KAPLAN

AP® Biology Prep Plus 2020 & 2021

ACKNOWLEDGEMENTS

Lead Editor
Brandon Deason, MD

Special thanks to our faculty writers and editors on this edition: Maria Cuffe, Chris Culbertson, Sumir Desai, Elizabeth Minten, Bonnie Wang, Allison Wilkes St. Clair, Valerie Yeung

Special thanks to the following for their contributions to this text: Laura Aitcheson, Matthew Callan, Andrea Carver, Christopher Durland, M. Dominic Eggert, Tim Eich, Tyler Fara, Elizabeth Flagge, Dan Frey, Bella Furth, Adam Grey, Allison Harm, Katy Haynicz-Smith, Duncan Honeycutt, Rebecca Knauer, Jeffrey Koetje, Liz Laub, Terry McMullen, Mark Metz, Jenn Moore, Kristin Murner, Nicholas Nguyen, Aishwarya Pillai, Elijah Schwartz, Jill Sherman, Linda Brooke Stabler, Oscar Velazquez, Shayna Webb-Dray, Lee Weiss, Lauren White, Dan Wittich, Michael Wolff, Jessica Yee, and Nina Zhang.

Published by Kaplan Publishing, a division of Kaplan, Inc.
750 Third Avenue
New York, NY 10017

10 9 8 7 6 5 4 3 2 1
Retail ISBN: 978-1-5062-4808-0

Kaplan Publishing print books are available at special quantity discounts to use for sales promotions, employee premiums, or educational purposes. For more information or to purchase books, please call the Simon & Schuster special sales department at 866-506-1949.

10 9 8 7 6 5 4 3 2 1
Course ISBN: 978-1-5062-4940-7

AP® Biology Prep Plus

2020 & 2021

ACKNOWLEDGEMENTS

Lead Editor
Brandon Deason, MD

Special thanks to our faculty writers and editors on this edition: Maria Cuffe, Chris Culbertson, Sumir Desai, Elizabeth Minten, Bonnie Wang, Allison Wilkes St. Clair, Valerie Yeung

Special thanks to the following for their contributions to this text: Laura Aitcheson, Matthew Callan, Andrea Carver, Christopher Durland, M. Dominic Eggert, Tim Eich, Tyler Fara, Elizabeth Flagge, Dan Frey, Bella Furth, Adam Grey, Allison Harm, Katy Haynicz-Smith, Duncan Honeycutt, Rebecca Knauer, Jeffrey Koetje, Liz Laub, Terry McMullen, Mark Metz, Jenn Moore, Kristin Murner, Nicholas Nguyen, Aishwarya Pillai, Elijah Schwartz, Jill Sherman, Linda Brooke Stabler, Oscar Velazquez, Shayna Webb-Dray, Lee Weiss, Lauren White, Dan Wittich, Michael Wolff, Jessica Yee, and Nina Zhang.

Published by Kaplan Publishing, a division of Kaplan, Inc.
750 Third Avenue
New York, NY 10017

10 9 8 7 6 5 4 3 2 1
Retail ISBN: 978-1-5062-4808-0

Kaplan Publishing print books are available at special quantity discounts to use for sales promotions, employee premiums, or educational purposes. For more information or to purchase books, please call the Simon & Schuster special sales department at 866-506-1949.

10 9 8 7 6 5 4 3 2 1
Course ISBN: 978-1-5062-4940-7

TABLE OF CONTENTS

PART 6: INTERACTIONS OF LIFE

PART 7: PRACTICE

PART 1

Getting Started

CHAPTER 1

What You Need to Know About the AP Biology Exam

Congratulations—you have chosen Kaplan to help you get a top score on your AP Biology exam. Kaplan understands your goals and what you're up against: conquering a tough exam while participating in everything else that high school has to offer.

You expect realistic practice, authoritative advice, and accurate, up-to-the-minute information on the exam. And that's exactly what you'll find in this book.

ABOUT THIS BOOK

In preparing for the AP exam, you certainly will have built a solid foundation of knowledge about the biological basis of life. While this knowledge is critical to your learning, keep in mind that just being able to recall isolated structures, processes, and facts does not ensure success on the exam. Biology is about big ideas: how evolution drives diversity, how organisms use energy to grow and reproduce, how information processes are vital to life, and how biological systems cooperate and compete within an ecosystem. The College Board (the maker of the AP exam) asks you to apply the knowledge you've learned at a higher level in order to show evidence of college-level abilities.

That's where this book comes in. This guide offers much more than a review of basic content. We'll show you how to put your knowledge to brilliant use on the AP exam through structured practice and efficient review of the areas you need to work on most. We'll explain the ins and outs of the exam structure and question formats so you won't experience any surprises. We'll even give you test-taking strategies that successful students use to earn high scores.

Are you ready for your adventure in the study and mastery of everything AP Biology? Good luck!

EXAM STRUCTURE

The AP Biology exam is 3 hours and 10 minutes long and is divided into two, 90-minute sections with a 10-minute break in between. Section I has 60 multiple-choice questions with four answer choices each. Section II has 6 free-response questions. The 8 official AP Biology course units tested, along with the frequency of questions on each topic, are as follows:

1. Chemistry of Life (8-11%)
2. Cell Structure and Function (10-13%)
3. Cellular Energetics (12-16%)
4. Cell Communication and Cell Cycle (10-15%)
5. Heredity (8-11%)
6. Gene Expression and Regulation (12-16%)
7. Natural Selection (13-20%)
8. Ecology (10-15%)

Our book has taken these 8 units and organized them into the 5 parts: Foundations of Life, Structures of Life, Processes of Life, Transformations of Life, and Interactions of Life. This approach to scientific discovery is about learning concepts and how they relate, not just facts. *The College Board has increased the emphasis on themes and concepts and placed less weight on specific facts in both the AP Biology course and exam*. The chapters in this book are intended to take advantage of this design by focusing on specific learning objectives and synthesizing information from different concepts to help you better understand, and learn, the AP Biology course and exam content.

Section II begins with a 10-minute, recommended "reading period" in which you're given 10 minutes to pore over Section II of the exam, which consists of two long free-response questions and four short free-response questions. You then have 80 minutes to answer all of these questions. The phrase "long free-response" means roughly the same thing as "large, multi-step, and involved." Although the two long free-response questions are worth a significant amount each and are broken into multiple parts, they usually don't cover an obscure topic. Instead, they take a fairly basic biology concept and ask you several questions about it. Sometimes diagrams are required or experiments must be set up properly. It's a lot of biology work, but it is fundamental biology work. The short free-response items are typically illustrative examples or concepts that you are expected to explain or analyze, providing appropriate scientific evidence and reasoning. A typical response for the short free-response items will be about a paragraph or two in length. You have approximately 20 minutes for each of the long free-response questions and 10 minutes for each of the short-response questions.

EXAM SCORING

Once you complete your AP exam, it will be sent to the College Board for grading. Student answer sheets for the multiple-choice section (Section I, Part A) are scored by machine. Scores are based on the number of questions answered correctly. No points are deducted for wrong answers, and no points are awarded for unanswered questions.

The free-response section (Section II) is evaluated and scored by hand by trained AP readers. Rubrics based on each specific free-response prompt are released on the AP central website after the exams are administered.

The score from the multiple-choice section of the exam counts for 50% of your total exam score. The other 50% is the combined score from the two long and four short free-response questions.

After your total scores from Sections I and II are calculated, your results are converted to a scaled score from 1 to 5. The range of points for each scaled score varies depending on the difficulty of the exam in a particular year, but the significance of each value is constant from year to year. According to the College Board, AP scores should be interpreted as follows:

5 = Extremely well qualified

4 = Well qualified

3 = Qualified

2 = Possibly qualified

1 = No recommendation

Colleges will generally not award course credit for any score below a 3, with more selective schools requiring a 4 or 5. Note that some schools will not award college credit regardless of your score. Be sure to research schools that you plan to apply to so you can determine the score you need to aim for on the AP exam.

Registration and Fees

To register for the exam, contact your school guidance counselor or AP Coordinator. If your school does not administer the AP exam, contact the College Board for a listing of schools that do.

There is a fee for taking AP exams. The current cost can be found at the official exam website listed below. For students with acute financial need, the College Board offers a fee reduction equal to about one third of the cost of the exam. In addition, most states offer exam subsidies to cover all or part of the remaining cost for eligible students. To learn about other sources of financial aid, contact your AP Coordinator.

For more information on all things AP, contact the Advanced Placement Program:

Phone: (888) 225-5427 or (212) 632-1780
Email: apstudents@info.collegeboard.org
Website: https://apstudent.collegeboard.org/home

CHAPTER 2

How to Get the Score You Need

HOW TO GET THE MOST OUT OF THIS BOOK

Kaplan's *AP Biology Prep Plus* contains precisely what you'll need to get the score you want in the time you have to study. The unique format of this book allows you to customize your prep experience to make the most of your time.

Start by going to kaptest.com/moreonline to register your book and get a glimpse of the additional online resources available to you.

Book Features

Specific Strategies

This chapter features both general test-taking strategies and strategies tailored specifically to the AP Biology exam. You'll learn about the types of questions you'll see on the official exam and how to best approach them to achieve a top score. In addition, chapter 20 of this book is devoted exclusively to free-response question strategy and sample questions.

Customizable Study Plans

We recognize that every student is a unique individual, and there is no single recipe for success that works for everyone. To give you the best chance to succeed, we have developed three customizable study plans. Each offers guidance on how to make the most of your available study time. With the instructions offered in the following section of this chapter, you'll be able to select and customize the study plan that is right for you, maximizing your chances of earning the score you need.

A Review of ALL the Relevant Subjects Based on Specific Test-like Questions

The 17 content chapters of this book (chapters 3 through 19) are designed to cover every concept tested on the AP Biology exam. However, unlike the textbook used in your class, this book focuses exclusively on the material you are required to know and examples you can use in free-response questions. Each content chapter includes a series of Learning Objectives that identify key takeaways on each topic and help to organize the chapter. Throughout the text, important terms are highlighted in bold; these terms are compiled in the Glossary, which you can find online. The most commonly tested topics, ones that appear on virtually every AP Biology exam, are demarcated with High Yield icons to help you recognize when information is absolutely essential to know.

Pre- and Post-Chapter Quizzes

Each chapter begins and ends with a mini-quiz—one to test what you already know about the given topic ("Test What You Already Know") and one to test what you learned by working through the chapter ("Test What You Learned"). You can use the results of these quizzes to guide your studying and chart your progress in mastering the Learning Objectives.

AP Biology Lab Investigations

Several of the content chapters feature sections that begin with "AP Biology Lab" followed by a number and a title. These sections review information that is relevant to the 13 official laboratory investigations that the College Board recommends for students in all AP Biology classes. Whether or not you completed these labs as part of your course, you'll find the information in these sections helpful; they're full of high-yield content applicable to multiple-choice and free-response questions.

Full-Length Practice Exams

In addition to all of the exam-like practice questions featured in the chapter quizzes, we have provided three full-length practice exams. These full-length exams mimic the multiple-choice and free-response questions on the real AP exam. Taking a practice full-length exam gives you an idea of what it's like to answer exam-like questions for about three hours. Granted, that's not exactly a fun experience, but it is a helpful one. And the best part is that it doesn't count; mistakes you make on our practice exams are mistakes you won't make on your real exam.

After taking each practice exam, you'll score your multiple-choice and free-response sections using the answers and explanations. Then, you'll navigate to the scoring section in your online resources and input your raw scores to see what your overall score would be with a similar performance on the official exam.

Online Quizzes

While this book contains hundreds of exam-like multiple-choice questions, you may still find yourself wanting additional practice on particular topics. That's what the online quizzes are for! Your online resources contain additional quizzes for each content chapter. Go to kaptest.com/moreonline to find them all.

CHOOSING THE BEST STUDY PLAN FOR YOU

There's a lot of material to review before the AP exam, so it's essential to have a solid game plan that optimizes your available study time. The tear-out sheet in the front of the book consists of three separable bookmarks, each of which covers a specific, customizable study plan. You can use one of these bookmarks both to hold your place in the book and to keep track of your progress in completing one of these study plans. But how do you choose the study plan that's right for you?

Fortunately, all you need to know to make this decision is how much time you have to prep. If you have two months or more with plenty of time to study, then we recommend using the Two Months Plan. If you only have about a month, or if you have more than a month but your time will be split among competing priorities, you should probably choose the One Month Plan. Finally, if you have less than a month to prep, your best bet is the Two Weeks Plan.

Regardless of your chosen plan, you have flexibility in how you follow the instructions. You can stick to the order and timing that the plan recommends or tailor those recommendations to fit your particular study schedule. For example, if you have six weeks before your exam, you could use the One Month Plan but spread out the recommended activities for Week 1 across the first two weeks of your studying.

After you've made your selection, tear out the perforated study plan page, separate the bookmark that contains your choice of plan, and use it to keep track of both your place in the book and your progress in the plan. You can further customize any of the study plans by skipping over chapters or sections that you've already mastered or by adjusting the recommended time to better suit your schedule. Don't forget to also use the guidelines in the Rapid Review chapters to further customize how you study.

STRATEGIES FOR EACH QUESTION TYPE

The AP Biology exam can be challenging, but with the right strategic mindset, you can get yourself on track for earning the score you need to qualify for college credit or advanced placement. Let's review some strategies that, along with the content review and practice questions in the rest of this book, will help you succeed on the AP exam.

✔ AP Expert Note

General Test-Taking Strategies

1. **Pacing.** Because many tests are timed, proper pacing allows you to attempt every question in the time allotted. Poor pacing causes students to spend too much time on some questions to the point where they run out of time before completing all the questions.
2. **Process of Elimination.** On every multiple-choice test you ever take, the answer is given to you. If you can eliminate answer choices you know are incorrect and only one choice remains, then that must be the correct answer.
3. **Knowing When to Guess.** The AP Biology exam does not deduct points for wrong answers, while questions left unanswered receive zero points. That means you should always guess on a question you can't answer any other way.
4. **Recognizing Patterns and Trends.** The AP Biology exam doesn't change greatly from year to year. Sure, each question won't be the same, and different topics will be covered from one administration to the next, but there will also be a lot of overlap from one year to the next. Because of this, certain patterns can be uncovered. Learning about these trends and patterns can help students taking the test for the first time.
5. **Taking the Right Approach.** Having the right mindset plays a large part in how well you do on a test. Those students who are nervous about the exam and hesitant to make guesses often fare much worse than students with an aggressive, confident attitude.

These points are generally valid for standardized tests, but they are quite broad in scope. The rest of this section will discuss how these general ideas can be modified to apply specifically to the AP Biology exam. These test-specific strategies and the factual information reviewed in this book's content chapters are the one-two punch that will help you succeed on the exam.

Multiple-Choice Questions

The multiple-choice questions are numbered, but that does not mean you must answer the questions in the given order. In fact, it's highly unlikely that the questions will be presented to you in a confidence-inspiring, point-building, time-saving order. There is good news, though: you are free to navigate the section in a manner that highlights your strengths and downplays your weaknesses. All of the multiple-choice questions carry the same weight, so you don't get extra credit for correctly answering a hard question. In the end, colleges will never know which questions you answered correctly. All they will know is that you were smart enough to spend your time where it was more likely to turn into points.

Of the 60 multiple-choice questions, there are two distinct question types: Stand-Alone Questions and Data Question Sets.

Stand-Alone Questions

These questions typically make up a little over half of the AP Biology exam. Each Stand-Alone question covers a specific topic, and then the next Stand-Alone hits a different topic. The question stem may be as short as a single sentence, but it is not uncommon to see multiple paragraphs for a single question! In addition to words, many question stems will be accompanied by an equation, table, graph, or figure.

You get some information to start with, and then you're expected to answer the question. The number of the question makes no difference because there's no order of difficulty on the AP Biology exam. Tough questions are scattered between easy and medium questions.

The overarching goal is to correctly answer the greatest possible number of questions in the time available. To do this, focus on your strengths during the first pass through the section. Some questions might be very difficult, even in a subject you're familiar with. Take a minute or so on a tough question, and if you can't come up with an answer, make a mark by the question number in your test booklet and move on. The first pass is about picking up easy points.

Once you've swept through and snagged all the easy questions, take a second pass and try the tougher ones. These tougher questions may cover subjects you're not strong in, or they may just be very difficult questions on subjects you are familiar with. Odds are high that you won't know the answer to some of these questions, but don't leave them blank. You should always take a stab at eliminating some answer choices, and then make an educated guess.

> ✔ **AP Expert Note**
>
> **Educated Guessing**
>
> **Many times you can eliminate at least one answer choice from a problem. It may seem insignificant, but it gets you closer to the correct answer and it can significantly increase your chances of guessing correctly. You won't get every guess right, but over the course of the test, this form of educated guessing will improve your score.**

Data Question Sets

Just as the name suggests, a set of two to five questions is preceded by data in one form or another. The data may be a simple sentence or two, but usually it is something more complex, such as:

- A description of an experiment (50–200 words), often with an accompanying illustration
- A graph or series of graphs
- A large table
- A diagram

Data Sets require a slightly greater initial time investment, but don't let that intimidate you! Once the data is understood, you may find that you can answer the questions rather quickly. Because most of the new information is in the shared introduction, you'll probably notice that the question stems are actually a little shorter than those of the average Stand-Alone. As you navigate through the multiple-choice questions, treat the Data Sets in much the same way you would Stand-Alones, with awareness of your strengths and weaknesses and your overall goal of getting more questions correct. If you see a Punnett square and you love heredity, then dive right in. However, if the topic is one that is more likely to induce anxiety than correct answers, then skip it and return after your first pass through the section.

The key to getting through Data questions on the exam is to be able to quickly analyze and draw conclusions from the data presented.

✔ **AP Expert Note**

At least one—and most likely several—of the Data questions you see on Test Day will deal with experiments. Make sure you understand all the basic points of an experiment—testing a hypothesis, setting up an experiment properly to isolate a particular variable, and so on—so that you will be able to breeze through these questions when you come to them.

Questions with Graphs

Most graph questions require a bit of biology knowledge to determine what the right answer is, but some graph questions only test whether or not you can read a graph properly. If you can make sense of the vertical and horizontal axes, then you can determine what the correct answer is. Granted, very few graph questions are this easy but, even so, it's nice to have a slam-dunk question or two. Therefore, if you see a graph, look at the problem and see if you can answer the question just by knowing how to read a graph.

Computational Questions

To help with the more quantitative problems on the AP Biology exam, you will be allowed to use a basic four-function, scientific, or graphing calculator on both sections of the exam. In addition, because these questions focus on applying math and not just recalling information, a formula list will be given to you on Test Day. A sample formula list is provided in the Appendix of this book.

Those are the two types of questions you'll see in Section I. Combine this knowledge with the fact that you must answer 60 multiple-choice questions in 90 minutes, and you'll see why it is imperative to take control of the test. Section management may not come to mind when you think about Biology, but it will be tested on the AP Biology exam!

There's more to it than just tackling questions in the right order, however. The more you know about the question types, the better equipped you will be to handle them. To select the correct answer on an AP Biology exam question, you will need to know the relevant science, but even if certain facts elude you, you can still increase your odds of choosing correctly by keeping the following key ideas in mind.

✔ **AP Expert Note**

1. The AP Biology test is NOT a sneaky test. The test works hard to be as comprehensive as it can be, so that students who only know one or two biology topics will soon find themselves struggling.
2. Focus on "good science" by crossing out extreme answer choices, choices that are factually inaccurate, and choices that are out of place.
3. Lastly, remember that you don't need to get every multiple-choice question right on the AP Biology exam. To get a 4 or 5, you need to get a large portion, but not all, of the questions right. If you don't have enough time to get to every question, make sure that the questions you skip are the longest, most involved ones. That's a great use of your limited resource: time.

Free-Response Questions

Of course, the multiple-choice questions only account for 50 percent of your total score. To get the other 50 percent, you have to tackle the free-response questions. Because free-response questions are so distinctive, a separate chapter in the book is devoted to free-response strategy and sample questions.

✔ **AP Expert Note**

See chapter 20 for information on making the most of the 10-minute reading period, planning your free responses effectively, and scoring as many points as possible. Chapter 20 also includes several sample free-response questions for you to practice with, along with grading rubrics that allow you to determine how many points you would earn on Test Day.

Be sure to use all the strategies discussed in this chapter (and chapter 20) when taking the chapter quizzes and practice exams. Trying out the strategies during practice will get you comfortable with them and make it easier for you to put them to good use on the real exam.

COUNTDOWN TO THE EXAM

This book contains detailed review, guidance, and practice for you to utilize in the weeks leading up to your AP exam. In the final few days before your exam, we recommend the following steps.

Three Days Before the Exam

Take a full-length practice exam under timed conditions. Use the techniques and strategies you've learned in this book. Approach the exam strategically, actively, and confidently. (Note that you should *not* take a full-length practice exam with fewer than 48 hours left before your real exam. Doing so will probably exhaust you and hurt your score.)

Two Days Before the Exam

Go over the results of your latest practice exam. Don't worry too much about your score or whether you got a specific question right or wrong. Instead, examine your overall performance on the different topics, choose a few of the topics where you struggled the most, and brush up on them one final time.

Know exactly where you're going to take the official exam, how you're getting there, and how long it takes to get there. It's probably a good idea to visit your testing center sometime before the day of your exam so that you know what to expect: what the rooms are like, how the desks are set up, and so on.

The Night Before the Exam

Do not study! You cannot cram for a test as extensive as the AP Exam. Worse, pulling an all-nighter will simply deplete your stamina ahead of the exam. If you feel you must review some AP material, only do so for a little while and stick to broad review (such as the Essential Content sections of this book). The best, most effective way to prepare for the AP Exam at this point is to rest the night beforehand.

Get together an "AP Exam Kit" containing the following items:

- A few No. 2 pencils (Pencils with slightly dull points fill the ovals better; mechanical pencils are NOT permitted.)
- A few pens with black or dark blue ink (for the free-response questions)
- Erasers
- A watch (as long as it doesn't have Internet access, have an alarm, or make noise)
- Your 6-digit school code (Home-schooled students will be provided with their state's or country's home-school code at the time of the exam.)
- Photo ID card
- Your AP Student Pack
- If applicable, your Student Accommodation Letter verifying that you have been approved for a testing accommodation such as braille or large-type exams

Make sure that you don't bring anything that is *not* allowed in the exam room. You can find a complete list at the College Board's website (https://apstudent.collegeboard.org/home). Your school may have additional restrictions, so make sure you get this information from your school's AP Coordinator prior to the exam.

Again, try to relax. Read a good book, take a hot shower, watch something you enjoy. Go to bed early and get a good night's sleep.

The Morning of the Exam

Wake up early, leaving yourself plenty of time to get ready without rushing. Dress in layers so that you can adjust to the temperature of the testing room. Eat a solid breakfast: something substantial, but nothing too heavy or greasy. Don't drink a lot of coffee, especially if you're not used to it; bathroom breaks cut into your time, and too much caffeine is a bad idea. Read something as you eat breakfast, such as a newspaper or a magazine; you shouldn't let the exam be the first thing you read that day.

Leave extra early so that you can ensure you are on time to the testing location. Allow yourself extra time for any traffic, mass transit delays, and/or detours.

During the Exam

Breathe. Don't get shaken up. If you find your confidence slipping, remind yourself how well you've prepared. You know the structure of the exam; you know the material covered on it; you've had practice with every question type.

If something goes really wrong, do not panic! If you accidentally misgrid your answer page or put the answers in the wrong section, raise your hand and tell the proctor. He or she may be able to arrange for you to regrid your exam after it's over, when it won't cost you any time.

After the Exam

You might walk out of the AP exam thinking that you blew it. This is a normal reaction. Lots of people—even the highest scorers—feel that way. You tend to remember the questions that stumped you, not the ones that you knew. Keep in mind that almost nobody gets everything correct. You can still score a 4 or 5 even if you get some multiple-choice questions incorrect or miss several points on a free-response question.

We're positive that you will have performed well and scored your best on the exam because you followed the Kaplan strategies outlined in this chapter and reviewed all the content provided in the rest of this book. Be confident and celebrate the fact that, after many hours of hard work and preparation, you have just completed the AP Biology exam!

PART 2

Foundations of Life

CHAPTER 3

Biological Molecules

LEARNING OBJECTIVES

In this chapter, you will review how to:

- **3.1** Recall the elements and molecules important to life
- **3.2** Explain why life requires free energy
- **3.3** Recognize the role of water in biological processes
- **3.4** Differentiate between common monomers and polymers
- **3.5** Determine how structure affects properties of polymers

TEST WHAT YOU ALREADY KNOW

1. The diagram below shows part of the carbon cycle as it occurs on land. Which of the following processes is occurring at the box labeled "A" ?

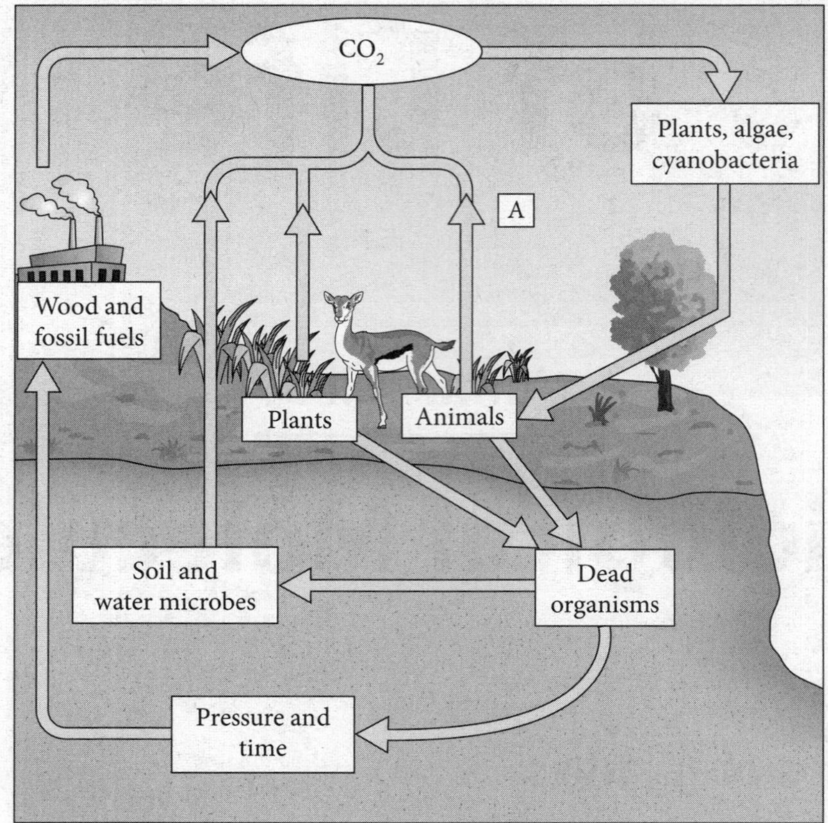

(A) Organic matter is broken down into simpler molecules by microorganisms.

(B) Carbon dioxide reacts with water to produce glucose as the primary food source for these organisms.

(C) Organic matter from the remains of dead animals is converted by extreme pressure into coal and oil.

(D) Nutrients inside a cell, such as glucose, are converted into usable forms of energy, such as ATP.

2. When a person exercises, muscles in the body convert energy stored in complex molecules into kinetic energy. This action releases a large amount of heat, which is transferred to the surroundings. Which law of thermodynamics best explains why this dissipation of heat occurs?

(A) The first law of thermodynamics, because this is an example of energy being created

(B) The second law of thermodynamics, because this is an example of energy being created

(C) The first law of thermodynamics, because this is an example of increasing entropy

(D) The second law of thermodynamics, because this is an example of increasing entropy

3. Researchers are comparing capillary action in two different species of plants by monitoring water flow with water-based dyes. The first species is a flowering plant in which the walls of the xylem are lined with negatively charged molecules. The second species is a wilted *Myrothamnus flabellifolia,* a plant that becomes brown and shriveled under drought conditions and is revived to its original green color and appearance once water becomes available. The researchers discovered that the walls of the water-conducting vessels in desiccated *Myrothamnus* are lined with hydrophobic molecules. Which of the following capillary activities will most likely be observed by the researchers?
   (A) The dye will reach the same height in both plants because the chemical nature of the vessel walls does not affect capillary action.
   (B) The dye will reach a higher point in *Myrothamnus* because the hydrophobic walls increase the surface tension of water.
   (C) The dye will reach a higher point in the flowering plant because the vessel walls in *Myrothamnus* do not interact with water and have lower adhesion forces.
   (D) No movement of water will be observed in the wilted *Myrothamnus* because the hydrophobic lining interferes with the cohesion of water molecules.
4. A scientist finds a novel nucleic acid in a seawater sample and hopes to determine its origin. The nucleic acid consists of a single strand and does not have characteristics of a viral nucleic acid. Which of the following is most likely the composition of the molecule?
   (A) The molecule contains a pentose sugar, a phosphate group, and the nitrogenous bases adenine, thymine, cytosine, and guanine.
   (B) The molecule contains a pentose sugar, a phosphate group, and the nitrogenous bases adenine, uracil, cytosine, and guanine.
   (C) The molecule contains a tetrose sugar, a phosphate group, and the nitrogenous bases adenine, uracil, cytosine, and guanine.
   (D) The molecule contains a pentose sugar, a fatty acid, and the nitrogenous bases adenine, uracil, cytosine, and guanine.
5. A researcher examines the structure of a protein that has been extracted from the membrane of a eukaryotic cell. She finds that the surface of one domain of the protein consists entirely of hydrophobic amino acids, whereas the surface of another domain consists entirely of hydrophilic amino acids. What can she most reasonably conclude about this protein?
   (A) This protein is probably found deep within the cell and must have been mixed with the membrane material in error during sample preparation.
   (B) The protein is probably found embedded completely within the phospholipid bilayer.
   (C) The protein is probably bound to the outer surface of the cell membrane, facing the extra cellular matrix.
   (D) The protein is probably embedded in the cell membrane with portions extending either into the extracellular matrix or the cytoplasm.

Answers to this quiz can be found at the end of this chapter.

MATTER AND ENERGY

3.1 Recall the elements and molecules important to life

3.2 Explain why life requires free energy

The world around us follows a hierarchy of organization. All life on Earth is connected, from the smallest individual units of matter (atoms and molecules) to complex organisms. In this chapter, we will discuss the basic building blocks that compose the living world. Although some organisms are more complex than others, different levels of organization do not correlate with levels of complexity. An individual **cell** with its multitude of chemical reactions is just as dynamic and complex as an entire community of species. This chapter will review the most crucial concepts of the building blocks of life.

Elements Essential to Life

Ninety-nine percent of all living matter is made up of only four elements. These elements are nitrogen (N), carbon (C), hydrogen (H), and **oxygen** (O). Phosphorus (P) and sulfur (S) account for almost all of the remaining 1 percent of living matter. All six elements are important in biochemistry, especially carbon (C).

> ✔ **AP Expert Note**
>
> **The mnemonic device N'CHOPS (for nice chops!) can be used to remember the main elements of life.**

Most molecules that contain carbon (C) are known as **organic molecules** or organic compounds. (There are some exceptions: for instance, carbon dioxide and hydrogen cyanide have the chemical formulas CO_2 and HCN, respectively, but are generally not considered to be organic.) Life wouldn't exist without organic compounds. Carbon earned its place as a staple in biology because it can bond to other atoms or to itself in four different equally spaced directions, allowing complex molecules of almost unlimited size and shape to be formed. The structural properties of the carbon atom have allowed for the formation of molecular compounds—such as DNA and enzymes—that have unique chemical identities and functions. The message to take home is that there are only a few elements important to biology (N'CHOPS) and that carbon is the main element of life because of the variety of organic compounds it can form.

> ✔ **AP Expert Note**
>
> **Macromolecules such as carbohydrates, lipids, proteins, and nucleic acids are examples of organic compounds that serve as building blocks in all living organisms. These compounds are discussed in detail later in this chapter.**

Free Energy in Living Systems

High-Yield

All living things require the capture or harvest of free energy from the environment to grow, reproduce, and maintain dynamic homeostasis. However, living systems must also follow the laws of thermodynamics, which means that entropy is always increasing. To balance the movement toward entropy, organisms must take in more energy than is used by the organism.

Though you do not need to memorize the specific steps of the reactions that help living things to capture and use free energy, it is important to remember that endergonic (energy-consuming) and exergonic (energy-releasing) reactions are typically connected. For example, the catabolic breakdown of ATP to ADP is an energetically favorable exergonic reaction. This release of free energy is coupled to endergonic reactions such as the synthesis of glutamine, an essential amino acid.

THE IMPORTANCE OF WATER

3.3 Recognize the role of water in biological processes

A water molecule is formed between two H atoms bonded covalently to a single O atom. The oxygen molecule tends to control all of the electrons by keeping them away from the hydrogen atoms, giving the oxygen a slightly negative charge, which is balanced by slightly positive charges on the hydrogen atoms. This is an example of a polar covalent bond. The water molecule looks like Mickey Mouse because it has a big head (the oxygen atom) and two big ears (the hydrogen atoms).

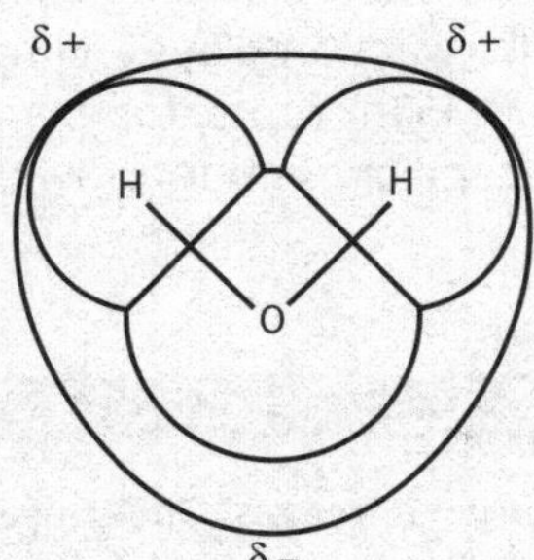

The head has a slightly negative charge and the ears have a slightly positive charge, which makes water a polar compound. The shape and charge of this molecule give it unique properties.

Water is the only substance on Earth that commonly exists in all three physical states (gas, liquid, and solid). The substance has a high specific heat so it serves as a temperature stabilizer for other compounds and vaporizes at a relatively high temperature.

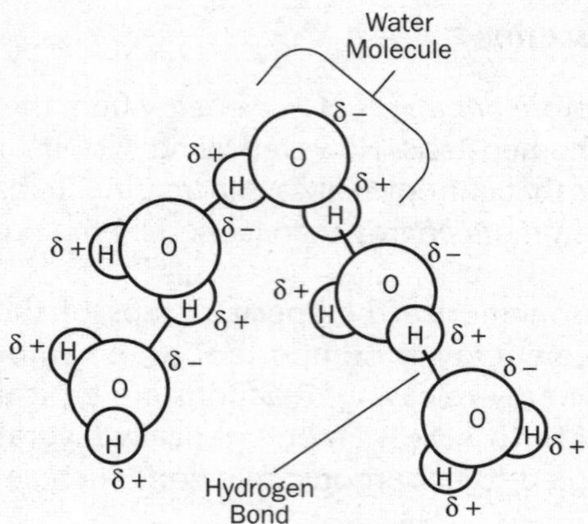

Water plays a key role in hydrolysis, condensation, and other chemical reactions that are essential to life. It is also fundamental to the biological activity of nucleotides, carbohydrates, lipids, and proteins. There are two specific characteristics of water that are particularly important to remember. First, the polar nature of water makes it "sticky." The positive charges on the hydrogen atoms cling to the negative charges on the oxygen atoms between molecules; that is, water molecules attract one another, causing water to have high surface tension. This is what makes water bead on windshields and form round raindrops. Even matter with greater density can float on top of water if it doesn't break the surface tension.

This surface tension is also the force behind capillary action, by which water (and anything dissolved in it) will climb up a thin tube or move through the spaces of a porous material until it is overcome by gravity. It is as if each water molecule drags along the one behind it, as well as any **nutrients** dissolved in the water. Capillary action plays an important role in moving nutrients and other metabolites through living things. Plant roots, for example, take in water from the soil, full of **minerals** and dissolved nutrients; capillary action draws the water and its load through the plant against gravity.

✔ AP Expert Note

Water expands rather than contracts when it freezes. One consequence of this is that solid ice floats on liquid water.

Second, the polar nature of water makes it a good solvent. Capillary action would be a lot less useful if water lacked this property. Fortunately, water molecules are happy to surround positive and negative ions and readily dissolve other polar compounds. In addition, hydrogen bonds can form between the hydrogen and oxygen atoms of the water molecule and surrounding molecules.

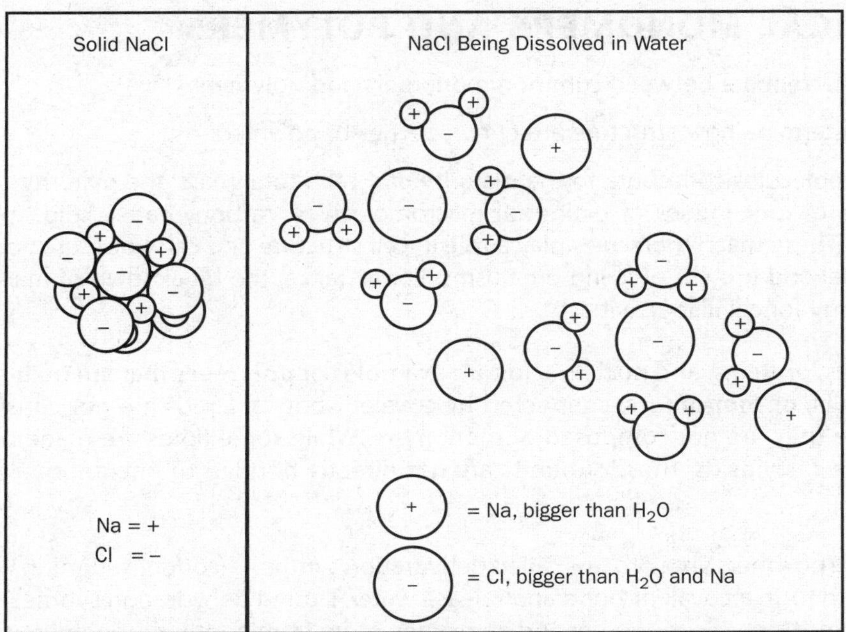

Water's role as a solvent is another reason it is so important for living things. Chemicals that are dissolved by the water in blood can be carried around the body rapidly and easily; this is how sugar gets to muscles, for example. Each living cell is, in large part, a **membrane** surrounding chemicals dissolved in water. It is only by being dissolved in water that these chemicals can participate in many of the biological reactions that keep living things alive.

> ✔ **AP Expert Note**
>
> **Remember that many organic compounds are nonpolar or have nonpolar regions and won't mix with water. Such nonpolar regions are consequently called hydrophobic (literally meaning "water-fearing"), while polar and ionic parts of molecules are hydrophilic ("water-loving").**

BIOLOGICAL MONOMERS AND POLYMERS

3.4 Differentiate between common monomers and polymers

3.5 Determine how structure affects properties of polymers

While water molecules contribute to the majority of a cell's total mass, the majority of a cell's dry mass consists of four classes of biological macromolecules: carbohydrates, lipids, proteins, and nucleic acids. These macromolecules play a role in cell structure and carry out functions necessary for the survival and growth of living organisms. For instance, the breakdown of macromolecules provides energy for cellular activities.

Carbohydrates, proteins, and nucleic acids are examples of **polymers** that are built from smaller molecular units, or **monomers**, connected by covalent bonds. Lipids are not considered polymers because they are not composed of monomers. While some lipids are made up of smaller molecules like fatty acids, those subunits are not directly bonded to one another as in proper polymers.

Cells assemble polymers via a process called dehydration synthesis (condensation), in which energy is consumed to form a covalent bond and release water. During dehydration synthesis, either two hydrogen atoms from one monomer and an oxygen atom from another monomer are combined, or a hydrogen atom from one monomer and a hydroxyl (OH) group from another monomer are combined. The reverse process is hydrolysis. Hydrolysis reactions use water to break down polymers and release energy—one monomer gains a hydrogen atom and the other monomer gains a hydroxyl group.

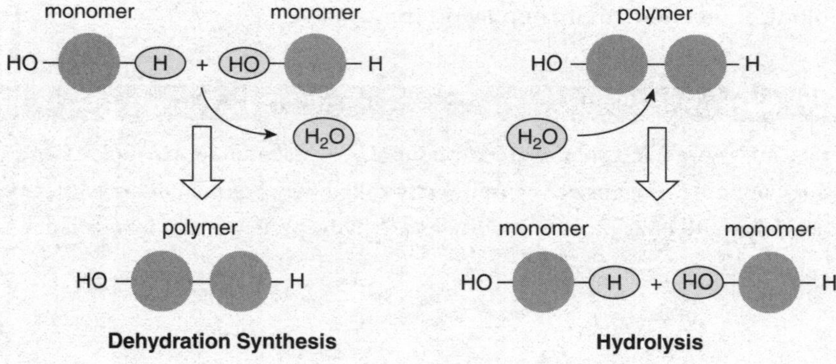

Carbohydrates

Carbohydrates are a source of immediate energy in living systems and serve as structural components. The three primary classes of carbohydrates are monosaccharides, disaccharides, and polysaccharides.

Monosaccharides are simple sugars that have the formula $(CH_2O)_n$, where *n* represents the number of carbons in the backbone and ranges from three to seven. Monosaccharides are classified based on the number of carbon atoms (triose, tetrose, pentose, hexose, or heptose), as well as on the position of their carbonyl group (an oxygen double-bonded to a carbon). A monosaccharide with an aldehyde group (carbonyl group at the end of the carbon chain) is an aldose, and a monosaccharide with a ketone group (carbonyl group in the middle of the carbon chain) is a ketose. Glucose and galactose are examples of aldoses, and fructose is an example of a ketose; all three are hexoses because they each have six carbons.

Disaccharides are formed when two monosaccharides undergo dehydration synthesis to form a glycosidic bond. Common disaccharides include lactose (glucose and galactose), maltose (two glucose molecules), and sucrose (glucose and fructose).

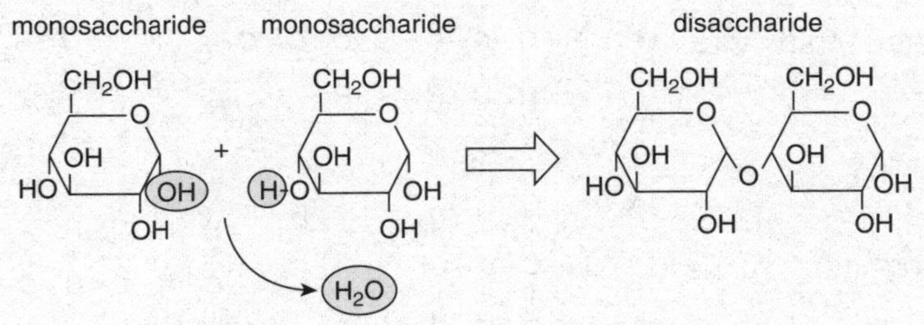

monosaccharide + monosaccharide → disaccharide + H_2O

Polysaccharides are composed of many monosaccharides linked by glycosidic bonds. In plants, both starch and cellulose are polysaccharides composed of glucose molecules, but they differ in structure and function. The difference is the linkage of the glucose bonds formed during dehydration synthesis: starch contains alpha linkages and cellulose contains beta linkages. Starch serves as the main energy storage material in plants, while cellulose provides structural support. To use energy stored in starch, organisms must first break it down into glucose. Animals, however, are unable to hydrolyze cellulose because they do not have the enzymes that break down the beta linkages. In animals, glycogen (a glucose polymer similar to starch, but containing branches) is a form of energy storage and chitin (analogous in structure to cellulose) provides support.

Lipids

Lipids store energy, contribute to cell membrane structure, protect against desiccation, and provide building blocks for hormones. A subgroup of lipids is fatty acids, which consist of a long hydrocarbon chain attached to a carboxyl group. The length of fatty acid chains ranges from 4–36 carbons but is usually 12–18 carbons. If the fatty acid contains only single bonds in the hydrocarbon chain, it is saturated with hydrogen atoms and thus called a saturated fatty acid. A double bond between carbons in the hydrocarbon chain causes the fatty acid to become unsaturated. Fatty acids may differ in their degree of unsaturation. Monounsaturated fatty acids like oleic acid (the primary fatty acid in olive oil) have one double bond and polyunsaturated fatty acids like linoleic acid have two or more double bonds.

The configuration of the double bonds affects the behavior of fatty acids (and the fats that contain them). In the cis configuration, hydrogen atoms at a double bond are in the same plane (on the same side), and in the trans configuration, they are in two different planes (on opposite sides). Unlike single bonds, which allow atoms to rotate freely, double bonds restrict rotation and lock the orientation of atoms. A cis double bond causes a bend in the chain, which prevents fatty acids from packing tightly against one another, so they are liquid at room temperature and have a relatively low melting point. Saturated fatty acids on the other hand pack tightly together, so they are solid at room temperature and have a relatively high melting point.

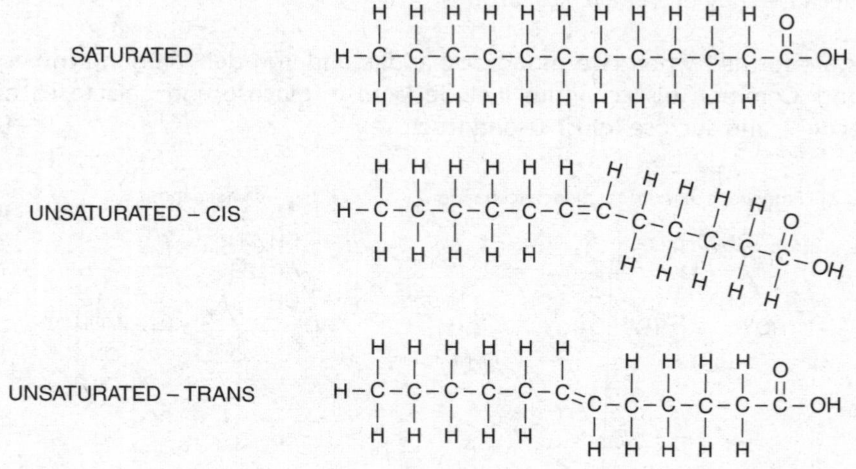

SATURATED

UNSATURATED – CIS

UNSATURATED – TRANS

Fats called triglycerides are another subgroup of lipids. Triglycerides are formed via dehydration synthesis. Hydroxyl groups on a glycerol backbone react with the carboxyl groups of three fatty acids to form ester bonds and release three molecules of water. Primarily stored in **adipose** (fat) tissue, triglycerides are a major form of energy storage.

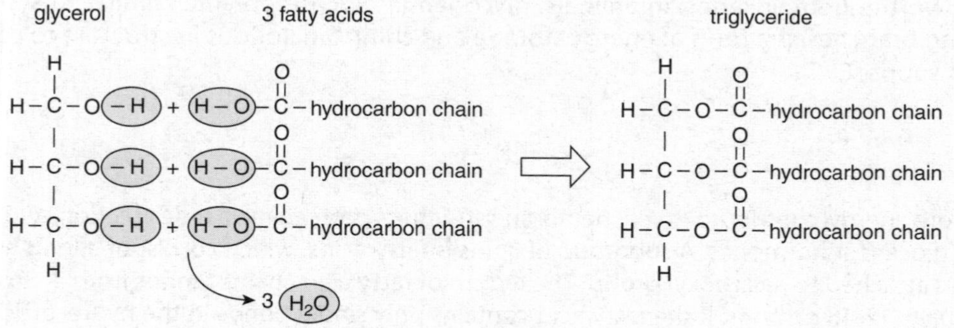

glycerol

3 fatty acids

triglyceride

hydrocarbon chain

$3\ H_2O$

Phospholipids, like fats, are composed of glycerol and fatty acid chains. However, instead of three fatty acids, phospholipids have two fatty acids and a modified phosphate group. The phosphate group or head is negatively charged, polar, and hydrophilic, whereas the fatty acids or tails are unsaturated, uncharged, nonpolar, and hydrophobic. This makes phospholipids amphipathic (containing both hydrophilic and hydrophobic regions).

As the major components of the **plasma membrane**, phospholipids are arranged in a bilayer with the heads facing outward toward the intracellular and extracellular fluids and the tails facing inward to prevent them from contacting water. The phospholipids' chemical and physical characteristics cause the membrane to be fluid, semipermeable, and to require low energy to remain stable. The separation of the intracellular and extracellular fluids is vital for cell communication and metabolism.

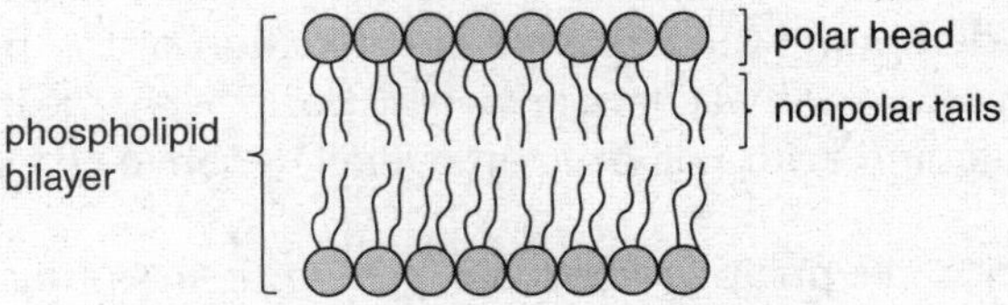

phospholipid bilayer — polar head — nonpolar tails

Waxes are lipids formed via dehydration synthesis, in which an ester bond is formed between a long-chain alcohol (12–32 carbons) and a fatty acid. Since waxes have hydrophobic properties, plants use them as protective coatings to prevent losing excessive amounts of water and drying out.

Steroids, though different in structure than triglycerides and phospholipids, are also considered lipids. Consisting of four fused rings of carbon atoms and insoluble in water, sterol lipids form the basis of steroid hormones. Cholesterol, for example, absorbs fat and is used to produce hormones such as estrogen, testosterone, and vitamin D.

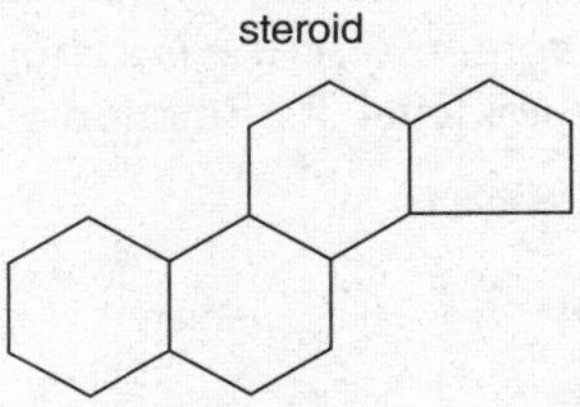

steroid

Proteins

Making up half the total dry weight of cells, **proteins** function as enzymes, structural materials, membrane transport, and signaling molecules. Examples include **lactase**, **lipase**, and **pepsin**, which are enzymes that break down lactose, lipids, and proteins, respectively; keratin and collagen, which provide structural support; hemoglobin, which transports oxygen and iron; and insulin, which is a hormone that regulates blood glucose levels.

Proteins are formed from amino acids via dehydration synthesis. An amino acid consists of a central carbon, an amino group, a carboxyl group, a hydrogen atom, and a side chain R group. The R group determines the amino acid's properties (size, polarity, and pH). In organisms, there are twenty α-amino acids; each is either acidic, basic, polar, or nonpolar at physiological pH (7.2–7.4). During dehydration synthesis, the amino group of one amino acid reacts with the carboxyl group of another to form a peptide bond and release a water molecule.

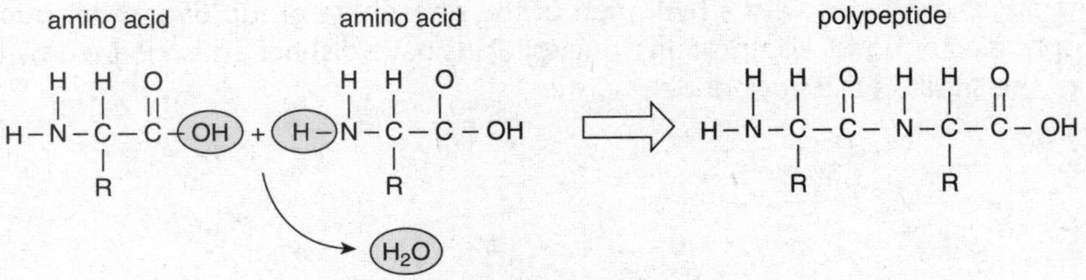

amino acid + amino acid → polypeptide + H_2O

A chain of amino acids is a polypeptide, which typically ranges from 50 to 1,000 amino acids. Since polypeptides have two distinct ends, they have directionality. The amino acid sequence is written and read from the amino end (N-terminus) to the carboxyl end (C-terminus). The properties and order of the amino acids determine the structure and function of the polypeptide. For instance, proteins with more basic amino acids will have an overall positive charge in a neutral solution, while proteins with more acidic amino acids will have an overall negative charge.

The sequence of amino acids in a polypeptide chain is the protein's primary structure and is determined by the DNA of the gene that encodes the protein. The protein's secondary structure refers to local folds along the chain due to interactions between the atoms of the backbone. The most common types of secondary structures are the α helix and the β pleated sheet, which involve hydrogen bonds between the carbonyl O of one amino acid and the amino H of another. A protein's shape and stability are influenced by the interactions between a protein and its immediate environment. For example, a protein in an aqueous environment will fold so that the hydrophilic R groups are at the surface and the hydrophobic R groups are on the inside of the protein. This three-dimensional folding, due to interactions between the R groups of amino acids, is a protein's tertiary structure. R group interactions include hydrogen bonds, ionic bonds, dipole-dipole interactions, London dispersion forces, hydrophobic interactions, and covalent disulfide bonds. The orientation and arrangement of multiple polypeptide chains or subunits constitute a protein's quaternary structure. Hemoglobin, which consists of two α and two β subunits, is an example of a protein with quaternary structure. Protein structure is important to its function: if a protein loses its shape (is denatured) at any structural level, its function will likely be diminished and possibly lost entirely.

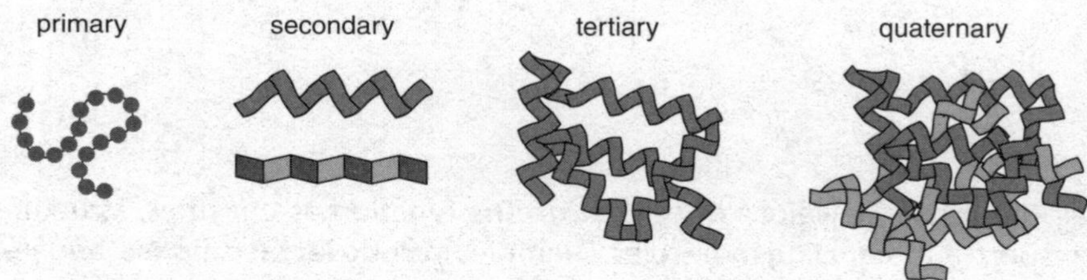

Nucleic Acids

Nucleic acids encode biological information in nucleotide sequences and provide information for **protein synthesis**. Nucleotides are composed of three components: a pentose sugar (**deoxyribose** or **ribose**), a **nitrogenous base** that is a purine (**adenine** and **guanine**) or pyrimidine (**thymine**, **cytosine**, or **uracil**), and one or more phosphate groups attached to the 5′ carbon of the sugar.

Nucleotide monomers join by dehydration synthesis to form nucleic acids: the 3′ carbon hydroxyl group of the sugar combines with a hydrogen of the phosphate group of another nucleotide to form a phosphodiester bond. Like proteins, nucleic acids have distinct ends, defined by the 3′ and 5′ carbons of the sugar in the nucleotide.

There are two types of nucleic acids: deoxyribonucleic acid (**DNA**) and ribonucleic acid (**RNA**). DNA carries the genetic information living organisms need to function, grow, and reproduce. Living organisms then use RNA to carry instructions from DNA for protein synthesis. The structural differences between DNA and RNA account for their differing functions. DNA contains deoxyribose and thymine and is double-stranded, while RNA contains ribose and uracil and is single-stranded. However, remember that viruses, unlike organisms, can have double- or single-stranded DNA or RNA.

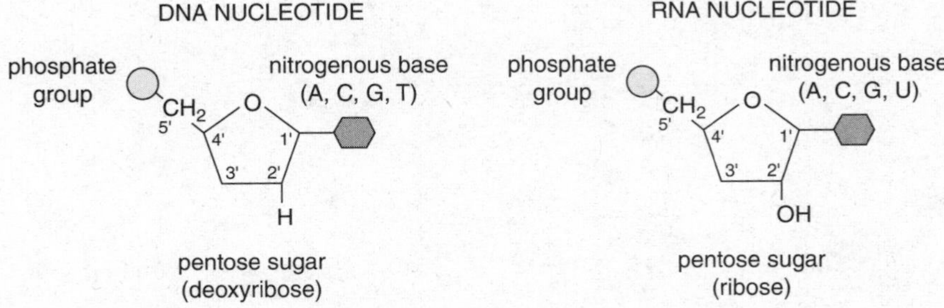

In the DNA double helix, sugars and phosphate form the backbone and the nitrogenous bases of each strand extend inward and are bound by base pairing rules (adenine with thymine and cytosine with guanine). The two strands in a DNA double helix are antiparallel (run in opposite directions). Directionality determines how complementary nucleotides are added during DNA synthesis and the direction in which DNA is transcribed to RNA (from 5′ to 3′). RNA occurs in different forms including messenger RNA, ribosomal RNA, and transfer RNA.

✓ AP Expert Note

Nucleic acids are discussed in further detail in chapter 13, Molecular Genetics.

RAPID REVIEW

If you take away only 4 things from this chapter:

1. Organic compounds are molecules that contain carbon. All living matter is made up of nitrogen, carbon, hydrogen, oxygen, phosphorus, and sulfur (N'CHOPS).
2. Living systems require free energy and matter from the environment to grow, reproduce, and maintain homeostasis. Organisms survive by coupling chemical reactions that increase entropy with those that decrease entropy.
3. The water molecule's polar nature leads to surface tension (enabling capillary action) and makes it an effective solvent. It expands rather than contracts when it freezes. These properties make water essential to life on Earth.
4. The four most common types of biological molecules are carbohydrates, lipids, proteins, and nucleic acids. Carbs, proteins, and nucleic acids include polymers composed of simpler monomer subunits, which give the molecules their distinctive properties: monosaccharides make up disaccharides and polysaccharides, amino acids make up monopeptides and polypeptides, and nucleotides make up RNA and DNA. Lipids are not technically polymers, but some of them are composed of smaller molecules: triglycerides (fats and oils) contain glycerol and three fatty acids, while phospholipids contain glycerol, two fatty acids, and a modified phosphate group.

TEST WHAT YOU LEARNED

1. Adenosine triphosphate (ATP) can be synthesized from adenosine diphosphate (ADP) according to the following reaction:

$$ADP + P_i \rightarrow ATP + H_2O$$

The energy change occurring during this reaction is shown in the graph:

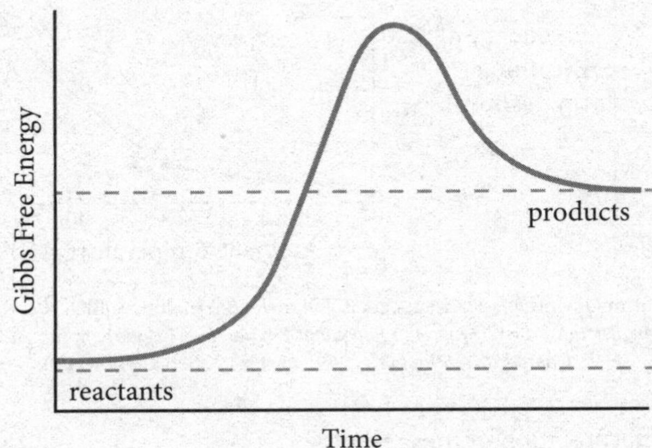

Which statement best describes the formation of ATP?

(A) The formation of ATP is endergonic, and energy is released during the process.

(B) The formation of ATP is exergonic, and energy is absorbed during the process.

(C) The formation of ATP is endergonic, and energy is absorbed during the process.

(D) The formation of ATP is exergonic, and energy is released during the process.

2. Reindeer live in arctic regions and have large bodies covered by thick, insulating fur. A scientist studying how reindeer adapt to warm temperatures observed a change in respiratory patterns as temperature increased. The frequency of reindeer panting as a function of ambient temperature is shown in the graph.

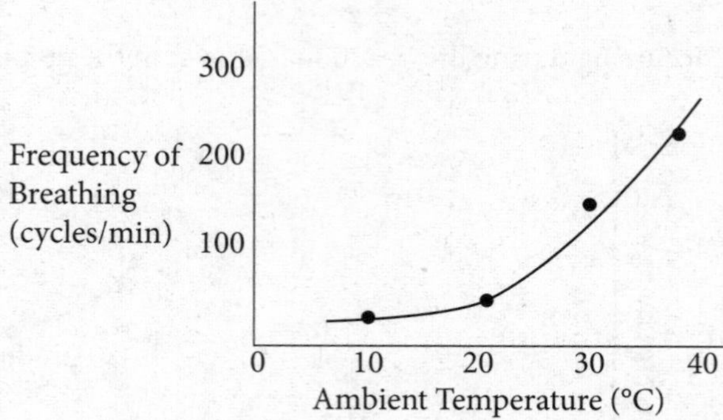

Adapted from Øyvind Aas-Hansen, Lars P. Folkow, and Arnoldus S. Blix, "Panting in reindeer (*Rangifer tarandus*)," *American Journal of Physiology—Regulatory, Integrative and Comparative Physiology* 279, no. 4 (October 2000): R1190.

What is the best interpretation of these results?

(A) Panting dissipates heat from the body when that heat radiates from the open nose and mouth.

(B) Panting decreases the core temperature of the reindeer when moisture from the lining of the nose and mouth evaporate into the air.

(C) Water in saliva has a high heat capacity and absorbs excess heat from the environment.

(D) Panting removes water from the surface of the tongue, allowing reindeer to absorb more heat from the water they drink.

3. Animals store glucose in the liver as a polymer called glycogen. Plants store glucose as a different polymer called starch. Cellulose is another polymer that is an important part of producing strong plant cell walls. Although cellulose is also a polymer of glucose, plant foods with high cellulose content, such as celery, do not provide many calories when eaten by humans. Which statement best explains why humans cannot effectively extract energy from cellulose, despite it being a polymer of glucose?

(A) The glucose monomers in cellulose are connected by different bonds than those in glycogen and starch. Human digestive enzymes cannot break down these bonds.

(B) The monomers in glycogen, starch, and cellulose differ slightly. Humans cannot break down the form of glucose found in cellulose.

(C) Cellulose is structured in a way that makes it impossible for mammals to digest.

(D) Cellulose is a larger polymer than glycogen and starch, with more associated water molecules that protect it from digestion.

4. Glycogen and triacylglycerols are both used by animals to store energy. If scientists were examining a sample of storage molecules, what is the best way to distinguish whether those molecules consisted of glycogen or triacylglycerols?

    (A) Use spectrometry and spectroscopy to determine whether the molecules in the tissue contained oxygen.

    (B) Examine whether the molecules were hydrophilic or hydrophobic and determine how many calories per gram they contain.

    (C) Determine the identity of the material by noting whether it was found in skeletal muscle.

    (D) Determine the mass of the material and compare it to reference values.

5. Photosynthesis is a chemical process by which plants use sunlight to convert carbon dioxide and water to an energy source, as shown in the reaction below:

$$6\ CO_2 + 12\ H_2O + \text{sunlight} \rightarrow C_6H_{12}O_6 + 6\ O_2 + 6\ H_2O$$

    This reaction usually requires the pigment chlorophyll. Which of the following best explains what occurs during this reaction?

    (A) Carbon dioxide undergoes reduction and water undergoes oxidation. In the process, glucose is made as the primary food source for the plant and oxygen gas is released as a byproduct.

    (B) Carbon dioxide undergoes oxidation and water undergoes reduction. In the process, glucose is made as the primary food source for the plant, and oxygen gas is released as a byproduct.

    (C) Carbon dioxide undergoes reduction and water undergoes oxidation. In the process, oxygen gas is made as the primary food source for the plant, and glucose is released as a byproduct.

    (D) Carbon dioxide undergoes oxidation and water undergoes reduction. In the process, oxygen gas is made as the primary food source for the plant, and glucose is released as a byproduct.

Questions 6–7

All of life as we know it needs water. Water's ability to form up to 4 hydrogen bonds gives it many unique properties. Hydrogen bonds form when a hydrogen bound to a nitrogen (N), oxygen (O), or fluorine (F) interacts with a lone pair of electrons on another nitrogen, oxygen, or fluorine. The oxygen atoms shown on both a fatty acid (Figure I) and a phospholipid (Figure II) have a lone pair of electrons available for hydrogen bonding. The long chains of carbon and hydrogen shown toward the bottom of both figures do not contain such lone pairs, nor do they contain any N, O, or F. Phospholipids form phospholipid bilayers and are the primary components of membranes.

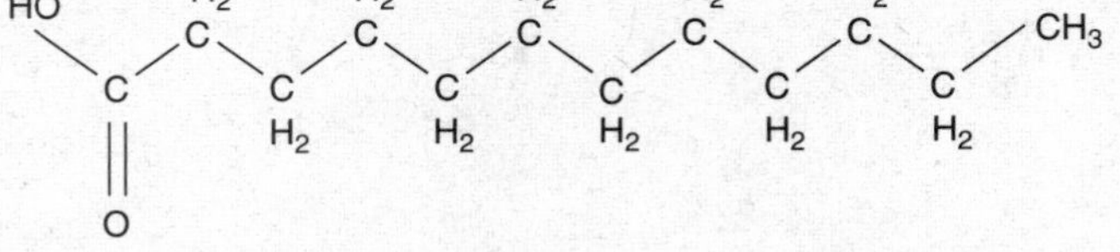

Figure I The structure of a fatty acid

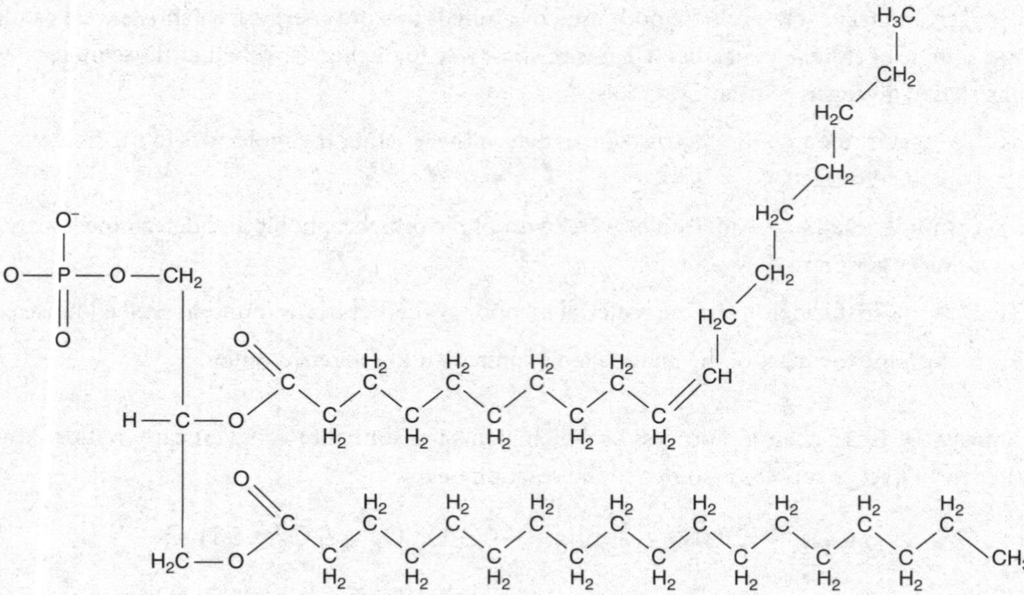

Figure II The structure of a phospholipid, which is a major component of membranes

6. Phospholipids have a polar end that can form hydrogen bonds and is hydrophilic, and a nonpolar end that is hydrophobic. Based on this information and the information given in the preceding material, which of the following is true about the fatty acid shown in Figure I?

    (A) The fatty acid is unlike the phospholipid because the fatty acid does not have a hydrophilic end that interacts with water. (The fatty acid is completely hydrophobic.) Therefore, phospholipids can form bilayers in water while fatty acids cannot.

    (B) Like the phospholipid, the fatty acid has both a hydrophilic end and a hydrophobic end. Therefore, fatty acids can also form bilayers in water.

    (C) The fatty acid is unlike the phospholipid because the fatty acid does not have a hydrophobic end that will not interact with water. (The fatty acid is completely hydrophilic.) Therefore, phospholipids can form bilayers in water while fatty acids cannot.

    (D) Translation of mRNA by ribosomes produces fatty acids. The sequence of carbon atoms incorporated in the polymer chain determines whether or not the fatty acid will be able to form a bilayer.

7. One hypothesis for an evolutionary pathway that would lead to modern membranes includes a step in which early membranes were composed of fatty acids. Which of the following best explains why fatty acids are a good candidate for a component of early membranes?

   (A) Since fatty acids are capable of catalyzing reactions and are simpler than protein-based enzymes, they are likely to have evolved first.

   (B) Phospholipids likely evolved into fatty acids because the complex structure of the phospholipid was unnecessary.

   (C) Both fatty acids and phospholipids are capable of storing genetic information. Fatty acids likely evolved first because they contain hereditary material.

   (D) Fatty acids are simpler in structure than phospholipids, but they still retain the ability to form a barrier that defines the inside and outside of a cell.

Questions 8–10

A team of researchers is studying a new species of bacteria that was recently discovered in Rio Tinto, a river in Southwestern Spain whose waters flow with a red tint due to a high concentration of iron. The river is also known for its extremely low pH and its high concentrations of heavy metals such as copper, gold, and silver. Though the waters have been deemed too dangerous for humans to swim in, there are a variety of prokaryotic and eukaryotic communities that have made the river their home.

While scientists initially believed that the large concentration of copper and mining operations were the cause of the low pH, new information led to the discovery that the decrease in pH was actually due to the oxidation of sulfur by chemolithotrophic organisms, which are a group of organisms characterized by their ability to oxidize inorganic compounds to obtain energy. The remaining organisms that inhabit this extreme ecosystem fall into the following three groups: photosynthetic primary producers, consumers, and decomposers. Further investigation led them to discover that the two main microbial processes that define this environment are the oxidation of iron and sulfur.

8. While the river itself is uninhabitable for most organisms, the climate surrounding this large body of water tends to be extremely moderate and pleasant to live in. Which of the following statements best explains the hospitable climate of the surrounding area?

   (A) The low specific heat of water helps to moderate air temperatures.

   (B) A great deal of heat is absorbed as hydrogen bonds form following water condensation.

   (C) Water releases a lot of heat to the environment as it cools.

   (D) Very little solar heat is absorbed during the gradual rise in water temperature.

9. A sample of chemolithotrophic bacteria *L. ferrooxidans* is isolated and cultured from a sample of the Rio Tinto water. Researchers determine the rates of synthesis of several organic compounds found in the *L. ferrooxidans* in water from the Rio Tinto and in a nutrient-rich water solution, as show in the graph below.

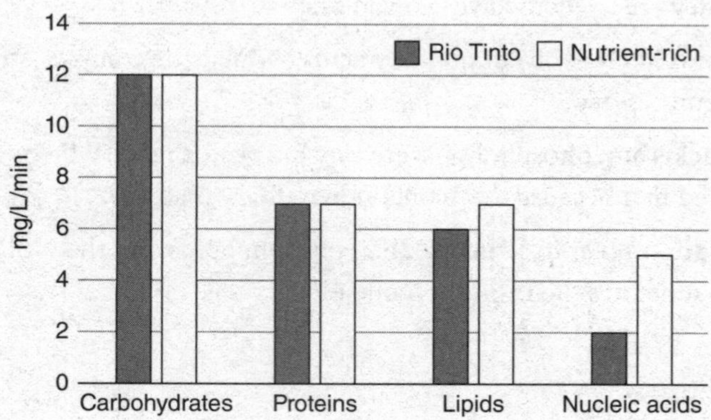

Based on the results shown, which of the following elements is most likely deficient in the river?

I. Carbon

II. Phosphorus

III. Nitrogen

(A) I only

(B) II only

(C) II and III only

(D) I, II, and III

10. Researchers begin to study the metabolic strategies used by the various eukaryotic and prokaryotic organisms in the river. Which of the following could be a possible explanation for their observation that not all of the organisms use the same strategies to meet their energy requirements for growth, reproduction, and homeostasis?

(A) Prokaryotes do not possess introns within their DNA, while most eukaryotes contain many of these noncoding sections of DNA.

(B) Algae have cell walls that protect them from the harsh environment, while bacteria use their flagella to transport themselves to pockets of water with higher pH.

(C) The protozoa were forced to evolve at a more rapid rate than their bacterial counterparts to adapt to the low pH in the river.

(D) Some organisms are able to generate ATP via chemiosmosis, while others are limited to what they can create through glycolysis.

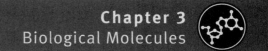

11. Cellulose and starch both have the same empirical formula. However, an alteration in the location of their linkages between monomers causes them to have completely different functions. Which of the following best explains the differences between these two molecules?
    (A) Starch is hard to digest, while cellulose is easy to digest.
    (B) Starch is used for energy storage, while cellulose is used for structural support.
    (C) Starch is broken down into glucose, while cellulose is broken down into fructose.
    (D) Starch has 1-4 glycosidic linkages of beta glucose monomers, while cellulose has 1-4 glycosidic linkages of alpha glucose monomers.
12. To differentiate between two unique carbohydrates, scientists added equal concentrations of both of them to a bag with selective permeability and submerged it in a container of water, as shown in the image below.

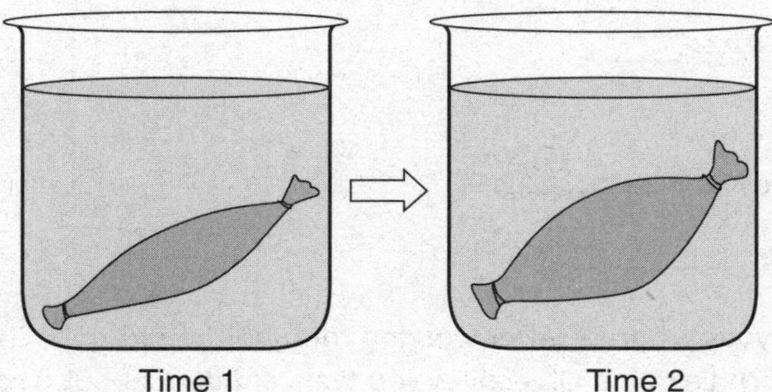

Both molecules are composed of carbon, oxygen, and hydrogen, but only one is able to move through the membrane. The mass of each substance at the beginning, middle, and end of the experiment is shown in the table below.

	Mass Inside Bag (G)			Mass Outside Bag (G)		
Substance	**Time 1**	**Time 2**	**Time 3**	**Time 1**	**Time 2**	**Time 3**
Carbohydrate 1	10	10	10	0	0	0
Carbohydrate 2	10	7	5	0	3	5

Given this information, which of the following molecules could the nonmotile carbohydrate be?
(A) Starch
(B) Fructose
(C) Ribose
(D) Glucose

Answer Key

Test What You Already Know

1. D
2. D
3. C
4. B
5. D

Test What You Learned

1. C
2. B
3. A
4. B
5. A
6. B
7. D
8. C
9. B
10. D
11. B
12. A

REFLECTION

Test What You Already Know score: ___________

Test What You Learned score: ___________

Use this section to evaluate your progress. After working through the pre-quiz, check off the boxes in the "Pre" column to indicate which Learning Objectives you feel confident about. Then, after completing the chapter, including the post-quiz, do the same to the boxes in the "Post" column. Keep working on unchecked Objectives until you're confident about them all!

Pre	Post	
❑	❑	**3.1** Recall the elements and molecules important to life
❑	❑	**3.2** Explain why life requires free energy
❑	❑	**3.3** Recognize the role of water in biological processes
❑	❑	**3.4** Differentiate between common monomers and polymers
❑	❑	**3.5** Determine how structure affects properties of polymers

FOR MORE PRACTICE

Complete more practice online at kaptest.com. Haven't registered your book yet? Go to kaptest.com/booksonline to begin.

ANSWERS AND EXPLANATIONS

Test What You Already Know

1. D Learning Objective: 3.1

The process labeled "A" shows carbon moving from animals to carbon dioxide in the atmosphere, which must be cellular respiration. Animals use cellular respiration to synthesize ATP from molecules in food, giving off CO_2 as a byproduct. **(D)** is thus correct. (A) is incorrect because it describes decomposition, in which dead plants and animals are broken down by bacteria. (B) is incorrect because it describes the process of photosynthesis, which actually removes carbon dioxide from the atmosphere and which is not performed by animals. (C) is incorrect because it describes the formation of fossil fuels.

2. D Learning Objective: 3.2

Living systems require a continuous input of energy to maintain their ordered state. During the process of maintaining order (dynamic homeostasis), some energy is transferred to the surroundings, thereby increasing the total entropy of the universe. The tendency of the universe to increase in entropy over time is described in the second law of thermodynamics, making **(D)** correct. (A) and (B) are incorrect because energy can never be created or destroyed, but only converted into other forms (according to the first law of thermodynamics). (C) is incorrect because the first law does not pertain to entropy.

3. C Learning Objective: 3.3

Water will reach a higher point if adhesion is greater in the vessel. Adhesion would be greater in the negatively-charged lining of the flowering plant's vessels, while adhesion would be minimal to a hydrophobic surface, such as the lining of *Myrothamnus*. **(C)** is thus correct. (A) and (B) are incorrect because the dye will reach a higher level in the flowering plant. (D) is incorrect because the hydrophobic lining only affects adhesion and not cohesion, which describes the attractive forces between the water molecules themselves.

4. B Learning Objective: 3.4

Since the molecule is single-stranded, it is likely RNA. DNA is usually double-stranded. Although certain viruses have single-stranded DNA, the scientist has already determined that the nucleic acid does not have viral origins. Because RNA contains ribose (a pentose), a phosphate group, and the bases adenine, uracil, cytosine, and guanine, **(B)** is correct. (A) is incorrect because only DNA contains thymine. (C) is incorrect because ribose is a pentose (5-carbon sugar), not a tetrose (4-carbon sugar). (D) is incorrect because nucleic acids do not contain fatty acids.

5. D Learning Objective: 3.5

The hydrophilic portion of the protein would be found toward the outside of the plasma membrane extending into the cytoplasm of the cell or outward into the extracellular matrix. In the intact cell, the hydrophobic portion would be contained within the hydrophobic lipid region of the membrane. **(D)** is thus correct. (A) is incorrect because the interior of the cell contains cytoplasm, an aqueous solution, which would repel the hydrophobic portion of the protein. (B) is incorrect because the interior of the phospholipid bilayer is hydrophobic and would repel the hydrophilic portion of the protein. (C) is incorrect because the hydrophobic portion of the protein would most likely be embedded in the membrane, not on its surface, and because the hydrophilic portion would not necessarily extend to the outside of the cell but could also extend inside.

Test What You Learned

1. C Learning Objective: 3.2

According to the graph, the energy of the products is higher than the energy of the reactants, which means that energy is absorbed during the reaction. Such a reaction is described as endergonic, making **(C)** correct. (A) is incorrect because energy is absorbed, not released, in an endergonic reaction. (B) and (D) are incorrect because the reaction is not exergonic.

2. B Learning Objective: 3.3

Panting works in an analogous fashion to sweating, with both processes using the evaporation of water as a cooling mechanism. Because water has a high heat of vaporization, a lot of energy is absorbed (and removed from the reindeer) in the process of converting liquid water to vapor. **(B)** is therefore correct. (A) is incorrect because the surfaces of the nose and mouth are too small to allow much heat to radiate away. (C) and (D) are incorrect because absorbing heat from outside the reindeer would actually increase body temperature, not decrease it.

3. A Learning Objective: 3.5

Starch and glycogen are more similar in structure than either is to cellulose. Humans do not have the necessary enzyme to break the bonds connecting glucose molecules in cellulose, even though some organisms do have these enzymes. **(A)** is thus correct. (B) is incorrect because the glucose monomers are identical in all three polymers. (C) is incorrect because some herbivorous mammals (e.g., cows, horses, and goats) are capable of digesting cellulose, due to the symbiotic bacteria that inhabit their digestive tracts. (D) is incorrect because cellulose does not have more associated water molecules than other glucose polymers and because it is not necessarily larger than the others.

4. B Learning Objective: 3.4

Triacylglycerols are hydrophobic and compact, containing more energy per gram than glycogen. Glycogen is hydrophilic and takes up more space. Thus, determining whether the unknown molecules are hydrophobic or hydrophilic and ascertaining their energy density would be sufficient to distinguish between the two types of compounds. **(B)** is correct. (A) is incorrect because both types of compounds contain oxygen. (C) is incorrect because both types of compounds can be found in skeletal muscle. Even though knowing the molar mass of an unknown substance can help to identify it, (D) is incorrect because determining the mass of the material would be insufficient without also knowing the number of moles present.

5. A Learning Objective: 3.1

Reduction is the gain of electrons whereas oxidation is the loss of electrons. (You can also think of oxidation as an increase in the number of bonds to oxygen and reduction as a decrease in the number of bonds to oxygen.) During photosynthesis, electrons are given off from water and accepted by derivatives of carbon dioxide. This results in the production of glucose, which supplies the plant with food, and oxygen gas is a byproduct. **(A)** is thus correct. (B) is incorrect because carbon dioxide is reduced and water oxidized. (C) is incorrect because oxygen is not a food source. (D) is incorrect because it combines the errors from (B) and (C).

6. B Learning Objective: 3.5

A fatty acid molecule has a carboxylic acid group (COOH) that can hydrogen bond with water molecules and a hydrocarbon chain that is hydrophobic. The phosphate group (PO_4), amino group (NH3), and oxygen atoms on the fatty acid tails of phospholipids are all hydrophilic, while the hydrocarbon chains are hydrophobic. As a result, both of these molecules are capable of forming bilayers when placed into water. Thus, **(B)** is correct. (A) is incorrect because fatty acids *do* have a hydrophilic end. (C) is incorrect because fatty acids are *not* completely hydrophilic and *can* form bilayers in water. (D) conflates fatty acids with proteins.

7. D Learning Objective: 3.5

Living organisms must have a boundary, such as a membrane, that defines the limits of the organism. There must have been a stepwise pathway, from nonliving things to the complex organisms of today, that involved simpler molecules performing the functions of more complex ones. One such function is the formation of a barrier by fatty acids, making **(D)** correct. Neither fatty acids nor phospholipids are capable of catalyzing reactions. Therefore, (A) is incorrect. (B) reverses the likely evolutionary sequence. Fatty acids were likely to be major components of early membranes and were replaced by phospholipids later. (C) is incorrect because fatty acids and phospholipids are *not* capable of storing genetic information.

8. C Learning Objective: 3.3

Living systems depend on properties of water that result from its polarity and hydrogen bonding, such as its specific heat and heat of vaporization. Due to its high specific heat, water is able to absorb a large amount of solar heat while undergoing only a slight change in temperature. This allows water to act as both an effective coolant during a hot day, due to the large amount of heat needed to heat up water, and as a warming mechanism during the fall and winter, since a large amount of heat release is required to cause the water to become cold. As a result, winds that blow off large bodies of warm water make the temperature less frigid than it would otherwise be. Therefore, **(C)** is the correct answer, and (A) and (D) can be eliminated. Similarly, due to the high heat of vaporization, water molecules must absorb a large amount of energy to break bonds and change state. Therefore, (B) is also incorrect because heat is released, not absorbed, during hydrogen bond formation.

9. B Learning Objective: 3.1

The graph shows that nucleic acid synthesis is the only thing that is significantly decreased in the Rio Tinto and that the remaining organic compounds are largely unaffected. Therefore, the river must be deficient in the element that appears only in nucleic acids. Nitrogen is found in nucleic acids and proteins, while carbon is found in all four organic compounds, meaning that phosphorus must be the only deficient element. Because only element II is deficient, **(B)** must be the correct answer. Phosphorus is found in nucleic acids, ATP, and phospholipids, which may explain the minor difference in lipid generation between both water sources as well.

10. D Learning Objective: 3.2

Metabolic strategies are the means by which organisms obtain the energy and nutrients they need to live and reproduce. The specific metabolic properties of different organisms are the major factors in determining their ecological niches. Organisms are often divided based on (1) how they obtain carbon (autotrophic vs. heterotrophic), (2) how they obtain reducing elements used in energy conservation or biosynthetic reactions (lithotrophic vs. organotrophic), and (3) how they obtain energy for living or growing (chemotrophic vs. phototrophic). Autotrophs are organisms that produce their own food via photosynthesis or chemosynthesis, while heterotrophs, which are unable to do so, must rely on other organisms for nutrition. Lithotrophs obtain reducing elements from inorganic compounds, while organotrophs obtain reducing elements from organic compounds. Chemotrophs obtain energy from external chemical compounds, while phototrophs obtain energy from light. When used to describe an organism these terms are typically combined, such as the "chemolithotrophic organisms" referenced in the stimulus. Different metabolic strategies yield different levels of nutrients and energy, thereby leading to variation in unrelated organisms' abilities to grow, reproduce, and maintain homeostasis. **(D)** highlights one of these metabolic differences and thus provides a reasonable explanation for why growth, reproduction, and homeostasis requirements may be met in different ways. (A) is incorrect, despite being a true statement, because the presence of introns has nothing to do with these organisms meeting their energy needs through different metabolic strategies. (B) is incorrect because both bacteria and algae are capable of having cell walls. Additionally, cell walls mainly function to provide structural support and protect cells from osmotic lysis. They offer little protection with respect to extreme pH conditions like those that exist in the Rio Tinto. While chemotaxis may direct bacterial movement toward a region of higher pH or away from a region of lower pH, it cannot be assumed that these pockets exist in the river. Finally, neither of the structures mentioned in (B) provides any type of metabolic advantage. (C) is incorrect because prokaryotes usually evolve faster than eukaryotes due to their usage of asexual reproduction and short generation times.

11. B Learning Objective: 3.5

Both starch and cellulose are polymers of glucose, eliminating (C). Structurally, starch contains 1-4 glycosidic linkages of alpha glucose monomers, while cellulose contains 1-4 glycosidic linkages of beta glucose monomers, eliminating (D). While most organisms have enzymes that can hydrolyze the alpha linkages of starch, very few have enzymes that can hydrolyze the beta linkages of cellulose, eliminating (A). By process of elimination, **(B)** is the correct answer. Indeed, due to the beta linkages in cellulose, the polymer forms a straight line, making it very effective for structural support. Conversely, the alpha linkages in starch cause it to form a spiral, making it less optimal for structural support but perfect for storing a large quantity of glucose in a limited space. This structure also allows short chains of glucose to branch off of the main helix, and, as a result, when plants need energy, they can quickly break down these smaller branches.

12. A Learning Objective: 3.4

All carbohydrates are composed of carbon, oxygen, and hydrogen and follow the empirical ratio $Cm(H_2O)n$ (where m and n are whole numbers). There are three different types of carbohydrates, based on the number of subunits they have. Monosaccharides have a single subunit or monomer. Common monosaccharides are fructose, glucose, ribose, and galactose. Disaccharides are composed of two sugar subunits, which are joined together through dehydration synthesis, also known as a condensation reaction. Common disaccharides are maltose, lactose, and sucrose. The third type of carbohydrates, polysaccharides, are formed when more than two monosaccharides are joined together. Common polysaccharides are cellulose, starch, glycogen, and chitin. In this question, three of the answers, (B), (C), and (D), are monosaccharides, while the fourth, **(A)**, is a polysaccharide. The question stem states that the membrane is selectively permeable, but since there are no active transport mechanisms built into the bag, passive transport is the only way for molecules to enter or leave. As a result, small molecules like water and monosaccharides will be able to cross, while a large polysaccharide like starch will be unable to do so. Therefore, **(A)** is correct.

CHAPTER 4

The Origin of Life and Natural Selection

LEARNING OBJECTIVES

In this chapter, you will review how to:

- **4.1** Evaluate evidence for origin hypotheses
- **4.2** Recall Darwin's theory
- **4.3** Explain how natural selection impacts evolution

TEST WHAT YOU ALREADY KNOW

1. One important hypothesis that addresses the origin of life on Earth is the RNA world hypothesis. Which of the following statements provides the greatest support for the RNA world hypothesis?
   (A) RNA is a more stable molecule than DNA.
   (B) The first self-replicating structures were RNA-based protobionts.
   (C) RNA was the first chemical polymer to form on Earth.
   (D) The first living cells used reverse transcription to produce a DNA genome from RNA.

2. White sand lizards vary in their color, which may be an adaptation to their environment. Those living in lighter environments may blend in better when their color is lighter, and those with darker coloration may fare better on darker backgrounds. Researchers used a nontoxic paint to color the backs of lizards with similar natural colorations and placed them in environments that matched their color ("matched") or that conflicted with their color ("mismatched"). The figure below shows the proportion of lizards in each treatment group that were recaptured (i.e., still alive) after 16 days.

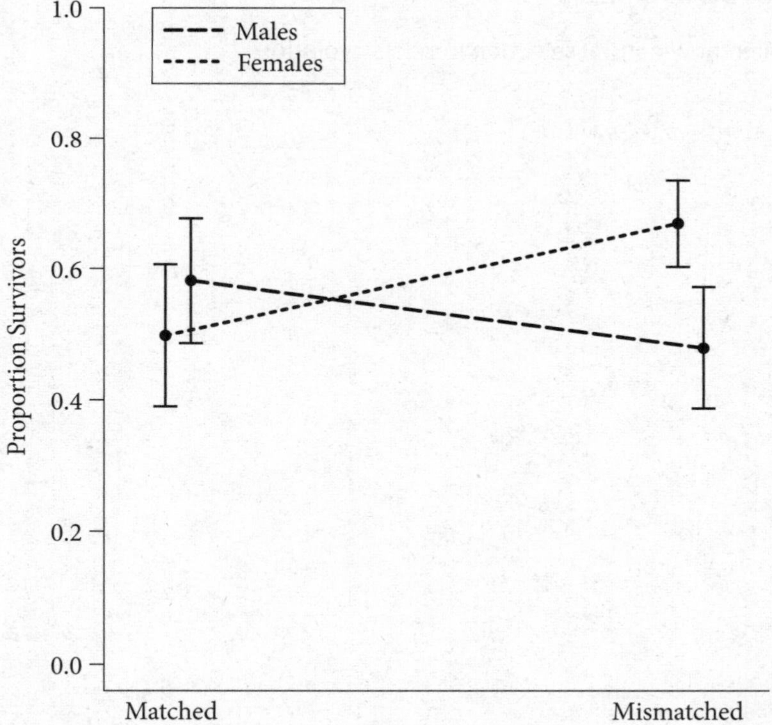

Adapted from Kayla M. Hardwick et al., "When Field Experiments Yield Unexpected Results: Lessons Learned from Measuring Selection in White Sands Lizards," *PLoS ONE* 10, no. 2 (2015): e0118560.

Assume that males and females reproduce at a similar rate. According to Darwin's theory of natural selection, which group of individuals most likely had the highest fitness?

(A) Female lizards painted to resemble their background had the highest fitness.

(B) Male lizards painted to resemble their background had the highest fitness.

(C) Female lizards painted a different color from their background had the highest fitness.

(D) Male lizards painted a different color from their background had the highest fitness.

3. A scientist studying bird migration places bands around the legs of each bird that visits her research site each year in order to examine whether increased water temperatures affect survival. In this way, she is able to document a pattern that larger birds are increasingly more likely to return than smaller birds. What conclusion can most reasonably be drawn from these data?

(A) Larger birds probably have access to more food, allowing them more energy to migrate successfully.

(B) The birds may become larger in future generations due to selective pressure.

(C) Smaller birds may reproduce at younger ages so that they can reproduce successfully despite having shorter lifespans.

(D) The birds may cease migrating so that more of them can survive.

Answers to this quiz can be found at the end of this chapter.

ORIGIN HYPOTHESES

4.1 Evaluate evidence for origin hypotheses

One of the most enduring debates in biology concerns the origin of life. While evolution is universally accepted among biologists, there is still disagreement concerning how the whole process began. Evolution, as we currently understand it, depends not merely on the existence of heritable information in the form of genes, but also on the numerous enzymes and other cellular structures that facilitate the replication of those genes and the transmission of them to offspring. The question then emerges: how did genes, enzymes, and cells come about in the first place?

There are some generally accepted points of agreement. Scientists estimate that the Earth is about 4.5 billion years old and that, in the beginning, it was a very inhospitable place. When Earth first came into existence, there was very little or no atmospheric oxygen, and the surface of the Earth was bombarded by intense ultraviolet radiation. Around 3.9 billion years ago, there were heavy rains and violent storms, which led to the production of basic inorganic chemical building blocks from the soil and the accumulation of energy needed to drive reactions for producing simple organic molecules.

The prevailing theory of the origin of life, sometimes known as the "organic soup" model, is that these organic molecules became more and more complex until amino acids and nucleic acids were formed. Once strings of nucleic acids formed, they could self-replicate within the "soup." These self-replicating structures organized into **protobionts**, which were droplets of segregated chemicals. Chemicals continued to organize until the first identifiable cells, the first unicellular organisms, came into being.

Of course, there are points of dispute concerning the details of this process, such as whether the initial reactions that produced organic molecules actually occurred in solution, as proposed by the organic soup model, or on solid reactive surfaces, such as clay particles. There is also uncertainty over whether the first self-replicating molecules were deoxyribonucleic acid (DNA) or ribonucleic acid (RNA). The RNA world hypothesis theorizes that RNA molecules were the first self-replicators, serving as both heritable information and as functional enzymes. The current paradigm of life, in which DNA is the primary source of heritable information and proteins the primary biological catalysts, only emerged later, according to this view.

Evidence for origin theories comes from biochemistry, laboratory experiments, and models of early Earth environments. However, the strongest evidence can be found in the fossil record. The first prokaryote fossils have been found in geological deposits thought to be 3.5 billion years old. The oldest eukaryotic cells are from deposits that are about 2 billion years old, while the oldest fossils of multicellular organisms are from deposits that are 1.25 billion years old.

Molecular and genetic evidence demonstrates that all life on Earth shares a common ancestor. For example, all eukaryotic cells share common traits, such as the presence of a cytoskeleton, a nucleus, membrane-bound organelles, linear chromosomes, and endomembrane systems, elements that have likely been conserved from some of the earliest life-forms.

DARWIN AND NATURAL SELECTION

4.2 Recall Darwin's theory

4.3 Explain how natural selection impacts evolution

Early Evolutionary Ideas and Darwinism

Although some early scientists believed that species are immutable and do not evolve or change, others believed that evolution occurs. Theories of evolution developed long before the time of **Charles Robert Darwin** (1809–1882), some even emerging in the ancient world. However, prior to Darwin, even the scientists who accepted the idea that species change over time did not have a solid idea for a mechanism that could explain the changes they observed.

The most well-known hypothesis prior to Darwin was that of **Jean-Baptiste de Lamarck** (1744–1829), who suggested that organisms pass on acquired traits in an attempt to reach a more perfect form. For example, by Lamarck's logic, if a mother works out at the gym and becomes strong and healthy, she will pass her acquired strength and health to her children, and so on. Lamarck's ideas concerning the inheritance of acquired characteristics have generally been discredited, though some of them are being reconsidered in light of research into epigenetics, the study of heritable factors other than genes (such as molecular modifications to DNA that influence gene expression).

Darwin presented his postulates for his **theory of evolution** by **natural selection** in his work ***The Origin of Species*** (1859). There have been slight modifications to the theory based on more up-to-date knowledge about genetics and molecular biology but, for the most part, Darwin's theories are accepted today. Virtually all scientists consider evolution by means of natural selection to be an established fact.

Darwin had several original postulates for his theory of evolution by natural selection:

1. Individuals vary in their characteristics within a population. This means that all giraffes have long necks, but their necks aren't all the exact same length.
2. The variations observed in populations are inherited. When a big dog has puppies, they tend to be big, and a little dog's puppies tend to be little.
3. A considerable number of individuals in a population seem to die as they compete for limited resources in the environment. This is where the term "survival of the fittest" emerged; "fit" organisms are simply the ones that don't die off, those with characteristics that make them more likely to survive to reproductive age within their specific environments. These characteristics could include being bigger or stronger, but they also could include being smaller and smarter—what counts as fitness depends on the environment and the role that the organism plays within it.
4. Individuals who have more resources because of their particular characteristics tend to produce more offspring that survive. For example, if a bird with a long beak can get more food from holes in trees because of its long beak, it will be more likely to survive and provide more food for its offspring. If beak length is a result of genetics, that bird's offspring are more likely to have long beaks, and the following generations of offspring are more likely to have long beaks, until every bird in the population has a long beak.

The selection for more "adaptive" traits tends to narrow a population of individuals down to those who are best suited for a particular environment. If changes occur in the environment, selection favors individuals best suited to the new environment. The theory of natural selection could explain the differences Darwin observed in species he studied and helps to explain biodiversity in organisms today.

Natural Selection and Variation

Natural selection is the differential survival and reproduction of individuals based on variation in genetically controlled traits. These differing rates of survival and reproduction are due to forces in the environment and/or to forces exhibited by other species. Evolutionary **fitness** is measured by the reproductive success of a species. To understand natural selection, it is necessary to understand variation.

Ultimately, all variation originates in the mutation of DNA in an organism's genome. For a mutation to have an impact on evolution, it must occur in a gamete and be passed on to offspring. If a mutation occurs in a gamete that forms a zygote, the offspring will inherit that mutation and pass it on to its offspring. Genetic variation can also occur during recombination in meiosis I. Most mutations are harmful, but occasionally a mutation exists in viable offspring. These offspring may then exhibit a phenotype that differs from the rest of the population.

✔ **AP Expert Note**

Variation that occurs in a population will have a distribution based upon the kind of natural selection that is taking place in the population.

The three types of variation that can occur in a population are stabilizing selection, directional selection, and disruptive (diversifying) selection. If a population is subject to **stabilizing selection**, extremes at both ends of a phenotype are eliminated, resulting in less genetic variability. For example, if the variation in color of a bird species ranges from dark gray to white and the population is subject to stabilizing selection, the medium-gray phenotype will be most common. If the population is under **directional selection**, one extreme is preferentially selected against the other (e.g., white birds being more easily spotted by predators leading to the selection of dark gray), so that the average in the population moves in one direction. **Disruptive (diversifying) selection** favors both extremes while selecting against the average, which would mean dark gray and white are selected over medium-gray in the case of the birds.

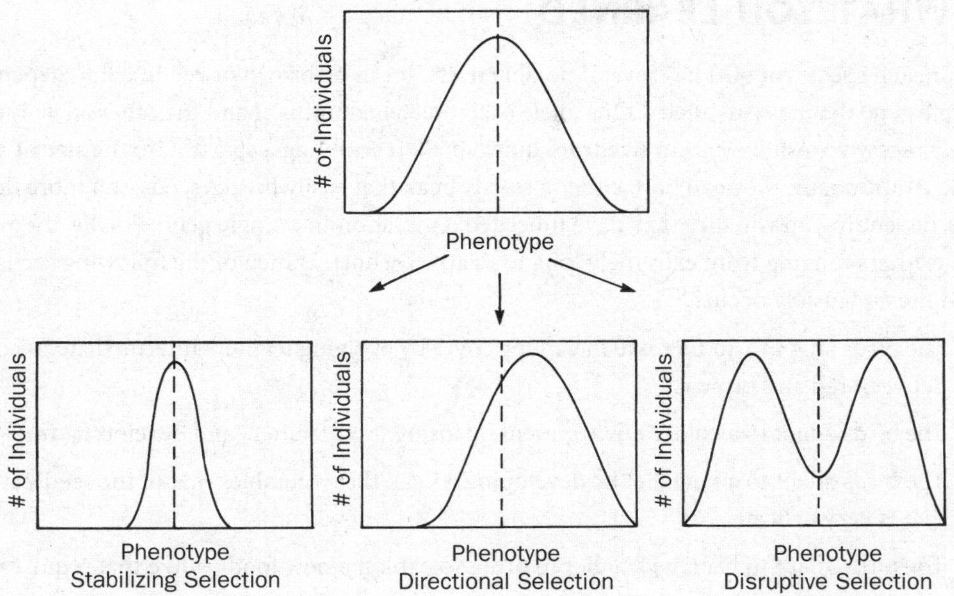

Modes of Natural Selection

RAPID REVIEW

If you take away only 3 things from this chapter:

1. It is thought that the chemical components of life on Earth originated through radiation and storms. These compounds became increasingly complex, forming protobionts and, ultimately, living organisms.
2. According to Darwin, species evolve via natural selection, in which animals with certain traits are more likely to survive and reproduce, passing on those traits.
3. Selection can be stabilizing (median is encouraged), directional (the norm shifts toward one extreme), or disruptive (extremes are favored over the norm).

TEST WHAT YOU LEARNED

1. An unusual species of bird has several notable traits. It can be brown or red in color, depending on a single gene that has two alleles. One allele is completely dominant and the other allele is completely recessive. Adults vary in size from quite small, resembling a sparrow, to the size of a small hawk. Additionally, they can have either a sturdy beak that easily breaks seeds or a more delicate beak, depending on whether they have inherited a mutation in a single gene. Finally, they can have feathers ranging from extremely long to relatively short. Which of the following scenarios could most plausibly occur?
   (A) The birds adapt to an increase in canopy cover by evolving feathers intermediate in color between red and brown.
   (B) The birds adapt to a colder environment by losing their feathers and developing fur.
   (C) The birds adapt to a new diet by developing a beak that resembles that of the seed-eaters but that is very delicate.
   (D) The birds adapt to become smaller in order to exploit a new food source that requires them to enter small crevices.

2. Which of the following statements is most consistent with scientists' understanding of the origin of life on Earth?
   (A) The first cells must have been composed of simple monomers, rather than more complex polymers.
   (B) Early life evolved in an anaerobic environment, meaning that the first cells must have been able to grow without oxygen.
   (C) The first cells grew in a relatively cold environment, requiring enzymes that could function at low temperatures.
   (D) Early life appeared multiple times, meaning that living organisms have evolved from several different original ancestors.

3. In order to study possible effects of changing ocean conditions, researchers caught Atlantic cod larvae and grew them in tanks. The researchers wanted to examine how well the cod could survive with carbon dioxide levels greater than those found naturally (i.e., high levels of "ambient CO_2"). Increased CO_2 levels lead to the phenomenon of ocean acidification because carbon dioxide acts as an acid when dissolved in water. The figure below shows the survival curves for varying food conditions and carbon dioxide levels.

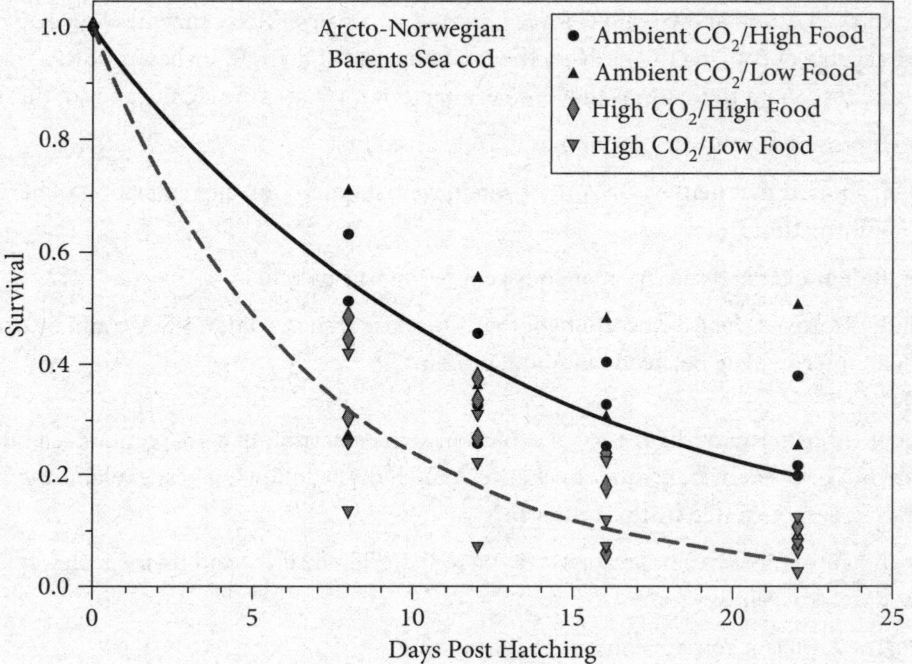

Adapted from Martina H. Stiasny et al., "Ocean Acidification Effects on Atlantic Cod Larval Survival and Recruitment to the Fished Population," *PLoS ONE* 11, no. 8 (August 2016): e0155448.

When conditions change too rapidly, organisms may not be able to adapt and may become extinct. If change is slower, animals may be able to adapt. What would be required for the cod to be able to adapt to the changing conditions described above?

(A) The cod would need genetic variation in genes associated with water acidity.

(B) The cod would need genetic variation in genes allowing them to metabolize carbon dioxide more efficiently.

(C) The cod would need to learn how to consume food more efficiently.

(D) The cod would need to move to cooler locations in order to survive.

4. RNA is well known as an information molecule. However, RNA can also have enzymatic functions. Ribozymes are enzymes made of RNA. The RNA world hypothesis states that early life used RNA as both information molecules and catalysts. The hypothesis also says that DNA and protein evolved later. DNA is a more durable information molecule than RNA. Protein is a more versatile structural and catalytic molecule than RNA. Except for some RNA viruses, DNA is the molecule of heredity. It is replicated and passed to offspring. The RNA viruses that use RNA as their genomes use RNA-dependent RNA polymerases (RdRPs) made of protein to copy RNA from RNA. In 2013, Antonio C. Ferretti and Gerald F. Joyce reported on a pair of RNA enzymes they had designed that were capable of copying themselves. The doubling rate of these RNA-based RdRPs was about 20 minutes (about the same as that of many bacteria). What is the significance of this finding?

   (A) It proved beyond any doubt that RNA evolved first.

   (B) It showed that neither DNA nor protein were required for the replication of heritable information.

   (C) It demonstrated that spontaneous generation was possible.

   (D) It disproved the RNA world hypothesis because it showed that RNA could be useful only as an intermediate between DNA and protein.

5. A popular criticism of the theory of evolution is the assertion that adaptations seen in nature could not have evolved by random chance. Which of the following best explains the misunderstanding demonstrated by this criticism?

   (A) Evolution involves organisms striving to better themselves and their species. It is a nonrandom process.

   (B) The criticism contains no misunderstandings.

   (C) If given enough time, random events can make anything happen.

   (D) Evolution requires both the random variability of mutation and the nonrandom process of natural selection.

6. Which of the following examples does not illustrate a principle from Darwin's theory of natural selection?

   (A) Wild animals, such as crocodiles and wolves, produce excess offspring because only a fraction will survive to adulthood.

   (B) The number of dark-colored insects in a population increased after a forest fire caused the landscape to become darker.

   (C) Rabbits that can run faster are better able to outrun predators, increasing the relative fitness of their genotype.

   (D) Some reptiles and amphibians reproduce asexually, which impacts their ability to survive by decreasing the occurrence of significant variation.

7. The deep sea vents model for the origin of life proposes that life began at submarine hydrothermal vents rich in organic molecules. The extreme heat found at deep sea vents provided conditions favorable for chemosynthesis, the conversion of carbon and nutrients into organic molecules (the building blocks of life) in the absence of sunlight. Which of the following statements would provide the best evidence for this model regarding the origin of life?

   (A) Scientists discovered fossils of plants from 3.6 billion years ago with organelles that had photosynthetic properties.

   (B) Fossils of a particular species of fish that is indigenous to only the Atlantic ocean were found in three different oceans across the globe.

   (C) Chemical analysis of fossils showed that the earliest forms of life were thermophilic, meaning they lived in systems composed of hot water.

   (D) Chemical analysis of fossils revealed that early yeast cells had RNA molecules with enzymatic properties.

8. Farmers routinely use antibiotics to treat livestock in an attempt to prevent and treat infections that arise due to overcrowding and poor living conditions. One example of an antibiotic used since 1995 is ciprofloxacin in poultry production. However, studies have have shown that overuse of antibiotics has led to antibiotic-resistant bacteria. By 1999, about 20 percent of chicken breasts tested by the FDA contained ciprofloxacin-resistant *Camplobacter*, a bacterium known to cause diarrhoeal disease. By 2005, nearly 30 percent of the *Camplobacter* found in chicken breasts across the United States were ciprofloxacin resistant. During that year, hospitals reported that after ciprofloxacin use, ciprofloxacin resistance in *Camplobacter* populations occurred in weeks, rather than months.

   Which of the following statements best explains these observations concerning ciprofloxacin resistance in *Camplobacter*?

   (A) Natural selection favors ciprofloxacin-resistant *Camplobacter* that were already in the population when ciprofloxacin exposure occurred.

   (B) The proportion of ciprofloxacin-resistant *Camplobacter* remained constant in the population as more patients with diarrhoeal disease were treated with ciprofloxacin.

   (C) Treatment with ciprofloxacin caused a decrease in *Camplobacter* growth and division, resulting in a greater proportion of ciprofloxacin resistance in the population.

   (D) New strains of *Camplobacter* unexposed to ciprofloxacin were constantly introduced to the population.

9. Sickle cell disease is an inherited group of blood disorders that results in the production of abnormal hemoglobin, which causes red blood cells to become distorted into a sickle shape. These cells are very fragile and rupture easily, leading to anemia and eventually death, if not managed properly. Additionally, these irregularly shaped cells can cause blockages within blood vessels, leading to pain and organ failure. To develop the disease, a person must inherit two mutated copies of the hemoglobin gene. If only one faulty gene is inherited, the offspring is often asymptomatic or may display a very mild sickling of cells.

   Researchers who observe that the incidence of sickle cell anemia is significantly higher in Americans whose grandparents or great-grandparents immigrated from the Middle East, India, or Africa, begin to investigate possible reasons for this occurrence. They come across the map below, displaying the risk of malaria worldwide, and find that it matches the increased frequency of the sickled hemoglobin gene observed in different American populations.

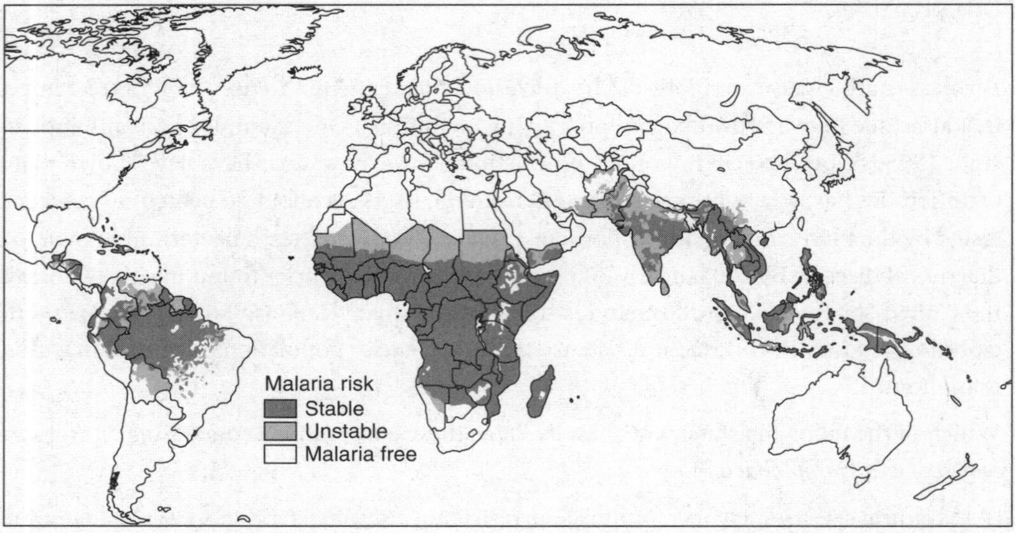

   Which of the following conclusions is best supported by this data?

   (A) There is no correlation between the increased frequency of the sickled hemoglobin gene in locations known for a high risk of malaria.

   (B) Individuals who are carriers of the sickled hemoglobin gene have an advantage over individuals without it in countries known for a high risk of malaria.

   (C) Individuals who are carriers of the sickled hemoglobin gene provide a better target for malaria-carrying mosquitoes to feed on.

   (D) Being infected with malaria causes individuals to develop mutations that result in sickle cell disease if both copies of the hemoglobin gene are mutated.

10. Ribozymes are RNA molecules that have enzyme-like characteristics. They have been found to be involved in catalyzing chemical reactions such as the synthesis of short complementary strands of RNA. There has even been evidence of ribozymes that are capable of self-replication, given ample amounts of nucleotide building blocks in their environment. The existence of these ribozymes best supports which of the following models for the origin of life?

    (A) The deep sea vents model, because the submarine hydrothermal vents rich in organic molecules provided the building blocks for ribozymes to self-replicate

    (B) Oparin-Haldane's primordial "soup" model, because the nucleotide building blocks necessary for RNA replication were formed as a result of Earth's reducing atmosphere

    (C) The RNA world hypothesis, because ribozymes provide evidence that RNA could replicate, store genetic information, and catalyze chemical reactions without DNA

    (D) The volcanic origin of life model, because the extreme temperature conditions coupled with the volcanic gas formed the precursors for amino acids, which ribozymes used as building blocks to self-replicate

11. DDT is a pesticide that was commonly used in the 1950s that targets sodium channels in insects. The World Health Organization (WHO) recommended its use to help decrease the rates of mosquito-transmitted malaria in many countries around the world. However, the WHO was unable to sustain the use of DDT in areas with tropical climates, such as sub-Saharan Africa, due to the rapid life cycle of mosquitoes. Today, after receiving more funding, the WHO attempted again to use the same concentration of DDT as in the 1950s to better control mosquitoes in sub-Saharan Africa; however, the pesticide was found to be ineffective. Assuming that the mosquitoes today are descendants of those who survived DDT exposure in the 1950s, which of the following is the best explanation for the ineffectiveness of DDT?

    (A) The concentration of DDT used in the 1950s is not potent enough to kill the mosquitoes today.

    (B) The DDT-resistant mosquitoes that survived in Northern Africa naturally migrated to sub-Saharan Africa, mating with local mosquitoes, passing on the trait for DDT resistance.

    (C) The malarial parasite found in mosquitoes underwent genetic mutations over the last 60 years that allowed the mosquitos to gain DDT resistance.

    (D) The mosquitoes in sub-Saharan Africa today are descendants of the DDT-resistant mosquitoes that survived exposure to DDT in the 1950s.

Answer Key

Test What You Already Know

1. B

2. C

3. B

Test What You Learned

1. D

2. B

3. A

4. B

5. D

6. D

7. C

8. A

9. B

10. C

11. D

Test What You Already Know score: ___________

Test What You Learned score: ___________

Use this section to evaluate your progress. After working through the pre-quiz, check off the boxes in the "Pre" column to indicate which Learning Objectives you feel confident about. Then, after completing the chapter, including the post-quiz, do the same to the boxes in the "Post" column. Keep working on unchecked Objectives until you're confident about them all!

Pre	Post	
❑	❑	**4.1** Evaluate evidence for origin hypotheses
❑	❑	**4.2** Recall Darwin's theory
❑	❑	**4.3** Explain how natural selection impacts evolution

FOR MORE PRACTICE

Complete more practice online at kaptest.com. Haven't registered your book yet? Go to kaptest.com/booksonline to begin.

ANSWERS AND EXPLANATIONS

Test What You Already Know

1. B Learning Objective: 4.1

The RNA world hypothesis proposes that self-replicating RNA molecules arose early in Earth's history and became increasingly organized. These self-replicating structures were called protobionts. **(B)** is thus correct. (A) is incorrect because DNA is actually more stable than RNA, which may explain why most organisms now use DNA genomes rather than RNA genomes. (C) is incorrect because a number of polymers are believed to have preceded the existence of cells and their protobiont precursors, but which one happened to be first does not necessarily have any impact on whether protobionts were RNA-based or DNA-based. (D) is incorrect because the RNA world hypothesis maintains that the first self-replicating molecules were purely RNA-based, not requiring DNA at any point in their replication.

2. C Learning Objective: 4.2

According to Darwin's theory of natural selection, fit individuals are those that are able to survive and reproduce successfully, passing their genes on to the next generation. Because the question stem says to assume that rates of reproduction are effectively equal, the only relevant criterion for fitness will be the rate of survival. Because mismatched females had the highest survival rate, they have the greatest fitness, making **(C)** correct. (A), (B), and (D) are incorrect because all of these groups demonstrated lower rates of survival.

3. B Learning Objective: 4.3

Because larger birds are more likely to survive, there is selection for larger size. This means that birds may become larger in future generations unless other selective pressures against larger size arise. **(B)** is thus correct. (A) is incorrect because no information is given about the availability of food. (C) is incorrect because no information is given concerning the reproductive cycles of the birds. (D) is incorrect because non-human animals do not simply decide to change their behavior as a species in response to environmental circumstances; if behavior in a species of non-human animals changes, it is almost always the result of selective pressures that favor one kind of behavior over another, with those organisms that behave in the way that is selected against eventually being removed from the gene pool.

Test What You Learned

1. D Learning Objective: 4.3

Given the great variation in size among the species, it is plausible to believe that multiple genes control the size of the birds. Consequently, it is quite probable that the species would respond relatively quickly to selective pressure for a particular size, making **(D)** correct. (A) and (C) are incorrect because the question stem states that differences in color and beak depend on single genes with only two distinctive alleles. Thus, it is improbable that intermediate forms would develop easily. (B) is incorrect because none of the birds currently possess fur and it would be highly unlikely that this completely novel trait would arise in the species simply as a result of random mutations. It is far more probable that the species would develop thicker feathers in a colder environment rather than an entirely different type of protective covering.

2. B Learning Objective: 4.1

The atmosphere of the Earth was very low in oxygen content when life arose, so early life had to be able to thrive in an anaerobic environment. **(B)** is thus correct. (A) is incorrect because life requires self-replicating molecules (such as DNA or RNA), but monomers like nucleotides and amino acids are too simple to replicate themselves. (C) is incorrect because the early Earth is actually believed to have had higher temperatures than are experienced today, so the first cells had to function in a hot environment. (D) is incorrect because biologists widely agree that life originated on Earth only once, so that all living organisms share a common ancestor.

3. A Learning Objective: 4.2

In order for adaptation to a particular environmental circumstance to occur, genetic variation in a relevant trait must exist or arise through mutation. Because the increased carbon dioxide levels make their environment more acidic, the cod should possess variability in their ability to survive in acidic conditions, so that genes promoting greater survival in an acidic environment can be selected for. **(A)** is thus correct. (B) is incorrect because animals do not metabolize carbon dioxide; rather, CO_2 is a waste product of the metabolism of glucose and other food sources. (C) is incorrect because learning plays a relatively small role in the ability of simple animals like fish to survive. (D) is incorrect because no data is provided about temperature, so there is no basis for concluding that a cooler environment would help the cod to survive.

4. B Learning Objective: 4.1

RNA is capable of acting as both an information molecule, like DNA, and a catalyst, like protein. Therefore, it seems reasonable to believe that RNA performed these functions in early living things, making **(B)** correct. DNA and protein evolved later as specialized molecules that were better than RNA at their functions. However, neither DNA nor protein could carry out both functions. (A) should be rejected because the phrase "proved beyond any doubt" is an unscientific statement. Although this experiment demonstrated that it was possible to engineer self-replicating RNA molecules, and we may speculate that such a molecular system could have worked in early life, that's the extent of what we can conclude. Spontaneous generation, as mentioned in (C), is the formation of living organisms without descent from similar organisms (such as mice spontaneously forming from rags or maggots spontaneously forming from meat). An RNA molecule that can self-replicate in a test tube lacks many of the fundamental properties of life, such as a membrane. Therefore, this is not a correct choice. (D) is incorrect because the experiment *supported* the RNA world hypothesis. It showed that RNA could act as an information molecule while simultaneously acting as an enzyme capable of copying that information.

5. D Learning Objective: 4.3

The misunderstanding demonstrated by this criticism is that evolution relies *entirely* on random processes, but natural selection is nonrandom. Therefore, **(D)** is correct. (A) is incorrect because evolution is not goal directed. Organisms are not trying to evolve into something else. Organisms vary, and some varieties are advantageous in a particular environment. (B) maintains that there is no misunderstanding. However, this is not true as explained in the information provided above. (C) accepts the false premise that evolution relies on random chance alone and is therefore incorrect. The amount of time for a modern eukaryotic cell to arise by random chance is likely more than the time that the universe has existed (although this is difficult to estimate).

6. D Learning Objective: 4.2

Darwin's theory of natural selection depends on three requirements: (1) the overproduction of young in each generation, (2) a high chance of heritable variations in each generation, and (3) competition due to limited resources in the environment. **(D)** is correct because it provides an example that opposes the second requirement by decreasing the rate of mutations and, therefore, heritable variations in each generation. (A) provides an example of the first requirement, while (B) and (C) provide examples of the third requirement.

7. C Learning Objective: 4.1

(C) provides evidence that the earliest forms of life lived in water under extreme conditions of heat. This supports the deep sea vents model, which maintains that the extreme temperatures found at vents in the sea created the organic molecules that were the precursors to life. (A) and (B) are both incorrect because they provide evidence that is irrelevant to the question; photosynthetic organelles and evidence that a particular fish species was once found in several locations around the world do not support the idea that the building blocks of life were synthesized due to the extreme conditions found at deep sea vents. Similarly, (D) provides evidence for the RNA world hypothesis, not the deep sea vents model.

8. A Learning Objective: 4.3

A proportion of *Camplobacter* populations in hospitals that were treated with ciprofloxacin was already ciprofloxacin-resistant due to overuse of antibiotics in chickens. As a result, these ciprofloxacin-resistant *Camplobacter* were selected for, surviving the treatment of ciprofloxacin and continuing to multiply, allowing for ciprofloxacin resistance to occur more rapidly than normal. **(A)** is thus correct. (B) is incorrect because the proportion of ciprofloxacin-resistant *Camplobacter* would increase after treatment with ciprofloxacin. (C) is incorrect because there is no evidence that ciprofloxacin would affect growth and division; the role of antibiotics is to eradicate disease-causing bacteria, not to slow persistence in the population. (D) is incorrect because it does not address the issue of rapid ciprofloxacin resistance in the population; the new strains of *Camplobacter* may or may not be ciprofloxacin-resistant since they are unexposed to the antibiotic, yielding no additional information on increased ciprofloxacin resistance.

9. B Learning Objective: 4.2

Darwin's theory of natural selection states that individuals with more favorable phenotypes are more likely to survive and produce more offspring, thus passing traits to subsequent generations. If people in a specific population have an increased frequency of sickled hemoglobin genes, as compared to the rest of the world, it is therefore likely that the faulty gene confers some advantage onto them that makes them more likely to survive. Thus, **(B)** is correct. (A) can be eliminated because, from the information provided, it is clear that there is a correlation between frequency of sickled hemoglobin gene and risk for malaria, though the causation is unknown. If individuals who are carriers of the sickled hemoglobin gene provided a better target for malaria-carrying mosquitoes, under Darwin's theory, a lower frequency of the sickled hemoglobin gene in this population would then be expected, eliminating (C). Similarly, unless the sickled hemoglobin gene confers some evolutionary advantage, even if it were caused by mutations resulting from a malarial infection, the frequency of this gene and the number of individuals with the disease would decrease after a few generations, eliminating (D).

10. C Learning Objective: 4.1

The RNA world hypothesis proposes that RNA was the precursor to forms of life on Earth. Ribozymes provide evidence that there are strands of RNA molecules that have functions similar to catalytic enzymes. The fact that they are able to self-replicate short strands of complementary RNA (as long as they are supplied with nucleotides) indicates that ribozymes were a primitive form of enzymes that could both store genetic material and catalyze chemical reactions before the evolution of DNA. **(C)** is thus correct. (A) and (B) are incorrect because they only take into account how the organic molecules (nucleotide building blocks) needed by ribozymes are made; they do not however explain how the ribozymes relate to the origin of all life on Earth. (D) is incorrect because it mentions amino acids, which are not the building blocks for ribozymes to self-replicate (nucleotides are the building blocks.)

11. D Learning Objective: 4.3

It is important to understand that DDT-resistant mosquitoes in sub-Saharan Africa were the only ones that were able to survive after DDT exposure in the 1950s and have been continually reproducing, passing on the trait for DDT resistance. This mechanism of natural selection allowed mosquitoes in sub-Saharan Africa to maintain this trait for DDT resistance for more than 60 years. **(D)** is thus correct. The concentration of DDT is irrelevant, since the mosquitoes already have the trait for DDT resistance, which indicates that the DDT is unable to attack sodium channels. Therefore, no matter how much DDT is used, it will never be able to affect its target site, making (A) incorrect. Recall that mosquitoes fit the criteria for *r*-selected species: they have short lifespans, small body sizes, and reproduce quickly, producing many offspring. Therefore, it is highly unlikely that they would survive long enough to be able to migrate such long distances without the influence of external factors, eliminating (B). (C) is incorrect because there is no evidence of selective pressure acting against the malarial parasite. DDT targets the mosquitos' sodium channels, not the malarial parasite found within the mosquito, so it is unlikely that the parasite evolved to confer DDT resistance to the mosquito.

PART 3

Structures of Life

CHAPTER 5

Cells

LEARNING OBJECTIVES

In this chapter, you will review how to:

5.1 Differentiate between prokaryotic and eukaryotic structures

5.2 Explain the function of common organelles

5.3 Recall membrane composition and function

5.4 Explain transport mechanisms

5.5 Investigate effects of tonicity and gradients

TEST WHAT YOU ALREADY KNOW

1. Biologists isolated a giant cell from the gut of a surgeonfish. The cell is 600 μm long and can be seen with the naked eye. Electron microscope images revealed convoluted membranes and tangles of DNA not surrounded by membranes on the margins of the cell. Ribosomes were visible in the cytoplasm. Which of the following conclusions can the scientists most reasonably draw from their observations?

   (A) The organism is a eukaryote because it is visible to the naked eye and contains loose membranes.

   (B) The organism cannot be yet classified because there are too few observations to make an informed decision.

   (C) The organism is probably a virus because it was found inside the gut.

   (D) The organism is a prokaryote because it does not contain a true nucleus.

2. Scientists measured the density of mitochondria in several human tissues. They estimated the mitochondrial content by measuring the activity of a marker enzyme, citrate synthase, which is present only in mitochondria. Their results are plotted in the bar graph below.

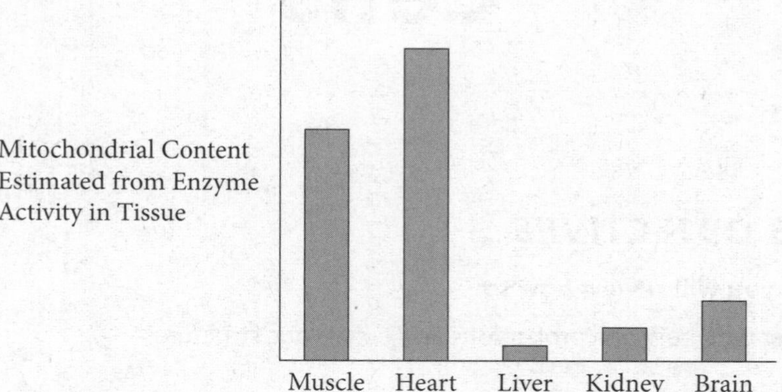

Adapted from G. Benard et al., "Physiological Diversity of Mitochondrial Oxidative Phosphorylation," *Am J Physiol Cell Physiol* 291, no. 6 (June 2006): C1172–C1182.

Which of the following hypotheses would provide the strongest justification for the scientists' observations?

   (A) The heart muscle uses a large supply of ATP because it contracts and relaxes continuously.

   (B) Skeletal muscle needs a large amount of mitochondria to support protein synthesis.

   (C) The liver does not need mitochondria to function.

   (D) The relative amount of mitochondria in each tissue does not have physiological significance.

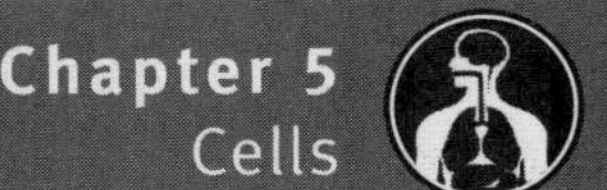

3. Saturated fats have a high melting point and remain solid at room temperature. Unsaturated fats are liquid at room temperature and solidify at much lower temperatures than saturated fats. In order for the membrane to function properly, membrane fluidity must stay within a certain physiological range. Chemical analysis of membrane lipids in bacteria adapted to different temperature ranges is shown in the table below.

Organism	**Temperature Range (°C)**	**Average Proportion of Unsaturated Fatty Acids**
Bacillus psychrophilus	0–28	Large (17–28%)
Bacillus cereus	10–50	Small (7–12%)
Geobacillus stearothermophilus	35–80	Small to insignificant (~3%)

Adapted from S. E. Diomandé et al., "Role of Fatty Acids in Bacillus Environmental Adaptation," *Frontiers in Microbiology* 6 (August 2015): 813.

Which of the following conclusions can be reasonably drawn from the data provided in the table?

(A) The composition of phospholipids in the membrane is not correlated to ambient temperature.

(B) Bacteria that thrive in cold environments adapt by increasing membrane rigidity.

(C) The membranes of *G. stearothermophilus* are adapted to increase fluidity at high temperatures.

(D) Fatty acids in *B. psychrophilus* allow the membrane to maintain its fluidity at low temperatures.

Rate of Oxygen Diffusion across the Alveoli

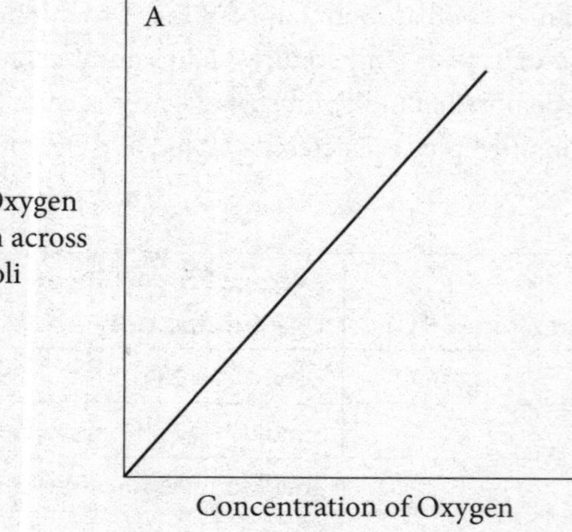

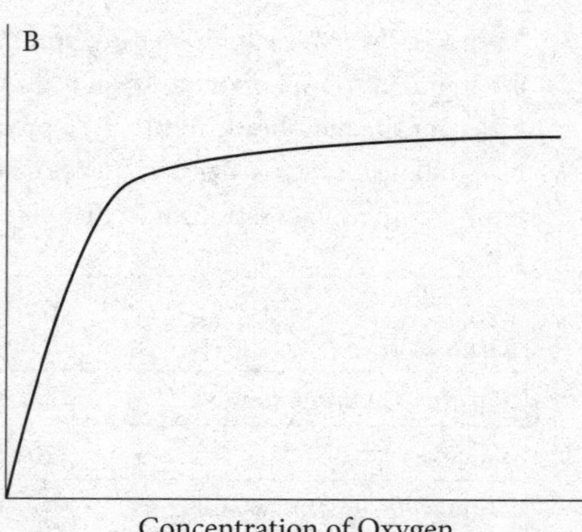

4. The alveoli of the lungs are sac-like structures where gas exchanges take place during respiration. Oxygen and carbon dioxide are small, hydrophobic gases. Two different groups of investigators report different kinetics for oxygen movement across the epithelium. Which of the above graphs is most likely to represent their results correctly?
   (A) Graph A, because the membrane is impermeable to small, hydrophobic molecules; oxygen uses active transport to flow against its concentration gradient
   (B) Graph A, because the membrane is permeable to small, hydrophobic molecules; oxygen diffuses through the membrane down its concentration gradient
   (C) Graph B, because the membrane is impermeable to small, charged molecules; oxygen binds to a carrier that allows it to flow down its concentration gradient
   (D) Graph B, because the membrane is permeable to all gas molecules; oxygen flows by osmosis down its concentration gradient

5. In an experiment to study diffusion across a semi-permeable membrane, students fill a length of dialysis tubing with a solution of 0.9% NaCl. The filled dialysis tube is soaked in a beaker containing 5% albumin (MW 64,500) in distilled water.

The students tested the content of the dialysis tube and the beaker before soaking the dialysis tube in the beaker. After letting the dialysis tube sit in the beaker for half an hour at room temperature, the students recorded that the dialysis tube had swelled and tested the contents of the beaker and the tube for NaCl and albumin. Testing results for the experiment are summarized in the table below, in which a + indicates the presence of the compound and a − indicates the absence.

	Start of Experiment		**End of Experiment**	
Compound	**In Bag**	**In Beaker**	**In Bag**	**In Beaker**
NaCl	+	−	+	+
Albumin	−	+	−	+

Which of the following is a valid comparison between the dialysis tube and the plasma membrane of a red blood cell?

(A) NaCl can never exit a red blood cell because it is composed of ions, but it can cross the dialysis tube membrane.

(B) Albumin can permeate a red blood cell membrane, but it cannot cross the dialysis tube membrane because of its size.

(C) Both membranes are similar in that both are impermeable to albumin but permeable to NaCl.

(D) The membranes are similar in that both are impermeable to albumin, but NaCl can permeate only across the dialysis tube membrane.

Answers to this quiz can be found at the end of this chapter.

PROKARYOTES VERSUS EUKARYOTES

High-Yield

5.1 Differentiate between prokaryotic and eukaryotic structures

There are two main types of cells—**prokaryotes** and **eukaryotes**. Prokaryotes are much simpler than eukaryotes; they are composed of a plasma membrane, cytoplasm, cell wall, DNA, ribosomes, and simple microtubules. Bacteria are an example of prokaryotic cells. Eukaryotes include both plant and animal cells. These cells are much more complex than prokaryotes and contain numerous **organelles**.

Prokaryotes

Prokaryotes are the more basic and simpler of the two major cell types. These cells are considered to be the older form of cells. There are three major regions of prokaryotic cells. The inside region is called the cytoplasmic region. It contains the circular DNA that makes up the genetic material of the cell. Because prokaryotes do not contain a nucleus, DNA is condensed inside an area of the cytoplasm, the nucleoid. Prokaryotes are useful in science because they carry extrachromosomal DNA elements called plasmids. Plasmids are circular bits of DNA that can be added or changed to allow for the addition or suppression of certain functions based on the coding of inserted DNA sections.

The cell envelope usually consists of a cell wall that covers a plasma membrane and may sometimes also include another protective layer called the capsule. The envelope provides structure as well as a protective filter for the cell. Most prokaryotes contain a cell wall that acts as yet another protective barrier from the cell's external environment.

Many prokaryotes have external projections known as flagella and pili. Flagella (singular: flagellum) are long projections or appendages that protrude from the cell body. The primary function of flagella is locomotion, but they can also function as a sensory structure. Pili (singular: pilus) are shorter structures, some of which are used for locomotion and others of which play a role in bacterial conjugation, discussed in the following chapter.

Eukaryotes

All multicellular organisms (such as you, a tree, or a mushroom) and all protoctists (such as amoebas and paramecia) are eukaryotic. The eukaryotes include the protoctists (protists), fungi, animals, and plants. Eukaryotic cells are enclosed within a lipid bilayer cell membrane, as are prokaryotic cells. Unlike prokaryotes, eukaryotic cells contain membrane-bound organelles (see figure). An organelle is a structure within the cell with a specific function that is separated from the rest of the cell by a membrane. The presence of membrane-bound organelles in eukaryotes allows eukaryotic cells to compartmentalize activities in different parts of the cell, making them more efficient. Compartments within a cell allow the cell to carry out activities, such as ATP production and consumption, within the same cell and to control each independently.

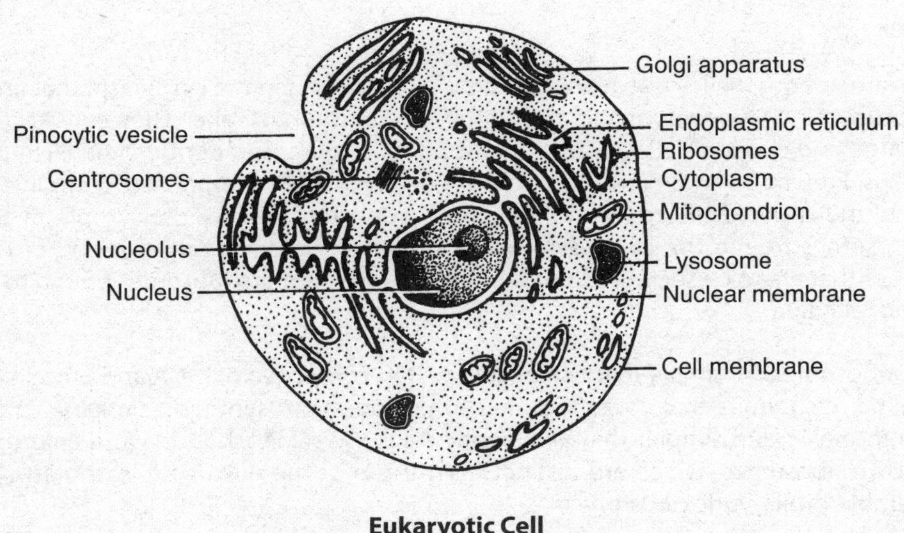

Eukaryotic Cell

ORGANELLES

High-Yield

5.2 Explain the function of common organelles

Nucleus

The genetic material, the DNA genome, is found in the largest organelle of a eukaryotic cell, the **nucleus**. The nucleus is separated from the rest of the cell by the nuclear envelope, a double membrane that has a large number of nuclear pores for communication of material between the interior and exterior of the nucleus. The pores are large enough to allow proteins to pass through but are also selective in the proteins that are transported into the nucleus or excluded from the nucleus. Special sequences in proteins signal a protein to be imported into the nucleus.

While the prokaryotic genome is generally found in a single circular piece of DNA, the eukaryotic genome in each cell is split into chromosomes. Chromosomes contain the DNA genome complexed with structural proteins, called histones, that help package the large strands of DNA in each chromosome within the limited space of the nucleus. Genes in the DNA genome are read (transcribed) to make RNA, which is processed in the nucleus before it is exported to the cytoplasm, where the RNA is read in turn (translated) to make proteins. The basic information flow of the cell is DNA to RNA to protein. The DNA genome is replicated in the nucleus when the cell divides. Other metabolic activities, such as energy production, are excluded from the nucleus. The structure and function of the eukaryotic genome will be presented later in more detail.

The dense structure within the nucleus where ribosomal RNA (rRNA) synthesis occurs is known as the nucleolus. The nucleolus is not surrounded by a membrane but is the site of assembly of ribosomal subunits from RNA and protein components. After assembly, the ribosomal subunits are exported from the nucleus to the cytoplasm to carry out protein synthesis.

Ribosomes

Ribosomes are not organelles but are large, complex structures in the cytoplasm that are involved in protein production (translation) and are synthesized in the nucleolus. They consist of two subunits, one large and one small. Each ribosomal subunit is composed of ribosomal RNA (rRNA) and many proteins. Free ribosomes are found in the cytoplasm, while bound ribosomes line the outer membrane of the endoplasmic reticulum. Proteins that are destined for the cytoplasm are synthesized by ribosomes free in the cytoplasm, while proteins that are bound for one of the several membranes or that are to be secreted from the cell are translated on ribosomes bound to the rough endoplasmic reticulum.

Prokaryotic ribosomes are similar to those of eukaryotes, composed of rRNA and proteins that form two different size subunits that come together to perform DNA synthesis. Prokaryotic ribosomes are, however, smaller and simpler than eukaryotic ribosomes. Mitochondria and chloroplasts also have their own ribosomes, which are distinct from those of the eukaryotic cytoplasm and more closely resemble prokaryotic ribosomes.

Endoplasmic Reticulum

The **endoplasmic reticulum** (ER) is an extensive network of membrane-enclosed spaces in the cytoplasm. The interior of the ER between membrane layers is called the lumen, and at points in the ER, the lumen is continuous with the nuclear envelope. If a region of the ER has ribosomes lining its outer surface, it is termed rough endoplasmic reticulum (rough ER); without ribosomes, it is known as smooth endoplasmic reticulum (smooth ER). Smooth ER is involved in lipid synthesis and the detoxification of drugs and poisons, and it has the appearance of a network of tubes, while rough ER is involved in protein synthesis and is a series of stacked plates.

Proteins that are secreted or found in the cell membrane, the ER, or the Golgi are made by ribosomes on the rough ER. Proteins synthesized on the rough ER cross into the lumen of the rough ER during synthesis. The presence of a hydrophobic sequence of amino acids at the amino terminus of proteins determines whether the protein will be sorted into the secretory pathway starting at the rough ER or synthesized in the cytoplasm. Proteins that are secreted will have only one hydrophobic signal sequence, the signal peptide, and will be inserted into the ER lumen when they are synthesized, then released from the cell later. Proteins that are destined to be membrane bound have hydrophobic transmembrane domains that are threaded through the rough ER membrane as the protein is synthesized. When the protein reaches the correct membrane destination along the secretory pathway, additional signals in the protein sequence and structure will cause the protein to stay localized at the current location.

Small regions of ER membrane bud off to form small round membrane-bound vesicles that contain newly synthesized proteins. These cytoplasmic vesicles are then transported to the Golgi apparatus, which is the next stop along the secretory pathway.

Golgi Apparatus

The **Golgi apparatus** is a stack of membrane-enclosed sacs, usually located in the cell between the ER and the plasma membrane (see the figure under the Eukaryotes heading above). The stacks closest to the ER are called the cis Golgi and the stacks farthest from the ER, closer to the plasma membrane, are called the trans Golgi. Vesicles containing newly synthesized proteins bud off of the ER and fuse with the cis Golgi. In the Golgi, these proteins are modified and then repackaged for delivery to other destinations in the cell. For example, the Golgi carries out post-translational modification of proteins through glycosylation, the process of adding sugar groups to the proteins to form glycoproteins. Many proteins destined for the plasma membrane have carbohydrate groups added to the surface of the protein facing the exterior of the cell.

After processing in the cis Golgi, proteins are packaged in vesicles that move to the next layer in the stack, where they fuse and release their contents. Proteins proceed in this manner from one stack to the next until they reach the trans Golgi. In the trans Golgi, proteins are sorted into vesicles based on signals in different proteins that indicate their final destination. The nature of the signal varies but includes the protein's primary sequence, structure, and post-translational modifications. Once packaged into vesicles, the vesicles move on to their final destination.

The final destination for a protein may include the lysosome, the plasma membrane, or the exterior of the cell. Some proteins are retained in the Golgi or the ER. Proteins that are destined for the plasma membrane as transmembrane proteins are inserted in the membrane in the ER as they are synthesized, and they maintain their orientation in the membrane as they move from the ER to the Golgi to the vesicle to the plasma membrane. Proteins that are secreted from the cell are inserted in the ER lumen during protein synthesis and remain in the lumen of the ER to the Golgi, where they form secretory vesicles. The last step in secretion is the fusion of the secretory vesicle with the plasma membrane, releasing the contents of the vesicle to the cellular exterior.

Lysosomes

Lysosomes contain hydrolytic enzymes involved in intracellular digestion that break down proteins, carbohydrates, and nucleic acids. For white blood cells, the lysosome may degrade bacteria or damaged cells. For a protist, lysosomes may provide food for the cell. They also aid in renewing a cell's own components by breaking them down and releasing their molecular building blocks into the cytosol for reuse. A cell in injured or dying tissue may rupture the lysosome membrane and release its hydrolytic enzymes to digest its own cellular contents.

The lysosome maintains a slightly acidic pH of 5 in its interior, a pH at which lysosomal enzymes are maximally active. The contents of the lysosome are isolated from the cytoplasm by the lysosomal membrane, keeping the pH distinct from the more neutral pH of the cytoplasm. The optimal pH and compartmentalization of lysosomal enzymes prevent the rest of the cellular contents from degrading.

Peroxisomes

Peroxisomes contain oxidative enzymes that catalyze reactions in which hydrogen peroxide is produced and degraded. Peroxisomes break fats down into small molecules that can be used for fuel; they are also used in the liver to detoxify compounds, such as alcohol, that may be harmful to the body. The peroxides produced in the peroxisome would be hazardous to the cell if present in the cytoplasm, because these molecules are highly reactive and could covalently alter molecules such as DNA. Compartmentalization of these activities within the peroxisome reduces this risk.

Mitochondria

Mitochondria are the source of most energy in the eukaryotic cell as the site of aerobic respiration. Mitochondria are bound by an outer and inner phospholipid bilayer membrane. The outer membrane has many pores and acts as a sieve, allowing molecules through on the basis of their size. The area between the inner and outer membranes is known as the intermembrane space. The inner membrane has many convolutions called cristae, as well as a high protein content that includes the proteins of the electron transport chain. The area bounded by the inner membrane is known as the mitochondrial matrix and is the site of many of the reactions in cellular respiration, including electron transport, the Krebs cycle, and ATP production.

Mitochondria are somewhat unusual in that they are semiautonomous within the cell. They contain their own circular DNA and ribosomes, which enable them to produce some of their own proteins. The genome and ribosomes of mitochondria resemble those of prokaryotes more than eukaryotes. In addition, they are able to self-replicate through binary fission. Mitochondria are believed to have developed from early prokaryotic cells that began a symbiotic relationship with the ancestors of eukaryotes, with the mitochondria providing energy and the host cell providing nutrients and protection from the exterior environment. This theory of the origin of mitochondria and the modern eukaryotic cell is called the endosymbiotic hypothesis.

Specialized Plant Organelles

Plants also have some organelles that are not found in animal cells. **Chloroplasts** are found only in plant cells and some protists. With the help of one of their primary components, chlorophyll, chloroplasts function as the site of photosynthesis, using the energy of the Sun to produce glucose. Chloroplasts have two membranes, an inner and an outer membrane. Additional membrane sacs called thylakoids inside the chloroplast are derived from the inner membrane and form stacks called grana. The fluid inside the chloroplast surrounding the grana is the stroma. The thylakoid membranes contain the chlorophyll of the cell.

Like mitochondria, chloroplasts contain their own DNA and ribosomes and exhibit the same semiautonomy. They are also believed to have evolved via symbiosis of an early photosynthetic prokaryote that invaded the precursor of the eukaryotic cell. In this arrangement, the chloroplast precursor cell provided food and received protection. Photosynthetic prokaryotes today carry out photosynthesis in a manner similar to the chloroplast.

Vacuoles are membrane-enclosed sacs within the cell. Many types of cells have vacuoles, but plant vacuoles are particularly large, taking up 90 percent of the cell volume in some cases. Plants use the vacuole to store waste products, and the pressure of liquid and solutes in the vacuole helps the plant to maintain stiffness and structure as well.

All plant cells have a cellulose cell wall that distinguishes them from animal cells, which lack a cell wall. The cell wall of plants is also distinct from the peptidoglycan cell wall of bacteria and the chitin cell wall of fungi. The cell wall provides structure and strength to plants.

Cilia and Flagella

Cilia and flagella are both anchored into the cell membrane by arrangements of microtubule triplets, which are called basal bodies. Because the microtubules in cilia and flagella must be rebuilt often, tubulin dimers use these basal bodies as the foundation to make new microtubules, which are used to maintain cilia and flagella.

As you can see in the following figure, cilia and flagella are composed of long stabilized microtubules arranged in a "9 + 2" structure (nine pairs of microtubules surrounding two central microtubules for added stability). These nine doublets slide past each other as dynein proteins grab neighboring tubules and pull them. This rapid sliding generates the force needed for the cilia or flagella to quickly beat back and forth and cause movement.

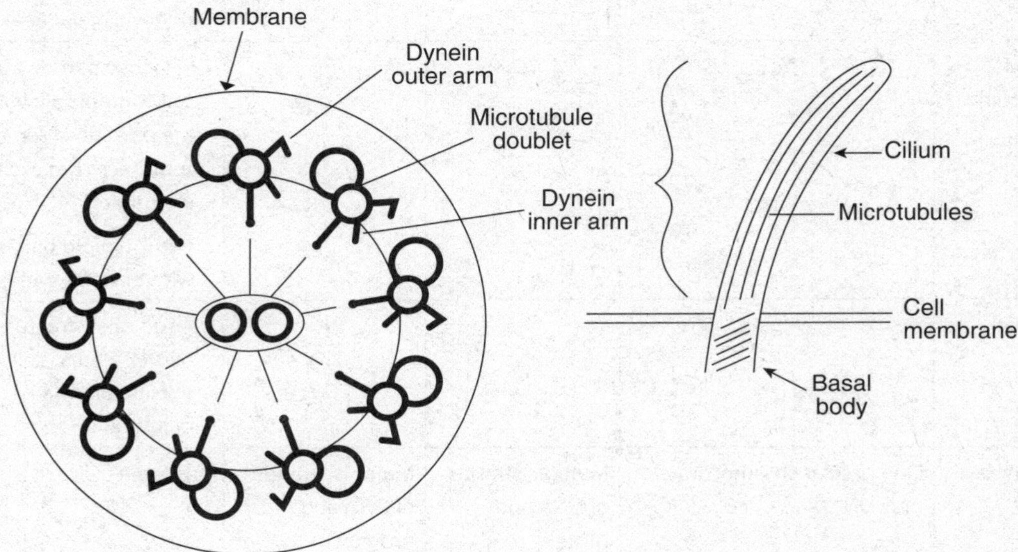

Cilium Cross-Section

CHARACTERISTICS OF CELLS				
		Eukaryotes		
	Prokaryotes	**Plant Cells**	**Animal Cells**	
Size	**0.2–500μm, most 1–10μm**	**Most 30–50μm**	**Most 10–20μm**	
Structure				**Properties**
Cytoplasm	Yes	Yes	Yes	• Intracellular matrix outside of nucleus
Nucleus	No	Yes	Yes	• Contains DNA • Pores allow communication with cellular matrix
Plasma Membrane	Yes	Yes	Yes	• Selective barrier around cell contents allowing the passage of some substances but excluding others • Phospholipid bilayer with proteins embedded
Cell Wall	Most	Yes	No	• Additional structural barrier around cell outside plasma membrane
Chromosomes	One circular chromosome, only DNA	Multiple strands of DNA and protein	Multiple strands of DNA and protein	• The cell's DNA
Ribosomes	Yes	Yes	Yes	• Site of protein synthesis (translation)
Endoplasmic Reticulum (ER)	No	Yes	Yes	• Site of attachment for ribosomes • Protein and membrane synthesis • Formation of vesicles for transport
Golgi Apparatus	No	Yes	Yes	• Synthesis, accumulation, storage, and transport of products

(Continued)

CHARACTERISTICS OF CELLS				
		Eukaryotes		
	Prokaryotes	**Plant Cells**	**Animal Cells**	
Size	**0.2–500μm, most 1–10μm**	**Most 30–50μm**	**Most 10–20μm**	
Structure				**Properties**
Lysosomes	No	Some vacuoles function as lysosomes	Usually	• Vesicle containing hydrolytic enzymes
Vacuoles or Vesicles	No	Yes	Some	• Membrane-bound sacs in the cytoplasm
Mitochondria	No	Yes	Yes	• Site of cellular respiration
Plastids	No	Yes	No	• Group of plant organelles that includes chloroplasts • Site of photosynthesis • Carbohydrate storage
Micro-tubules (Cilia or Flagella)	Simple	On some sperm	Complex (9 + 2 arrangement)	• Tubes of globular protein, tubulins • Provides structural framework for cell • Provides motility
Centrioles	No	No	Yes	• Cell center for microtubule formation

MEMBRANE TRAFFIC

5.3 Recall membrane composition and function

5.4 Explain transport mechanisms

Membrane Structure

All cells are surrounded by a **plasma membrane**. In eukaryotic cells, the nucleus and most of the organelles are also surrounded by plasma membranes. Membranes are composed mostly of **lipids**, which are full of nonpolar covalent bonds, so they are **hydrophobic** and do not dissolve in water. Because membranes are composed primarily of lipids, most of the material in membranes will not mix well with water.

Most of the lipids in membranes have a phosphate group attached to one end, so they are called **phospholipids**. The charged end (phosphate group) is polar and happy to be in water, which is why it is termed **hydrophilic**. The other end of the phospholipid is a tail that is nonpolar, and it turns in toward the center of the membrane. The attraction between the nonpolar regions of these phospholipids creates the foundation for the bilayer of the membrane. The lipid ends group together like the insides of a sandwich, surrounded by polar barriers.

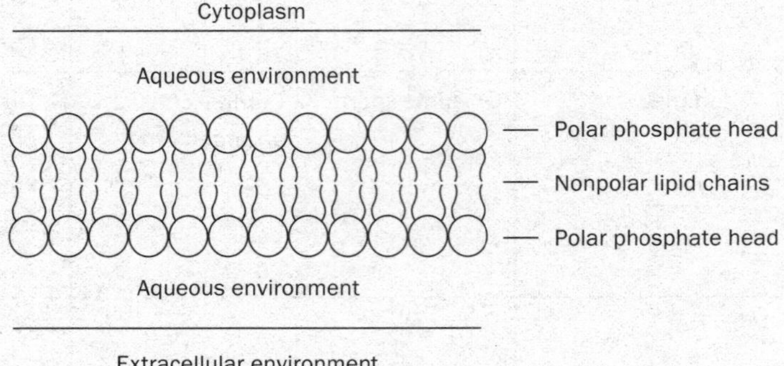

Phospholipid Bilayer

> **✔ AP Expert Note**
>
> **Make sure you are familiar with the structure and function of macromolecules as they relate to the plasma membrane structure. Be able to identify the *general* structures and functions of phospholipids, proteins, and cholesterol.**

Embedded among all of these membrane lipids are **proteins**, **carbohydrates**, and **sterols** (like cholesterol). Some proteins are embedded on the outer surface, some on the inner surface, and some span the entire width of the plasma membrane (these usually function as transport proteins). Some surface proteins have sugar groups attached to them, called glycoproteins.

Each component of a cell membrane contributes to how the membrane functions. Proteins act as transport molecules, receptor sites, attachments to the **cytoskeleton**, and surface enzymes. Carbohydrates on the surface of the cell and glycoproteins contribute to cell recognition, particularly in immune response. Cholesterol helps to maintain the fluidity of the membrane.

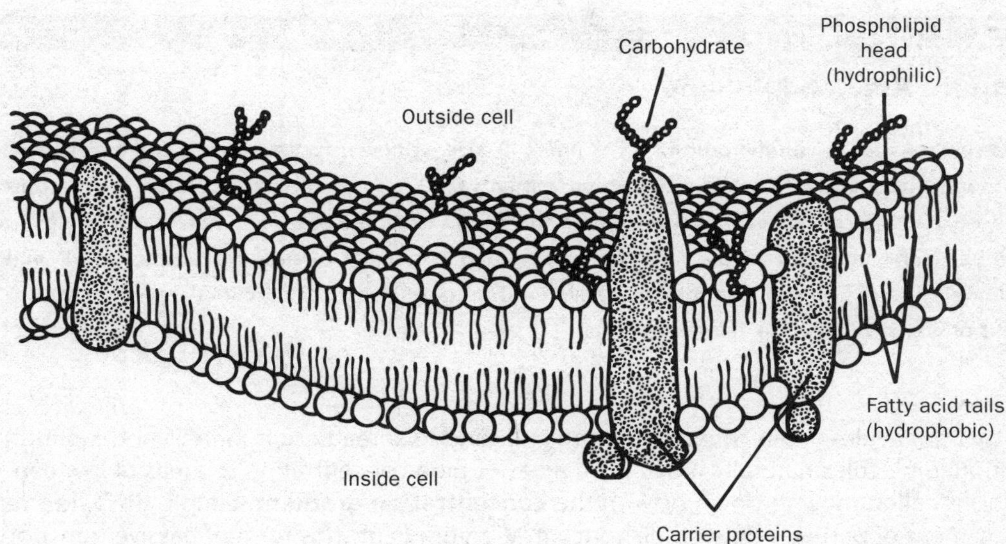

Cell Membrane

Membrane Proteins

Membrane proteins, like the membrane phospholipids, usually have carbohydrate groups attached to them so that the outside surface of the plasma membrane is extremely sugar-rich.

Membrane-spanning proteins have regions that are hydrophobic as well as regions that are hydrophilic, with the nonpolar (hydrophobic) regions passing through the nonpolar interior of the membrane and the polar areas sticking both into the cytoplasm and out into the extracellular space. Other proteins can be located completely intracellularly or extracellularly, anchored to the cell membrane by a variety of special lipids. These proteins are made in the cytosol and bind into the cell membrane only because they subsequently have a lipid molecule attached to their structure.

Transmembrane proteins are involved not only in carrying materials across the membrane, but also in cell recognition, cell adhesion, cell signaling, and enzymatic reactions. Recall that most proteins sticking up from the surface of the membrane are covered in carbohydrates on the extracellular surface. The term used to describe the protein- and carbohydrate-rich coating on the cell surface is glycocalyx. Keep in mind that these sugars reside exclusively on the exterior of the membrane.

Membrane Transport Mechanisms

Membranes are **selectively permeable**, which means that they allow some things to pass through, but not others. The main limiting factors that determine whether a molecule will be able to pass through a cell's membrane are the *size of the particle* and *its charge* (polarity). Simply stated, the molecules quickest to pass through the lipid bilayer are those that are *small* and *nonpolar*, because the interior of the membrane is far too hydrophobic for others to make it through without assistance. This assistance can come in the form of membrane-spanning proteins, which either can bind to extracellular molecules and bring them inside the cell via a conformational change or can open up a temporary tunnel through the membrane lipids so that the molecules can pass through.

AP Expert Note

Surface Area and Volume

The surface area-to-volume ratio of a cell limits its size. While surface area affects the transport of nutrients into and waste out of a cell, volume affects the consumption and production of resources and waste. As a cell grows, the surface area-to-volume ratio decreases. At a certain point, the cell will be unable to maintain the rate of transport that is needed to support the increased cellular volume. Thus, the cell will not obtain sufficient nutrients and waste will accumulate within the cell—both of which will lead to cell death.

Recall that molecules are in constant motion and tend to spread out if there is nothing in the way. In solution, molecules naturally move from areas of high concentration to areas of low concentration. This is called moving down or with the **concentration gradient**. Simple **diffusion** refers to the movement of particles down their concentration gradient. This form of passive transport takes place directly through the cell membrane lipid bilayer without using any form of energy or membrane proteins in order to move particles. Again, small nonpolar molecules move most freely by simple diffusion. Examples include water (small but polar), carbon dioxide (nonpolar), and oxygen (nonpolar).

The simple diffusion of water is referred to as **osmosis** and occurs from a region of higher water concentration to a region of lower water concentration. For water to be in high concentration, the amount of dissolved solute (salts, sugars, etc.) must be low, and vice versa for water in low concentration. So, although water diffusion works like any other passive diffusion in terms of movement from high → low concentration, it is generally stated that water moves from an area of lower *solute* concentration to one of higher *solute* concentration.

AP Expert Note

Water Potential

The flow of water in a system is determined by its water potential (Ψ), which is a measure of the potential energy in water. Water potential can be described as the sum of pressure potential (Ψ_P) and solute potential (Ψ_S). At atmospheric pressure, the pressure potential of water in an open container is zero and equal to the solute potential. Adding solute decreases solute potential (makes it more negative) and in turn decreases water potential. Negative hydrostatic pressures will also decrease water potential. The gradient of water potential causes water to move (from high to low water potential). In plants, water potential is regulated by osmosis and transpiration to transport water from the roots to the leaves for photosynthesis.

Solutions low in solute concentration relative to other solutions are said to be **hypotonic**, whereas solutions higher in solute than others are **hypertonic**. When two solutions have the same solute concentration as each other, they are said to be **isotonic**. The cell membrane effectively separates two distinct solutions: the extracellular environment and the cytoplasm. If the outside of a cell is hypertonic (e.g., a cell has just been moved from fresh water to salt water), water will move out of the cell into the high solute solution. If possible, some of those solutes will also move into the cell until a balance has been established so that both areas are equivalent in solute concentration. In hypotonic solutions, cells generally take on water, sometimes until they burst.

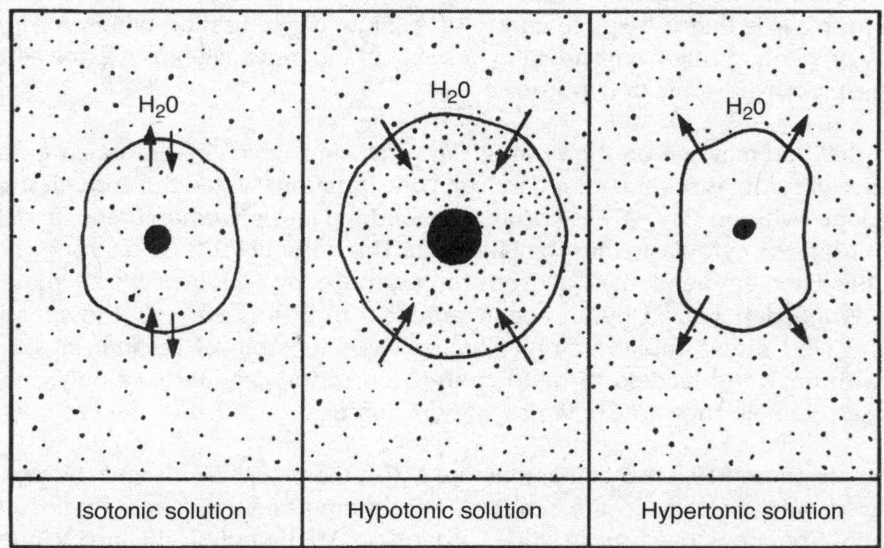

Tonicity and Osmosis

> **✔ AP Expert Note**
>
> **In isotonic solutions, water and solutes can and do move across the membrane, yet movement inward is always balanced by reciprocal movement outward. This means that overall concentrations of water and solute on opposite sides of the membrane do not change.**

Facilitated diffusion, another type of passive transport, involves the use of channel or carrier proteins embedded in the membrane to allow impermeable molecules to diffuse down a gradient. The structures of the proteins involved in this type of transport are very similar, and amino acid sequences are highly conserved across many species. In some cases, these proteins act as pores for ions; in other cases, they may open and close in response to external signals. Keep in mind that, because cells naturally have a negatively charged cytoplasm, the opening of ion channels favors the movement of positively charged ions into the cytoplasm. The combination of differential solute concentrations and an electrical gradient is called an **electrochemical gradient** and is the key determinant of what moves into and out of membranes when passive transport channels are open.

In **active transport**, membrane proteins use the energy of ATP to change the protein's conformation so that molecules can be brought into and out of the cell against their concentration gradients. These **ATPase** pumps are found in every membrane of cells, and they are extremely important in the maintenance of unequal concentrations of certain ions across the lipid bilayer—something essential for processes like nerve signal conduction.

A commonly cited example of active transport is the Na^+–K^+ ATPase membrane pump, whose conformational change uses the energy of ATP breakdown to pull 2 K^+ (potassium) ions into a cell while kicking out 3 Na^+ (sodium) ions at the same time. This transport of molecules in opposite directions is known as antiport, and it can be contrasted with pumps that pull two different molecules in the same direction (symport). Because the pump, which is present in all cell membranes, pumps out three positive charges for every two it brings in, the inside of the cell remains negatively charged compared with the outside of the cell under normal conditions. Yet, the more important role the pump plays is that it helps to control the solute concentration within the cytoplasm of cells, thereby preventing cells from shrinking or swelling too much when the extracellular environment becomes too hypertonic or hypotonic.

Another ion that you may see on the exam is Ca^{2+} (calcium), which is kept in extremely low concentrations in cell cytoplasm, yet is stored in high concentrations within the endoplasmic reticulum (ER). This is done by using Ca^{2+}-ATPase pumps, embedded in the ER membrane, to actively transport calcium from the cytoplasm into the ER lumen. This naturally sets up a strong calcium gradient across the ER membrane that is used, for example, by muscle cells to regulate muscle contraction. When depolarized by an action potential from a nerve cell, the specialized ER of muscle cells (called the sarcoplasmic reticulum) releases its store of calcium ions, flooding the cytoplasm with Ca^{2+} and leading to rapid contraction of the cell. Because only one ion moves through these channels, they are known as uniport pumps.

As a last example, those ATPases that manufacture ATP in the mitochondria and chloroplasts, as part of the **electron transport chain** are simply ATPase membrane-transport proteins working in a reverse manner from how they usually work. Rather than ATP hydrolysis driving changes in protein structure so that ions can pass through the membrane, it seems that these pumps are driven by the flow of H^+ ions moving through them to synthesize ATP from ADP and an inorganic phosphate.

Endocytosis and **exocytosis** are two mechanisms of transport that can move large molecules and even entire cells through the cell membrane. To accomplish this, the cell membrane actually invaginates, or pinches inward, to form a pocket into which the material to be transported can fall. In the case of endocytosis, this invagination pinches off completely, forming a vesicle that contains the transported material and can move freely within the cytoplasm. In exocytosis, a vesicle containing material to be expelled simply merges with the lipid bilayer, and the material is pushed off into the extracellular space.

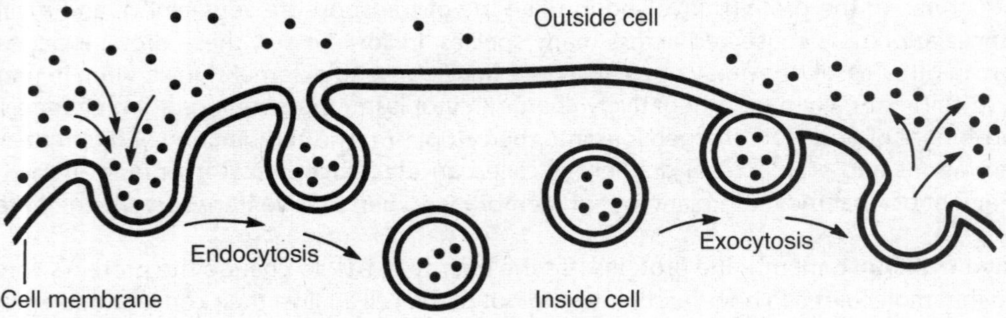

Endocytosis and Exocytosis

There are three distinct types of endocytosis. In phagocytosis, sometimes known as "cellular eating," the membrane invaginates around solid particles. In pinocytosis, or "cellular drinking," the membrane surrounds a mass of liquid. Finally, in receptor-mediated endocytosis, a carrier protein binds to a specific substance, which then triggers the endocytotic response.

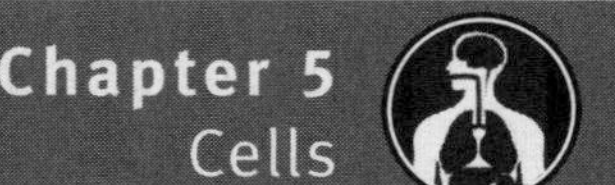

> ✔ **AP Expert Note**
>
> **As you prepare for the exam, be familiar with similarities and differences between the different types of transport across the cell membrane. Pay close attention to which ones require energy and which ones do not.**

TYPES OF MEMBRANE TRANSPORT		
Type of Transport	**Requires Energy?**	**Concentration Gradient**
Diffusion	No	Down
Osmosis	No	Down
Facilitated Diffusion	No	Down
Active Transport	Yes	Against
Exocytosis	Yes	N/A
Endocytosis	Yes	N/A

AP BIOLOGY LAB 4: DIFFUSION AND OSMOSIS INVESTIGATION

5.5 Investigate effects of tonicity and gradients

> ✔ **AP Expert Note**
>
> **This and the other lab sections of this book provide supplemental information for the 13 laboratory investigations officially recommended by the College Board for all AP Biology students. Many of these lab sections include high-yield content not covered in other parts of the text, so they are worth reviewing irrespective of whether you completed the labs in class.**

The properties of diffusion and osmosis, discussed earlier in this chapter, will be explored in this investigation. Recall that selectively permeable membranes let some molecules through but not others. For this investigation, it is important to know that:

- Isotonic solutions are two solutions that have the same concentration of solute.
- A hypotonic solution has a lower solute concentration than another solution.
- A hypertonic solution has a higher solute concentration than another solution (think *hypo* = "below" and *hyper* = "above").

Keep in mind that water potential is the measure of force a solution has for pulling or drawing water into it. The more negative the water potential, the stronger its pulling force.

Dialysis bags allow for the movement of water but not ions. A common osmosis experiment is to fill dialysis bags with different solutions. These bags are tied at each end and put into hypertonic, hypotonic, and isotonic solutions. The water will move in the direction of the hypertonic solution. Suppose you tied your dialysis bags and put them in solutions to complete the experiment but didn't get the results you expected. Skewed results can be obtained by not tying the knot on your bag tight enough, not getting all the solution washed off the outside of your bag, or by some other slight oversight. Often, one group gets usable results for one part of the lab and another group gets another part right, so everyone shares the "good" results. Occasionally, classes have really good luck and all the groups get "good" results. Either way, you need to know what happened in order to score well on the exam. You can expect to see an osmosis question in one form or another on the exam. It may be in the multiple-choice section or it may be a free-response question.

✔ AP Expert Note

Two key skills of the diffusion and osmosis lab are:

- Measuring the effects (e.g., weight change) of osmosis
- Determination of the osmotic concentration/water potential of an unknown tissue or solution using solutions of known concentrations

The first skill uses experimental methods to obtain results. Even if you know how osmosis works, it is important to know how to measure it. You can't really watch water molecules or sugar and starch molecules move because they're too small. In this investigation, you learn how to measure things you don't see by observing things you can. To obtain results, the weight change of a dialysis bag or a piece of potato can be measured. An indicator color will appear if glucose or starch occurs in a solution. If a bag or piece of potato increased in weight, it gained water and had a more negative water potential than (was hypertonic to) the solution it was placed into, and vice versa. If a solution produced an indicator color, the semipermeable membrane allowed the passage of molecules that the dye is an indicator for.

The second key skill deals with using a standard to figure out the "identity" of an unknown and using observed data to interpolate expected data. Part two of the experiment begins with a dialysis bag full of solution (the concentration of which is unknown). Questions to ask during the experiment include: What is the osmolarity/water potential of the unknown, and how do you measure it? The first parts of the investigation teach you to observe the effects of osmosis by measuring the change in weight. Samples are weighed and placed into solutions with known concentrations, usually ranging from distilled water to a high concentration like 1 M. Percentage weight change can be plotted on a graph, as shown on the following page.

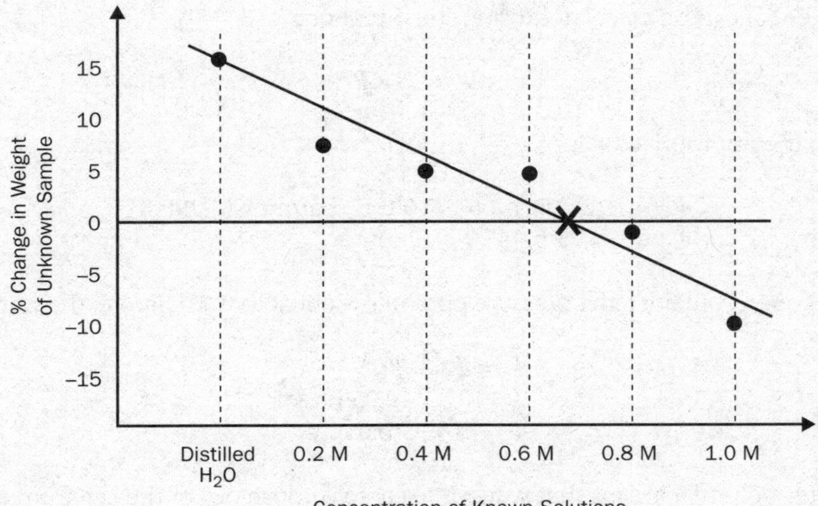

An imaginary line can be drawn that nearly bisects all data points. Note where the line crosses the *x*-axis. At this point, there is no change in weight of the unknown sample, which represents the point at which the osmolarity/water potential of the unknown is the same as a known solution equal to that point on the *x*-axis (on the graph, this is about 0.675 M). By assuming change in weight is linear with respect to change in concentration, the expected concentration of a theoretical solution can be determined. The concentration of the unknown can be estimated by comparing it to the concentration of the theoretical solution (as we saw in the graph above, this is approximately 0.675 M). This type of experimental design and interpolation is very common in biological research. The College Board expects you to be able to design simple experiments like this to test simple hypotheses.

Water potential [represented by the Greek letter psi (Ψ)] predicts which way water diffuses. Water potential is calculated from the solute potential (Ψ_S), which is dependent on solute concentration, and the pressure potential (Ψ_P), which results from the exertion of pressure on a solution. When a solution is open to the atmosphere, the pressure potential is equal to 0 because there is no tension on that solution.

$$\Psi = \Psi_P + \Psi_S$$

$$\text{water potential} = \text{pressure potential} + \text{solute potential}$$

In an open beaker of pure water, the water potential is equal to zero. There is no solute and no tension on the solution, so the solute and pressure potentials are zero. If you add solute in a measured concentration to the beaker, you can calculate the solute potential using the following formula:

$$\Psi_S = -iCRT$$

where *i* is equal to the number of particles the molecule will dissolve into in water, *C* is the molar concentration, *R* is the pressure constant (equal to 0.0831 L·bar/mol·K), and *T* is the temperature of the solution in Kelvins.

For example, suppose a plant cell is placed in an open container of 0.1 M NaCl solution at 25°C. What would the water potential be?

The solute potential can be calculated using the equation:

$$\Psi_S = -iCRT$$

Substituting in the appropriate values,

$$\Psi_S = -(2)\ (0.1\ \text{mol/L})(0.0831\text{L}\cdot\text{bar/mol}\cdot\text{K})(298\ \text{K})$$
$$\Psi_S = -4.95\ \text{bars}$$

Because it is an open container, the pressure potential is equal to 0. Therefore, the water potential is:

$$\Psi = \Psi_P + \Psi_S$$
$$\Psi = 0 + -4.95\ \text{bars}$$
$$\Psi = -4.95\ \text{bars}$$

A negative water potential means that water is likely to osmose out of the cell from a place of high water potential to a place of low water potential.

RAPID REVIEW

If you take away only 5 things from this chapter:

1. Cells are the basic structural and functional units of all known living organisms. There are two major types: prokaryotic cells and eukaryotic cells.
2. Prokaryotic cells have no real nucleus, have circular DNA, and reproduce using binary fission. Prokaryotes include archaea and bacteria. These cells are small and move via flagella.
3. Eukaryotic cells have a nucleus, linear DNA, highly structured cell membranes, and organelles, and they reproduce via mitosis and meiosis. Eukaryotes include protists, fungi, plants, and animals. These cells are large and move via a variety of methods including flagella and cilia.
4. Cell membranes are made up of a lipid bilayer that includes a hydrophobic and a hydrophilic region. Specific structures embedded within the membrane help to facilitate transport.
5. The cell membrane is selectively permeable, meaning it allows certain things through while keeping others out. Water diffuses across the membrane from areas of lesser to greater solute concentration (osmosis). While certain things can cross the membrane in the processes of diffusion or facilitated diffusion, which do not require energy, others require the expenditure of energy for active transport against the concentration gradient.

TEST WHAT YOU LEARNED

1. A group of investigators isolated a cDNA from the plant *Alonsoa meridionalis* and expressed the protein in yeast cells. They observed that sucrose accumulated in the yeast cells only if the sense cDNA is expressed. The group then followed the accumulation of radioactive sucrose by yeast cells under several conditions as summarized in the table. Sucrose is made of a molecule of fructose linked to a molecule of glucose. Maltose is a disaccharide made of two molecules of glucose. Raffinose is a trisaccharide made of galactose, glucose, and fructose. Chlorophenyl hydrazone is a metabolic poison that prevents synthesis of ATP.

Compound in Solution	Transport Rate as Percent of Control
^{14}C-Sucrose (Control)	100
^{14}C-Sucrose + Maltose	45
^{14}C-Sucrose + Raffinose	96
^{14}C-Sucrose + Chlorophenyl Hydrazone	12

Adapted from Christian Knop et al., "AmSUT1, a Sucrose Transporter in Collection and Transport Phloem of the Putative Symplastic Phloem Loader *Alonsoa meridionalis*," *Plant Physiology* 134, no. 1 (January 2004): 204–214.

What conclusion can the researchers most justifiably draw from this data?

(A) The cloned protein is a carrier protein that facilitates passive movement along a concentration gradient.

(B) The cloned protein is a channel protein that is not specific for sucrose.

(C) Transport of sucrose by the cloned protein does not require energy.

(D) The cloned protein is involved in the active transport of sucrose.

2. In an emergency, a doctor can replace a patient's blood plasma with a solution of 5% human albumin and 0.9% NaCl. A pharmacist does not have ready-made 5% albumin solution and decided to use a 25% albumin solution instead. He mixed 1 part of the 25% solution with 4 parts of sterile distilled water.

The administration of the solution caused severe damage to the patient. How can the consequence of the administration of substitute plasma be explained?

(A) The dilution was incorrect, resulting in an excessive concentration of albumin that led to the shriveling of the red blood cells.

(B) The dilution was incorrect, resulting in a deficient concentration of albumin that led to the bursting of the red blood cells.

(C) The dilution of albumin in distilled water caused the bursting of red blood cells because the solution was hypotonic to the red blood cells.

(D) The dilution of albumin in distilled water caused the shriveling of the red blood cells because the solution was hypertonic to the red blood cells.

3. The energy required for protein segments to span a plasma membrane was calculated and presented in the graphs shown. The higher the free energy required, the less likely the amino acid will be found in that environment. The grayed rectangle shows the region associated with the membrane lipids. Graphs A and B below represent results obtained for different types of amino acids, but are not labeled.

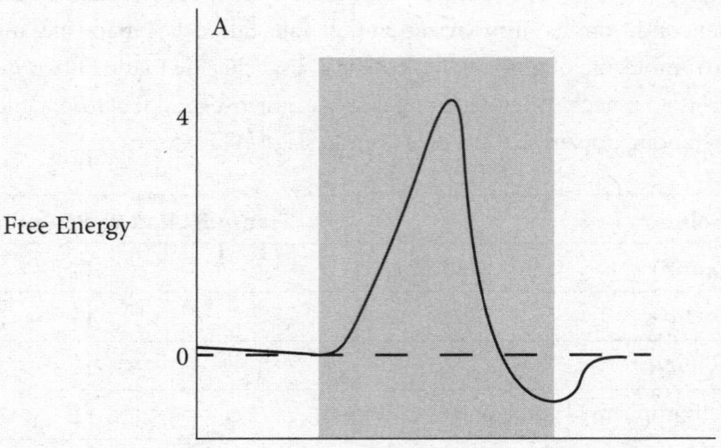

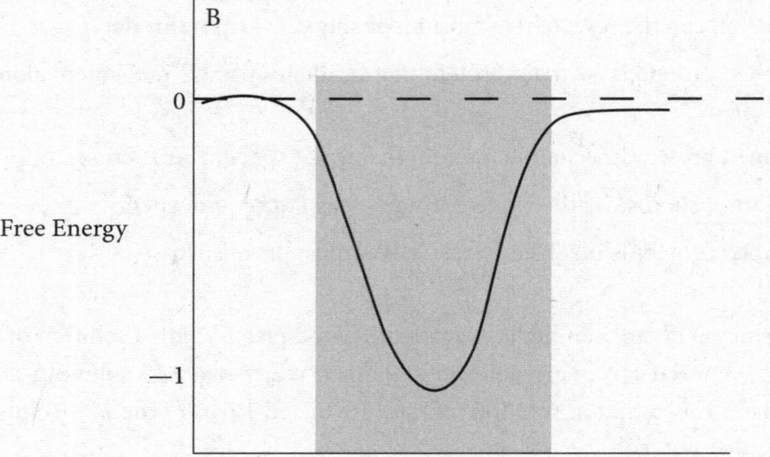

Adapted from Gunnar von Heijne, "Membrane-Protein Topology," *Nature Reviews Molecular Cell Biology 7* (December 2006): 909–918.

Which of the following best characterizes how the graphs should be labeled?

(A) Graph A represents nonpolar amino acids and Graph B represents charged amino acids.

(B) Graph A represents amino acids that have short side chains and Graph B represents amino acids that have long side chains.

(C) Graph A represents charged amino acids and Graph B represents nonpolar amino acids.

(D) Graph A represents amino acids that have long side chains and Graph B represents amino acids that have short side chains.

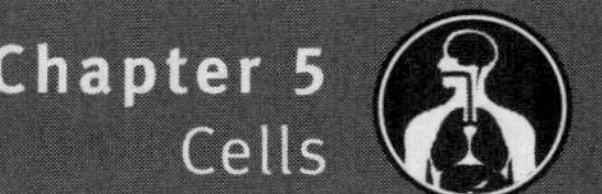

4. A pharmaceutical company is evaluating new types of drugs to treat bacterial infections. They have several candidates with different mechanisms of action. Which of these candidates would most likely be the safest for use in humans?

   (A) A molecule that prevents the formation of pores in the nuclear membrane

   (B) A molecule that inhibits the crosslinking of subunits in the biosynthesis of cell walls

   (C) A molecule that prevents the modification of proteins in the Golgi apparatus

   (D) A molecule that inhibits biosynthesis of the phospholipids in the plasma membrane

5. An investigator supplies radioactive carbon in the form of $^{14}CO_2$ to plants and harvests tissues at regular intervals. The tissue is fractionated and analyzed for radioactive carbon content in phosphoglyceric acid (PGA), a 3-carbon sugar, which is an early product of carbon fixation. The results are plotted in a graph that shows the content of radioactive PGA as a function of time of exposure to radioactive carbon.

Labeling Curve of Phosphoglyceric Acid in Barley Cells

^{14}C Incorporated in Phosphoglyceric Acid (ng C/mg Chlorophyll)

40

20

Total cell fraction

Chloroplasts

Cytosol

Mitochondria

100

200

Duration of Exposure to ^{14}C (s)

Adapted from Olav Keerberg et al., "Quantitative Analysis of Photosynthetic Carbon Metabolism in Protoplasts and Intact Leaves of Barley. Determination of Carbon Fluxes and Pool Sizes of Metabolites in Different Cellular Compartments," *Biosystems* 103 (October 2010): 291–301.

After analyzing the graphs, the scientist can reasonably conclude that

   (A) carbon fixation occurs in all compartments

   (B) carbon fixation takes place mostly in the cytosol

   (C) PGA produced in the mitochondria comes from CO_2 in the air

   (D) the first steps of carbon fixation take place in the chloroplast

6. Researchers created phospholipid bilayer spheres into which they incorporated purified protein as part of the bilayer. If the protein incorporated was a water channel (aquaporin), which of the following should be true?

   (A) The proteins should dissociate from the membrane.

   (B) If the researchers added salt to the solution, the average diameter of the spheres should stay the same.

   (C) If the researchers added salt to the solution, the average diameter of the spheres should increase.

   (D) If the researchers added salt to the solution, the average diameter of the spheres should decrease.

Questions 7–8

The neurotransmitter acetylcholine has been shown to be an important target for a large number of synthetically produced drugs, in addition to numerous venoms and toxins produced by plants, animals, and bacteria, which interfere with acetylcholine release or reuptake.

In the brain, acetylcholine can function as both a neurotransmitter and a neuromodulator. As a neurotransmitter, it is stored in the axon terminal until an action potential triggers its release into the synaptic cleft, where it can bind to activate skeletal muscles.

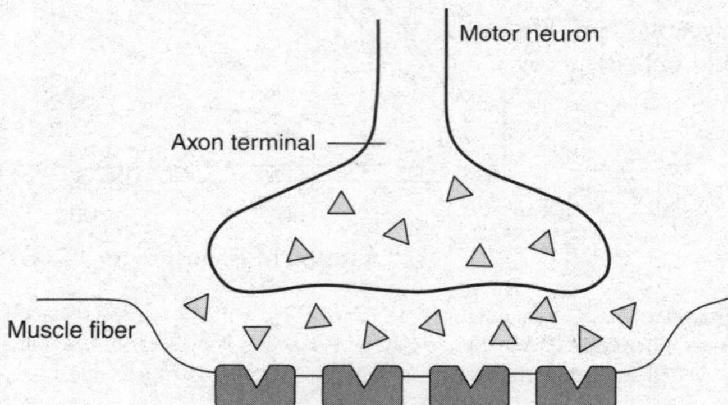

There are two main classes of acetylcholine receptors: nicotinic and muscarinic. Nicotinic receptors are ligand-gated ion channels while muscarinic receptors function as G protein-coupled receptors and exert their effects through a second messenger system.

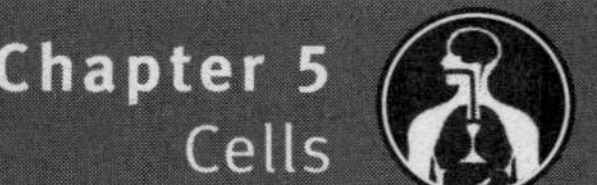

7. In order to propagate an action potential, neurons utilize different types of transport mechanisms. Which of the following correctly link the type of transport used to its function?

    I. Simple diffusion occurs when sodium ions flow through leak channels in the cell membrane.
    II. Facilitated diffusion occurs when acetylcholine binds to a nicotinic receptor to open an ion channel, allowing sodium to enter the cell.
    III. The sodium/potassium pump utilizes passive transport to draw potassium into the cell and push sodium out of it.

    (A) I only
    (B) II only
    (C) I and II only
    (D) II and III only

8. Researchers have determined that patients with Alzheimer's produce increasingly less acetylcholine as the disease progresses. They hypothesize that this is due to oxidative stress that alters the cell membrane and disrupts the function of affected brain cells. Which of the following provides the best explanation for how this occurs?
    (A) The oxidative stress increases the amount of channels in the membrane, disrupting normal ion levels.
    (B) The disruption of the cell membrane causes fewer muscarinic receptors to be embedded on the cell surface.
    (C) The oxidative stress increases the amount of cholesterol in the membrane, making it difficult for integral proteins to stay secured in the membrane.
    (D) The disruption of the cell membrane causes fewer nicotinic receptors to be embedded on the cell surface.

Questions 9–10

The questions that follow refer to the three transport processes shown below.

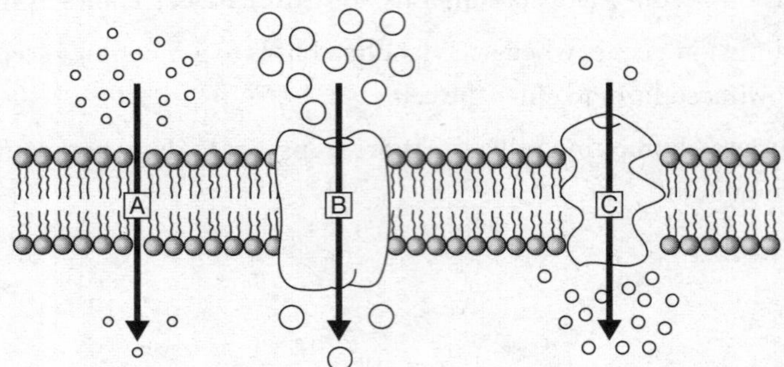

9. Which of these transport processes requires the input of energy?

(A) Process A requires the input of energy because molecules must move rapidly to cross membranes without protein channels.

(B) Process B requires the input of energy because the molecules are flowing up their concentration gradient.

(C) Process C requires the input of energy because the molecules are moving from an area of low concentration to an area of high concentration.

(D) None of these processes require the input of energy because they all occur spontaneously.

10. Which of the following most likely depicts the transport of a nonpolar substance?

(A) The molecules in process A are able to cross without aid through the nonpolar fatty acid tails of the membrane.

(B) The molecules in process B are able to cross through a channel protein, which is necessary because the fatty acid tails are polar.

(C) The molecules in process C are nonpolar because only nonpolar substances can be moved against their concentration gradient.

(D) None of the molecules shown in these processes can be nonpolar.

11. Hypernatremia is defined as the condition of having too much sodium in the blood plasma. When this occurs, the sodium diffuses into the interstitial fluid, causing water to flow out of the cells through osmosis. As a result, the cells shrink. When neurons shrink, people can experience different symptoms, ranging from confusion to hallucinations to death.

Researchers wish to observe the effects of hypernatremia in the lab. They take a synthetic cell originally submerged in distilled water and place it in a solution containing 0.5 M NaCl. Using the information provided and assuming the experiment is being carried out in an open container at room temperature (25°C), calculate the difference in the water potential before and immediately after the cell is placed in the salt solution. Round your answer to the nearest tenth.

(A) −24.8 (B) −12.4 (C) −2.1 (D) −1.0

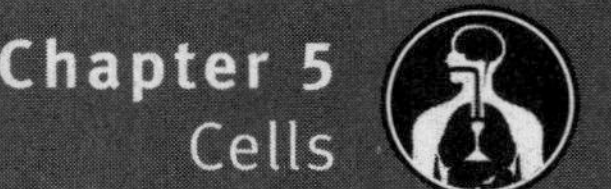

12. Although prokaryotes lack the membrane-bound organelles that are found in eukaryotes, they are still able to carry out most of the same functions. Which of the following provides an example of a mechanism prokaryotes use to carry out the function of the associated eukaryotic organelle?
    (A) Rather than using chloroplasts to synthesize food using light energy, prokaryotes either prepare food through chemosynthesis or depend on other substances for nutrition.
    (B) Rather than hydrolyze foreign material in lysosomes, prokaryotes use peroxisomes for detoxification.
    (C) Rather than synthesize ATP in mitochondria, specialized molecules in the cell wall use an electron gradient to generate ATP.
    (D) Rather than generate an mRNA transcript in the nucleus, prokaryotic transcription of the bacterial chromosome occurs in the cytosol.

13. Researchers are testing the efficacy of a drug known to target and inhibit the functioning of the endoplasmic reticulum. Which of the following tables provides the most accurate representation of the expected effects?

(A)

	Protein production	**Lipid production**	**RNA production**	**Carbohydrate production**
Effect	decreased	normal	decreased	normal

(B)

	Protein production	**Lipid production**	**RNA production**	**Carbohydrate production**
Effect	decreased	decreased	decreased	normal

(C)

	Protein production	**Lipid production**	**RNA production**	**Carbohydrate production**
Effect	decreased	decreased	normal	normal

(D)

	Protein production	**Lipid production**	**RNA production**	**Carbohydrate production**
Effect	decreased	decreased	normal	decreased

Structures

Answer Key

Test What You Already Know

1. D
2. A
3. D
4. B
5. D

Test What You Learned

1. D
2. C
3. C
4. B
5. D
6. D
7. B 
8. A
9. C
10. A
11. A
12. D
13. C 

REFLECTION

Test What You Already Know score: __________

Test What You Learned score: __________

Use this section to evaluate your progress. After working through the pre-quiz, check off the boxes in the "Pre" column to indicate which Learning Objectives you feel confident about. Then, after completing the chapter, including the post-quiz, do the same to the boxes in the "Post" column. Keep working on unchecked Objectives until you're confident about them all!

Pre	Post	
❑	❑	**5.1** Differentiate between prokaryotic and eukaryotic structures
❑	❑	**5.2** Explain the function of common organelles
❑	❑	**5.3** Recall membrane composition and function
❑	❑	**5.4** Explain transport mechanisms
❑	❑	**5.5** Investigate effects of tonicity and gradients

FOR MORE PRACTICE

Complete more practice online at kaptest.com. Haven't registered your book yet? Go to kaptest.com/booksonline to begin.

ANSWERS AND EXPLANATIONS

Test What You Already Know

1. D Learning Objective: 5.1

All prokaryotes lack a true nucleus, while all eukaryotes contain one, so the cell isolated by the biologists must be a prokaryote. Therefore, **(D)** is correct. (A) is incorrect because eukaryotes possess a true nucleus, but the isolated cell did not. In addition, some prokaryotes are quite large, so size does not immediately reveal the identity of a cell. (B) is incorrect because the lack of a nucleus is enough to conclude that the cell is prokaryotic. (C) is incorrect because viruses are not, properly speaking, organisms and because viruses lack their own ribosomes, instead using those of the host cells that they occupy.

2. A Learning Objective: 5.2

The contractions of the heart are powered by ATP, so **(A)** is correct. (B) is incorrect because muscle cells need large amounts of mitochondria to contract, not to synthesize proteins. While muscles do contain large amounts of proteins, turnover is slow and proteins are continuously made. (C) is incorrect because liver cells do contain some mitochondria, just fewer than other tissue types, as can be seen in the graph. (D) is incorrect because the amount of mitochondria varies in accordance with the function of the tissue.

3. D Learning Objective: 5.3

Based on the information provided, cell membranes require a significant proportion of unsaturated fatty acids in order to maintain fluidity at low temperatures. Thus, *B. psychrophilus* needs to contain a larger proportion of unsaturated fatty acids because it tends to be found at lower temperatures, making **(D)** correct. (A) is incorrect because there is a correlation found in the table: temperature is inversely correlated with unsaturated fatty acid content. (B) is incorrect because bacteria in colder environments require additional fluidity, not additional rigidity. (C) is incorrect because the high temperatures that *G. stearothermophilus* is exposed to require greater rigidity, not greater fluidity, which is why the species contains such a small proportion of unsaturated fatty acids.

4. B Learning Objective: 5.4

Molecular oxygen is a small, hydrophobic molecule that diffuses through the layer of surfactants and the thin epithelium in an alveolus. The higher the concentration of oxygen, the greater will be the rate of diffusion, as seen in Graph A. **(B)** is thus correct. (A) is incorrect because the membrane is permeable to molecules like oxygen and because oxygen diffuses down its concentration gradient. (C) and (D) are incorrect because they indicate the wrong graph and because oxygen neither binds to a carrier nor flows by osmosis (osmosis specifically describes the diffusion of water).

5. D Learning Objective: 5.5

Albumin does not permeate either type of membrane due to its size. The dialysis tube selects only according to size, so NaCl is small enough to cross it. A red blood cell membrane, in contrast, selects according to both hydrophobicity and size; because NaCl is polar, it is not permeable across the plasma membrane, though it can be transported with carrier proteins. Hence, **(D)** is correct. (A) is incorrect because carrier proteins can allow NaCl to exit a red blood cell. (B) is incorrect because albumin is not permeable across a cell membrane (it would require an alternative mechanism of transport). (C) is incorrect because a red blood cell membrane is not permeable to NaCl.

Test What You Learned

1. D Learning Objective: 5.4

The results in the table for chlorophenyl hydrazone suggest that the protein requires energy in the form of ATP to transport sucrose. Thus, the most justifiable conclusion is that the protein is involved in the active transport of sucrose, **(D)**. (A) and (C) are incorrect because the protein requires ATP to be effective. (B) is incorrect because the protein is inhibited by a similar sugar (maltose) but not by a dissimilar sugar (raffinose), which suggests that the protein is specific to sucrose.

2. C Learning Objective: 5.5

The pharmacist's dilution of albumin was correct: when a solution of 25% albumin is diluted to contain five times as much water, it will have one-fifth of the original concentration, or 5% albumin. However, the pharmacist should have used sterile saline solution instead of the distilled water, to ensure that there was an adequate concentration of NaCl. Because the resulting solution was hypotonic to the red blood cells, water from the solution osmosed into the cells, causing them to burst, as in **(C)**. (A) and (B) are incorrect because the pharmacist did not err in his dilution of the albumin. (D) is incorrect because the solution is hypotonic to the red blood cells, not hypertonic.

3. C Learning Objective: 5.3

Because the center of the cell membrane consists of hydrophobic fatty acid tails, amino acids with nonpolar (hydrophobic) side chains would be far more likely to be found there than amino acids with charged (hydrophilic) side chains. A hydrophobic compound would be more stable (have a lower free energy) while a hydrophilic compound would be less stable (have a higher free energy) within a membrane. Thus, Graph A must correspond to charged amino acid side chains and Graph B to nonpolar amino acid side chains, making **(C)** correct. (A) is incorrect because it reverses the results. (B) and (D) are incorrect because the length of the side chains is irrelevant; only the polarity of the side chain matters.

4. B Learning Objective: 5.1

Animal cells do not contain cell walls, but many prokaryotes do contain such structures. Thus, a drug specifically targeting cell walls (such as penicillin) would be relatively safe to use in humans. Therefore, **(B)** is correct. (A) and (C) are incorrect because only eukaryotic cells contain organelles such as the nucleus and Golgi apparatus, so these drugs would be harmless against bacteria while being toxic to humans. (D) is is incorrect because both eukaryotes and prokaryotes contain phosopholipids in their plasma membranes, so this drug would be harmful to both bacteria and human cells.

5. D Learning Objective: 5.2

The graph indicates that most of the labelled carbon ends up in the chloroplasts, suggesting that carbon fixation primarily takes place there. **(D)** is thus correct. (A) is incorrect because the proportion of labeled carbon varies greatly between compartments, with very little found in the cytosol and mitochondria. (B) is incorrect because the fraction of labeled carbon found in the cytosol is quite low. (C) is incorrect because PGA is not produced in the mitochondria and because very little labeled carbon is found there.

6. D Learning Objective: 5.5

If salt was added to the solution surrounding the artificial cell described, water would flow through the aquaporins out of the cell. As a result, the diameter of the cell would decrease, **(D)**. There is no reason to believe that the proteins would dissociate from the membrane, so (A) is incorrect. Since the diameter of the cell should decrease, the diameter wouldn't stay the same, (B), or increase, (C).

7. B Learning Objective: 5.4

Simple diffusion refers to the process by which a substance passes through a membrane without the aid of an intermediary, such as an integral membrane protein. Since leak channels are a type of integral membrane protein, statement I cannot be true, eliminating (A) and (C). Facilitated diffusion is the process of spontaneous transport of molecules across a membrane via specific membrane-spanning integral proteins. When acetylcholine binds to a nicotinic receptor, it opens a ligand-gated ion channel that allows sodium, calcium, and potassium to flow through it along their respective gradients. As a result, this is a clear example of facilitated diffusion and II is true. The sodium/potassium pump is an active transport process since it involves the hydrolysis of ATP to provide the necessary energy to push three sodium molecules out of the cell and draw two potassium molecules in. Therefore, statement III is false, eliminating (D) and making **(B)** the correct answer.

8. A Learning Objective: 5.3

Oxidative stress is an imbalance between the generation of free radicals and the ability of the body to neutralize their harmful effects by using antioxidants. However, as this is outside of the scope of necessary knowledge for AP Biology, the best approach for this question is process of elimination. The question stem states that less acetylcholine is being produced due to the disease. As a result, (B) and (D) are incorrect because they do not play a role in the production of acetylcholine (instead, both describe effects on acetylcholine receptors) and so do not provide any connection between oxidative stress and the decrease in the neurotransmitter. (C) is incorrect because increasing the amount of cholesterol would make integral proteins more secure in the membrane by forming "rafts" around them. Thus, **(A)** must be the correct answer. In fact, oxidative stress essentially "punches holes" in the plasma membrane, allowing ions to flow in and out without regulation. This affects the membrane potential of the cell and could prevent the cell from responding to or conducting an action potential. As a result, affected cells are more likely to atrophy and die, decreasing the quantity of acetylcholine-producing cells.

9. C Learning Objective: 5.4

Energy must be used to move molecules up their concentration gradient, as shown in process C. Therefore, **(C)** is correct. Processes A and B show molecules moving down their concentration gradients. Thus, no input of energy is required, and (A) and (B) are both incorrect. Since process C does require the input of energy to move molecules up their concentration gradient, (D) is incorrect.

10. A Learning Objective: 5.3

Nonpolar substances are not blocked by the nonpolar fatty acid portion of the membrane and can therefore pass through the membrane without a channel protein, making **(A)** correct. (B) is incorrect because fatty acid tails are nonpolar. Any substance can be moved against its concentration gradient, so (C) is incorrect. Since the molecules in process A can be nonpolar, (D) is incorrect.

11. A Learning Objective: 5.5

The water potential is equal to the solute potential of a solute in an open container because the pressure potential of the solution in an open container is zero. Furthermore, because distilled water does not contain any solutes, the solute potential, and therefore the water potential, is equal to zero.

In the salt solution, the solute potential can be calculated using the following equation (found in the formula sheet provided to you on Test Day): $\Psi_S = -iCRT$. The ionization constant, i, is 2 because NaCl ionizes in water to become Na^+ and Cl^-. The molar concentration, C, is 0.5 M, based on the value provided by the question stem. The pressure constant, R, is equal to 0.0831. (This value is included in the formula sheet provided to you on Test Day.) The temperature, T, is 298 Kelvins, based on the 25°C temperature from the question stem. Plugging in these values results in the following: $\Psi_S = -(2)(0.5)(0.0831)(298) = -24.76$. Thus, the value of the solute potential for the salt solution is -24.76. Since the cell is being considered immediately after being placed in the salt solution, water has not had a chance to leave the cell to equilibrate with its surroundings. Thus, the pressure potential is still zero, making the water potential equal to -24.76. Therefore, the difference in water potential is $-24.76 - 0 = -24.76$, or -24.8 (A), when rounded to the nearest tenth.

(B) is the result of inputting 1 for the ionization constant. (C) is the result of inputting 25°C for temperature. (D) is the result of making the mistakes of both (B) and (C).

12. D Learning Objective: 5.1

(D) is correct because, for prokaryotes, all aspects of protein generation, including transcription, translation, and post-translational processing, occur in the cytosol. Various types of bacteria, such as purple phototrophic bacteria, heliobacteria, and cyanobacteria utilize photosynthesis to generate a usable form of energy, eliminating (A). They use proteins called photosynthetic reaction centers, which contain chlorophyll or chlorophyll-like molecules, to absorb light energy, thereby kickstarting the photosynthetic process. In eukaryotes, these proteins are called chloroplasts; however, in prokaryotes, these proteins are embedded in the plasma membrane or on surfaces known as photosynthetic membranes. (B) is incorrect because peroxisomes are an example of a membrane-bound organelle, so it is impossible for a prokaryote to contain one. Additionally, lysosomes are responsible for the digestion of cells, using the enzyme hydrolase, while peroxisomes utilize oxidation reactions and are involved in a variety of metabolic functions, such as catabolism of very long chain fatty acids, branched chain fatty acids, D-amino acids, and polyamines, reduction of reactive oxygen species, and biosynthesis of plasmalogens. (C) is incorrect because the electron transport chain is located in the prokaryotic cell membrane, not the cell wall.

13. C Learning Objective: 5.2

The endoplasmic reticulum (ER) is an important organelle involved in manufacturing lipids and proteins. The ER can be divided into the rough endoplasmic reticulum (rough ER) and smooth endoplasmic reticulum (smooth ER). The rough ER is covered with ribosomes on its outer surface that are responsible for translating mRNA into proteins. The smooth ER functions mainly to manufacture lipids. Though they are often discussed as separate entities, they are, in fact, sub-compartments of the same organelle. As a result, a drug targeting the ER as a whole would cause a decrease in protein and lipid production. RNA production would be unaffected because transcription of DNA to RNA occurs in the nucleus, prior to any interaction with the ER. While carbohydrates can be added to proteins made in the rough ER and then transported to areas of the cell where they are needed or sent to the Golgi apparatus for further processing and modification, they are not actually produced in the ER. Additionally, metabolism, but not production, of carbohydrates through hydrolysis occurs in the smooth ER. Thus, because only protein and lipid production would be decreased, but not RNA or carbohydrate production, **(C)** is correct.

CHAPTER 6

Viruses and Bacteria

LEARNING OBJECTIVES

In this chapter, you will review how to:

6.1 Recognize common virus types

6.2 Recall virus replication pathways

6.3 Explain why viruses mutate easily

6.4 Explain plasmid function in bacteria

6.5 Recall gene transfer processes in bacteria

6.6 Investigate bacterial transformation gene transfer

TEST WHAT YOU ALREADY KNOW

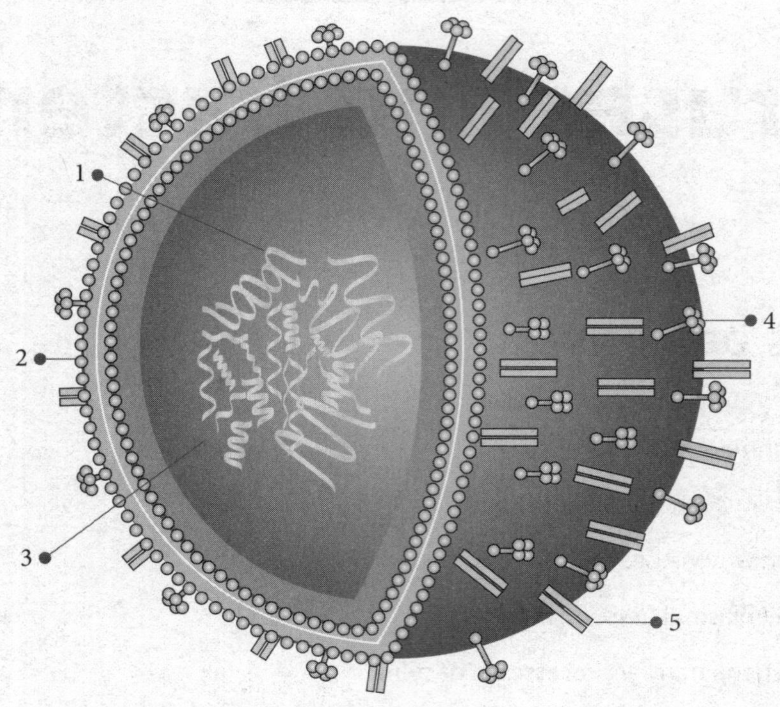

1. In the above illustration of an influenza virus, what are structures 2 and 3 and how are their functions related?
   (A) Structure 2 is the nucleoprotein that acts as a protective coating for structure 3, the capsid, which contains the genetic material of the virus.
   (B) Structure 3 is the lipid envelope that acts as a protective coating for structure 2, the capsid, which contains the genetic material of the virus.
   (C) Structure 2 is the lipid envelope that acts as a protective coating for structure 3, the neuraminidase, which contains the genetic material of the virus.
   (D) Structure 2 is the lipid envelope that acts as a protective coating for structure 3, the capsid, which contains the genetic material of the virus.

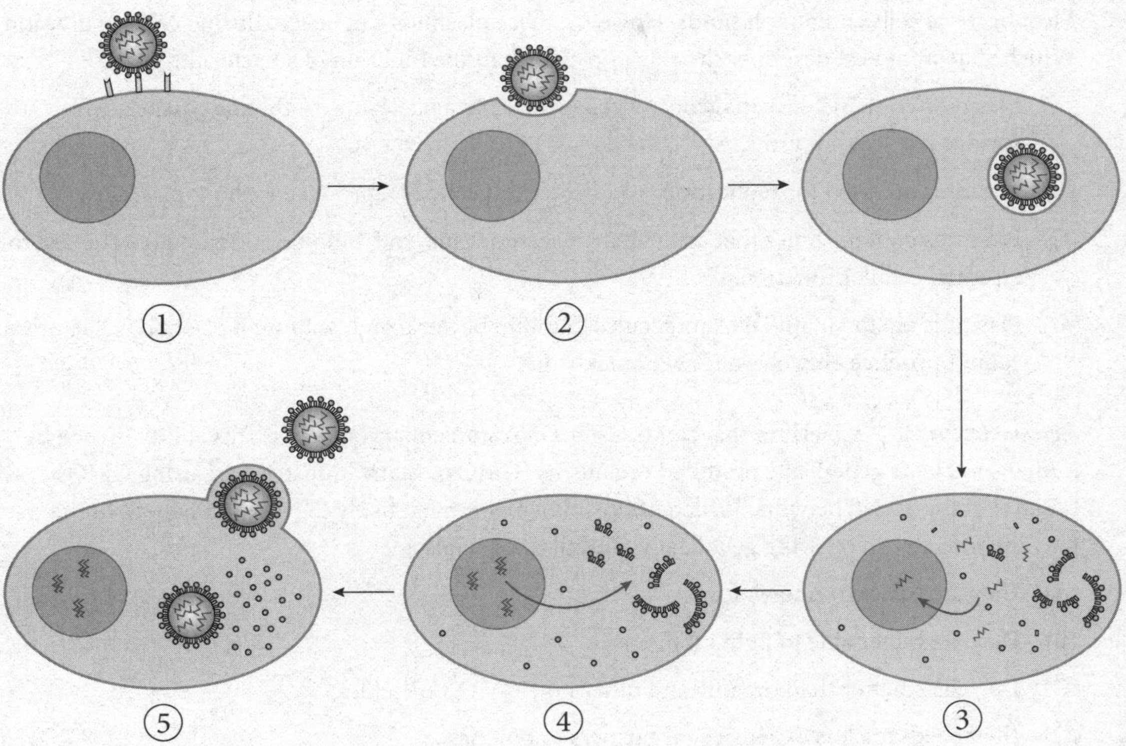

2. The diagram above illustrates the infection of an epithelial cell by an influenza virus. During which step does the viral mRNA start synthesizing new viral proteins?

   (A) 1

   (B) 2

   (C) 3

   (D) 4

3. When a human being is infected with the influenza virus, his or her immune system produces antibodies that help fight off viral antigens. Vaccines for influenza work by stimulating the body to produce the antibodies that recognize the antigens of the virus. Which of the following is the most plausible explanation of why previously inoculated people have to receive the influenza vaccine annually?

   (A) The influenza virus will likely undergo minor mutations in the genes that encode the influenza antigens.

   (B) The influenza virus will likely undergo major mutations that are caused by genetic recombination resulting from two different influenza strains infecting a host cell at the same time.

   (C) Natural selection causes only the most successful viruses from the previous year to survive to the current year.

   (D) Crossing over of the viruses' chromosomes recombines genes, creating the variation required for new strains to emerge.

4. Most bacteria cells contain plasmids. However, these plasmids can be lost during cellular division. Which statement best describes the role of plasmids in the function of a bacterium?

    (A) Plasmids contain genes that control cell growth, so a bacterium without a plasmid will survive but will not grow.

    (B) Plasmids are used for replication, so a bacterium cannot reproduce without plasmids.

    (C) Plasmids contain genes that can enhance bacterial survival, but plasmids are not required for essential cellular functions.

    (D) Plasmids are the main DNA molecules found in bacteria and, without a plasmid, a bacterium cannot produce enzymes and will quickly die.

5. *Agrobacterium* is a bacterium that can transfer DNA into eukaryotic cells. This ability is used by humans to create genetically modified organisms (GMOs). Many industries are using GMOs because of the added benefits. Which of the following is most likely the greatest benefit to the farming industry in growing genetically modified vegetables?

    (A) They are easier to transport.

    (B) They are better able to fight off diseases.

    (C) They taste better than organic and other non-GMO vegetables.

    (D) Their seeds are less expensive for farmers to purchase.

6. A researcher wants to study the effects of penicillin on the growth of *Escherichia coli.* Two Luria broth agar plates, one containing penicillin and the other only nutrient agar, were both inoculated with *E. coli.* Which of the following is the most likely outcome for the experiment?

    (A) The agar plate containing penicillin will show the same amount of bacteria growth as the agar plate without penicillin.

    (B) The agar plate containing penicillin will show more bacteria growth than the agar plate without penicillin.

    (C) The agar plate containing penicillin will show less bacteria growth than the agar plate without penicillin.

    (D) Both agar plates will show reduced bacterial growth.

Answers to this quiz can be found at the end of this chapter.

VIRAL STRUCTURE

6.1 Recognize common virus types

Viruses defy much of the logic with which we approach our study of life on Earth. Their genomes often differ strikingly from those of other living organisms, their life cycles depend upon their ability to enter and replicate within a living host cell, and they have no cell membranes or organelles as part of their structure.

Virus particles are surrounded by a capsid, or protein shell, which can come in a variety of shapes (cones, rods, and polyhedrons). Capsids are built out of proteins, many of which have various sugars attached to them, poking upward from the viral surface. These glycoproteins are used to gain entry into a living cell by binding with surface proteins on the living cell's membrane. Most viral capsids are made up of only one or two different types of protein. In addition, some viruses are able to surround themselves with an envelope of cell membrane as they burst out of a cell they have just infected. This viral envelope can help them avoid detection by the host's immune system, because the viral particles resemble (at least on the outside) the host's own cells.

Bacteriophages

The viruses that infect bacteria are known as bacteriophages, or phages for short. They are a diverse group of organisms that are best characterized by their head and tail structure, which is unique to phages:

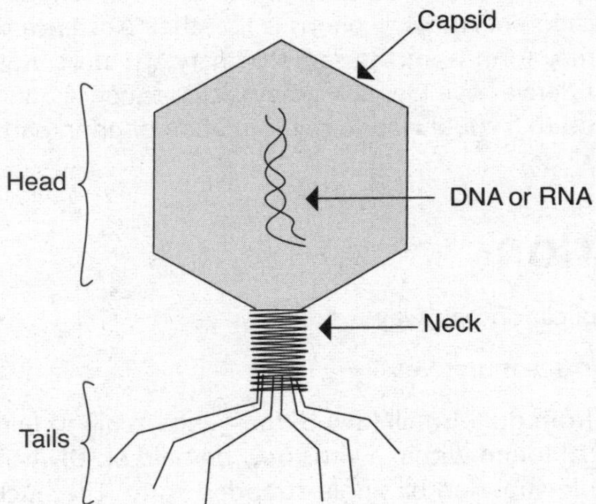

Bacteriophage Structure

The tails are used to latch on to the host cell surface, after which enzymes digest a portion of the cell membrane to insert viral genetic material (either DNA or RNA) into the host cell. The capsid and the tails are left outside of the host cell membrane, differentiating phage infection from that of other viruses. Often, the viral DNA or RNA is rapidly destroyed by powerful bacterial enzymes called restriction enzymes (which are described in detail in the lab section of chapter 16). These enzymes are a primitive type of immune system in bacteria and chew up foreign genetic material.

For the viral genomes that do survive, they will either cause the host cell to produce new viral particles (as described later in the chapter) or will integrate into the bacterial chromosome. Phages that infect bacteria in a lytic way and cause active viral replication are called lytic phages, while those whose DNA gets integrated into the host cell's DNA (in a lysogenic fashion) are called temperate phages. While integrated, the bacteriophage is known as a prophage.

Atypical Virus-like Forms

There are a few exceptions to the typical virus described above, and these particles (which certainly cannot be considered living organisms according to the standard definition of life) are infectious to many living cells:

Viroids are virus-like particles that are composed of a single molecule of circular RNA without any surrounding capsid or envelope. The RNA can replicate using a host's machinery, but it does not seem to code for any specific proteins. Despite this, viroids have been implicated in some plant diseases, though not in diseases in any other organisms.

Prions are simply pieces of protein that are infectious. They have gained fame through the recent spread of mad cow disease (bovine spongiform encephalopathy), and they are connected with diseases that are typically slow to form, taking years before symptoms develop. Despite once being called slow viruses, prions are not viruses at all and do not contain any DNA or RNA. Recent evidence suggests that prions are infectious because they change the structure of an organism's normal proteins, a switch not encoded in genes at all, by coming into contact with the proteins. They are not recognized by the immune system, probably because their structure is similar to the structure of normal proteins. Perhaps the most fascinating implication of studying prions is that their existence shows that more than one tertiary protein structure may form from the same primary structure. Both prions and the normal proteins from which prions derive have identical amino acid sequences, and no traceable DNA mutation has yet been discovered that could lead to the formation of prion particles.

VIRAL REPLICATION

6.2 Recall virus replication pathways

6.3 Explain why viruses mutate easily

Viral genomes can range from quite small (five to ten genes in all) to fairly large (several hundred genes). The genetic material found within a virus may be held on DNA or RNA, both of which can be found in either a double-stranded or single-stranded state. The nucleic acid can be linear or circular, and although the DNA is always found together in a single chromosome, the RNA can be in several pieces. Viruses are considered to be obligate intracellular parasites, meaning that they must live and reproduce within another cell, where they act as a parasite, using the host cell's machinery to copy themselves or to make proteins encoded for by their DNA or RNA. Independent of host cells, viruses conduct no metabolic activity of their own.

Lytic Cycle

Some viruses contain the enzymes they need for replication of their genome, stuffing these proteins into each "baby" virus as it is produced by the host cell. Others do not travel with their proteins but rather make them only when they are inside a host. The typical viral growth cycle includes the ordered events of attachment and penetration, uncoating, viral mRNA and protein synthesis, replication of the genome, and assembly and release.

Attachment and penetration by parent virion (viral particle): The specificity of the proteins coating the capsid of the virus determines the host range of the particular virus, or how many kinds of cells the virus can infect. Those viruses with a wide range of surface proteins (unusual) or with proteins that can bind to many kinds of cell surface receptors (more common) are said to have a wide host range.

Uncoating of the viral genome: As the viral particle penetrates, the cell traps it in a vesicle, at which point the virus will break open its capsid. When the vesicle breaks due to viral uncoating, the inner core of the virus with its genetic material can dump itself into the cytoplasm.

Viral mRNA and protein synthesis: DNA viruses replicate their DNA in the nucleus of the cell and use the host cell's RNA polymerases to make their proteins. There are a few exceptions to this (e.g., the pox family of viruses that causes smallpox, chicken pox, etc.) that cannot enter the nucleus and actually carry the necessary RNA polymerase with them. RNA viruses replicate in the cytoplasm, sometimes using their RNA as mRNA directly, and sometimes using their RNA as a template for mRNA synthesis. One group of RNA viruses, the retroviruses (e.g., HIV), converts its RNA into DNA first, using the enzyme *reverse transcriptase*. The DNA copy is then transcribed into mRNA by the host's RNA polymerases. Many of the viral proteins that are first synthesized on the host's ribosomes are the ones needed for replication of the genome.

Replication of the genome: All virus particles will first replicate the DNA or RNA they came into the host cell with; the complementary copy that is created is then used as a template for all new copies of the genome. This is analogous to making a negative from a photograph and using the negative for all subsequent copies rather than using the photograph itself. Replication proceeds using host cell DNA and RNA polymerases as cells normally would.

Assembly and release: Viral nucleic acid is packaged within a protein capsid, and "baby" virus particles are released from the cell either en masse or slowly. *En masse* release ruptures and kills the host cell; *slowly* means the particles bud out of the host cell surface and envelop themselves in the host cell membrane as they push their way out. Budding in a slow and deliberate fashion will usually allow the host cell to live.

Lysogenic Cycle

The cycle described in the previous section is typical of most viruses: it is called a lytic cycle because the host cells usually lyse, or burst, in the end. Some viruses, however, use a lysogenic cycle, whereby the viral DNA gets integrated into the host cell's DNA and can remain there indefinitely, allowing viral DNA to be replicated for generations alongside the host cell DNA. The infection continues so that all offspring of the host cell carry the viral DNA. This DNA can simply remain a part of the host cell's genome forever. For example, certain bacteria (e.g., *Corynebacterium diphtheriae*, which causes diphtheria, and *Clostridium botulinum*, which causes botulism) secrete toxins that are coded for by viral genes that were acquired from a bacteriophage (a virus that infects bacteria). In some cases, however, certain environmental events may cause this integrated viral DNA to begin a full replicative cycle (lytic cycle), accompanied by the production of new viruses and the eventual death of the cell.

BACTERIA

6.4 Explain plasmid function in bacteria

6.5 Recall gene transfer processes in bacteria

Bacteria are **prokaryotes**, micrometers in size, that live in diverse environments. Single-celled organisms composed of an outer cell wall and inner cell membrane, bacteria contain a free-floating DNA chromosome (not enclosed in a nucleus). Unlike eukaryotes, bacteria do not have membrane-bound organelles. Bacteria exist in three basic shapes: cocci (round), bacilli (rod), and spirilla (spiral). Some bacteria also contain plasmids and have flagella (whiplike **appendages**) for motility and/or have pili (hair-like appendages) for surface attachment.

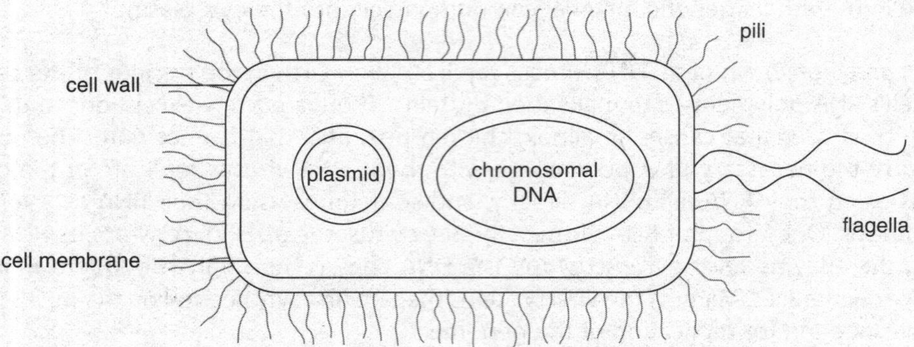

> ✔ **AP Expert Note**
>
> **The study of bacteria falls under a branch of microbiology known as bacteriology. Bacteriology is a growing field with a rapidly expanding knowledge base, but this section is limited only to material that is highly tested on the AP Biology exam.**

In response to changes in their environment, bacteria may undergo chemotaxis and quorum sensing. Chemotaxis involves bacteria responding to a chemical stimulus by moving in a certain direction. For example, bacteria move toward higher concentrations of food molecules like glucose and move away from higher concentrations of toxins like phenol. Quorum sensing involves cell-to-cell communication in which bacteria release signal molecules into the environment to monitor population density. At high cell densities, bacteria regulate gene expression to produce the most beneficial phenotypes.

Bacteria reproduce via **binary fission**, which produces genetically identical copies of the parent bacterium. Aside from random mutations, binary fission does not promote genetic recombination or genetic diversity like sexual reproduction. Genetic variation, which allows organisms to adapt to changes in their environment, is advantageous for evolution and survival. Thus, bacteria have evolved mechanisms to ensure genetic variation and to yield new phenotypes. Transfer of genetic material from one bacterium to another is called horizontal gene transfer and can occur via three mechanisms: 1) transformation, 2) transduction, or 3) conjugation. Bacteria also use transposition to increase genetic variation. In some bacterial species, the generation time is only a few minutes, which allows bacteria to evolve quickly.

Plasmids

High-Yield

Mainly found in bacteria, plasmids also naturally occur in some archaea and eukaryotes such as yeast and plants. Plasmids are small, double-stranded circular DNA molecules that are separate from the chromosomal DNA. Plasmids may carry genes that benefit the survival of the organism by providing a selective advantage. For example, plasmids may encode for **antibiotic** resistance, the production of toxins in a competitive environment, or the utilization of particular organic compounds when nutrients are scarce. Possessing its own origin of replication, a plasmid can replicate independently from its host's chromosomal DNA.

In genetic engineering, recombinant plasmids are used as vectors to clone and amplify or express particular genes. The advantages of using plasmids as vectors include the ability to easily modify and mass-produce them.

Bacterial Gene Transfers

Transformation

Upon cell death, some bacterial species release pieces of their chromosomal DNA and/or plasmids to be taken up by other bacteria. The uptake and incorporation of the exogenous (foreign) DNA into a host (recipient) bacterium is called transformation. For transformation to occur, the host cell must be competent—the cell membrane becomes more permeable to allow the DNA to pass from the environment into the cell. Transformation works best between closely-related species and results in a stable genetic change. In a laboratory setting, artificial competence can be obtained by exposing cells to calcium chloride and heat shock or by using electrical pulses to create pores in the cell membrane.

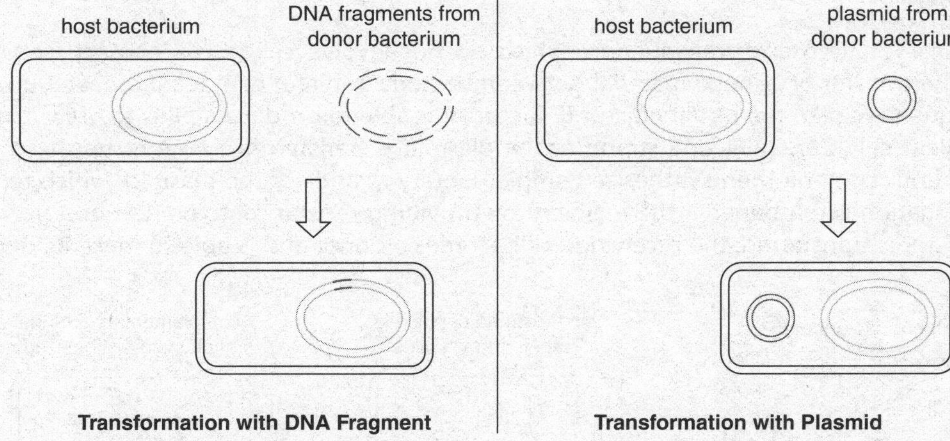

Transformation with DNA Fragment

Transformation with Plasmid

Transduction

High-Yield

Introducing exogenous DNA into a bacterium by a virus is called transduction. The bacteriophage (a virus that infects bacteria) injects viral DNA into a host bacterium and breaks down the host DNA. Infected by the phage, the host cell (donor) creates new phages that may contain host DNA packaged in the viral capsid. Bacterial lysis releases the new phages, which inject the host cell DNA into recipient bacterial cells. In the recipient bacterium, the donor DNA may be recycled for spare parts, it may recircularize and become a plasmid if it was originally one, or it may be exchanged via recombination to create a bacterium with a different genotype.

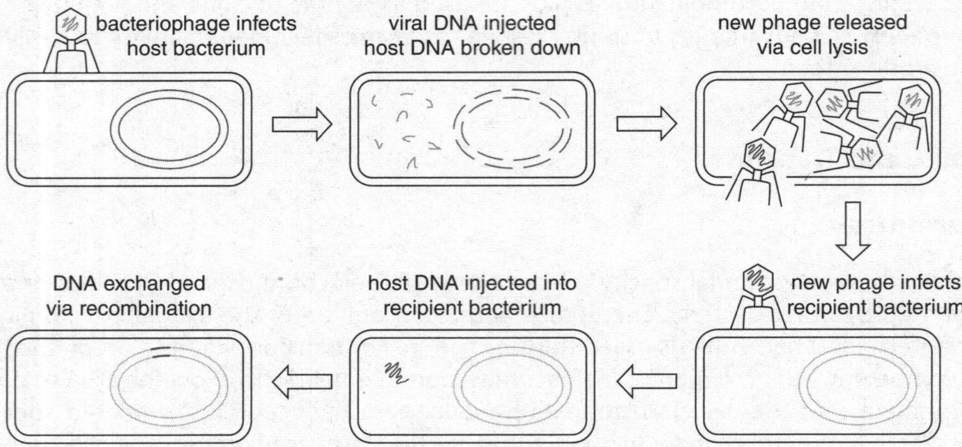

Transduction with DNA Fragment

Conjugation

Unlike transformation and transduction, which do not involve cell-to-cell contact, conjugation involves the transfer of genetic material between bacteria in direct contact. If a bacterium carries a DNA sequence called the fertility factor (F-factor), it is able to produce a pilus to draw itself close to a recipient cell. Generally, one strand of the plasmid is transferred through the thin tube-like structure. Both bacteria then synthesize complementary strands of the plasmid, which recircularizes. Conjugation often benefits the recipient by providing resistance, tolerance, or a new ability. If the F-factor is transferred, the receiving cell becomes a donor and is able to make its own pilus.

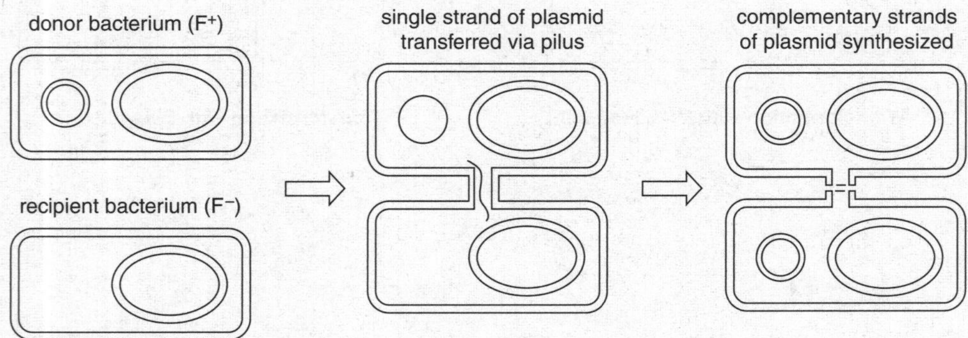

Conjugation with Plasmid

Transposition

Transposition involves the movement of DNA segments within a genome or between the chromosome and plasmid of a bacterium. Transposable elements (transposons), sometimes known as jumping genes, can change their position from one place to another by cutting and inserting themselves in new spots. In bacteria, transposable elements may carry antibiotic resistance or virulence. After transposition, bacteria may then spread the advantageous new genes to the population via some kind of horizontal transfer.

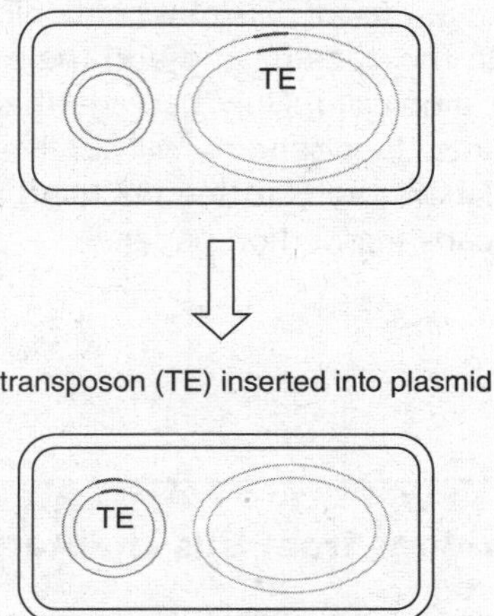

Transposition between Chromsomal DNA and Plasmid

AP BIOLOGY LAB 8: BIOTECHNOLOGY: BACTERIAL TRANSFORMATION INVESTIGATION

High-Yield

6.6 Investigate bacterial transformation gene transfer

This is a difficult investigation to conceptualize. It is also difficult to perform because it requires a considerable amount of specialized materials, expensive equipment, and aseptic conditions. Some companies have even prepared software that replaces the lab with a virtual experiment.

✔ AP Expert Note

You will most likely see questions on the AP Biology exam about the concepts covered in the two biotechnology investigations (the other one, AP Biology Lab 9, is covered in chapter 16)—along with other techniques such as polymerase chain reaction (PCR) and restriction fragment length polymorphism (RFLP) analysis.

Be sure that you familiarize yourself with the general procedures and purposes of using these techniques.

Structures

This investigation deals with the transformation of bacteria using plasmids that contain known genes. Bacteria can incorporate plasmids after being shocked by a chemical, temperature, or electrical pulses. Heat shock followed by an ice bath allows the plasmid DNA to penetrate the cell wall more easily. As long as the gene from the plasmid inserts after a promoter region, it will be expressed as part of the new, recombined bacterial genome.

The common example is to take plasmids that contain an ampicillin-resistance gene and transform bacteria that are affected by ampicillin. After applying the treatment, you can test for transformation by looking for bacterial growth on a medium that contains ampicillin. You start with a stock of ampicillin-sensitive bacteria and a stock of plasmids containing the gene for ampicillin resistance. You apply the shock treatment to the bacteria and add the plasmids, then apply the bacteria to growth media with and without ampicillin. Only bacteria that have incorporated the ampicillin-resistant gene from the plasmid into their genome will survive in the medium containing ampicillin. Your goal in this part of the laboratory is to see firsthand the products of recombination and transformation that you have already learned on paper.

RAPID REVIEW

If you take away only 4 things from this chapter:

1. Viruses are obligate intracellular parasites that must replicate inside the cells of living organisms. Viral genomes can range from small to large and be DNA or RNA and double-stranded or single-stranded.
2. There are multiple types of viruses and virus-like structures. Bacteriophages, or phages, are viruses that infect bacteria. Viroids and prions are not viruses in the traditional sense but are still infectious. One of the most famous prions is responsible for causing bovine spongiform encephalopathy, or mad cow disease.
3. Viruses can replicate in several ways, including the lytic and lysogenic cycles.
4. Bacteria use four distinctive mechanisms to increase genetic variation: transformation, transduction, conjugation, and transposition.

TEST WHAT YOU LEARNED

1. It is useful for scientists to study whether genetically transformed organisms can pass on new traits to future offspring. Which of the following organisms would be the most practical to use for the study of genetic transformations and why?

   (A) *Escherichia coli*, because they have short generation times

   (B) Insects, because most use external fertilization

   (C) Laboratory mice, because they reproduce sexually

   (D) Chimpanzees, because they are very genetically similar to humans

2. Which of the following conclusions can most reasonably be drawn if a host cell's function has been impaired following a viral infection?

   (A) The host cell will lyse or undergo apoptosis regardless of the severity of damage.

   (B) The host cell will lyse or undergo apoptosis if the severity of damage exceeds the cell's ability to repair itself.

   (C) The host cell will only lyse.

   (D) The host cell will only undergo apoptosis.

3. Some bacteria, such as *Methicillin-resistant Staphylococcus aureus* (MRSA), are known to carry genes for antibiotic resistance. Which of the following structures could best facilitate the inheritance of antibiotic-resistance genes by a bacterium?

   (A) Flagellum

   (B) Pilus

   (C) Plasmid

   (D) Chlorosome

4. There has been an outbreak of the influenza virus at a school. A student, who happens to work in the cafeteria, has been infected. Which of the following would be the best advice to give to the student to help reduce the spread of infection?

   (A) Stay at home until all of the symptoms are gone.

   (B) Go see the doctor immediately and get a prescription for antibiotics.

   (C) Take over-the-counter flu medication and go back to work.

   (D) Use antibiotics from a previous illness.

Questions 5–6

Currently, research is being performed to determine the feasibility of enabling diabetes patients to produce insulin autonomously, rather than relying on the administration of strictly-timed synthetic insulin. One idea that has been proposed is genetically engineering bacteria to reprogram cells in the intestine to secrete insulin in response to changing glucose levels in the body. Another option is to engineer the bacteria to produce insulin themselves, thereby simplifying the process.

To test the second idea, scientists perform a transformation procedure on a culture of *E. coli* cells with a plasmid containing a gene coding for insulin secretion. The plasmid also contains a gene (*amp*^{r}) that confers ampicillin resistance to the bacteria. A second culture of *E. coli* was grown, and the plasmid was not added to this sample. Both samples were plated on nutrient agar plates, half of which were supplemented with ampicillin. The results of *E. coli* growth are summarized below. The shaded area represents extensive growth of bacteria, while dots represent individual colonies of bacteria.

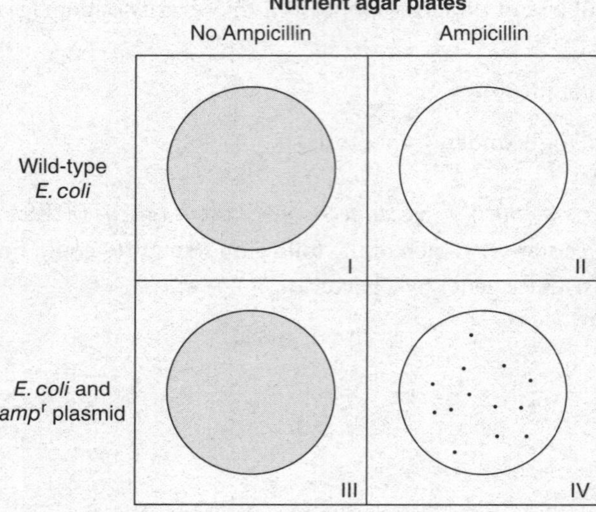

5. Scientists must be careful when inserting plasmids into bacteria in case they mutate or spread accidentally. For example, if a sample of the modified bacterial culture were to leave the lab and infect another animal, which of the following provides an example of the adverse effects that could result?

   (A) The transformed bacteria will continue secreting insulin, which will cause the animal to develop chronic high blood sugar problems.

   (B) The transformed bacteria will continue secreting insulin, which will cause the animal to stop producing its own insulin.

   (C) The transformed bacteria will continue secreting insulin, which will cause the animal to develop chronic low blood sugar problems.

   (D) The transformed bacteria will eliminate the plasmid with the gene for insulin secretion since it is no longer necessary.

6. It is also important for the scientists to consider the different ways that the transformed bacteria can share genetic information with other strains and species of bacteria. Which of the following correctly matches a way bacteria can increase their genetic variation with its corresponding term?

(A) Bacteria can transfer small fragments of DNA between individual bacteria of the same species or different species via vertical transfer.

(B) Bacteria can take up fragments of DNA from their external environment that are usually present due to the death and lysis of another bacterium via transposition.

(C) Bacteria can transfer plasmids to other bacteria that may or may not be closely related through direct cell-to-cell contact via transformation.

(D) Bacteria can transfer DNA when a bacteriophage accidentally takes up bacterial DNA from its host before infecting another closely-related bacterium via transduction.

7. Horizontal gene transfer between bacteria can occur through several mechanistic pathways as shown in the diagram below.

How does pathway 2 differ from pathway 3?

(A) Pathway 2 is transformation and involves the exchange of DNA using bacteriophages, whereas pathway 3 is conjugation and involves the transfer of genetic material using a pilus.

(B) Pathway 3 is transduction and involves the exchange of genetic material using bacteriophages, whereas pathway 2 is conjugation and involves the transfer of genetic material using a pilus.

(C) Pathway 2 is transduction and involves the exchange of genetic material using a pilus, whereas pathway 3 is conjugation and involves the transfer of genetic material using bacteriophages.

(D) Pathway 2 is transduction and involves the exchange of genetic material using bacteriophages, whereas pathway 3 is conjugation and involves the transfer of genetic material using a pilus.

8. Certain viruses, such as Ebola, have higher mutation rates than other viruses, such as the virus that causes chicken pox. Which of the following statements best explains this difference?

   (A) Ebola is a DNA virus, while chicken pox is an RNA virus.

   (B) Ebola is an RNA virus, while chicken pox is a DNA virus.

   (C) Ebola is a DNA virus, while chicken pox is a retrovirus.

   (D) Ebola is an RNA virus, while chicken pox is a retrovirus.

9. Some viruses, such as the dengue virus and the tobacco mosaic virus, have RNA genomes that are replicated by RNA-dependent RNA polymerases (RdRPs). These RdRPs have no proofreading ability. As a result, RdRPs have very high mutation rates. Which of the following explains the adaptive significance of this?

   (A) The high mutation rate is inconsequential because viruses produce enough offspring to maintain a stable genome.

   (B) Since their genomes are large, RNA viruses typically encode several additional proteins to proofread the RNA before packaging.

   (C) Since RNA viruses use the host's DNA-editing mechanisms on their RNA, error-prone RdRPs are acceptable.

   (D) The high mutation rate leads to increased variability and rapid evolution.

10. While viruses that infect bacteria may seem beneficial with respect to curing bacterial illnesses, they can also cause adverse effects to occur by providing harmless bacteria with genetic information that allows them to become pathogenic. For example, *Escherichia coli*, *Streptococcus pyogenes*, *Vibrio cholerae*, and *Shigella* only become harmful when viruses transfer the genes that produce toxic substances to them. This allows them to infect humans and cause diarrhoeal disease, in addition to other, possibly fatal, symptoms.

    Which of the following best explains how a specialized gene encoding a particular pathogenic toxin can be consistently incorporated into new bacterial genomes?

   (A) The virus integrates its genetic information into a specific location on a host cell's chromosome until stress causes it to be excised, taking some of the bacterial chromosome with it.

   (B) The virus produces enzymes that cause the host DNA to be broken down into fragments and pieces of host DNA are transferred to subsequent bacteria through bacterial transformation.

   (C) The virus contains RNA that has mutated to code for the toxin, which it injects into the host cell to be translated, packaged, and released as a protein into the extracellular space.

   (D) The virus contains RNA that has mutated to code for the toxin and uses reverse transcriptase to convert the toxin gene to DNA, so it can be processed and distributed by the host cell.

11. A microbiologist studies proteins from a virus that he obtained from a monkey. He finds that the virus is an enveloped RNA virus that actually infected bacteria found in the samples, rather than the monkey cells themselves. Which of the following correctly identifies this type of virus and explains why?

(A) Retrovirus, since it contains RNA

(B) Bacteriophage, since it infects bacteria

(C) Provirus, since it was found in a bacterium

(D) Viroid, since it replicates within mammals

12. A student is performing a lab, in which she attempts to transfer a gene that confers ampicillin resistance to a culture of *E. coli*. She adds 10 microliters of the 0.005 μg/μL plasmid solution to a tube of *E. coli*. 150 microliters of the resulting 600 μL suspension are then placed on agar plates and incubated at 37°C for 24 hours. After the procedure is finished, the student identifies 35 individual colonies that have grown on the plate. She then calculates the transformation efficiency in order to determine how effective she was in getting the plasmids carrying ampicillin resistance into the host bacterial cells. Based on the information provided above, calculate the transformation efficiency of the experiment. Give your answer to the nearest whole number with units of colonies per microgram of plasmid.

(A) 7

(B) 14

(C) 229

(D) 2800

13. A student studying viruses in lab is surprised to find that RNA viruses tend to have a significantly higher mutation rate than DNA viruses. What is a potential explanation for this occurrence?

(A) RNA viruses do not have replication error-checking mechanisms, as DNA viruses do.

(B) RNA viruses assemble in both the nucleus and cytoplasm, while DNA viruses assemble solely in the nucleus.

(C) All RNA viruses contain reverse transcriptase, which commonly causes frameshift mutations.

(D) RNA viruses have genome RNA that forms mRNA, which is then transcribed, bypassing the error-checking mechanisms that occur during translation altogether.

Answer Key

Test What You Already Know

1. D
2. D
3. A
4. C
5. B
6. C

Test What You Learned

1. A
2. B
3. C
4. A
5. C
6. D
7. D
8. B
9. D
10. A
11. B
12. D
13. A

REFLECTION

Test What You Already Know score: __________

Test What You Learned score: __________

Use this section to evaluate your progress. After working through the pre-quiz, check off the boxes in the "Pre" column to indicate which Learning Objectives you feel confident about. Then, after completing the chapter, including the post-quiz, do the same to the boxes in the "Post" column. Keep working on unchecked Objectives until you're confident about them all!

Pre	Post	
❑	❑	**6.1** Recognize common virus types
❑	❑	**6.2** Recall virus replication pathways
❑	❑	**6.3** Explain why viruses mutate easily
❑	❑	**6.4** Explain plasmid function in bacteria
❑	❑	**6.5** Recall gene transfer processes in bacteria
❑	❑	**6.6** Investigate bacterial transformation gene transfer

FOR MORE PRACTICE

Complete more practice online at kaptest.com. Haven't registered your book yet? Go to kaptest.com/booksonline to begin.

ANSWERS AND EXPLANATIONS

Test What You Already Know

1. D Learning Objective: 6.1

Structure 2 is the lipid envelope and structure 3 is the capsid. The lipid envelope provides a protective coating for the capsid, which contains the RNA, the genetic material of the influenza virus. **(D)** is thus correct. (A) is incorrect because the nucleoprotein is actually structure 1, the genetic material of the virus. (B) is incorrect because it reverses the identities of structures 2 and 3. (C) is incorrect because the neuraminidase is structure 4, a surface enzyme that facilitates viral replication.

2. D Learning Objective: 6.2

In step 1, the influenza virus attaches to the host cell and prepares for infection. In step 2, the virus is engulfed by the epithelial cell and, in step 3, the virus is released and enters the nucleus, where it will begin replicating by using the resources of the host cell. Only in step 4 does the viral mRNA begin synthesizing new viral proteins. **(D)** is thus correct.

3. A Learning Objective: 6.3

Minor mutations that alter the antigens of the virus are quite common and, if enough of these mutations accumulate, can make an immune system exposed to an earlier form of the virus unable to recognize the new form. **(A)** is thus correct. (B) is incorrect because major mutations resulting from two strains infecting a host at the same time are quite uncommon. (C) is incorrect because the most successful viruses from a previous year would still be recognizable by an immune system that was inoculated against them. (D) is incorrect because viruses do not use meiosis to reproduce, so crossing over cannot occur in viruses.

4. C Learning Objective: 6.4

Plasmids may contain genes for antibiotic resistance or other characteristics that can help bacteria survive in adverse circumstances, but they are independent of the bacterial chromosome that contains genes for essential functions. **(C)** is thus correct. (A) is incorrect because genes for growth control are generally contained in the bacterial chromosome. (B) is incorrect because some bacteria contain no plasmids but are still able to reproduce. (D) is incorrect because plasmids are not the main DNA molecules found in bacteria; the circular bacterial chromosome contains genes for essential enzymes.

5. B Learning Objective: 6.5

Genetically modified vegetables can be bred to be more resistant to disease and to last much longer than organic and other non-GMO vegetables. **(B)** is thus correct. (A) is incorrect because transportation is generally the same for both GMO and non-GMO vegetables. (C) is incorrect because people disagree about taste, so there is no objective basis for saying that genetically modified vegetables taste better. (D) is incorrect because genetically modified seeds are typically much more expensive than other seeds.

6. C Learning Objective: 6.6

Penicillin is an antibiotic, so it would reduce the growth of *E. coli* unless some of the bacteria contained, for instance, plasmids with antibiotic resistance. In any case, it is safe to conclude that there would be less bacterial growth on the plate with penicillin. **(C)** is correct. (A) and (B) are incorrect because they suggest a different outcome. (D) is incorrect because there is no reason to suspect that the plate without penicillin would show less growth than normal.

Test What You Learned

1. A Learning Objective: 6.6

E. coli bacteria are single-celled organisms that reproduce easily with very short generation times, allowing for many generations of offspring to emerge in a relatively short time. This makes them prime candidates for the study of genetic transformations. **(A)** is thus correct. (B), (C), and (D) are all complex eukaryotic organisms that reproduce sexually with longer generation times than bacteria, making it more difficult to study the impact of a transformed gene across generations.

2. B Learning Objective: 6.2

If a cell loses its ability to function and cannot be repaired, then it could lyse (burst open) or undergo apoptosis (programmed cellular death). Whether it does either depends on the severity of the damage to the cell. **(B)** is thus correct. (A) is incorrect because the cell will remain intact if it is capable of repairing the damage. (C) and (D) are incorrect because they only present one of the two viable options for a damaged cell.

3. C Learning Objective: 6.4

Plasmids are small, circular DNA molecules that carry genes that can enhance bacterial survival. The exchange of plasmids between bacteria in processes of horizontal gene transfer allows for the spread of traits like antibiotic resistance. Plasmids can also be transferred vertically when a bacterium divides and replicates not only its chromosome but also any plasmids it contains. **(C)** is thus correct. (A) is incorrect because flagella are structures used for movement, not gene transmission. Pili are hair-like structures that allow for adhesion to other cells. Though they can be used in the process of conjugation, such horizontal gene transfer cannot occur without the actual genes, often contained in plasmids, which make plasmids the more important structure for spreading antibiotic resistance. Thus, (B) is incorrect. (D) is incorrect because chlorosomes are complexes used in photosynthesis.

4. A Learning Objective: 6.1

A sick student working in the school cafeteria could potentially infect many other students by contaminating food with the virus. The best advice for such a student would be to remain at home until all symptoms subside, at which point the student would likely no longer be infectious. Therefore, **(A)** is correct. (B) and (D) are incorrect because antibiotics are only effective against bacterial infections. (C) is incorrect because over-the-counter flu medications will at best alleviate symptoms but will not stop someone from being infectious.

5. C Learning Objective: 6.4

Bacteria that are manipulated to target a specific problem need to be heavily regulated due to the effects they may have on non-target populations. For example, one risk of an insulin-producing bacterial culture is that it may infect organisms that would be adversely affected by an increase in their internal insulin concentration. Too much insulin can lead to decreased blood sugar levels, as mentioned in **(C)**, that can eventually become fatal for an organism. (A) is incorrect because too much insulin would cause low blood sugar, not high blood sugar. Since insulin release is rapidly triggered in response to increased blood glucose levels, it is unlikely that the presence of bacteria that also release insulin in response to increased blood glucose levels would completely stop the animal's insulin production, eliminating (B). Finally, the bacteria would not be able to sense that their insulin secretion was killing their host, so they would have no cause for downregulating it. Moreover, bacteria are incapable of simply eliminating plasmids that contain "unnecessary" genes. As a result, (D) is also incorrect.

6. D Learning Objective: 6.5

Transduction is the process by which genetic material is transferred from one bacterium to another by a bacteriophage, either through generalized or specific transduction. Thus, **(D)** is correct. Transformation is the process by which a bacterial cell takes up foreign DNA from its surroundings. Naturally, this occurs when neighboring cells have died and been lysed. However, scientists utilize this occurrence to introduce specific genes of interest into bacteria for further study or for therapeutic processes. As a result, (B) and (C) are both incorrect. Conjugation, as described in (C), is the direct transfer of genetic material between bacteria while being temporarily joined. (A) is incorrect because it describes horizontal transfer, which includes transformation, transduction, conjugation, and transposition (the movement of DNA segments within and between DNA molecules). Vertical transfer is defined as the transmission of genetic material from parents to offspring during reproduction.

7. D Learning Objective: 6.5

Pathway 1 is transformation (genetic transfer from loose genes in the extracellular environment), pathway 2 is transduction (genetic transfer using a virus), and pathway 3 is conjugation (genetic transfer using a bacterial pilus). **(D)** is the appropriate match. (A), (B), and (C) each mislabel one or more of the pathways.

8. B Learning Objective: 6.3

RNA viruses lack the proofreading ability of DNA polymerases. As such, they are more likely to mutate than DNA viruses. **(B)** thus presents the best explanation for Ebola's higher mutation rate. (A) is incorrect because it reverses the two viruses. (C) is incorrect because retroviruses tend to mutate more quickly than DNA viruses. (D) is incorrect because retroviruses tend to mutate about as quickly as other RNA viruses.

9. D Learning Objective: 6.3

RNA viruses rely on their error-prone replicases (RNA-dependent RNA polymerases) to generate variety. This variability allows for rapid evolution. Hence, **(D)** is correct. A stable genome, as suggested in (A), is useful in *unchanging* environments. Adaptation to changing environments requires variability. Viral genomes are small (on the order of 10,000 base pairs) and typically do not encode proteins

Structures

beyond the absolute necessities. Therefore, (B) is incorrect. (C) is wrong since DNA-editing mechanisms would not work on RNA because DNA and RNA have significant structural differences.

10. A Learning Objective: 6.2

Transduction, the uptake of new genetic material by viral infection can occur in the lytic or lysogenic cycle. During the lysogenic cycle, the viral genome integrates into specific locations on the host chromosome. When certain environmental stimuli trigger the prophage to exit the lysogenic cycle and enter the lytic cycle, the phage DNA is excised from the host chromosome. Sometimes, the excision process is imprecise, and the phage DNA will contain a small part of bacterial DNA that was adjacent to the original insertion site. When the new virus infects a subsequent bacterium, it injects DNA from the previous bacterium into the host cell. This is known as specialized transduction, in which only genes from a particular region of the bacterial chromosome are transferred to another bacterium. The question stem asks how a single specialized gene could be consistently transduced; **(A)** correctly outlines the steps of this process. (B), which describes the transfer of a gene by transformation after viral infection, can be eliminated because the question stem states that the genes for toxins are transmitted through viral infection. (C) is incorrect because a virus would be unable to incorporate RNA into a host cell's chromosome. (D) is incorrect because it does not explain how the viral DNA would become a part of the bacterial genome, as posed in the question stem.

11. B Learning Objective: 6.1

This virus is best categorized as a bacteriophage, **(B)**, which is defined as a virus that infects and replicates within a bacterium. A retrovirus is a specific type of RNA virus that contains an enzyme called reverse transcriptase, which allows it to reverse transcribe its RNA into DNA after entering a cell. Since not all RNA viruses are retroviruses and there is nothing in the question stem to suggest that this virus uses reverse transcriptase, (A) can be eliminated. A provirus, as mentioned in (C), is not actually a type of virus, but rather refers to the genetic material of a virus that is incorporated into and able to replicate in the genome of a host cell. A viroid, as mentioned in (D), is a plant pathogen consisting of circular RNA. Since the virus from the question stem targeted bacteria and replicated within a monkey, (D) is incorrect.

12. D Learning Objective: 6.6

Transformation efficiency is a number that represents the total number of bacterial cells that express the gene for ampicillin resistance divided by the amount of DNA plasmid used in the experiment. It is calculated using the following formula:

$$\text{Transformation efficiency} = \frac{\begin{matrix}\text{Total colonies} \\ \text{growing on plate}\end{matrix}}{\begin{matrix}\text{DNA (in } \mu\text{g)} \\ \text{spread on plate}\end{matrix}}$$

For this problem, the student identifies that 35 colonies have grown on the agar plate, providing the value for the numerator. The denominator requires slightly more in-depth calculations. To determine the amount of DNA spread on the agar plate, the first step is to calculate how many micrograms of plasmid were added by multiplying the volume of plasmid solution added by the mass concentration of the plasmid: $10\ \mu\text{L} \times 0.005\ \mu\text{g}/\mu\text{L} = 0.05\ \mu\text{g}$. The question stem also states that 150 μL of a 600 μL suspension was plated, so the total amount of plasmid that was spread on the plate is equal to the plasmid mass multiplied by the fraction of the suspension that was spread: $0.05\ \mu\text{g} \times (150\ \mu\text{L}/600\ \mu\text{L}) = 0.0125\ \mu\text{g}$ of plasmid plated.

Plugging in these values to the equation for transformation efficiency gives the following result: $35/0.0125 = 2800$ colonies per microgram of plasmid. **(D)** is correct.

13. A Learning Objective: 6.3

Viral replication allows for mutations, such as those caused by environmental factors like UV radiation from the Sun and those resulting from mismatches between nitrogenous bases. However, the unique method of replication used by RNA viruses does not include the error-checking mechanisms that are present in DNA viruses, which can undo the effects of mutation. As a result, RNA viruses are observed to have significantly higher rates of mutation. This best matches **(A)**. (B) is incorrect because RNA viruses only assemble in the cytoplasm, while DNA viruses assemble in both the nucleus and the cytoplasm. (C) is incorrect because only a specific group of RNA viruses contain reverse transcriptase, and there is no proof that it leads to increased frameshift mutations. (D) is incorrect because the mRNA that is formed undergoes translation, not transcription.

CHAPTER 7

Plant Structure and Systems

LEARNING OBJECTIVES

In this chapter, you will review how to:

7.1 Describe structures that bring in nutrients

7.2 Describe the vascular system in plants

7.3 Explain how exposure to light affects plants

7.4 Recall methods of plant reproduction

7.5 Describe the effect of mutations across generations

7.6 Investigate water movement through plants

TEST WHAT YOU ALREADY KNOW

1. Mycorrhizae are fungi that live in symbiotic association with plant roots. The figure below shows the results of an experiment that examined the percent of petunia roots colonized by mycorrhizae in environments with different amounts of phosphate (PO_4^{3-}).

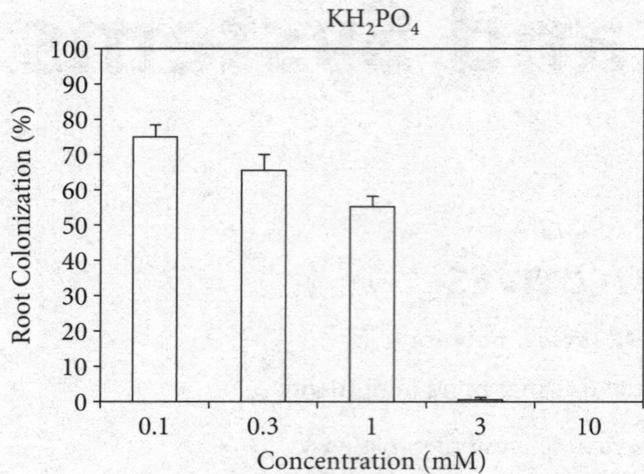

Adapted from Eva Nouri et al., "Phosphorus and Nitrogen Regulate Arbuscular Mycorrhizal Symbiosis in *Petunia hybrid*," *PLoS ONE* 9, no. 3 (March 2014): e90841.

Which of the following provides the most plausible interpretation of the graph?

(A) Petunias can take up more phosphate when they have mycorrhizae, so they are less likely to benefit from mycorrhizal colonization when there is plenty of phosphate available.

(B) Petunias under stress from high or low phosphate levels are less able to form associations with mycorrhizae than those at healthy phosphate levels.

(C) The pattern in the graph is unclear, so another abiotic factor not measured is probably responsible for the change in percent root colonization with changing phosphate concentration.

(D) Increased phosphate levels inhibit mycorrhizal growth, making them less available to colonize petunia roots.

2. Researchers are developing a way to produce clean drinking water using plant tissues. Which of the following strategies would probably be most successful, and why?
   (A) Because periderm forms an outer ring of tissue around a plant stem, it can be used to create a straw-like structure to transport water.
   (B) Because xylem transports water in plants, water could be pushed through it with the natural structure of the xylem acting as a filter.
   (C) Because phloem consists of dead tissue, it forms straw-like structures that can be used to filter water.
   (D) Because the cortex consists of living tissue, it can be used to actively pump water through a membrane.

3. To improve crop yields, a researcher decides to study the effects of light on factors related to photosynthesis in plants. In his study, he examines stomata under the microscope under light and dark conditions and measures changes in sugar concentrations. Which of the following provides the most plausible hypothesis that he could test with this experiment?
   (A) If the plant has receptors on guard cell membranes, it can increase its rate of photosynthesis by closing its stomata in response to light.
   (B) If the plant has receptors on guard cell membranes, it can increase its rate of photosynthesis by opening its stomata in response to light.
   (C) If the plant has receptors on cortical cell membranes, it can increase its rate of photosynthesis by closing its stomata in response to light.
   (D) If the plant has receptors on cortical cell membranes, it can increase its rate of photosynthesis by opening its stomata in response to light.

4. Researchers examined different methods of growing pineapples in Benin, Africa, to understand why some methods yielded larger pineapples than others. The figures below show two experiments. In both cases, plants were either artificially induced to flower (artificial flowering induction) or allowed to flower naturally (natural flowering induction). Additionally, the fruit was either artificially induced to mature (AMI) or allowed to mature naturally (NMI).

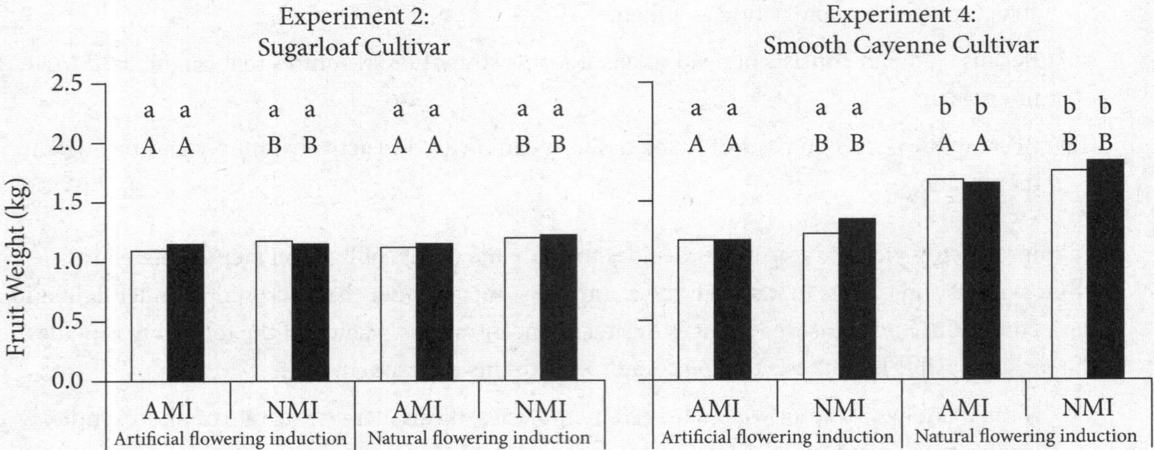

Adapted from V. Nicodème Fassinou Hotegni et al., "Trade-Offs of Flowering and Maturity Synchronisation for Pineapple Quality," *PLoS ONE* 10, no. 11 (November 2015): e0143290.

Which option produces the largest fruit, and why?

(A) The largest fruit is produced by the Sugarloaf cultivar when natural flowering induction is used, regardless of the maturation approach used. This technique allows the plant to invest maximum amounts of energy into fruit growth.

(B) The largest fruit is produced by the Sugarloaf cultivar with artificial flowering induction and by the Smooth Cayenne cultivar with natural flowering induction. This technique allows the plant to save more energy for future fruit production.

(C) The largest fruit is produced by the Smooth Cayenne cultivar when natural flowering induction is used, regardless of the maturation approach used. This technique allows the plant to invest maximum amounts of energy into fruit growth.

(D) The largest fruit is produced by the Sugarloaf cultivar when artificial flowering induction is used, regardless of the maturation approach used. This technique allows the plant to save energy to produce larger numbers of fruit.

5. If an error in meiosis in a diploid plant resulted in the production of diploid gametes, what would be the most likely consequences for the next generation?

   (A) The fusion of three diploid gametes would produce a hexaploid plant with six sets of chromosomes, which would then produce hexaploid gametes.

   (B) If the diploid gametes fused with diploid gametes from another plant, some chromosomes would be lost to produce a new diploid plant.

   (C) There would be no possibility that the individual could reproduce because polyploid organisms are not fertile.

   (D) If the diploid gametes were capable of self-fertilization, then a tetraploid plant with four sets of chromosomes would be produced. If this plant could self-fertilize, more tetraploid plants would be produced.

6. A researcher decides to compare the response of two plants to desiccation by growing them in a hot, dry environmental chamber and then returning the chamber to normal climate conditions. One plant showed a dramatic increase in transpiration relative to its initial transpiration rate as the environment became drier, but then reverted to a normal transpiration rate when the normal climate was restored. The other plant showed a much smaller change in transpiration relative to its initial transpiration rate under the same conditions. What conclusion can most likely be drawn from these results?

   (A) The plant that had a smaller change in transpiration probably normally lives in a moist environment and had water reserves. In contrast, the plant that had a larger change in transpiration is a xerophyte, a plant that is adapted to a dry environment.

   (B) The plant that had a smaller change in transpiration is probably a smaller plant and therefore has less transpiration under any conditions than the plant that had a larger change.

   (C) The plant that responded dramatically does not normally live in a dry environment. In contrast, the plant that had a smaller change in transpiration may be a xerophyte, a plant that is adapted to a dry environment.

   (D) The plant that responded dramatically is probably from a hot climate, whereas the plant that responded only slightly is probably from a cold climate.

Answers to this quiz can be found at the end of this chapter.

PLANT VASCULAR AND LEAF SYSTEMS

7.1 Describe structures that bring in nutrients

7.2 Describe the vascular system in plants

Plant Structure

Plants with vascular tissues usually have three types of structures, or organs: leaves, roots, and branches. The leaves provide most of the photosynthesis of the plant; the roots provide support in the soil, along with water and **minerals;** and the branches hold the leaves up to light and convey nutrients and water between the leaves and the roots. Each of these structures can be specialized in many ways.

Plants with taproots have long roots with a single extension deep into the soil, while other plants have highly branched roots. Cells on the surface of roots often have long extensions called root hairs, which increase the surface area of roots. Some plants without root hairs have a symbiotic relationship with fungi that increase the surface area of the root to absorb water and minerals. In legumes, nitrogen-fixing *Rhizobium* species of bacteria infect roots and form root nodules in symbiosis with the plant. The roots of plants often play an important role in preventing erosion. Tropical rain forest that is cleared is highly vulnerable to erosion of the thin soil if the plants and their root systems are absent.

Leaves can have a variety of shapes. Monocot leaves are usually very narrow with veins that run parallel to the length of the leaf, while dicot leaves are broad with veins that are arranged in a net in the leaf. Modified leaves form thorns in cacti, tendrils in pea plants, and petals in flowers. The leaves produce all of the energy of the plant and are specialized to gather sunlight. The broad shape helps to gather sunlight for themselves and in some cases to block sunlight from getting to competing plants. The shape and arrangement of leaves are key features used to distinguish plant species.

Terrestrial plants have broad leaves that maximize the absorption of sunlight but also tend to increase water loss by evaporation. Leaves of terrestrial plants have a waxy cuticle on top to conserve water. The lower **epidermis** of the leaf is punctuated by **stomata**: openings that allow diffusion of carbon dioxide, water vapor, and oxygen between the leaf interior and the atmosphere. A loosely packed spongy layer of cells inside the leaf contains chloroplasts with air spaces around cells. Another photosynthetic layer in the leaf, the palisade layer, consists of more densely packed elongated cells spread over a large surface area. A moist surface that lines the photosynthetic cells in the spongy layer is necessary for diffusion of gases into and out of cells. Air spaces in leaves increase the surface area available for gas diffusion by the cells. The size of stomata is controlled by guard cells that can open and close the opening. These cells open during the day to admit CO_2 for photosynthesis and close at night to limit the loss of water vapor through **transpiration**, the evaporation of water from leaves that draws water up through the plant's vascular tissues from the soil. The upper surface layer of cells in leaves has no openings, an **adaptation** that reduces water loss from the leaf.

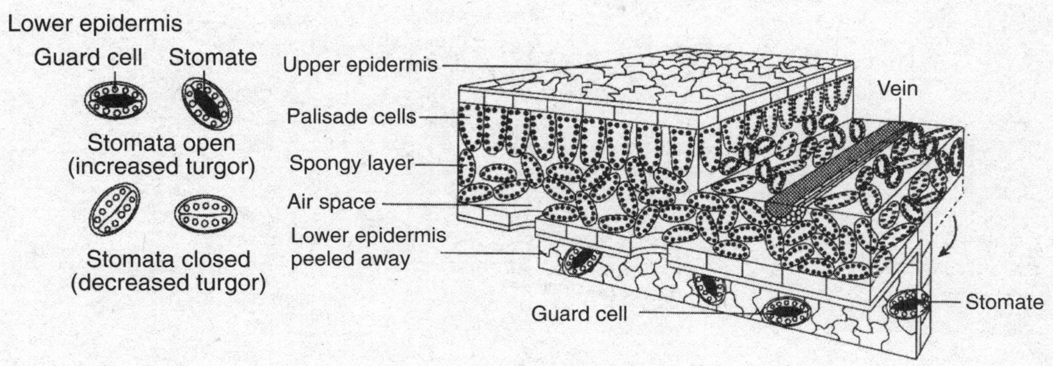

Leaf Structure

Guard cells are kidney-shaped cells in dicots and dumbbell-shaped cells in monocots that change their shape according to the amount of water that exists within them. This water exerts a pressure called turgor pressure. When water is abundant, the guard cells swell, and when water is sparse, they clamp down and shut the stomata. This makes perfect sense, as the guard cells allow transpiration when there is plenty of water in the leaves and water conservation when there is not. Guard cells absorb water in response to potassium ions that are driven into the cells. Water follows these ions due to the osmotic gradient that is created by the presence of extra particles inside the cell as compared with outside the cell.

The presence of blue-light receptors on the guard cell membranes is what drives the movement of K^+ into the guard cell. Sunlight, particularly blue wavelengths, causes K^+ ion channels to open and potassium to flood in. This explains why guard cells are usually open during the day and closed at night, helping photosynthesis to take place as carbon dioxide gas can then enter through the open stomata.

Growth in higher plants is restricted to areas of perpetually embryonic and undifferentiated tissue known as meristems. Meristems are self-renewing populations of cells that divide and cause plant growth either in height or width. Apical meristems exist at the tips of roots and stems, whereas lateral meristems (also known as cambium) are found within the stem between layers of **xylem** and **phloem** on the sides of the plant. The trunk of the plant thickens each year because the embryonic cells of the cambium produce more and more xylem and phloem, supporting the growth of a larger tree with more leaves. Upward growth that occurs as a result of cell division within apical meristems is called primary growth, while outward growth is called secondary growth. It is secondary growth that causes the well-known concentric tree rings that one sees when a large woody tree is cut down.

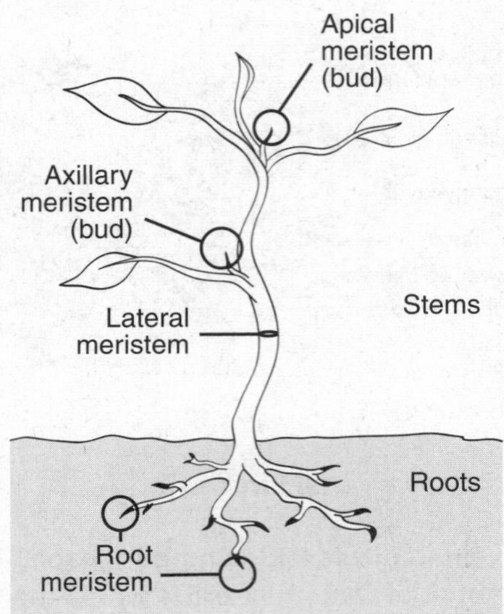

General Plant Anatomy

Plant Vasculature

Terrestrial plants left an aquatic environment during their evolution, and in the process lost some of the benefits that a liquid medium had previously provided. **Cell walls**, which provided simple rigidity for structures in small plants, became aligned, allowing for the growth of trees over 100 meters in height and creating channels for the delivery of important soil nutrients. Water loss became a constant challenge in environments where rainfall could be unpredictable or nearly nonexistent. The Sun, formerly relied upon solely for light-giving energy, also became a source of dehydration and overheating.

One of a plant's many challenges is providing all of its structures with the water needed to complete photosynthesis and to maintain an aqueous solution for all biological reactions. Water is heavy and a plant cannot rely on muscular contractions to move materials, as animals do. Instead, a plant has narrow, lifeless channels in its xylem tissue that take advantage of the cohesive properties of water. A plant loses the majority of its water during transpiration while its stomata are open for the exchange of CO_2 and O_2 in photosynthesis. As water is lost through the leaves, it creates a negative pressure in the xylem channels, just like sucking on a straw. The cohesive force of water keeps a steady flow of water moving through the plant, pulled by the negative pressure developed by transpiration occurring in the leaves. The force exerted in the water column due to cohesion is as strong as steel wire of the same diameter.

Plants need a mechanism to deliver the stored energy created during photosynthesis to all of their structures. While water travels upward from the roots through the xylem via the pull of transpiration and the cohesion of water, nutrients flow both up and down through the phloem via the pull of **osmotic pressure**. Osmotic pressure is built up between areas that have high concentrations of nutrients and those that have low concentrations. Translocation of materials in the phloem is like long-range diffusion. Similar to the cohesion of water, lower solute pressure on one end of a phloem vessel is translated along the narrow vessel to an area where the solute pressure is greater.

The adaptive significance of these transport systems is the colonization of a terrestrial environment. Plants also harnessed a physical disadvantage, water loss, and turned it into a benefit, with the force of water loss powering water transport through the xylem.

Xylem

The vascular tissue of the xylem contains a continuous column of water from the roots to the leaves, extending into the veins of the leaves. The leaves regulate the amount of water that is lost through transpiration. Water is transported from the roots to the shoots. In other words, from roots to stems, water is transported through two different mechanisms: root pressure and cohesion-tension.

Osmotic pressure in the roots tends to build up due to water absorption. This pressure pushes water up from the roots into the stem of the plant. Root pressure functions best in extremely humid conditions when there is abundant water in the ground or at night. The drops of water, or dew, that appear on blades of grass or other small plants in the morning result from this root pressure. During the day, transpiration occurs at a high enough rate that dew is generally not seen, because the water that leaves the tips of the grass is evaporating before it can build up.

Transpiration

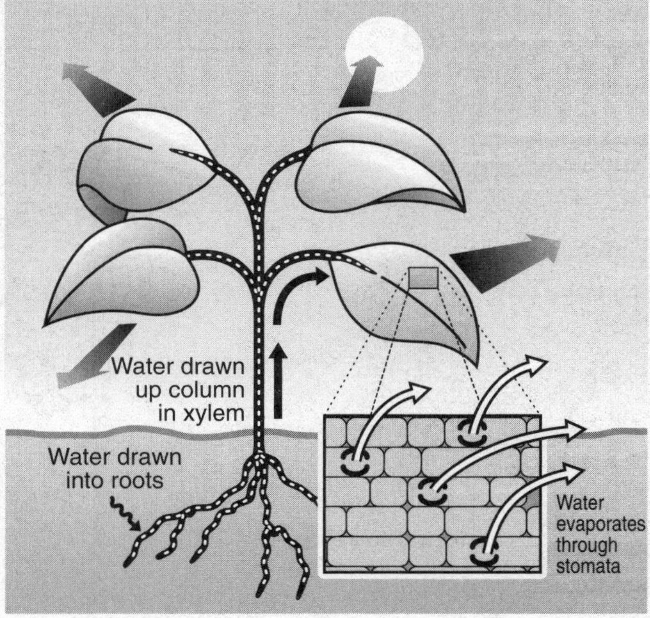

Water Gain and Loss in Plants

Phloem

Phloem tubes are much thinner-walled than xylem and are found toward the outside edges of stems. They begin up in the leaves, where sugars are made by photosynthesis and stored as starch within mesophyll cells. Much of this sugar and starch, though, is transported down phloem tubes from shoots to roots (from leaves and stems to the roots). Active transport between the mesophyll cells, where sugar is made, and the nearby phloem tubes is what allows sugar to move into the phloem's sieve tubes. The active transport pumps are symport pumps that move one H^+ ion along with every sucrose molecule. The H^+ that is used to push the sucrose across into the phloem is rapidly transported back out to the mesophyll cells by an H^+ ion ATPase pump, so that more sucrose can be moved. As sugar is loaded into the phloem, the water potential, or the amount of water pressure, in the phloem is effectively reduced. The influx of sugar into the phloem creates an osmotic potential that pulls more water into the phloem, generating a water pressure that forces sap (essentially sugar-rich water) down the phloem toward the roots. The xylem will recycle this water back from the roots.

As you can see in the following figure, vascular cambium (C) divides laterally to create new phloem (P) on the outer edges of the tree and new xylem (X) near the inner core. This results in the formation of annual tree rings.

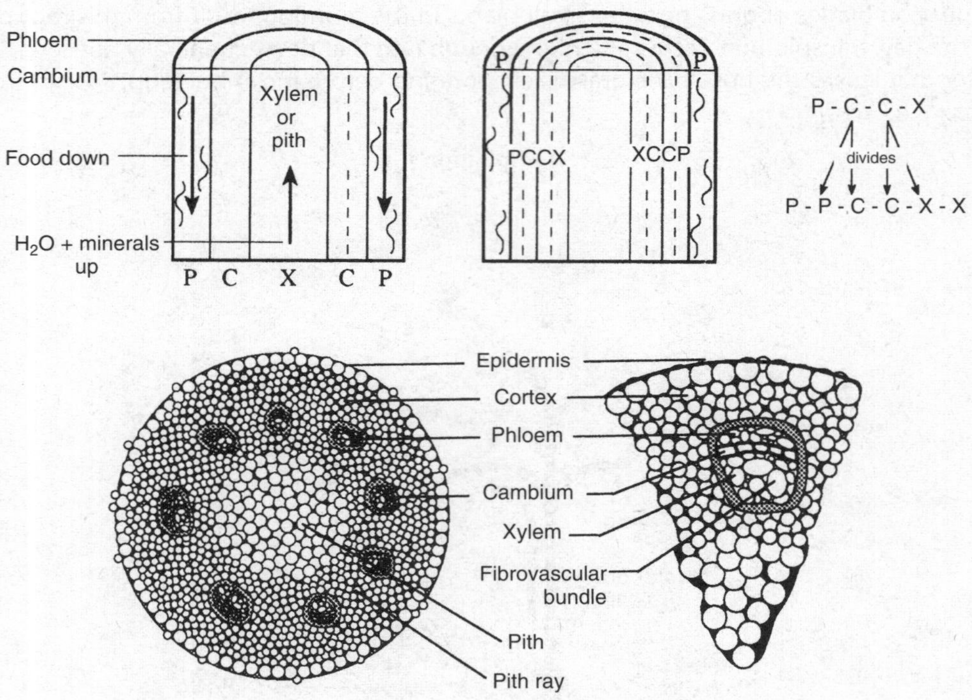

Internal Anatomy of a Plant

Plant Nutrients

Plants require certain nutrients and compounds that are important for normal physiological functions. For example, the photoreceptor phytochrome is an important pigment that plants use to detect light. Plants can use phytochrome to regulate flowering based on day length or to set up other daily rhythms. Large amounts of macronutrients such as potassium, calcium, magnesium, and phosphorous are required for normal plant functions.

PLANT NUTRIENTS			
Element/Compound	**Type**	**Origin**	**Action**
Phytochrome	Photopigment	Systemic	Detection of light to control photoperiodism
Potassium	Nutrient	Root uptake in soil	Protein synthesis, operation of stomata
Calcium	Nutrient	Root uptake in soil	Cell wall stability, enzyme activation
Magnesium	Nutrient	Root uptake in soil	Chlorophyll synthesis, enzyme activation
Phosphorus	Nutrient	Root uptake in soil	Nucleic acid and ATP synthesis

PLANT REPRODUCTION AND DEVELOPMENT

7.3 Explain how exposure to light affects plants

7.4 Recall methods of plant reproduction

7.5 Describe the effect of mutations across generations

Phototropism and Photoperiodism

The ability to respond to external stimuli allows plants to adjust to their environment and optimize growth, which is important for competition and survival. **Phototropism** is the directional growth of plants in response to a light stimulus. For example, if a plant on a window sill detects light coming from the window, it can respond by growing toward the light to obtain more light energy needed for photosynthesis. Shoots generally exhibit positive phototropism or growth toward a light source, while roots generally exhibit negative phototropism or growth away from light.

During phototropism, light is sensed at the coleoptile (the tip of a plant) and detected by blue-light photoreceptors called phototropins, which are composed of a protein bound to a light-absorbing pigment called the chromophore. The absorption of light by the chromophore causes the protein shape to change and activate a signaling pathway. The different levels of phototropin activation, in turn, cause unequal transport of auxin, a plant hormone that promotes cell elongation, down the sides of the coleoptile. More auxin is transported down the side that is away from the light than the illuminated side, which causes the plant to bend in the direction of light.

Photoperiodism is the response to changes in the photoperiod (relative lengths of day and night). Examples of photoperiodism in plants include regulating flowering, tuber formation, bud dormancy, and loss of leaves. To flower, some plants require a certain length of day or night. Long-day plants typically measure the continuous length of night (darkness) and flower when the night length is below a certain threshold (light-dominant). Short-day plants, on the other hand, typically measure the length of day and flower when the day length is below a certain threshold (dark-dominant).

Plants that do not depend on day length are day-neutral. Though it is unclear how plants determine day or night, models suggest the interactions between a plant's **circadian rhythms** (biological clock) and light cues from its environment govern a plant's photoperiodic response.

Plant Reproduction

Plants possess a reproductive advantage in that they can **reproduce vegetatively**, while very few groups of animals can reproduce without some form of sexual reproduction. Plants also live dual lives in that they have specific structures for reproduction (flowers, pollen, and fruit), while the rest of the plant continues life as usual. In contrast, female animals' entire bodies are affected by the reproductive process, so reproduction has to be an entire stage of existence for them. In fact, some animals' adult life stage is solely for the purpose of reproduction; they don't even have mouthparts for the **ingestion** of nutrients. Because of the specialization of certain plant structures, however, plants need not undergo such vast changes to reproduce.

✔AP Expert Note

You do not need to know the details of the sexual reproduction cycles in different plants and animals. Focus instead on the similarities among the processes and how this is important for genetic variation.

Plants can reproduce vegetatively as a function of their **indeterminate growth**. Plant cells can differentiate when isolated from the rest of the plant, so one could place the stem of a rose in the soil and it would grow adventitious roots and become a whole plant. By default, this produces a series of plants that are all genetically identical and reduces variation in the population, but it is advantageous when **sexual reproduction** is not suitable for a given environmental situation.

✔AP Expert Note

All plants exhibit a definite alternation of generations in which the plant life cycle alternates between haploid and diploid stages.

Most terrestrial plants live their lives as diploid **sporophytes** and produce haploid tissue in the form of ovules and **pollen** (**gametophytes**), sometimes within the same flower. Pollen is delivered to the female gametophyte either by an animal, such as a bird, or by the wind, and fertilization takes place. The flower functions as the entire reproductive organ, completely dependent on the sporophyte plant. This is a departure from aquatic plants, in which **spores** travel independently from the plant that produced them. Flowering plants seem to have evolved a specialized region where a plant can become "pregnant." These "seed babies" are housed within a seed coat and often a fruiting body. The seed coat and fruiting body aid in plant propagation by either acting as bait to an animal or providing a source of nutrition to the developing plant embryo.

Variation in Plants

The terrestrial environment provides a great degree of variation, in contrast to the constant temperatures and cycles of aquatic systems. Seasonal changes and the variability of weather requires plants to synthesize new chemical messengers that allow plant tissues to respond to both acute and chronic environmental cues. **Novel structures**, primarily for reproduction, require new forms of control mechanisms to function in time with a changing environment.

Within the plant kingdom are several phyla, with one of the key distinguishing characteristics between phyla being the presence or absence of vascular tissue. Within the tracheophytes, which have vascular tissue, two of the important modern phyla are the gymnosperms and the angiosperms. The gymnosperms, such as the conifers, have "naked seeds" that do not have endosperm and are not located in true fruits. The angiosperms are the flowering plants. They have flowers, true fruits, and a double fertilization system that creates endosperm to nourish the plant embryo. Angiosperms are the most successful form of plant life on the planet, and the following diagram distinguishes between two main classes of angiosperm: monocots and dicots.

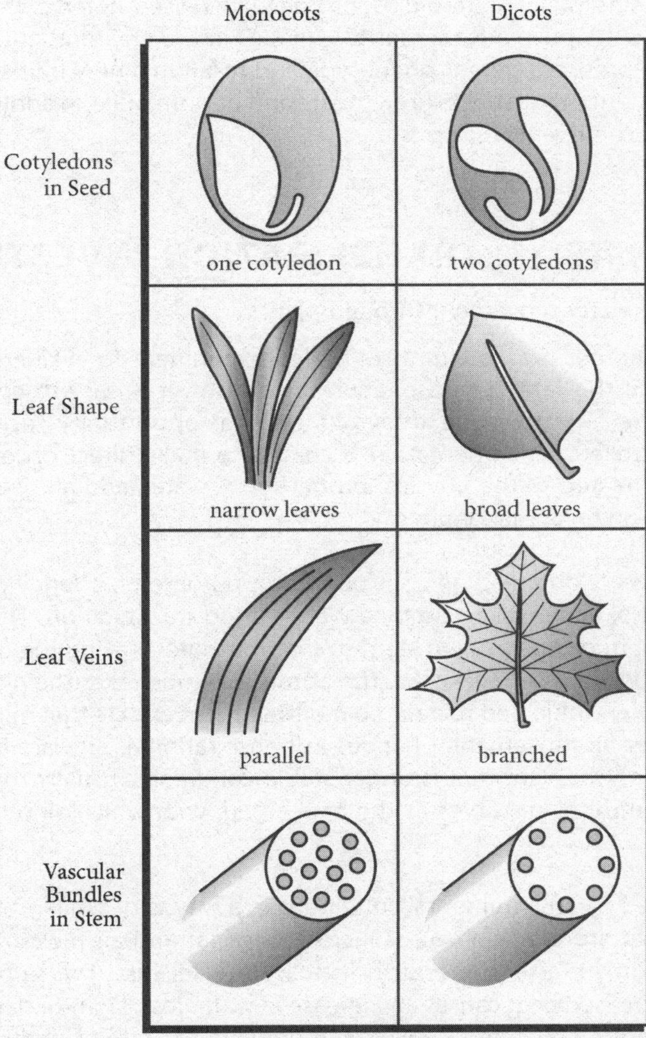

**Monocots vs. Dicots:
Two Angiosperm Classes**

Common in flowering plants, **polyploidy** drives adaptation and speciation. Most eukaryotes are diploid, having two sets of chromosomes (2*N*). Polyploids have three or more times the haploid (*N*) chromosome number. For example, triploids (3*N*) have three sets of chromosomes, tetraploids (4*N*) have four sets of chromosomes, and hexaploids (6*N*) have six sets of chromosomes. Polyploids occur spontaneously in nature and result from mutations during cell division, meiotic or mitotic failures, and the fusion of gametes. Combining two diploid cells, for instance, results in a tetraploid, and combining a diploid and haploid cell results in a triploid.

Polyploidy may arise from two closely related species (autopolyploidy) or two different species (allopolyploidy). Normally, **hybrids** are sterile because they do not have homologous pairs of chromosomes needed for gamete formation during meiosis. In some instances, polyploidization results in each chromosome having a homologue from the additional set of chromosomes, so meiosis can occur and the hybrid becomes fertile. Note that sterile polyploids can propagate via **asexual reproduction**. Polyploids like allotetraploids also have increased heterozygosity (varied sets of a chromosome) and thus adaptive advantages to changes in the environment. The different copies of a gene may each express a different phenotype and result in a new trait such as drought tolerance or pathogen resistance that affects growth rate and performance. In addition, some polyploids can interbreed and become a new species.

AP BIOLOGY LAB 11: TRANSPIRATION INVESTIGATION

7.6 Investigate water movement through plants

This investigation is another example of a controlled experiment to evaluate the effects of environmental variables on the rate of transpiration for a plant. It is easy to complete and requires simple equipment. The objectives are threefold: another opportunity to explore the scientific method through controlled experimentation, a chance to make direct observations on the phenomena immediately related to the physical properties of water, and an occasion to observe the environment's effects on an actual organism.

A suitable plant species is obtained and set up with a potometer, a fancy name for a tube that measures the amount of water taken up by a plant during transpiration. The water-filled tube is connected to the bottom of a plant's cut stem so that the water leaving the tube and entering the plant due to transpiration can be measured. The potometer tube takes the place of soil as a source of water. The plant is then subjected to a number of treatment effects that might include increased light, darkness, heat, or air movement. Changes in transpiration, if any, are compared to the rate of transpiration of a control plant that is under stable conditions, usually room temperature and room lighting. The control plant serves as the benchmark with which all other treatment effects are compared.

Transpiration is affected by environmental conditions in a way very similar to human skin. Increasing the amount of heat around skin causes increased sweat and air movement, which increases evaporation on the skin surface; the result is increased water loss. The same is true for the leaf surface of a plant—increased heat causes an increase in water loss. Light or darkness in the absence of changes in temperature only affects a leaf by modifying the rate of photosynthesis. Increasing the light intensity should increase the rate of photosynthesis and the opening of stomata, thereby increasing transpiration and water loss. As light levels decrease in intensity, water loss decreases. Water loss at the leaf surface causes greater uptake of water through the potometer, which can be measured as movement of the meniscus through the tube.

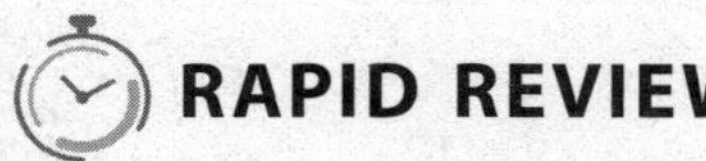

RAPID REVIEW

If you take away only 4 things from this chapter:

1. Plants with vascular tissues usually have three types of structures: leaves, roots, and branches.
2. Plants have specialized structures to deal with water and nutrients. These include stomata controlled by guard cells, a loosely packed spongy layer, the palisade layer, xylem, and phloem.
3. Plants produce energy through photosynthesis and lose water via transpiration. As water evaporates from the leaves, it pulls water up through channels in the xylem. The phloem carries nutrients throughout the plant.
4. Plants can reproduce asexually via vegetative propagation. Sexual reproduction in plants takes place in the flower.

TEST WHAT YOU LEARNED

1. A researcher uses a potometer to measure transpiration in plants. How is it possible to measure transpiration using a tube attached to the bottom of the plant?

   (A) Transpiration involves the loss of water through the leaves. This water must be replaced and the potometer measures the uptake of replacement water.

   (B) Transpiration involves the uptake of water through the roots. This can be measured directly using a potometer.

   (C) Transpiration involves the loss of water through the leaves. The tube must surround one or more leaves in order to properly measure transpiration.

   (D) Transpiration involves the uptake of water through the roots. This water is actively pushed out through the leaves and a potometer measures this pressure.

2. A researcher wants to understand what factors cause a particular species of plant to grow against buildings. He hypothesizes that this behavior is due to phototropism. Which of the following experimental designs would most effectively test this hypothesis?

   (A) 40 seeds are evenly divided into two groups. Half of the seeds are grown in pots placed against a wall, allowing the plants to grow in close proximity to the wall. The other half of the seeds are grown far from any walls. After two weeks, the average angles of plant growth are compared among the groups.

   (B) 200 seeds are evenly divided into two groups. Half of the seeds are grown with lighting directed from one side only. The other half of the seeds are grown with diffuse lighting from all directions. After two weeks, the average angles of plant growth are compared among the groups.

   (C) 10 seeds are evenly divided into two groups. Half of the seeds are grown with lighting above the pot. The other half of the seeds are grown with lighting below the pot. After two weeks, the average angles of plant growth are compared among the groups.

   (D) 400 seeds are evenly divided into two groups. Half of the seeds are grown in green light and half of the seeds are grown in red light. After two weeks, the average angles of plant growth are compared among the groups.

3. Mycorrhizae are fungi that grow in a symbiotic association with plant roots. Researchers were interested in knowing whether a particular species of fungus, *Tuber melanosporum*, obtained carbon from the soil or from its host plant (the hazel tree). They added ^{13}C to carbon dioxide in the air surrounding a hazel tree, which was enclosed in a cylinder. Which of the following experimental results most strongly supports the hypothesis that *T. melanosporum* obtains carbon primarily from its host rather than from the soil?

   (A) ^{13}C was found in high abundance in the soil surrounding the plant and in *T. melanosporum* cells.

   (B) ^{13}C was found in the leaves of the hazel tree, but not at all in the surrounding soil.

   (C) ^{13}C was found in the leaves of the hazel tree and in the surrounding soil.

   (D) ^{13}C was found in the leaves of the hazel tree and in *T. melanosporum* cells in much greater abundance than in the surrounding soil.

4. A farmer growing grape vines is testing out two subspecies of grapes in his vineyard. Some regions of the vineyard have unfertilized soils while other regions of the vineyard are artificially fertilized. In addition, he artificially induces some plants to flower while allowing others to flower naturally. Which of the following observations would be the most likely outcome of this experiment?

   (A) Fruit weight is lower for both subspecies when unfertilized because the plants have less energy to invest in fruit production.

   (B) Fruit weight is higher for both subspecies when unfertilized because the plants invest more energy in reproduction as they reduce expenditure on their own growth.

   (C) The naturally flowering plants in both subspecies have larger fruit because the artificially flowering plants would experience too much stress.

   (D) The artificially flowering plants in both subspecies have larger fruit because the naturally flowering plants produce smaller fruit to conserve energy.

5. To understand the spread of genes associated with herbicide resistance, researchers planted a herbicide-resistant strain of *Lolium rigidum* (annual ryegrass) plant surrounded by herbicide-susceptible plants to examine the spread of resistance genes by pollen. The table below shows the percent of pollen-mediated gene flow at varying distances (0, 1, 5, 10, and 15 meters) from the resistant plant.

	Pollen-Mediated Gene Flow (%)			
Distance (m)	**A**	**B**	**C**	**D**
2009				
0	2.3	5.9	6.7	5.5
1	3.6	2.5	3.7	2.8
5	3.2	1.8	0.9	3.0
10	1.1	—	—	—
15	0.1	—	—	1.5
2010				
0	7.1	11.2	7.1	9.0
1	4.9	7.0	6.1	8.4
5	0.9	3.1	4.2	3.6
10	0.7	1.1	1.7	2.5
15	0.5	0.4	0.6	0.6

Adapted from Iñigo Loureiro, María-Concepción Escorial, and María-Cristina Chueca, "Pollen-Mediated Movement of Herbicide Resistance Genes in *Lolium rigidum*," *PLoS ONE* 11, no. 6 (June 2016): e0157892.

Based on these findings, is it reasonable to expect that herbicide resistance may spread from resistant plants?

(A) No, because the percentage of pollen-mediated gene flow is too low to have a significant effect on future generations. This suggests that the amount of resistance will not increase in future years.

(B) No, because it does not matter what genes are in the pollen as long as the resistance genes are not in the seeds. The table would need to show data on resistance in seeds in order to draw conclusions about whether future generations will show resistance.

(C) Yes, because the trials consistently showed pollen-mediated gene flow within 5 meters of the resistant plant and less consistent gene flow within 15 meters. This suggests that more plants will be resistant in the next generation.

(D) Yes, because the trials consistently showed pollen-mediated gene flow within 5 meters of the resistant plant. This suggests that pollen is spreading to plants immediately adjacent to the resistant plant, although it is unlikely to be a problem over longer distances.

6. The diagram below illustrates the movement of water (indicated by the path of the arrows) through vascular tissue in angiosperms and gymnosperms.

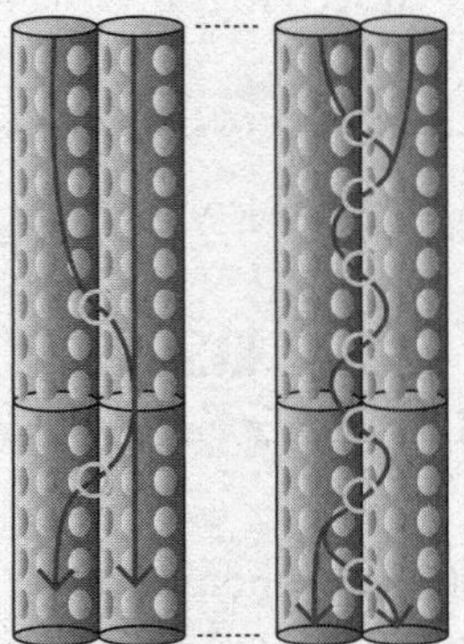

Adapted from Michael S. H. Boutilier et al., "Water Filtration Using Plant Xylem," *PLoS ONE* 9, no. 2 (February 2014): e89934.

Which of these structures would most effectively filter water moving through the plant?

(A) Angiosperm vascular tissue would be more effective because water can travel rapidly through these tissues, making it more efficient to use.

(B) Gymnosperm vascular tissue would be more effective because the water frequently moves from one cell to another through the vessel walls, which act as a filter.

(C) Angiosperm and gymnosperm vascular tissue would be equally effective in filtration because water travels over the same distance in both cases.

(D) It is impossible to determine which would make the better filter without knowing whether the plants are monocots or dicots.

Questions 7–8

In order to study phototropism, students placed plant seeds onto moistened soil in darkened chambers. Each chamber had a green, red, and blue light-emitting diode (LED) of the same intensity located equally distant from each other. Germinated seedlings invariably grew toward the blue LED.

7. Which of the following is a likely explanation of how plants sense differences in the color of light?
   (A) Different wavelengths of light affect rubisco differently. Rubisco is the enzyme that fixes carbon dioxide.
   (B) A light receptor protein changes shape when exposed to blue light. This conformation change exposes kinase sites, which activate a signal transduction cascade that changes gene expression.
   (C) Different colors of light have different amounts of energy. This leads to a change in soil temperature. Plant roots measure the kinetic energy in the soil and respond by bending toward the cooler blue light.
   (D) Light absorbed through the stomata affects the rate at which transpiration occurs. Hydrostatic receptors in the guard cells trigger the plant to bend in the direction of the most intense light.

8. Which of the following models would explain the mechanism by which the plants bended toward the blue light?
   (A) Cells detecting the blue light transmit cell-lengthening signals to the cells directly below them. These cells directly below expand, and the stem bends toward the light.
   (B) Cells detecting the blue light transmit cell-lengthening signals laterally while inhibiting cell lengthening locally. This causes cells stimulated by the least light to elongate, bending the plant toward the blue light.
   (C) Cells detecting the blue light transmit cell-lengthening signals locally while inhibiting cell lengthening laterally. This causes cells stimulated by the most light to elongate, bending the plant toward the blue light.
   (D) Blue light causes more photosynthesis in the cells on the side of the plant closest to the light. The increased sugar causes these cells to elongate, bending the plant toward the light.

9. Scientists have spent many years tracking the evolution of a certain flowering plant in southern Florida. Recent flooding in the area has caused regions of land that contain the plant to become submerged under water. Already, this environmental change has led to unexpected mutations that cause certain varieties of the plant to thrive in the new conditions.

Given that terrestrial plants and aquatic plants often use different mechanisms to reproduce, which of the following graphs best illustrates the most likely change to the plant population that will occur as a result of the flood?

(A)

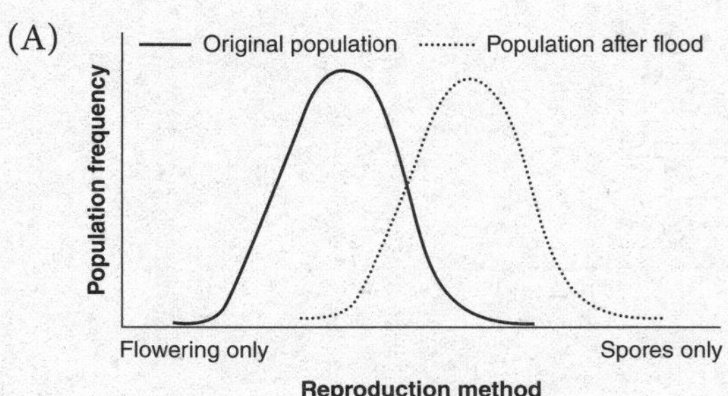

(B)

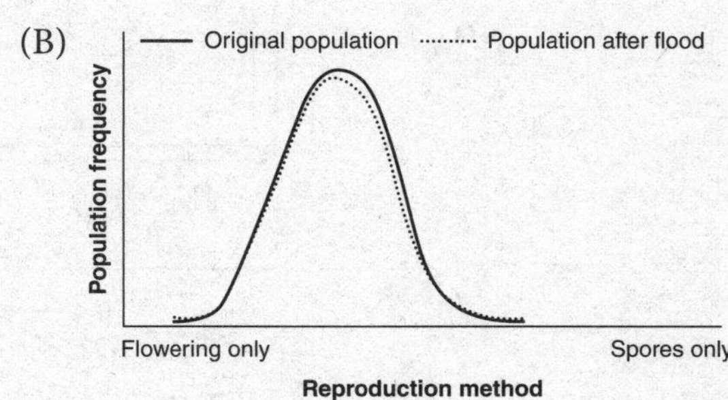

(B)

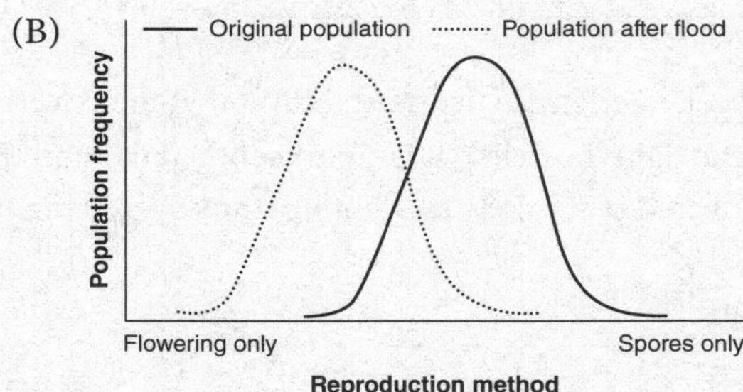

(D)

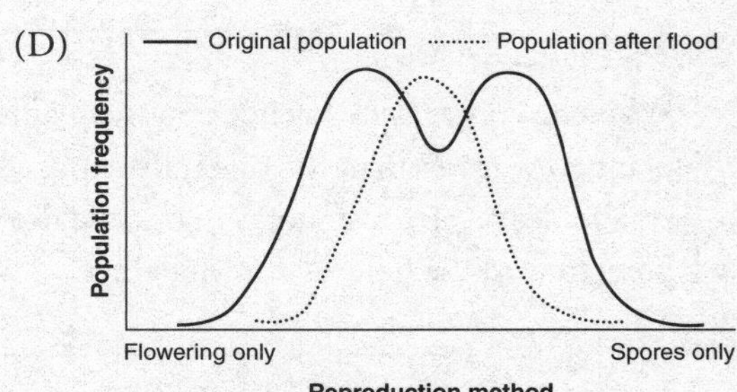

10. As plants evolved, moving from aquatic environments to terrestrial environments, they were forced to develop new mechanisms to acquire and transport necessary nutrients and resources. Which of the following provides an example of one of these structural alterations?

(A) Plants developed organelles known as lysosomes to breakdown macromolecules in their cells.

(B) Plants developed thick-walled phloem tubes to carry water from their roots to their leaves.

(C) Increased quantities of phytochrome were produced to set up daily rhythms in response to the presence of direct sunlight.

(D) Plants developed thin-walled xylem tubes to transport sugar and starch from their leaves and stems to their roots.

11. A botanist is interested in comparing the growth of bean plants growing in sunlight against beans growing in the dark. He places two bean plant seeds in two boxes, which he puts on a windowsill. Both boxes have a tiny hole on one side. To ensure that the hole does not affect the experiment, the botanist arranges the boxes so the hole is as far from the window pane as possible. He then covers one of the boxes with a lid. The experimental setup can be seen below:

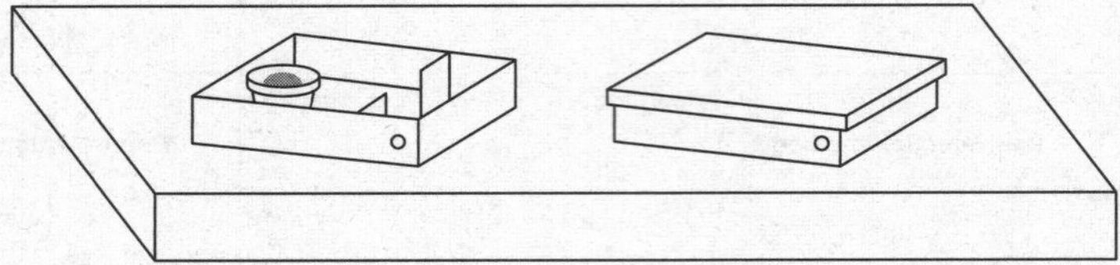

When he comes back to check on the plants a week later, he is surprised to find that the covered plant grew faster than the one fully exposed to the light; however, it is significantly paler and has smaller leaves. He also notes that while the uncovered plant grew straight up, the covered plant grew toward the hole, as shown below:

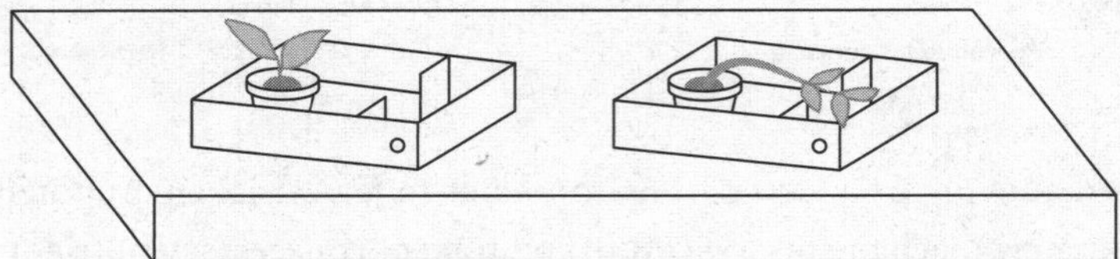

The botanist decides to redo the experiment, but this time, he removes the growing tips from both plants. When he returns one week later, he once again finds that the covered plant is pale with small leaves. However, this time, it did not grow toward the light, instead growing upward. Which of the following best explains what has occurred?

(A) By cutting the growing tip, the botanist blocked all photoperiodic effects by eliminating all of the plant's photoreceptors.

(B) By cutting the growing tip, the botanist blocked all phototropic effects by eliminating all of the plant's photoreceptors.

(C) By cutting the growing tip, the botanist eliminated all photoperiodic effects by blocking a signal to a growth repressor responsible for slower elongation on the lit side.

(D) By cutting the growing tip, the botanist eliminated all phototropic effects by blocking a signal to release a growth hormone responsible for faster elongation on the unlit side.

12. The transport of water from roots to shoots is dependent on differences in water potential, which are responsible for water movement from cell to cell or over long distances in the plant. Several factors, such as environmental pressure and solute concentration, contribute to water potential. For example, the process of transpiration, in which water evaporates from stomata, creates a lower osmotic potential in the leaf, which causes water to flow into it. However, to avoid dehydration, plants must constantly monitor and adjust their rates of transpiration in response to environmental conditions. Which of the following best explains what would be observed if the light intensity experienced by a plant were to be increased?
    (A) The rate of photosynthesis would increase, resulting in the closing of stomata and thereby decreasing water loss.
    (B) The rate of photosynthesis would increase, resulting in the opening of stomata and thereby increasing water loss.
    (C) The rate of photosynthesis would decrease, resulting in the closing of stomata and thereby increasing water loss.
    (D) The rate of photosynthesis would decrease, resulting in the opening of stomata and thereby decreasing water loss.

13. Plants utilize a variety of different structures to obtain necessary nutrients. Which of the following correctly matches the plant structure with how it contributes to the acquisition of nutrients from the environment?
    (A) Xylem transports sugar and other metabolic products made in the leaves downward through the rest of the plant.
    (B) Casparian strips block the passive flow of water and solutes into the central core of the stem via cell walls.
    (C) Leaf stomata are mainly responsible for the process of transpiration.
    (D) Root nodules are made up of multiple plant cells that contain nitrogen-fixing bacteria.

14. Some plants develop longer roots that allow them to adhere more tightly to rocks in the soil. To learn more about this mutation, scientists perform a cross between a long-rooted plant and an original short-rooted plant. When planted, the 38 seeds produced from the cross grew into 24 short-rooted plants and 14 long-rooted plants. Calculate the chi-squared value for the hypothesis that the short-rooted parent was heterozygous for the short-root gene, assuming the short-root allele is dominant. Give your answer to the nearest tenth.
    (A) 0.5
    (B) 1.3
    (C) 2.6
    (D) 2.8

Answers to this quiz can be found at the end of this chapter.

Answer Key

Test What You Already Know

1. A
2. B
3. B
4. C
5. D
6. C

Test What You Learned

1. A
2. B
3. D
4. A
5. C
6. B
7. B
8. B
9. A
10. C
11. D
12. B
13. D
14. C

REFLECTION

Test What You Already Know score: __________

Test What You Learned score: __________

Use this section to evaluate your progress. After working through the pre-quiz, check off the boxes in the "Pre" column to indicate which Learning Objectives you feel confident about. Then, after completing the chapter, including the post-quiz, do the same to the boxes in the "Post" column. Keep working on unchecked Objectives until you're confident about them all!

Pre	Post	
☐	☐	**7.1** Describe structures that bring in nutrients
☐	☐	**7.2** Describe the vascular system in plants
☐	☐	**7.3** Explain how exposure to light affects plants
☐	☐	**7.4** Recall methods of plant reproduction
☐	☐	**7.5** Describe the effect of mutations across generations
☐	☐	**7.6** Investigate water movement through plants

FOR MORE PRACTICE

Complete more practice online at kaptest.com. Haven't registered your book yet? Go to kaptest.com/booksonline to begin.

ANSWERS AND EXPLANATIONS

Test What You Already Know

1. A Learning Objective: 7.1

Mycorrhizae are important in helping plants obtain limiting nutrients, such as phosphate. The graph shows lower percent root colonization at higher phosphate levels, suggesting that mycorrhizae are less important to petunias under those conditions **(A)** is thus correct. (B) is incorrect because the graph shows higher colonization at low phosphate levels. (C) is incorrect because there is a clear pattern: mycorrhizae colonization increases as environmental phosphate levels decrease. (D) is incorrect because it provides a less plausible explanation of the results. There is no reason to suppose that higher levels of phosphate inhibit mycorrhizal growth, but there is reason to expect that petunias would be less accommodating to the mycorrhizae when they have an adequate supply of phosphate.

2. B Learning Objective: 7.2

Xylem is dead tissue that transports water in plants. Because it already contains pores, it can be used as a filter if water is pushed through it. Therefore, **(B)** is correct. (A) is incorrect because periderm, also known as bark, would have to be hollowed out to form a straw-like structure, and even then it would not contain pores useful for filtering. (C) is incorrect because phloem is living vascular tissue, not dead. (D) is incorrect because the cortex is not used to transport water, so it could not be used as a pump.

3. B Learning Objective: 7.3

Plants have receptors on guard cell membranes that respond to light. These allow the guard cells to become turgid as needed, opening the stomata and allowing more carbon dioxide to diffuse in and oxygen to diffuse out. This increased gas exchange allows for a higher rate of photosynthesis, which would be reflected in higher sugar concentrations measured in the leaves. **(B)** thus provides the most plausible hypothesis, making it correct. (A) is incorrect because closing the stomata would actually decrease the rate of photosynthesis since there would be less input of carbon dioxide and less output of oxygen. (C) and (D) are incorrect because cortical cells are not involved in the regulation of stomata.

4. C Learning Objective: 7.4

According to the graphs, significantly larger fruit is produced from natural flowering induction in the Smooth Cayenne cultivar, for both the AMI and NMI maturation approaches. Larger fruit growth means that more energy has been invested into growing the fruit. **(C)** is thus correct. The other choices are incorrect because they fail to identify the methods that produce the largest fruit.

5. D Learning Objective: 7.5

Some plants are capable of having a greater than normal number of chromosomes, a condition known as polyploidy. When these plants self-fertilize, they produce new plants with the same number of chromosomes. Thus, if two diploid gametes came together, they would form a tetraploid plant, from which other tetraploid plants could be produced, **(D)**. (A) is incorrect because a hexaploid plant would produce triploid gametes. (B) is incorrect because the loss of entire sets of chromosomes is an extremely rare event; two diploid gametes would normally fuse to form a tetraploid organism. (C) is incorrect because many polyploid plants are capable of reproduction.

6. C Learning Objective: 7.6

Plants that live in dry environments (xerophytes) have adaptations to reduce their water loss through transpiration because dry conditions cause greater evaporation. Thus, a xerophyte could be expected to see less dramatic changes in transpiration rates when placed in environments with differing humidity levels, making **(C)** correct. (A) is incorrect because it reverses the difference between the two plants. (B) is incorrect because a smaller plant would be expected to have a lower absolute rate of transpiration than a larger plant, but not a lower *difference* in transpiration rates at two distinct humidity levels. (D) is incorrect because humidity matters more than temperature for transpiration; for instance, plants would be expected to have lower transpiration rates in hot, moist environments like tropical rainforests than in hot, dry environments like deserts.

Test What You Learned

1. A Learning Objective: 7.6

In transpiration, plants lose water through their stomata. The only way for the plant to replace the lost water is by pulling it up through its roots and stem by capillary action. A potometer is able to measure the water taken up by a plant's cut stem indirectly by measuring the water lost from a tube connected to the stem. Thus, **(A)** is correct. (B) is incorrect because a potometer cannot directly measure water taken up by the roots, but instead indirectly measures how much is taken up by a stem. (C) is incorrect because a potometer does not attach to a plant's leaves. (D) is incorrect because a potometer measures the amount of water lost, not pressure, and because the water is not actively pushed through the plant's vascular system but drawn up using capillary action.

2. B Learning Objective: 7.3

To be an effective experiment, the sample size must be sufficiently large and the variable under investigation must be the only condition that differs between experimental groups. Because it uses a sizable sample of 200 and divides the group based on lighting direction, the experiment in **(B)** presents the most effective design for testing the impact of phototropism. (A) is incorrect because the sample size is too small and because including an actual wall can introduce complicating variables that make it harder to gauge the impact of phototropism alone. (C) is incorrect because the sample size is way too small and because the variation in lighting direction does not match the conditions of plants growing against a wall. (D) is incorrect because variation in the wavelengths of light used adds a complicating variable and does not mimic the conditions of plants growing against a wall.

3. D Learning Objective: 7.1

In order to support the hypothesis that *T. melanosporum* obtains its carbon from its host, rather than from the surrounding soil, experimental results should indicate that the radioactive carbon (which is taken up by the tree in photosynthesis) is found both in the tree itself and in the fungus. In addition, there should be very little of the ^{13}C in the surrounding soil, because otherwise the fungus might be obtaining the radioactive carbon from that. Thus, **(D)** is correct. (A) is incorrect because it suggests that the ^{13}C is somehow getting into the surrounding soil before being taken up by the fungus. (B) is incorrect because it does not indicate whether the fungus took up the radioactive carbon from the tree. (C) is incorrect because it says nothing about the carbon in the fungus. Even if the fungus did contain the ^{13}C, the results would not indicate whether this carbon came from the tree or the soil.

4. A Learning Objective: 7.4

The most likely outcome is that fruits are larger when the plants have access to a greater supply of nutrients, such as when the soil is fertilized. **(A)** is thus correct. (B) is incorrect because it suggests the opposite. (C) and (D) are incorrect because there is not enough information to determine whether natural or artificially induced flowering would enhance the availability of energy and nutrients for fruit growth.

5. C Learning Objective: 7.5

The results indicate significant levels of gene flow within 5 meters, as well as smaller amounts of gene flow at greater distances. Though the percentages are not huge, they are large enough to have an impact on the frequency of the herbicide resistance trait, especially if the application of herbicide provides a selective pressure. Therefore, **(C)** is correct. (A) is incorrect because even a small percentage of gene flow can have a significant effect, especially when the selective pressure of a herbicide is present. (B) is incorrect because the researchers planted seeds with the herbicide resistance genes, so it is already clear that such genes can be found in both pollen and seeds. (D) is incorrect because the trait can spread over large distances given enough time and the appropriate selective pressure.

6. B Learning Objective: 7.2

As water moves through pores from one cell to another, the size of the pores limits which dissolved materials can pass through. In this way, the tissue acts as a filter. Because they contain more cells, gymnosperm tissues provide more opportunities for filtration than angiosperm tissues with the same stem length. **(B)** is thus correct. (A) is incorrect because there is less passage through pores due to the smaller number of cells in angiosperm vascular tissue, so there would be less filtration. (C) is incorrect because the path followed by the water (and the number of pores it passes through) matters more than the distance traveled. (D) is incorrect because monocots and dicots are both angiosperms with only slight differences in vascular tissue.

7. B Learning Objective: 7.3

The mechanism in **(B)** would allow the plant to sense blue light as distinct from other colors and also produce a change in gene expression. (A) is incorrect because it does not explain how light changes gene expression. (C) is incorrect because it does not mention how the plant would sense changes in kinetic energy. In addition, blue light has more energy than other colors of visible light and is therefore not cooler. Light is not absorbed through the stomata, so (D) is incorrect.

8. B Learning Objective: 7.3

Of the mechanisms discussed, only that of **(B)** would bend the plant toward the blue light. (A), (C), and (D) would all cause the plant to bend away from the blue light.

9. A Learning Objective: 7.4

While most terrestrial plants live their lives as diploid sporophytes and produce haploid tissues in the form of ovules and pollen, sometimes within the same flower, aquatic plants typically reproduce through spores, which travel independently from the plant that produced them. Therefore, as the environment becomes more aquatic, a shift toward spore-producing plants would be expected, making **(A)** the correct answer. Given that a new environment poses a variety of new challenges, it is reasonable to expect that the process of pollination (whether by animal or by wind) would be affected too, making (B) a highly unlikely answer. Since, according to the question stem, the plant was already flowering, the line representing the "original population" on the graph should correspond to a higher population frequency of "flowering only" plants, thus reaching its peak earlier, than the line corresponding to the "population after flood." Thus, (C) can be eliminated. (D) is incorrect because, while it may be possible for the post-flood population to produce both flowers and spores, it is unlikely that the original population had equal numbers of flower-producing and spore-producing plants, especially since the question stem explains that it is a flowering plant.

10. C Learning Objective: 7.2

As competition for resources increased in aquatic environments, plants began to migrate to more terrestrial environments. On land, they faced new problems, like supporting the plant's entire structure and avoiding dehydration. As a result, they developed modifications that better enabled them to live on land. One of these modifications was the increased production of the photoreceptor phytochrome, which is a pigment that plants use to detect light, allowing them to regulate flowering based on day length and set up additional daily rhythms. Thus, **(C)** is the correct answer. (A) is incorrect because plant cells do not possess lysosomes; only animal cells do. Instead, plants use vacuoles to carry out the same function. (B) and (D) are incorrect because the terms used in each answer choice are switched. Phloem are thin-walled tubes used to transport sugar and starch from the leaves and stems of the plant to its roots and xylem are thick-walled tubes used to carry water from the roots of the plant to its leaves.

11. D Learning Objective: 7.3

Plants have many different responses to light. In phototropism, a plant bends or grows in a certain direction as a response to light. Shoots usually move towards the light while roots usually move away from it. In photoperiodism, developmental processes (such as flowering) are regulated in response to day length. During the initial experiment, the researcher observes that both the covered and uncovered plants grow toward the brightest available light source. After cutting the growth tips, this no longer occurs. Therefore, by cutting the tip, the botanist eliminated phototropic effects, since the covered plant was no longer able to grow toward the light. This occurred because the photoreceptors at the tip were no longer able to communicate with cells located lower in the plant responsible for plant growth, making **(D)** the correct answer. (A) and (C) can be immediately eliminated because the effects observed by the botanist are phototropic, not photoperiodic. (B) is incorrect because not all of the plant's photoreceptors are gone. Plants have photoreceptors in other locations as well, such as their roots, leaves, and flowers.

12. B Learning Objective: 7.6

Increasing the light intensity would increase the rate of photosynthesis. This would cause stomata to open, in order to take in more carbon dioxide to allow the reaction to occur. The act of opening the stomata would also result in increased transpiration and water loss, making **(B)** correct. (C) and (D) are incorrect because increased light intensity would cause photosynthesis to increase, not decrease. (A) is incorrect because an increase in the rate of photosynthesis would cause stomata to open, not close, thereby increasing the water loss.

13. D Learning Objective: 7.1

The question stem specifically refers to nutrient acquisition from the environment, which means that any answer choices discussing other functions can be eliminated. Xylem is vascular plant tissue that conducts most of the water and minerals upward from the roots to the shoots. The structure described in (A) is actually phloem. Since it incorrectly describes xylem and neither xylem nor phloem play a role in nutrient acquisition, (A) can be eliminated. Similarly, (B) can be eliminated because it discusses nutrients that have already been acquired. Transpiration is the process by which moisture changes to vapor and is released to the atmosphere. So rather than discussing nutrient acquisition, (C) is discussing the process by which plants give up water to the atmosphere, making it incorrect. Root nodules are composed of root cells that contain vesicles filled with nitrogen-fixing bacteria. These bacteria convert nitrogen from the Earth's atmosphere into ammonia that can then be used by the plant. Because this process involves the acquisition of nitrogenous nutrients from the environment, **(D)** is correct.

14. C Learning Objective: 7.5

The hypothesis is that the short-rooted parent is heterozygous. Therefore, the other parent's genotype must be homozygous recessive. Crossing the two should theoretically result in a population that is 50% long-rooted and 50% short-rooted. With 38 seeds, this means that 19 should develop into long-rooted plants and 19 should develop into short-rooted plants. These are the expected values. The question stem states that the observed values were 24 short-rooted and 14 long-rooted. The chi-squared value can be calculated as follows, where *o* stands for the observed values and *e* for the expected values:

$$\chi^2 = \sum \frac{(o-e)^2}{e}$$

phenotype	*o*	*e*	*o–e*	$(o-e)^2$	$(o-e)^2/e$
short-rooted	24	19	5	25	1.316
long-rooted	14	19	–5	25	1.316
					$\chi^2 = 2.632$

Rounding to the nearest tenth yields a value of 2.6. **(C)** is correct.

CHAPTER 8

Cell Communication

LEARNING OBJECTIVES

In this chapter, you will review how to:

- **8.1** Differentiate between stimulatory and inhibitory transduction signaling
- **8.2** Explain how selective pressure impacts pathway choice
- **8.3** Describe receptor recognition of ligands and protein shape change
- **8.4** Explain how blocking transduction or reception affects cells
- **8.5** Contrast distance and contact forms of cell-to-cell communication

TEST WHAT YOU ALREADY KNOW

1. Physiologists observe that epinephrine can have opposite effects on skeletal and smooth muscles. Epinephrine stimulates contraction in skeletal muscles, whereas it slows down the contraction of the smooth muscles found in bronchial and intestinal walls. Researchers have proposed the following signaling cascade to explain these differential effects.

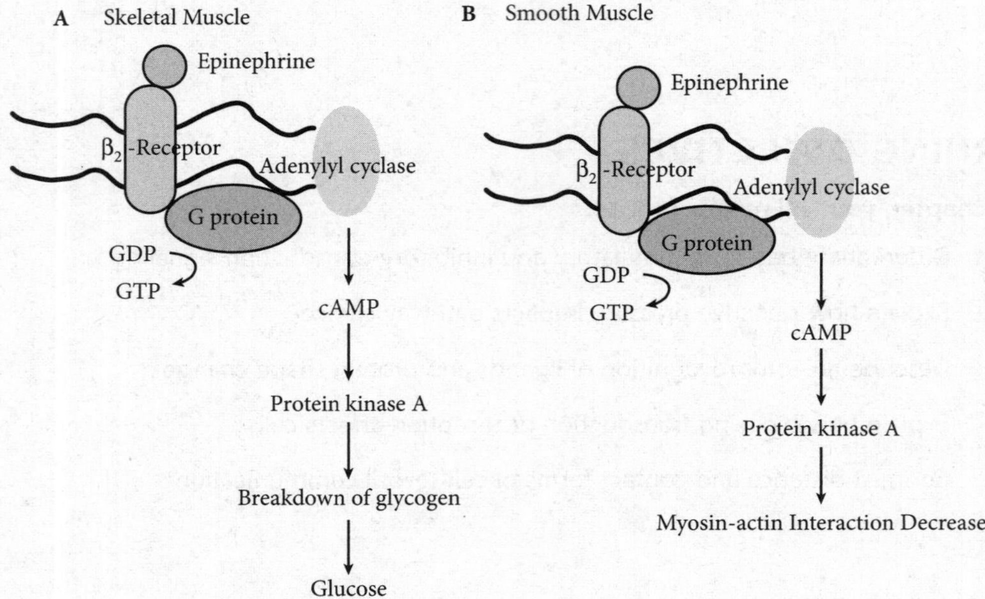

Based on the researchers' proposal, which statement best explains the differential effects of epinephrine on skeletal and smooth muscles?

(A) Epinephrine attaches to different types of receptors in skeletal and smooth muscles.

(B) Different second messengers transduce the signal in skeletal and smooth muscles.

(C) Glycogen metabolism is inhibited in smooth muscle and enhanced in skeletal muscle by epinephrine.

(D) The target of protein kinase A is different in skeletal and smooth muscles.

2. Infection by the bacterium *Clostridium difficile* (*C. diff*) results in extensive tissue damage caused by the secretion of a toxin. Microbiologists investigate secretion of the toxin by measuring its accumulation in the nutrient broth in which the bacteria grow.

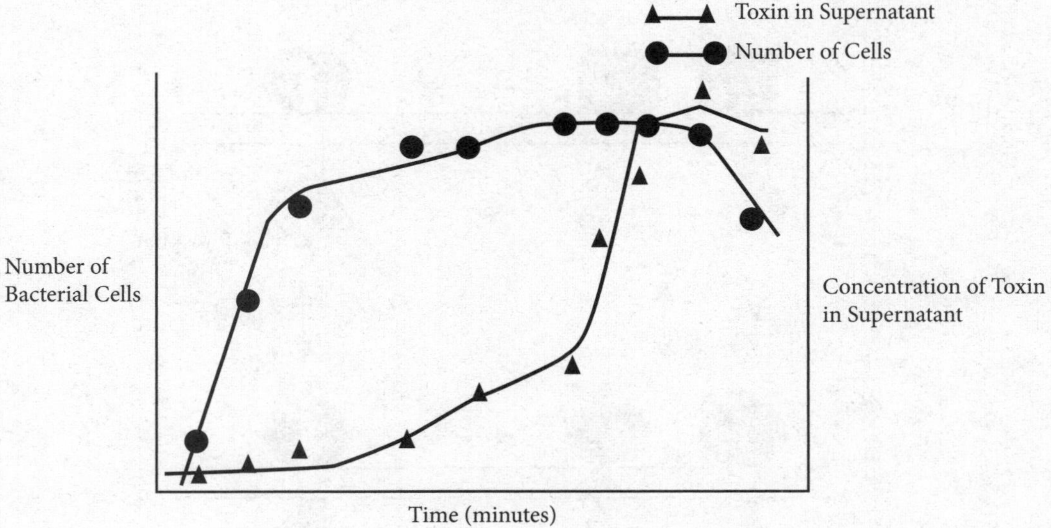

Adapted from Charles Darkoh et al., "Toxin Synthesis by *Clostridium difficile* Is Regulated through Quorum Signaling," *mBio* 6, no. 2 (February 2015): e02569–14.

They hypothesize that bacteria secrete the toxin only when the population reaches a certain cell density and that they coordinate expression of the protein through quorum sensing mediated by a diffusible inducer molecule.

Which of the following experiments would most effectively test their hypothesis?

(A) Resuspend cells in the late stationary phase of growth in the supernatant conditioned by cells in early stages of growth.

(B) Resuspend cells from the early stages of growth in the supernatant conditioned by cells in the late phase of growth.

(C) Resuspend cells in the late phase of growth in fresh medium.

(D) Resuspend cells from the early stages of growth in fresh medium.

3. Receptor tyrosine kinases (RTKs) are found mostly as monomers but also as dimers in the membrane. Signal transduction inside the cell is activated when the intracellular tyrosine kinase domain of the receptor is modified with four phosphate groups. Researchers studied ligand-receptor interactions and summarized their data in the diagram below.

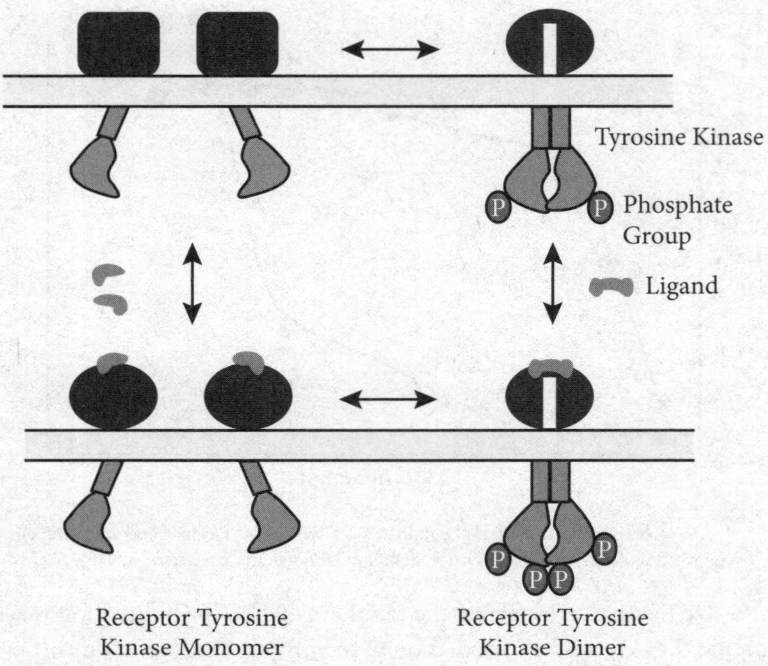

Adapted from Lijuan He and Kalina Hristova, "Physical-Chemical Principles Underlying RTK Activation, and Their Implications for Human Disease," *Biochim Biophys Acta* 1818, no. 4 (April 2012): 995–1005.

Based on the diagram, which of the following statements best describes the activation of an RTK?

(A) Ligands can bind either to monomers or dimers and, in each case, fully activate tyrosine kinase phosphorylation.

(B) Ligands can bind to monomers or dimers, but only the ligand-dimer complex can be phosphorylated.

(C) Only dimers bound to the ligand undergo extensive phosphorylation, initiating the signaling cascade.

(D) Monomers can form dimers without the ligand bound, can become phosphorylated, and can start the signaling cascade in the absence of a signal.

4. The cholera toxin secreted by the bacterium *Vibrio cholerae* enters the cells lining the intestinal tract and locks a G-protein in its GTP-bound form, which stimulates the continuous production of cAMP by adenylyl cyclase. In turn, high cAMP levels activate the cystic fibrosis transmembrane conductance regulator (CFTR), which pumps Cl^- out of the cell.

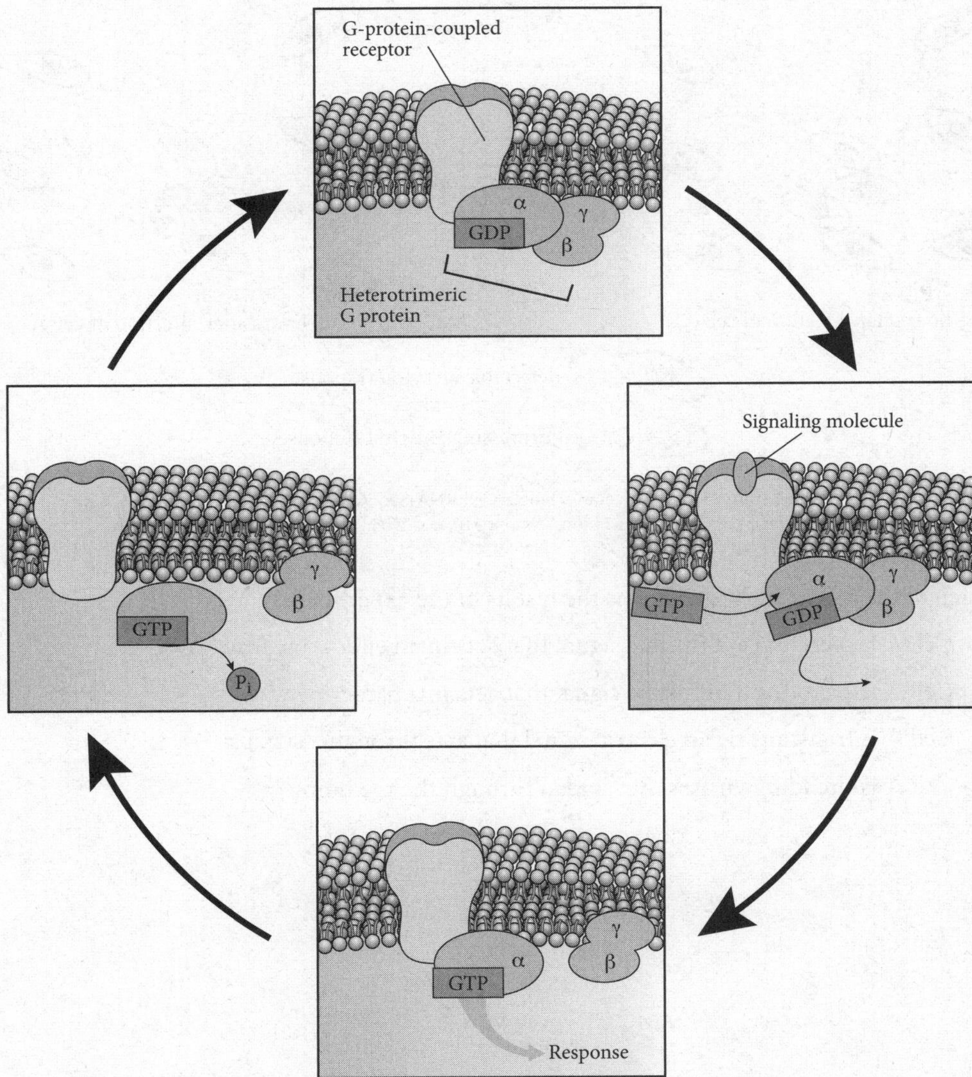

Which stage of the signaling cycle by a G-protein-coupled receptor is affected by the action of cholera toxin?

(A) Cholera toxin interferes with the binding of the signal molecule with its cognate receptor.

(B) The G-protein-coupled receptor activates dissociation of the G-protein complex into subunits.

(C) Activation of the enzyme adenylyl cyclase by a G-protein subunit leads to increased levels of cAMP.

(D) G-protein subunits reassemble to form an inactive G-protein that can reattach to the receptor.

5. Many viruses elicit the biosynthesis of the protein interferons in the cells they infect. The following diagrams show the synthesis of antiviral proteins in cells immobilized on a solid substrate in growth medium. Cell A is infected with a nonreplicating virus at time zero. The synthesis of mRNA encoding antiviral proteins is measured in all cells at the 1 and 2 hour time points.

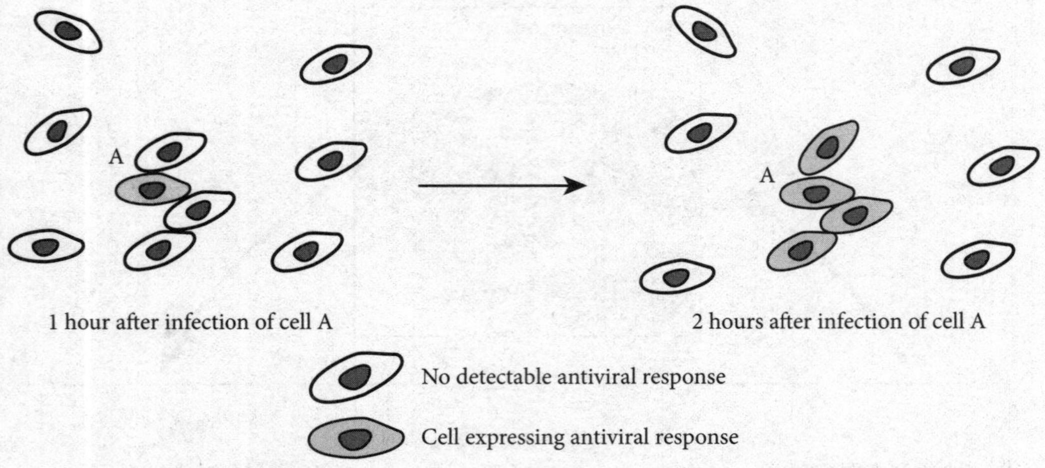

Adapted from A. K. Shalek et al., "Single-Cell RNA-Seq Reveals Dynamic Paracrine Control of Cellular Variation," *Nature* 510, no. 7505 (June 2014): 363–369.

Which of the following best explains the results of the experiment?

(A) Cell A is secreting a diffusible signal that acts in an endocrine way.

(B) Cell A is secreting a diffusible signal that acts in a paracrine way.

(C) Cell A is transmitting an electric signal that acts through a synapse.

(D) Cell A is shedding viruses that spread through the medium.

Answers to this quiz can be found at the end of this chapter.

SIGNAL TRANSDUCTION

8.1 Differentiate between stimulatory and inhibitory transduction signaling

8.2 Explain how selective pressure impacts pathway choice

8.3 Describe receptor recognition of ligands and protein shape change

8.4 Explain how blocking transduction or reception affects cells

General Principles of Signaling

Two main ways that animal cells communicate with one another are (1) via signaling molecules that are secreted by the cells and (2) through receptor molecules that rest on the cells' surfaces and remain attached to the cell even as they signal other cells. The target cell receives information through receptors on its surface, which are generally membrane-spanning proteins with binding domains on the outside of the plasma membrane.

Cells communicate with each other using a wide variety of molecules. Some can pass through the membranes of the target cell to act directly within the cell's cytoplasm or nucleus. Others bind to receptors on the cell membrane and act through *second messenger systems* to alter a process within the cell. All of these molecules can be considered hormones, circulating signal molecules that are released by specialized cells and travel throughout the bloodstream.

Responses to signal transduction (the conversion of an extracellular signal to a change in an intracellular process) may be stimulatory or inhibitory. Stimulatory responses typically involve activators or growth factors that promote gene expression, cell growth, and cell proliferation, whereas inhibitory responses involve inhibitors that may block signals from one molecule to another and/or lower enzyme activity. Epinephrine, also known as **adrenaline**, is an example that elicits both stimulatory and inhibitory responses. Epinephrine not only activates glycogen breakdown, but it also inhibits the enzyme that catalyzes glycogen synthesis. Another example is apoptosis. In a normal cell, apoptosis is prevented by inhibiting the expression of pro-apoptotic factors and by promoting the expression of anti-apoptotic factors. However, when DNA damage is beyond repair, the cell signals apoptosis to occur.

Selective pressures such as predators, diseases, and environmental threats also impact which signaling pathway that cells within an organism choose. If the selective pressure is neutral, the present signal transduction pathway persists because the pathway results in a stable population that is able to survive and reproduce. Otherwise, the selective pressure will cause the population to undergo natural selection and evolve. For instance, a mutation that alters the phenotype of an organism will spread in a population if the mutation enhances the behavior, fitness, and reproduction of the organism.

Signal transduction pathways of hormones (specifically peptide hormones) generally involve the binding of a hormone to a cell-surface receptor, which induces a conformational change in the receptor protein and sets off an intracellular cascade to alter the cell's behavior in some way. Cell-surface receptors come in three different forms: ion-channel-linked, G-protein-linked, and enzyme-linked.

Ion-channel-linked receptors are also called ligand-gated channels. These membrane-spanning proteins undergo a conformational change when a ligand binds to them so that a "tunnel" is opened through the membrane to allow the passage of a specific molecule. These ligands can be neurotransmitters or peptide hormones, and the molecules that pass through are often ions, such as sodium (Na^+) or potassium (K^+), which can alter the charge across the membrane. The ion channels, or pores, are opened only for a short time, after which the ligand dissociates from the receptor and the receptor is available once again for a new ligand to bind.

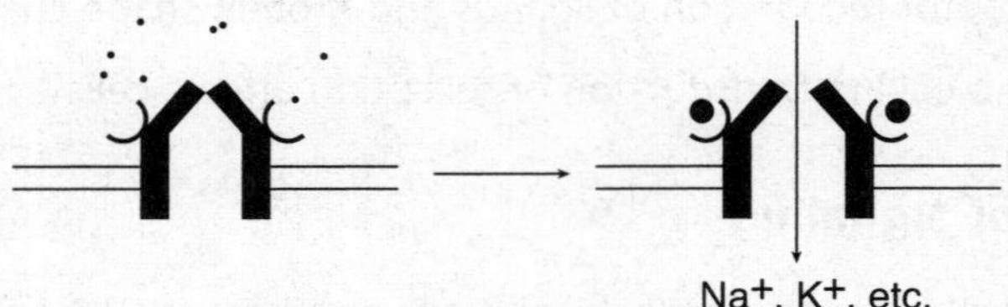

Ion-Channel-Linked Receptors

G-protein-linked channels cause G-proteins to dissociate from the cytoplasmic side of the receptor protein and bind to a nearby enzyme. This enzyme continues the signaling cascade by inducing changes in other intracellular molecules. In addition, it can also cause other membrane channels to open in areas some distance from the originating receptor.

Most G-proteins activate what are known as second messengers, small intracellular molecules like cyclic AMP (cAMP), calcium, and phosphates, which in turn activate key enzymes or transcription factors involved in essential reactions. Signaling cascades involving G-proteins can be very complex and involve many different enzymes and conversions, which prevents the reactions from running out of control.

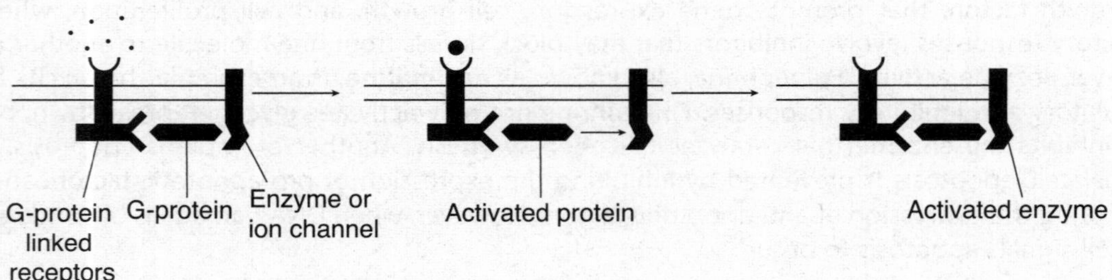

G-Protein-Linked Receptors

Enzyme-linked receptors can act directly as enzymes, catalyzing a reaction inside the cell, or they can be associated with enzymes that they activate within the cell. Most enzyme-linked receptors turn on a special class of enzymes called protein kinases, which add free-floating phosphate groups to proteins, regulating their activity. Protein phosphorylation is an essential means of intracellular signaling and control, as proteins become activated or deactivated simply by the addition or removal of phosphates. Protein kinases often target other protein kinases, initiating a cascade of kinase activity. Protein phosphatases reverse the action of protein kinases.

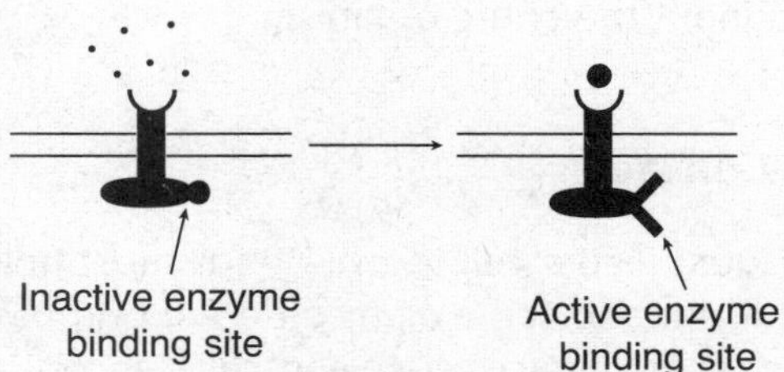

Enzyme-Linked Receptors

G-protein-linked and enzyme-linked receptors use complex relays of signal proteins to amplify and/or regulate their signal transduction. In some cases, a measure of safety requires that two different receptors on the cell surface become activated to turn on a particular intracellular protein. In other cases, signals at different receptors lead to the phosphorylation of different proteins that activate together only when *both* proteins have been phosphorylated. This signal integration leads to a measure of control over reactions and the ability to use multiple inputs to cause a certain effect that can vary in degree.

Signal transduction pathways that are blocked or defective can be deleterious (harmful) or prophylactic (preventative). For example, type I and type II diabetes arise from blocked signal transduction and blocked signal reception, respectively. Normally, the pancreas releases insulin, which signals liver, muscle, and fat cells to store sugar. In type I diabetes, pancreatic cells that produce insulin are destroyed by the immune system, so little to no insulin is produced. As a result, the effect is deleterious as sugar accumulates to toxic levels in the blood. In type II diabetes, cells are unable to respond to insulin, resulting in insulin resistance as well as toxic sugar levels in the blood. Some drugs such as antihistamines are prophylactic. Used to prevent and treat allergies, antihistamines inhibit the action of histamine (a compound released in allergic reactions) by suppressing the expression of histamine receptors in nasal mucosa.

CELL-TO-CELL COMMUNICATION

8.5 Contrast distance and contact forms of cell-to-cell communication

Cell Communication Over Distance

Although some signaling molecules may act far away from the cell that secreted them, many bind only to receptors on cells in the immediate vicinity. *Paracrine signaling* refers to the process of signaling only nearby cells, with signaling molecules quickly pulled out of the **extracellular matrix**. *Synaptic signaling* occurs between a nerve cell and either another nerve cell, a gland cell, or a muscle cell. Synaptic signaling itself occurs over extremely short distances, but the combined effect of multiple synaptic signals plus action potential transmission along lengthy axons can result in long distance communication. As electric signals reach axon terminals, they cause the release of chemicals called neurotransmitters. These chemical messengers bind to nearby receptors in post-synaptic cells and cause an electrical signal to be propagated or continued. By sending information from one neuron to another, nervous systems can produce rapid communication over long distances without a diminished effect (which can happen when sending individual molecules over long distances, such as through the bloodstream). *Endocrine signaling* refers to the secretion of chemical messengers into the bloodstream (or other liquid medium, such as is the case with plants) for widespread distribution throughout the entire organism.

Cell Communication via Contact

For cells to form complex **tissues**, secure cell-to-cell bonds must hold them together. These cell junctions come in a variety of forms and serve many different purposes. Occluding junctions, otherwise known as tight junctions, seal spaces between cells; anchoring junctions, which include desmosomes, connect one cell's cytoplasm to another through anchoring proteins; and communicating junctions, including gap junctions and plasmodesmata in plants, allow cells to directly exchange cytoplasmic material via channels that cross both cells' membranes. The contact form of communication enabled by communicating junctions is to be contrasted with the long-distance communication allowed by the endocrine system.

Communicating Junctions

The best known of these cell-to-cell connections are the gap junctions, formed by proteins called connexins, which build tubes or pores between two adjacent cells' cytoplasms. It is through these pores that ions and other material can pass from one cell to the other. In cells that rapidly transmit chemical or electrical signals across tissues, gap junctions are everywhere. Because chemical and electrical transmission is mediated through the movement of ions and other messengers, gap junctions *allow for undisrupted and very fast signal transmission* across wide areas of tissue. In heart cells, gap junctions allow for rhythmic contractions of large sections of the heart all at once. In the digestive tract, gap junctions allow for waves of muscle contraction such as those found in the esophagus. These junctions also allow for coordination of *rapid and complex movements,* such as a fish's tail-flip to escape an oncoming predator. The following diagram shows a generic gap junction with many connexin proteins that form a connexon.

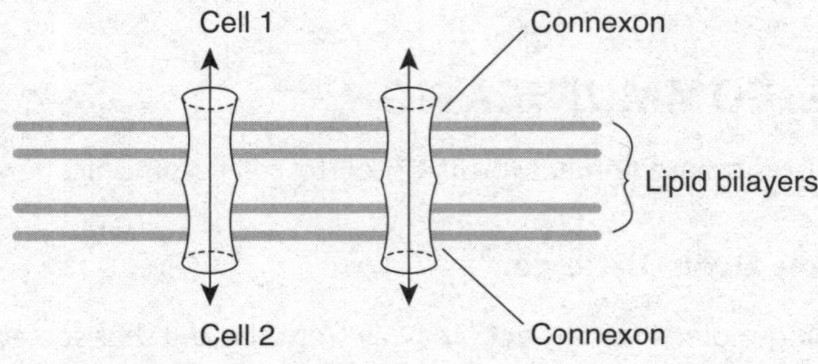

Gap Junctions

Communication in Plants

Plasmodesmata are plant cells' equivalent of gap junctions, which are particularly useful for the free flow of nuclei from one cell to another. These junctions in plants are not nearly as complex in structure as those in animals, but serve essentially the same purpose. Plant viruses, however, often exploit plasmodesmata because the openings allow the virus particles to spread rapidly from one section of the plant to others.

RAPID REVIEW

If you take away only 4 things from this chapter:

1. There are two major ways animal cells communicate: through signaling molecules secreted by cells and through receptor molecules that rest on the cells' surfaces.
2. Responses to signal transduction (the conversion of an extracellular signal to a change in an intracellular process) may be stimulatory or inhibitory.
3. *Endocrine signaling* refers to the secretion of chemical messengers for widespread distribution throughout the entire organism. *Paracrine signaling* refers to the process of signaling only nearby cells. *Synaptic signaling* occurs between a nerve cell and either another nerve cell, a gland cell, or a muscle cell.
4. The cells of many tissues contain pores known as gap junctions that allow for direct cell-to-cell communication, which allows for incredibly fast, undisrupted signal transmission across large areas. Plasmodesmata are plant cells' equivalent of gap junctions.

TEST WHAT YOU LEARNED

1. The sympathetic branch of the autonomic nervous system transmits signals through the release of the neurotransmitter acetylcholine. The muscarinic acetylcholine receptors associated with the sympathetic system are G-protein-coupled receptors usually found in the membrane of the target organ. The graph shows the response of a target organ to acetylcholine with or without atropine. The researchers observed that adding atropine shifted the dose-response curve to the right. An increased dose of acetylcholine was required to overcome the effect of atropine on the target organ.

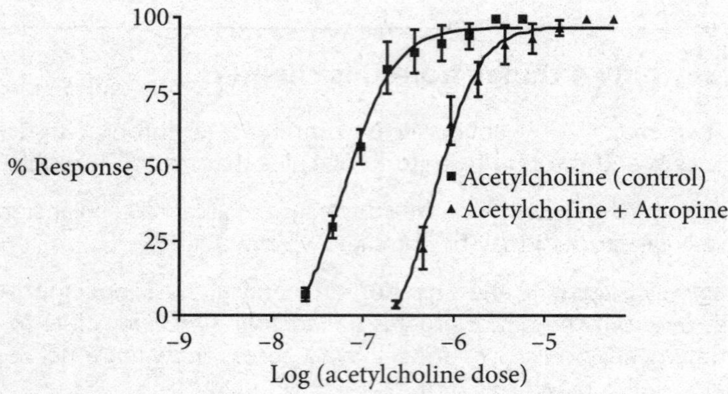

Which mechanism of action of atropine would produce the effects recorded in the experiment?

(A) Atropine and acetylcholine bind to the same site on the receptor.

(B) Atropine increases the effectiveness of acetylcholine.

(C) Atropine is an irreversible inhibitor of the muscarinic receptor.

(D) Atropine has no substantial effect on the response of the target organ to acetylcholine.

2. Acetylcholine is a known neurotransmitter that initiates muscle contraction when it binds to ion-gated channels. Epinephrine is a neurotransmitter of the sympathetic system, which decreases smooth muscle contraction. It does so by binding to a G-protein-coupled receptor that triggers adenylyl cyclase to release cAMP, which activates protein kinase A, causing less interaction between myosin and actin. In an experiment on smooth muscle tissue, a baseline of contraction in the absence of chemicals was recorded for 1 minute before introducing acetylcholine to the system. Epinephrine was applied 3 minutes after acetylcholine. The graph shows the response of the smooth muscle tissue to the addition of acetylcholine followed by epinephrine.

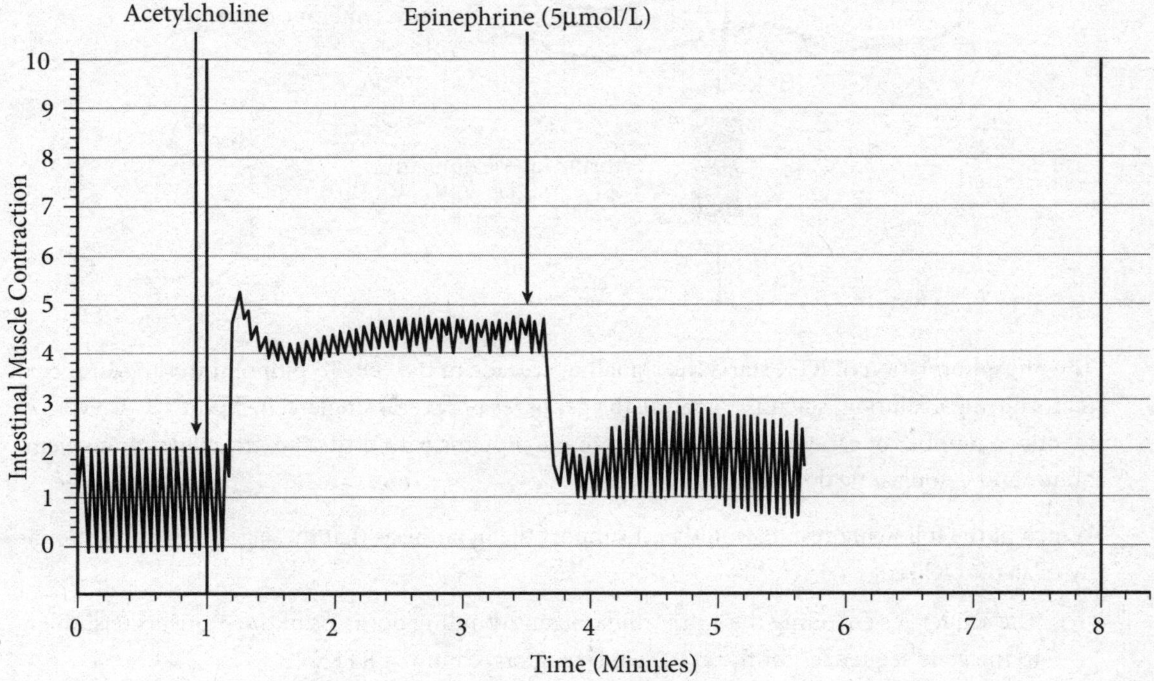

Which of the following hypotheses is best supported by the results shown in the graph?

(A) The contractions of intestinal muscle decrease after the addition of epinephrine because epinephrine competes with acetylcholine for a binding receptor in the membrane.

(B) The activation of the transducing cascade triggers activation of protein kinase A and inhibition of glucose synthesis.

(C) Epinephrine initiates a signaling cascade that activates protein kinase A, thereby indirectly decreasing muscle contraction.

(D) Epinephrine activates the membrane receptor, which inhibits the activation of adenylyl cyclase, preventing production of a second messenger.

3. Receptor tyrosine kinases (RTKs) are signaling receptors that consist of an extracellular region that binds the ligand, a hydrophobic region that spans the membrane, and a cytoplasmic region that acts as a tyrosine kinase enzyme.

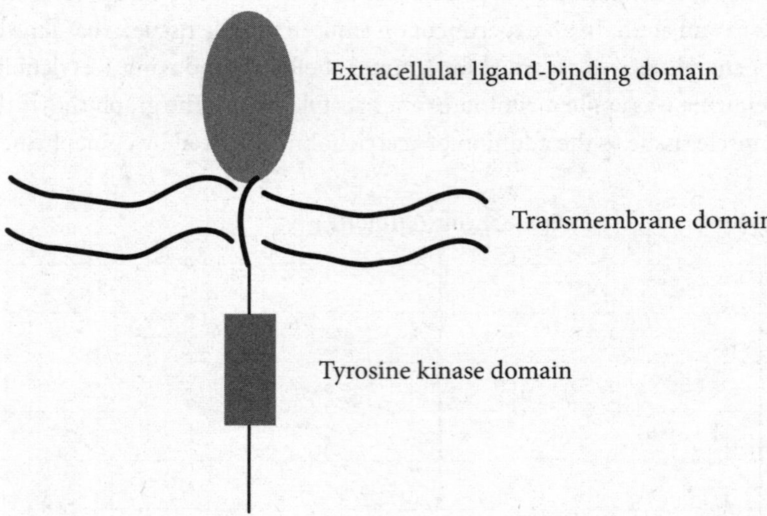

The phosphorylation of RTK starts the signaling cascade in the cell. To pinpoint the appearance of RTKs during evolution, scientists analyze the genomes of several single-celled protists. They identify a number of genes that encode membrane proteins with distinct extracellular, transmembrane, and cytoplasmic domains.

Which of the following results would best support the hypothesis that these genes are evolutionarily related to RTKs?

(A) The sequences encoding the extracellular ligand-binding domains of the protists are similar to the gene sequences for the extracellular region of human RTKs.

(B) The sequences encoding the membrane-spanning domains of the protists are similar to the gene sequences for the membrane-spanning region of human RTKs.

(C) The sequences encoding the tyrosine kinase domain of the protists are similar to the gene sequences for the tyrosine kinase regions of human RTKs.

(D) The entire gene sequences for the proteins of the protists are similar to the gene sequences for the entire coding region of human RTKs.

4. The hormone insulin is produced by the β cells of the pancreas. When the levels of glucose rise above a set threshold, insulin is released in the blood and promotes the uptake of glucose by cells. In type I diabetes, the body is unable to produce insulin. In type II diabetes, cells are insulin-resistant and do not respond to insulin stimulus. The graph shows levels of glucose and insulin in the blood of three patients after eating.

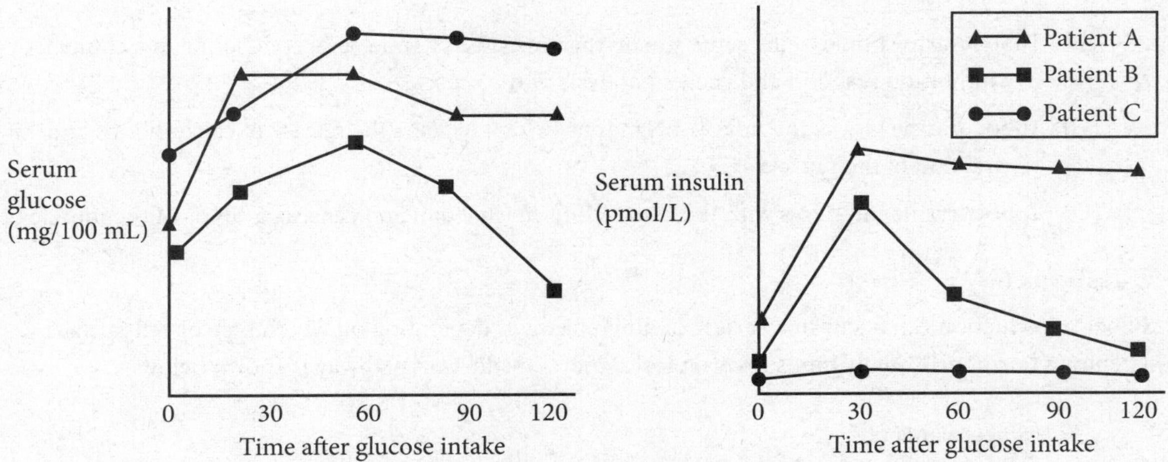

Adapted from N. Geidenstam et al., "Metabolite Profile Deviations in an Oral Glucose Tolerance Test—A Comparison between Lean and Obese Individuals," *Obesity* 22, no. 11 (November 2014): 2388–2395.

Based on the curves, which patient is the most likely to be diagnosed with type II diabetes?

(A) Patient A, because his insulin and glucose levels remain high after 2 hours

(B) Patient B, because her insulin levels show a strong peak after ingestion of glucose

(C) Patient C, because his insulin levels never increase

(D) All three patients, because the glucose levels in their blood increase sharply during the glucose tolerance test

5. Tubocurarine was first purified from curare, a poison obtained from the bark of a South American tree. Poisoning by tubocurarine can be reversed by the administration of a cholinesterase inhibitor, which results in an increased concentration of acetylcholine in the synaptic cleft. The inhibition of muscle contraction by tubocurarine is most likely explained by which of the following statements?
   (A) Tubocurarine prevents the release of acetylcholine from the nerve ending, causing paralysis of the muscle.
   (B) Tubocurarine binds to the same site on the ion-gated channel as acetylcholine in a competitive inhibition reaction and causes paralysis of the muscle.
   (C) Tubocurarine blocks the influx of Na^+ ions, which triggers the release of calcium ions and contraction of the muscle.
   (D) Tubocurarine interferes with the contraction mechanism and causes paralysis of the muscle.

Questions 6–7

Signal transduction can occur in a variety of different ways, depending on which type of cell-surface receptor a hormone initially binds to. A typical signal transduction pathway is shown below:

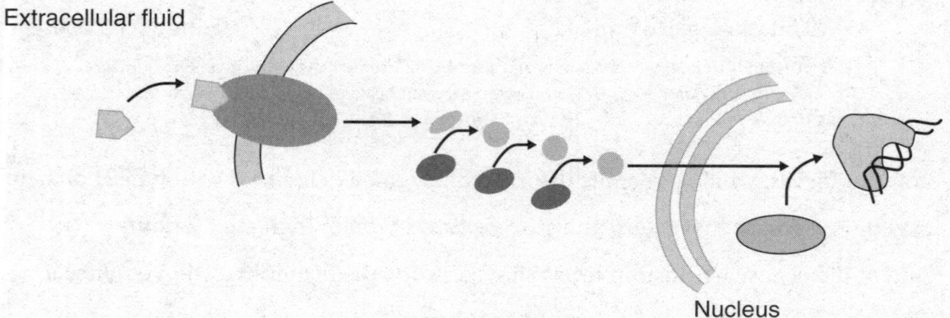

Blocking signal transduction pathways can be extremely harmful and even fatal. For example, cancer results when genetic and epigenetic changes adversely impact important cellular processes, such as by causing excessive cell division or by causing blood vessels to be attracted to growing tumors. These processes are usually regulated by growth factors, which are a type of hormone. As a result, many of the mutations that lead to cancer development can be found within signal transduction pathways.

6. Which of the following provides a possible explanation for what is occurring in the image as soon as the hormone binds to the receptor?
   (A) The ligand causes the membrane protein to change shape, allowing a specific extracellular molecule to enter.
   (B) The hormone causes an intracellular cascade, thereby amplifying the original signal.
   (C) The ligand targets the expression of a specific gene, causing its transcription to stop temporarily.
   (D) The hormone causes the membrane protein to undergo a conformational change, triggering second messengers.

7. Even though defective or obstructed pathways can be exceedingly dangerous, doctors and researchers have discovered a way to utilize the principles of signal transduction blockage to produce positive, therapeutic results. Which of the following does NOT provide an example of one of these uses?

    (A) Antihistamines bind to histamine receptors, thereby blocking histamine from binding and causing nasal allergy symptoms.

    (B) Trastuzumab, an antibody against the ErbB-2 receptor, binds to and initiates the destruction of cells overexpressing the receptor in patients with metastatic breast cancer.

    (C) Injectable insulin, provided to patients suffering from type I diabetes, acts as a replacement for their damaged insulin-producing pancreatic cells.

    (D) Felodipine, a drug used to treat high blood pressure, binds to voltage-gated calcium channels in smooth muscle cells to decrease contraction.

8. Different types of immune cells work together to capture and identify invading pathogens. Certain immune cells, known as antigen presenting cells, engulf pathogens and present small components of their unique proteins, known as antigens, on their cell surfaces. These antigens are then recognized when helper T cells and killer T cells contact them directly, triggering the start of the adaptive immune response. Which of the following cell junction functions is most similar to the method of communication used by immune cells?

    (A) Desmosomes, which allow two cells to become attached using intermediate filaments found in the cytoplasm of each cell, anchor cells that are exposed to high levels of shear stress, such as those in the heart.

    (B) Tight junctions, which form a total barrier between adjacent cells, such as those found in the intestines, prevent important nutrients from diffusing into areas where they should not be.

    (C) Anchoring junctions, which are found between cells that experience a large amount of stress such as epithelial cells in the skin, do not actually link cytoplasm, but rather physically join cells so they do not separate.

    (D) Connexins, a type of protein used to build pores between adjacent cells, allow cells to exchange ions and other material.

Questions 9–10

Saccharomyces cerevisiae, more commonly known as yeast, exist in two different genders that are often referred to as mating types. Unlike mammals, which are typically divided into either male or female, yeast are categorized as being either a cells or α (alpha) cells.

Interestingly, if a sample of a cells is placed into a test tube, the cells are able to divide indefinitely, provided there is an adequate supply of nutrients and space available. However, if an a cell detects the presence of an α cell nearby, it will immediately halt current and future mitosis and begin to prepare itself to mate with the α cell.

Scientists researching this phenomenon discovered that this occurs as a result of a 13-amino-acid peptide known as α-factor, which is secreted by α cells and diffuses from the α cell to the a cell. They also observed that the a cell responded in two notable ways to the α factor. First, the a cell radically reorganized its actin cytoskeleton to allow for the generation of a "germ tube" towards the source of the α factor. Second, the expression of several genes required for the mating process was activated. For example, production of FUS1, a protein necessary to allow the fusion of the nuclei, increased.

9. A sample of the yeast cells being studied is accidentally contaminated with a bacterium known for producing a toxin that inhibits DNA repair. Which of the following describes the most likely response of the cells in this culture?

   (A) An inhibitory response resulting in the upregulation of pro-apoptotic protein production

   (B) A stimulatory response resulting in the upregulation of pro-apoptotic protein production

   (C) An inhibitory response resulting in the upregulation of DNA repair protein production

   (D) A stimulatory response resulting in the upregulation of DNA repair protein production

10. Cellular interaction is an integral part of proper functioning in both single-celled and multi-celled organisms. Cells communicate by using a variety of molecules known as hormones. Although different hormones utilize different mechanisms of action, all of the subsequent responses that they produce can be labelled as either stimulatory or inhibitory. Which of the following correctly identifies an inhibitory effect of α factor recognition by an a cell?

    (A) The increased production of a protein that decreases the duration of interphase

    (B) The decreased production of a protein that increases the duration of interphase

    (C) The increased production of a protein that increases the duration of interphase

    (D) The decreased production of a protein that decreases the duration of interphase

The answer key to this quiz is located on the next page.

The answer key to this quiz is located on the next page.

Answer Key

Test What You Already Know

1. D

2. B

3. C

4. D

5. B

Test What You Learned

1. A

2. C

3. C

4. A

5. B

6. D

7. C

8. D

9. B

10. C

REFLECTION

Test What You Already Know score: __________

Test What You Learned score: __________

Use this section to evaluate your progress. After working through the pre-quiz, check off the boxes in the "Pre" column to indicate which Learning Objectives you feel confident about. Then, after completing the chapter, including the post-quiz, do the same to the boxes in the "Post" column. Keep working on unchecked Objectives until you're confident about them all!

Pre	Post	
❑	❑	**8.1** Differentiate between stimulatory and inhibitory transduction signaling
❑	❑	**8.2** Explain how selective pressure impacts pathway choice
❑	❑	**8.3** Describe receptor recognition of ligands and protein shape change
❑	❑	**8.4** Explain how blocking transduction or reception affects cells
❑	❑	**8.5** Contrast distance and contact forms of cell-to-cell communication

FOR MORE PRACTICE

Complete more practice online at kaptest.com. Haven't registered your book yet? Go to kaptest.com/booksonline to begin.

ANSWERS AND EXPLANATIONS

Test What You Already Know

1. D Learning Objective: 8.1

According to the diagrams, the same transducing cascade is employed in both skeletal and smooth muscles, up until the point when protein kinase A is activated. In skeletal muscles, it leads to the breakdown of glycogen, while in smooth muscles, it causes less interaction between myosin and actin. Thus, protein kinase A has different targets in the two muscle types, making **(D)** correct. (A) is incorrect because epinephrine attaches to the β_2 adrenergic receptor in both skeletal and smooth muscles. (B) is incorrect because cAMP is the second messenger in both types of muscles. (C) is incorrect because glycogen metabolism is not affected in smooth muscle.

2. B Learning Objective: 8.2

Cells in the late stage of growth secrete a signal molecule in the medium. If the hypothesis is correct, a supernatant conditioned by these cells would induce quorum sensing in cells that have not yet reached high density. Thus, **(B)** presents the most effective way to test the hypothesis. (A) is incorrect because early cells would not secrete much of the signal molecule into the supernatant. (C) and (D) are incorrect because a fresh medium would not contain the signaling molecule.

3. C Learning Objective: 8.3

According to the lower right image in the diagram, only dimers bound to ligands carry enough phosphate groups to be activated fully. **(C)** is thus correct. (A) is incorrect because full phosphorylation occurs only when the ligand is bound to the dimer. (B) is incorrect because the dimer without the ligand can be partially phosphorylated, as is clear from the image on the upper right. (D) is incorrect because the monomers cannot be phosphorylated and because the signaling cascade will not be initiated without a signal.

4. D Learning Objective: 8.4

The question stem states that the cholera toxin locks a G-protein into being bound to GTP. According to the top image in the diagram, the G-protein subunits cannot reform until GTP has been hydrolyzed back to GDP. Thus, the cholera toxin interferes with this stage, making **(D)** correct. (A) is incorrect because this is not a typical stage in the signaling cycle and because cholera does not bind to the receptor, but uses a different mechanism to enter the cell. (B) is incorrect because the G-protein complex is still able to dissociate into subunits in the presence of the toxin. (C) is incorrect because adenylyl cyclase is continuously activated in the presence of the cholera toxin, as stated in the question stem.

5. B Learning Objective: 8.5

According to the results, only cells in close proximity to the infected cell express an antiviral response. This suggests a type of paracrine signaling. **(B)** is thus correct. (A) is incorrect because endocrine signaling can act over large distances by traveling through the bloodstream, but cells outside of the immediate vicinity of cell A do not express the antiviral response. (C) is incorrect because electrical signals act quickly, but it takes hours for the antiviral response to be expressed by nearby cells. (D) is incorrect because the virus is nonreplicating, as noted in the question stem, so cell A cannot shed copies of the virus.

Test What You Learned

1. A Learning Objective: 8.3

As seen in the figure, if present at a high enough dose acetylcholine can reverse the effect of atropine, suggesting that the two compounds compete for the same site on the receptor. **(A)** is thus correct. (B) is incorrect because atropine actually has an inhibitory effect: a higher dose of acetylcholine is required to achieve the same response. (C) is incorrect because increased acetylcholine reversed the effects of atropine, indicating that it is not an irreversible inhibitor. (D) is incorrect because there was a substantial effect: the response curve moved to the right.

2. C Learning Objective: 8.1

As explained in the question stem, epinephrine binds to a G-protein-coupled receptor, which initiates a signaling cascade that activates protein kinase A, decreasing interaction between actin and myosin and thereby indirectly decreasing muscle contraction. Therefore, **(C)** is correct. (A) is incorrect because acetylcholine binds to an ion-gated channel while epinephrine binds to a G-protein-coupled receptor, so the two could not compete for the same receptor. (B) is incorrect because protein kinase A does not inhibit glucose synthesis, according to the question stem. (D) is incorrect because the binding of epinephrine to the receptor activates the adenylyl cyclase, as noted in the question stem.

3. C Learning Objective: 8.2

The tyrosine kinase domain is what makes RTKs distinctive. The presence of sequences encoding a putative tyrosine kinase domain in the protists would be a strong indication that the protein is in fact a receptor tyrosine kinase. **(C)** is thus correct. (A) is incorrect because ligand regions are determined by the ligands they bind, so they would not be conserved unless they bind to the same ligand. (B) is incorrect because transmembrane domains could be similar across a wide range of proteins, since spanning the membrane always requires hydrophobic residues. (D) is incorrect because only the tyrosine kinase domain is essential for determining the evolutionary relationship.

4. A Learning Objective: 8.5

A patient with type II diabetes is resistant to insulin, but the β cells of the pancreas still appropriately respond to high glucose levels. This suggests that both glucose and insulin levels would remain high in such a patient after eating. Only Patient A presents such a response, so he is the most likely to suffer from type II diabetes, making **(A)** correct. (B) is incorrect because Patient B shows a normal response, in which glucose and insulin levels spike after eating but then decline with time. (C) is incorrect because Patient C seems to have type I diabetes, since the consumption of glucose does not trigger an increase in his insulin levels. (D) is incorrect because only Patient A shows a response consistent with type II diabetes.

5. B Learning Objective: 8.4

The question stem indicates that a higher concentration of acetylcholine (resulting from the administration of a cholinesterase inhibitor) is able to reverse the effects of the poison. This suggests that tubocurarine and acetylcholine compete for the same site on the ion-gated channel. **(B)** is thus correct. (A) is incorrect because acetylcholine must be released for a cholinesterase inhibitor (a compound that inhibits cholinesterase, an enzyme that breaks down existing acetylcholine) to be capable of increasing acetylcholine concentration. (C) is incorrect because increased acetylcholine could reverse the effects of tubocurarine, which would not happen if the tubocurarine simply blocked the Na^+ influx. (D) is incorrect because the tubocurarine interferes with the effects of acetylcholine, not with the actual contraction mechanism of the muscle.

6. D Learning Objective: 8.3

During the first step of the signal transduction pathway, a ligand, in this case the hormone, must bind to the membrane receptor protein. When the correct ligand is bound, the receptor proteins undergoes a conformational change, which sets off an intracellular cascade. Thus, **(D)** is correct. (A) is incorrect because the image does not depict the membrane bound protein undergoing a conformational change to form a small pore for an ion to travel through when attached to the ligand. Rather, it shows a signaling cascade as the receptor activates a molecule which, in turn, has its own effects. (B) occurs after the conformational change occurs and is followed by (C), the activation or deactivation of certain genes or transcription factors. Both of these choices are incorrect because they do not occur immediately as the hormone binds to the receptor.

7. C Learning Objective: 8.4

While all four answer choices provide examples of drug therapies that are currently in use, **(C)** fails to demonstrate how insulin therapy utilizes the principle behind signal transduction pathway blockage, making it correct. (A) is incorrect because, by binding the receptors, antihistamines block the pathway initiated when histamine binds. Similarly, (B) is incorrect because, by binding to the ErbB-2 receptor, Trastuzumab limits the amount of actual ErbB-2 that are able to bind, helping to limit the excessive cell growth seen in breast cancer patients. (D) is incorrect because, by blocking voltage-gated calcium channels, Felodipine is not allowing the signal, and therefore the calcium, to enter the cell and affect cellular function.

8. D Learning Objective: 8.5

The communication that occurs between the immune cells in the question stem is dependent on the ability of the two cells to share particles. **(D)** presents the only cell junction that allows for the flow of materials between cells, making it correct. Desmosomes and anchoring junctions are integral for keeping cells in close proximity with each other, but they do not provide a mechanism for material exchange to occur, eliminating (A) and (C). Tight junctions bind cells together so tightly that a total barrier to transport and diffusion occurs in areas where they exist. They prevent molecules from going where they do not belong. Thus, (B) is also incorrect.

9. B Learning Objective: 8.2

The presence of selective pressures, such as predators, diseases, and environmental threats, impacts which signaling pathway cells choose. If the selective pressure is neutral, the current signal transduction pathway will continue since the resulting population is already capable of survival. However, if the selective pressure is not neutral, the species will experience natural selection for specific phenotypes, adapting to meet the new challenges. In a normal cell, apoptosis is prevented by inhibiting the expression of pro-apoptotic genes and increasing the expression of anti-apoptotic genes. However, when a cell undergoes excessive DNA damage and is unable to repair it sufficiently, the opposite occurs, causing the cell to actually promote the expression of pro-apoptotic factors. Thus, **(B)** is correct. (A) is incorrect because the upregulation of pro-apoptotic proteins would not be a result of an inhibitory response. (C) and (D) are incorrect because the question stem does not indicate how the toxin inhibits DNA repair. Increasing the concentration of DNA repair proteins would only be useful if the toxin were a competitive inhibitor.

10. C Learning Objective: 8.1

Inhibitory responses to signal transduction involve inhibitors that can block signals from one molecule to another or lower enzyme activity, while stimulatory responses usually involve activators or growth factors that increase gene expression, cell growth, or cell proliferation. Many processes can trigger both stimulatory and inhibitory responses. For example, epinephrine activates glycogen breakdown while simultaneously inhibiting the enzyme that catalyses glycogen synthesis. By increasing the production of a protein that increases the duration of interphase, that protein is subsequently inhibiting the entry of that cell into mitosis, making **(C)** correct. All forms of signal transduction, whether inhibitory or stimulatory, lead to the increased production of an effector protein, eliminating (B) and (D). (A) is incorrect because it is a false statement. As stated in the stimulus, if an a cell detects the presence of an α cell or α factor, it will halt entry into mitosis. Therefore, the time spent in interphase will be increased, not decreased.

PART 4

Processes of Life

CHAPTER 9

Enzymes

LEARNING OBJECTIVES

In this chapter, you will review how to:

9.1 Explain how enzymes and substrates interact

9.2 Explain how cofactors or coenzymes affect enzyme function

9.3 Contrast competitive and noncompetitive inhibition

9.4 Explain how concentration affects enzyme function

9.5 Investigate how enzyme reactivity is affected by concentration

TEST WHAT YOU ALREADY KNOW

1. Amplex red (AR) is an indicator used to monitor the activity of peroxidases. In the presence of peroxidases, AR reacts with hydrogen peroxide to produce a fluorescent red product, resorufin, which can be measured with a fluorometer. The figure shows an experiment that examined the rate of a reaction at different concentrations of H_2O_2 and AR.

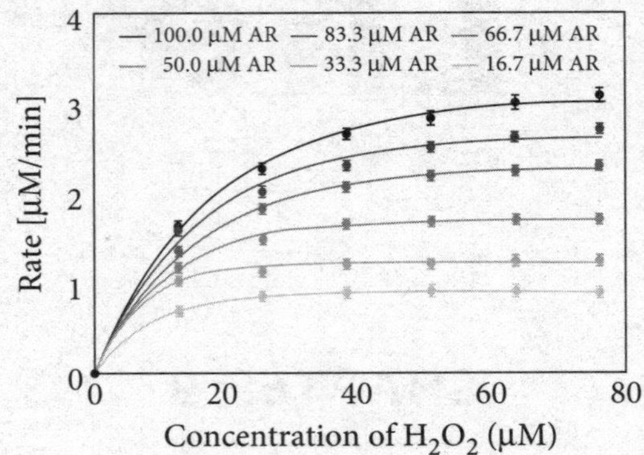

Adapted from Hoon Suk Rho et al., "Mapping of Enzyme Kinetics on a Microfluidic Device," *PLoS ONE* 11, no. 4 (April 2016): e0153437.

Why do the curves level off at a concentration slightly above 20 μM of H_2O_2?

(A) The reaction rate levels off as the temperature increases because most enzymes are proteins and denature at high temperatures.

(B) The reaction rate levels off as the substrate concentration decreases; as the H_2O_2 concentration increases towards the right of the graph, the concentration of AR decreases simultaneously.

(C) The induced fit model states that the active site changes shape as the substrate binds. This is not possible when the H_2O_2 concentration becomes too high.

(D) During an enzymatic reaction, the substrate binds to the active site of an enzyme. This means that the enzymes can only catalyze one reaction at a time, which limits the reaction rate.

Processes

2. Glucose-6-phosphate dehydrogenase (G6PDH) is an enzyme that can use NAD^+ or $NADP^+$ as a cofactor. The table shows the K_m and k_{cat} of this enzyme with the two cofactors. The K_m represents the substrate concentration at which the reaction rate is half of the maximum rate. k_{cat} is known as the turnover number and represents how quickly the enzyme processes substrate molecules.

Cofactor	**K_M (μM)**	**k_{cat} (s^{-1})**
$NADP^+$	243 ± 5.3	282 ± 1
NAD^+	8352 ± 307	298 ± 4

Adapted from M. Fuentealba et al., "Determinants of Cofactor Specificity for the Glucose-6-Phosphate Dehydrogenase from Escherichia coli: Simulation, Kinetics and Evolutionary Studies," *PLoS ONE* 11, no. 3 (March 2016): e0152403.

Based on the data in the table, which cofactor yields the more efficient enzyme, and why?

(A) $NADP^+$, because the reaction reaches maximum velocity at a lower substrate concentration and the turnover rate is higher

(B) $NADP^+$, because the reaction reaches maximum velocity at a lower substrate concentration and the turnover rate is lower

(C) NAD^+, because the reaction reaches maximum velocity at a lower substrate concentration and the turnover rate is higher

(D) NAD^+, because the reaction reaches maximum velocity at a lower substrate concentration and the turnover rate is lower

3. In a recent study, researchers examined the effects of an inhibitor to determine how it affected the activity of angiotensin I-converting enzyme. On the graph below, the maximum velocity of the reaction is represented by the y-intercept. The substrate concentration at which the reaction proceeds at half-maximum velocity is represented by the x-intercept.

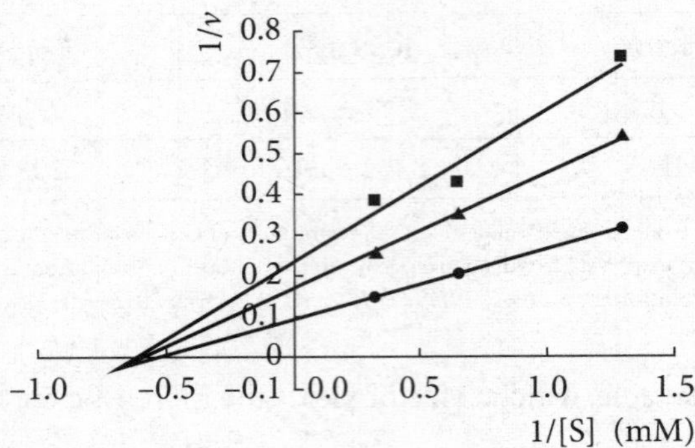

Adapted from He Ni et al., "Inhibition Mechanism and Model of an Angiotensin I-Converting Enzyme (ACE)-Inhibitory Hexapeptide from Yeast (*Saccharomyces cerevisiae*)," *PLoS ONE* 7, no. 5 (May 2012): e37077.

Based on this data, is this a competitive inhibitor or a noncompetitive inhibitor, and why?

(A) The x-intercept is the same for all of the reactions, so the maximum velocity must be the same with and without the inhibitor. This is competitive inhibition because high concentrations of substrate can outcompete a competitive inhibitor.

(B) The y-intercept is different for all of the reactions, so the maximum velocity must be lower with the inhibitor. This is noncompetitive inhibition because high concentrations of substrate can outcompete a competitive inhibitor and reach the same maximum velocity.

(C) The y-intercept is the same for all of the reactions, so the maximum velocity must be the same with and without the inhibitor. This is competitive inhibition because high concentrations of substrate can outcompete a competitive inhibitor.

(D) The x-intercept is different for all of the reactions, so the maximum velocity must be less with the inhibitor. This is noncompetitive inhibition because high concentrations of substrate can outcompete a competitive inhibitor and reach the same maximum velocity.

4. Researchers are studying the reaction rate of mevalonate-5-phosphate (Mev-P), an enzyme. The graph below shows the rate of the enzyme-catalyzed reaction as the initial concentration of Mev-P is increased.

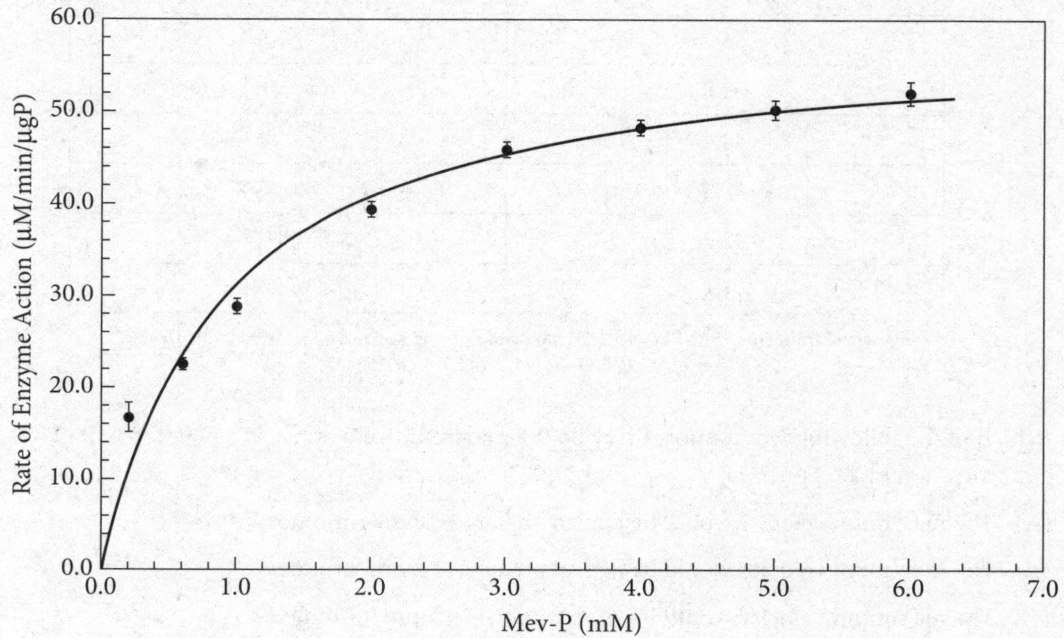

Adapted from David E. Garcia and Jay D. Keasling, "Kinetics of Phosphomevalonate Kinase from *Saccharomyces cerevisiae*," *PLoS ONE* 9, no. 1 (January 2014): e87112.

The graph is plotted using the initial concentration and reaction rate. Which of the following provides the most likely reason why the researchers used these initial values instead of values determined after the reaction had been allowed to progress?

(A) At the beginning of the reaction, the enzyme works more slowly. Over time, the reaction increases in rate as the concentration of the substrate goes down.

(B) The reaction will generate heat, which will increase the reaction rate as the reaction progresses. This will make the reaction appear to proceed faster than expected.

(C) At the start of the reaction, the researchers knew the exact concentrations of all species in the reaction. As the reaction proceeded, it would be more difficult to determine exact concentrations.

(D) To save time, the researchers measured the initial rate of reaction so they do not need to wait for the reaction to progress.

5. The table shows an experiment that examined the reaction rate of peroxidase with hydrogen peroxide. V_{max} refers to the maximum velocity of the reaction.

Concentration of H_2O_2 (μM)	**V_{max} (μM/minute)**
66.7	9.7 ± 1.2
55.6	8.3 ± 1.1
44.4	6.7 ± 0.7
33.3	5.3 ± 0.4
22.2	3.9 ± 0.2
11.1	2.3 ± 0.1

Adapted from Hoon Suk Rho et al., "Mapping of Enzyme Kinetics on a Microfluidic Device," *PLoS ONE* 11, no. 4 (April 2016): e0153437.

Which of the following would most likely be the maximum velocity of this reaction at an H_2O_2 concentration of 77.8 μM?

(A) The maximum velocity would be approximately 12.8 μM/minute.

(B) The maximum velocity would be approximately 11.2 μM/minute.

(C) The maximum velocity would be approximately 9.7 μM/minute.

(D) The maximum velocity would be approximately 8.9 μM/minute.

Answers to this quiz can be found at the end of this chapter.

ENZYME STRUCTURE

9.1 Explain how enzymes and substrates interact

9.2 Explain how cofactors or coenzymes affect enzyme function

Binding

Most **enzymes** are proteins with specific 3-D structures that allow them to bind to very particular molecules (called **substrate** molecules) and increase the rate of reactions between these molecules. In many cases, enzymes bind to larger molecules and break them into smaller ones. Enzymes can synthesize or break down molecules at a rate of thousands or millions per second—they are extremely fast-acting! They allow reactions to occur that either would not take place or would take place far too slowly under normal conditions to be useful.

Enzymes are very specific for the molecules they bind to and the reactions they catalyze. Each enzyme has a name that usually indicates exactly what it does, and the name often ends in *-ase*. For example, the lactase enzyme breaks the complex sugar lactose into the simple sugars glucose and galactose. The enzyme pyruvate decarboxylase removes a carbon from the three-carbon molecule pyruvate. Thinking about enzyme names in this way may be helpful on the AP Biology exam.

For the exam, you should be familiar with the term *catalyst*, which refers to any chemical agent that accelerates a reaction without being permanently changed in the reaction. Enzymes are biological catalysts, which can be used over and over again.

Enzyme specificity means that an enzyme will only catalyze one specific reaction. Molecules upon which an enzyme acts are called substrates, and the substrate binds to the active site on the enzyme, speeding up the conversion from substrate to product.

All enzymes possess an active site, a 3-D pocket within their structure in which substrate molecules can be held in a certain orientation to facilitate a reaction. The two models of enzyme-substrate interaction are shown below. In the *lock-and-key model*, the spatial structure of an enzyme's active site is exactly complementary to the spatial structure of the substrate so that the enzyme and substrate fit together like a lock and key. In the *induced fit model*, the active site has a flexibility that allows the 3-D shape of the enzyme to shift to accommodate the incoming substrate molecule.

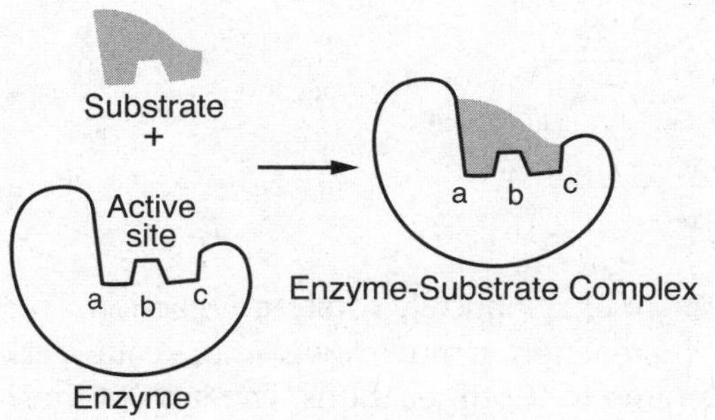

Lock-and-Key Model

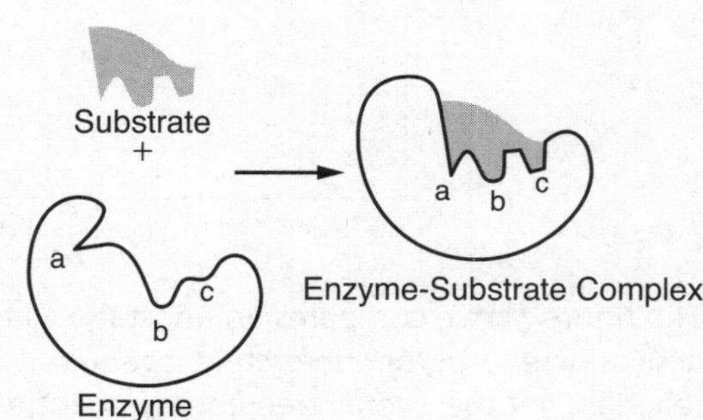

Induced Fit Model

Cofactors and Coenzymes

Most enzymes require cofactors to become active. **Cofactors** are nonprotein (inorganic) species that either play a role in binding to the substrate or stabilize the enzyme's active conformation. Two examples of cofactors are zinc and the iron in hemoglobin.

Coenzymes are other organic molecules that play a similar role. Most coenzymes cannot be synthesized by the body but are obtained from the diet as **vitamins**.

Without the appropriate cofactor or coenzyme, an enzyme will be less capable of catalyzing its reaction and may even lose its function entirely. As a consequence, compounds that must be synthesized to maintain homeostasis and growth could become deficient or waste products that, should they be broken down, could accumulate to toxic levels. This is why vitamins and minerals are such essential aspects of nutrition.

ENZYME REGULATION

9.3 Contrast competitive and noncompetitive inhibition

9.4 Explain how concentration affects enzyme function

Thermodynamics and Kinetics

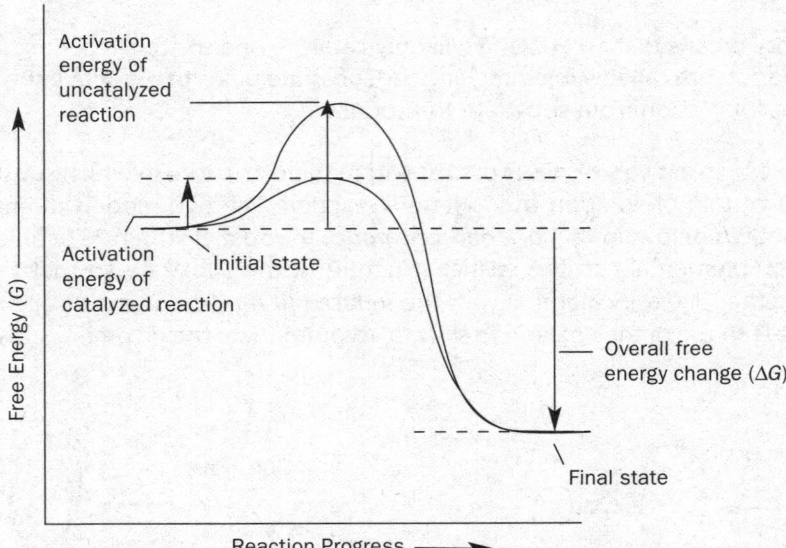

The figure above compares an uncatalyzed reaction with an enzymatically catalyzed reaction. The activation energy, or energy that is required to start up the reaction, is much lower in the catalyzed reaction, yet the overall free energy change (ΔG) is the same for both reactions. The laws of thermodynamics can be used to predict if a reaction will occur or not. If products have less free energy (G) than the reactants, the reaction has an overall negative change in G ($\Delta G < 0$) and will occur spontaneously. If products have more free energy than reactants, the reaction is an uphill one, needing a great deal of supplied energy to occur. Free energy is a measure of the potential energy

Processes

of the molecules in a reaction. Those starting out with high potential energy, or higher G, are more likely to react and lower their G through the reaction than vice versa. What that means is that reactions having a $\Delta G < 0$ are deemed favorable. Keep in mind that most biosynthetic reactions have $\Delta G > 0$ and will not occur spontaneously without the help of both enzymes and ATP.

Even though thermodynamics and ΔG alone may predict that a reaction is favorable or can occur spontaneously, the kinetics, or rate, of the reaction may be so slow that these reactions are not feasible for living systems. Sure, a hamburger will break down eventually if exposed to enough acid in your stomach, but without digestive enzymes to help speed up this breakdown, the hamburger might take months to break down sufficiently to be useful to you. Thus, although hamburger breakdown may be spontaneous, the limiting factor in the reaction is the reaction rate, which is dependent on energy being provided to start this breakdown. This energy is the activation energy, and enzymes provide a foundation on which molecules can react so that the energy needed to start a reaction is not as great as it would have to be without the enzymes' presence.

Inhibition

Cells must regulate enzyme action to keep these rapid reactions under control. Most enzymes are inactive most of the time, and enzyme pathways are regulated in complex fashions to ensure efficiency and safety. Enzymes can be regulated by inhibitors, molecules that bind to the enzyme either at the active site or the allosteric (regulatory) site.

Competitive inhibition occurs when the inhibitor and the substrate compete for the active site. This type of inhibition can be overcome by increasing the concentration of substrate.

Noncompetitive inhibition occurs when the inhibitor binds to the allosteric site, inducing a conformation change in the enzyme, rendering the active site inactive. This type of inhibition cannot be overcome by adding more substrate because the enzyme's shape has been altered. Therefore, noncompetitive inhibition may or may not be irreversible. Keep in mind that enzymes do not alter reaction equilibrium or affect the free energy of a reaction. They accelerate the forward and reverse reactions by the same factor.

Feedback inhibition occurs when the end product of an enzyme-catalyzed reaction works to block the activity of the enzymes that started the reaction in the first place. (This is an example of negative feedback, as distinct from positive feedback, which amplifies a particular bodily response.) In the following diagram, feedback inhibition would occur if product C could bind to the active site of enzyme 1 (this would be a kind of competitive inhibition), thereby preventing any more of compound A from becoming compound B.

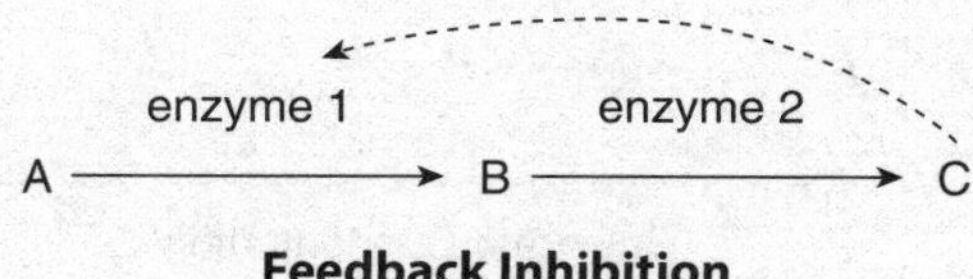

Feedback Inhibition

Physiological reactions can take place without enzymes, but they would take much longer to proceed. Though enzymes are neither changed nor consumed during the reaction, reaction conditions such as high temperatures, detergents, or acidic/basic conditions can cause enzymes to denature (lose their 3-D structure) and thereby lose their activity.

As a general rule, the rate of an enzyme will increase with increasing temperature but only up to a point. If the temperature increases too much, then the enzyme will become denatured. Moreover, enzymes are active only within a specific **pH** range. In the human body, most enzymes work best around neutral (pH = 7).

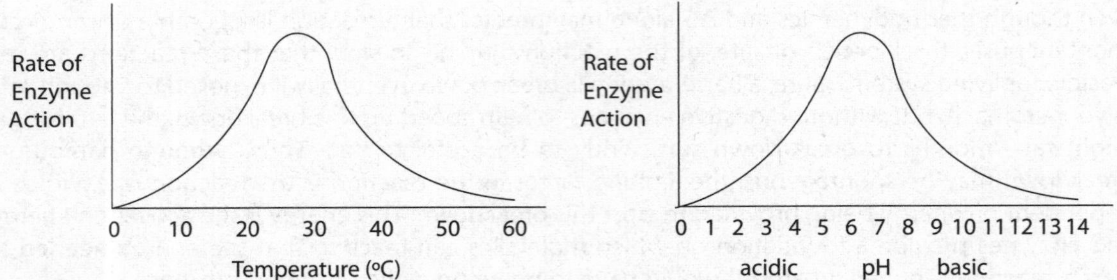

Temperature- and pH-Dependent Enzyme Reaction Rates

Effects of Concentration

Reaction rates increase as more and more enzyme is added to a particular environment. If the enzyme concentration is kept constant, the reaction rate will plateau at a maximum speed even as substrate concentration increases, because the enzymes can only work so fast. This maximum reaction rate is termed V_{max} and is illustrated in the following graph as the flat part of the rate versus substrate concentration curve for an enzymatic reaction. This point occurs when the enzymes become saturated with substrate. Enzymes become saturated at high substrate concentrations because substrate must bind to enzymes at a particular place: the active site. While the entire process is somewhat complex, the main idea is that increasing the substrate can produce the same V_{max} if competitive inhibitors are present, but noncompetitive inhibitors lower V_{max} regardless of the amount of substrate.

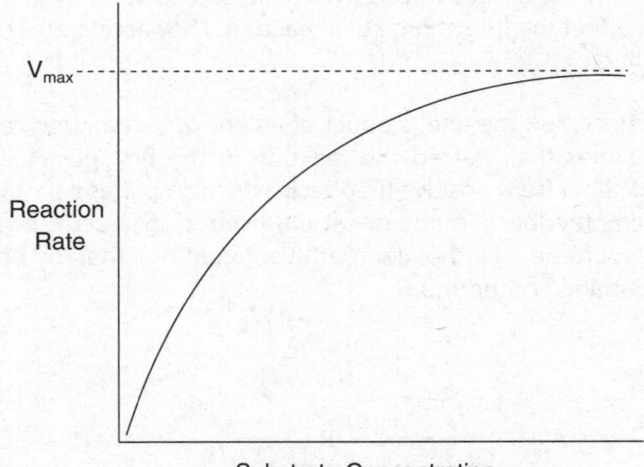

Concentration and Enzyme Reaction Rates

Increasing the concentration of the enzyme itself will also speed up the rate of the reaction. However, note that the synthesis of enzymes requires energy and resources that could be deployed elsewhere in an organism, so there is a trade-off to producing more copies of a particular enzyme.

AP BIOLOGY LAB 13: ENZYME ACTIVITY INVESTIGATION

High-Yield

9.5 Investigate how enzyme reactivity is affected by concentration

The enzyme catalysis lab is designed to allow you to explore structural proteins and environmental conditions that affect chemical activity in the natural world. In the lab, you will measure the chemical activity of the enzyme catalase as it breaks down hydrogen peroxide (H_2O_2) into water and oxygen gas. Specific experimental conditions—temperature, substrate and enzyme concentrations, pH, and so on—will be altered to help you understand the effects of environmental conditions on a specific chemical reaction. This experiment helps you to learn the experimental method through manipulation of experimental treatments compared with a control treatment.

✔ AP Expert Note

Graphing

For labs requiring graph construction and calculations, make sure that your graph is labeled thoroughly and accurately. Make sure you identify appropriate axes, show units, plot points accurately, and assign your graph a title. You should also follow these steps when answering AP Biology free response questions that require you to create graphs.

Understanding of the experiment itself can be encapsulated in just a few graphs. One or more of these graphs may be on the exam. The first graph is about free energy and enzyme activity.

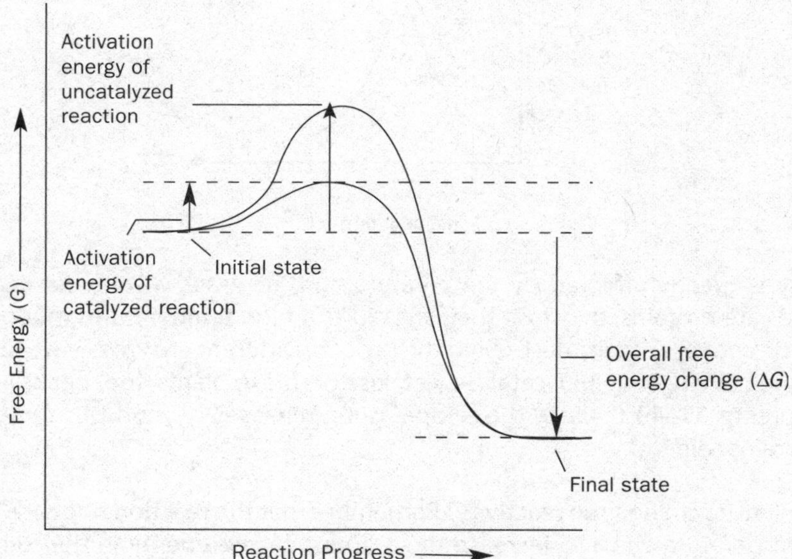

Before a reaction can take place, it must reach a point called its activation energy by receiving enough energy from the environment, termed free energy, in the form of heat or kinetic energy. An enzyme catalyzes a reaction by lowering the activation energy needed (it requires less free energy) to allow the reaction to take place. Hydrogen peroxide breaks down into water and oxygen gas on its own, but at an incredibly slow rate. The enzyme catalase lowers the activation energy of the reaction and the reaction happens very quickly.

Processes

✔ AP Expert Note

Many reactions that occur in biology can be sped up through the addition of enzymes that lower activation energy. Natural compounds readily metabolize and synthesize in the presence of certain enzymes, and biological systems control chemical activity largely through synchronizing the presence or absence of enzymes. Enzymes also control the release of energy so that cells don't combust during certain reactions from a sudden influx of energy.

Notice on the graph that the amount of free energy in the system changes over the progress of the reaction. Reactants of the reaction have an initial amount of energy, receive energy from the environment, and then energy is released with the products. Remember that the energy of the reactants and the energy of the products remain at the same values, whether the reaction is enzyme mediated or not. All the enzyme does is lower the activation energy for the molecule to do "what comes naturally" and make the reaction happen. When the reaction is complete, the enzyme is free and helps to speed up the reaction for another H_2O_2 molecule. This lab experiment demonstrates firsthand what occurs in all living systems on a regular basis and points out the importance of these specialized proteins.

The environment can also have a profound effect on the reaction. In this experiment, you measure the experimental effect of changing the reaction temperature and pH, as well as the effect of changing concentrations of the enzyme and substrate.

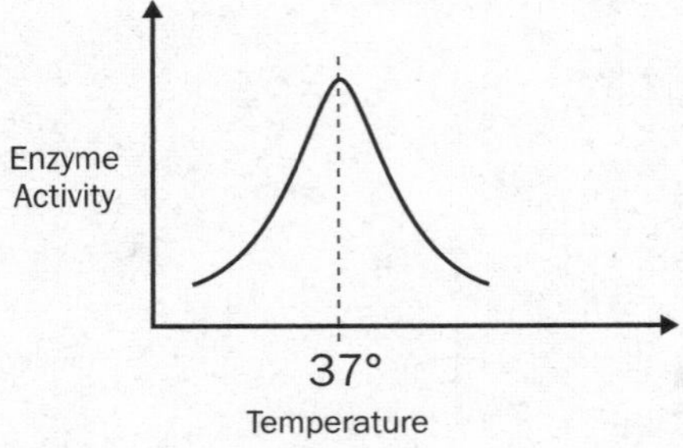

Catalytic activity is greatly affected by temperature and increases with increasing temperature. Because enzymes are proteins, they lose their structure at high temperatures, not only eliminating catalytic properties but essentially destroying the protein. Different enzymes have different specific temperatures. The activity of animal catalase (catalase occurs in plants, too) peaks at about normal body temperature, or 35–40°C. Once the temperature increases beyond this temperature range, the catalase proteins "die."

How does temperature change the reactivity? Remember that the reaction depends on free energy, so increasing temperature will also increase the amount of free energy in the form of both heat and kinetic energy (all the molecules will be moving faster). Because all of the molecules in the system will be moving faster at a higher temperature, there is an increased rate of collision between enzyme and substrate molecules. In essence, there are more chance meetings.

The pH of the environment affects the reactivity of enzymes in the same way that temperature does, in that there is a level of peak activity.

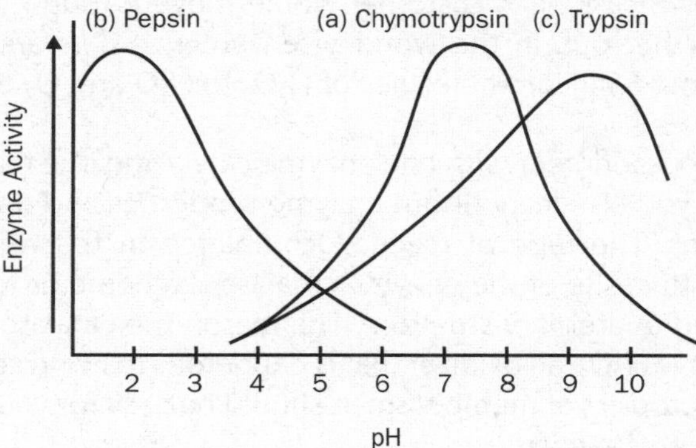

In the above graph, you can see the peak activity for different common digestive enzymes. Pepsin, for example, has a peak activity at a very low pH because it catalyzes reactions in the stomach, which is very acidic. The peak activity for chymotrypsin occurs at a pH of 7 to 8, which is fairly neutral. The pH of a system affects protein activity by altering the structure of the protein. Proteins have a tertiary structure based on the electrochemical properties of the amino acids in the chain. The charges on these amino acids can change as the pH changes because of the available H^+ and OH^- ions.

The last things you need to understand from this lab are the effects of changing the amount of enzyme and substrate available.

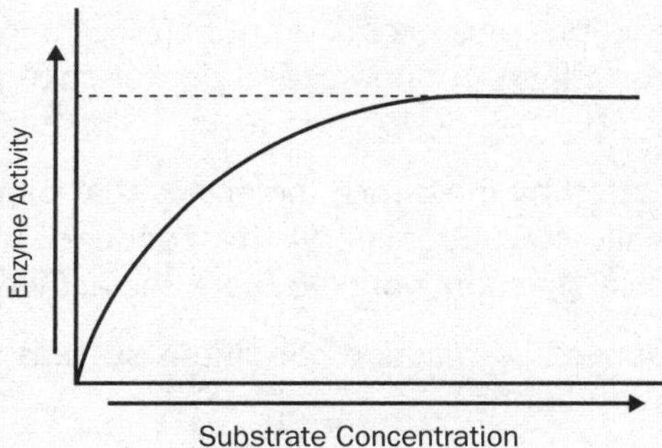

You may think that as you add more H_2O_2, you get more reaction. You would be right ... to a point. Eventually, no matter how much substrate you add, the reaction maxes out. This is because there are a limited number of enzyme molecules in the system. These enzymes can handle only one molecule of H_2O_2 at a time. Think of enzymes as cashiers at the grocery store. If several customers want to check out at the same time, they can only be helped one at a time by any single cashier. Once all of the cashiers are busy, a customer has to wait for the next available cashier. Thus, H_2O_2 has to wait for the next available enzyme to catalyze the reaction.

Changing the enzyme concentration affects the reaction in a similar way, except the reaction becomes almost instantaneous as the number of enzyme molecules becomes equal to the number of substrate molecules. An increase in enzyme concentration is like every customer in the grocery store having his or her own cashier. A customer would move through the line as quickly as the cashier can check him or her out. In the world of enzymes, orders are processed very quickly. Catalase can convert almost 6 million molecules of H_2O_2 to H_2O and O_2 each minute.

By completing this simple experiment with one enzyme catalyzing one reaction, you not only learn about catalase and H_2O_2, you also learn about enzyme properties and the effect the environment has on chemical reactions. The type of reaction completed in the experiment was oxidation-reduction, in which the active site of the enzyme catalase is a heme group like in hemoglobin, so it has a similar tertiary and quaternary structure. The reason the catalase enzyme exists in animal and plant tissue is to remove H_2O_2, an oxidizer that is dangerous to living cells and must be removed when it is created as a byproduct of metabolism. It should be easy for you to see now how biology is intricately connected with chemistry.

RAPID REVIEW

If you take away only 4 things from this chapter:

1. All enzymes possess an active site, a 3-D pocket within their structures, in which substrate molecules can be held in a certain orientation to facilitate a reaction. The two models of enzyme-substrate interaction are lock-and-key and induced fit.
2. Enzymes lower the activation energy of a reaction, do not get used up in the reaction, catalyze millions of reactions per second, do not affect the overall free-energy change of the reaction, increase the reaction rate, and do not change the equilibrium of reactions.
3. Enzymes can be regulated by inhibitors, molecules that bind to the enzyme either at the active site or the allosteric (regulatory) site. Feedback inhibition is when the end product of a biochemical reaction works to block the activity of the original enzyme.
4. Enzymes can be influenced by reaction conditions such as high temperatures, detergents, or acidic/basic conditions.

TEST WHAT YOU LEARNED

1. Researchers are studying the reaction rate of mevalonate-5-phosphate (Mev-P), an enzyme. The graph below shows the rate of the enzyme-catalyzed reaction as the initial concentration of Mev-P is increased.

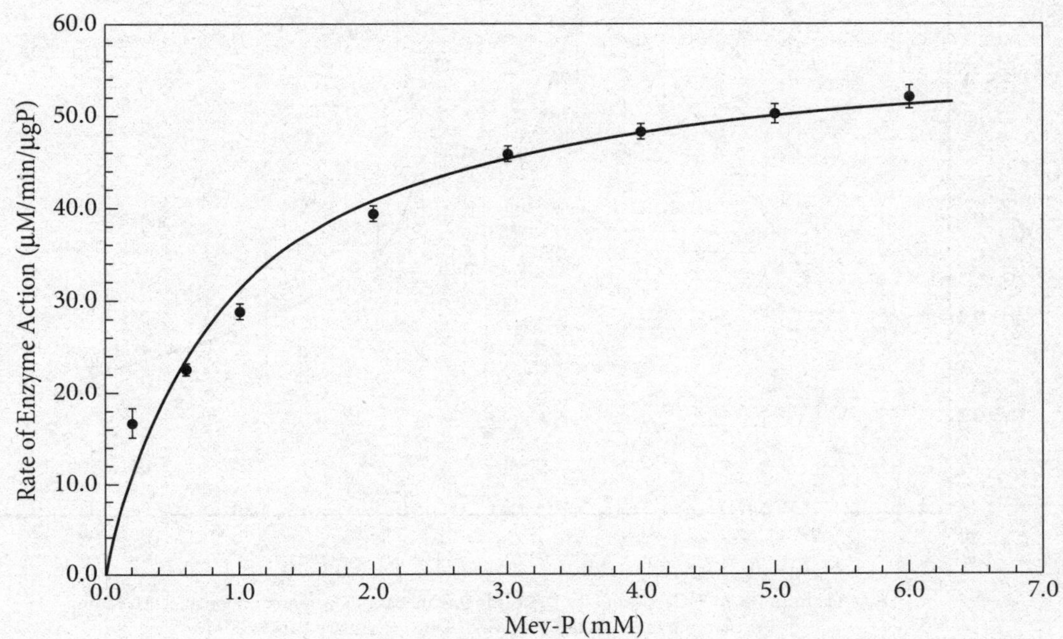

Adapted from David E. Garcia and Jay D. Keasling, "Kinetics of Phosphomevalonate Kinase from *Saccharomyces cerevisiae*," *PLoS ONE* 9, no. 1 (January 2014): e87112.

What would happen if the concentration of Mev-P was doubled from 6.0 mM to 12.0 mM?

(A) The reaction rate would increase significantly.

(B) The reaction rate would decrease sharply.

(C) The reaction rate would not change much.

(D) The reaction rate would change in an unpredictable way.

2. The figure below shows how the rate of an enzymatic reaction changed as the pH of the reaction changed.

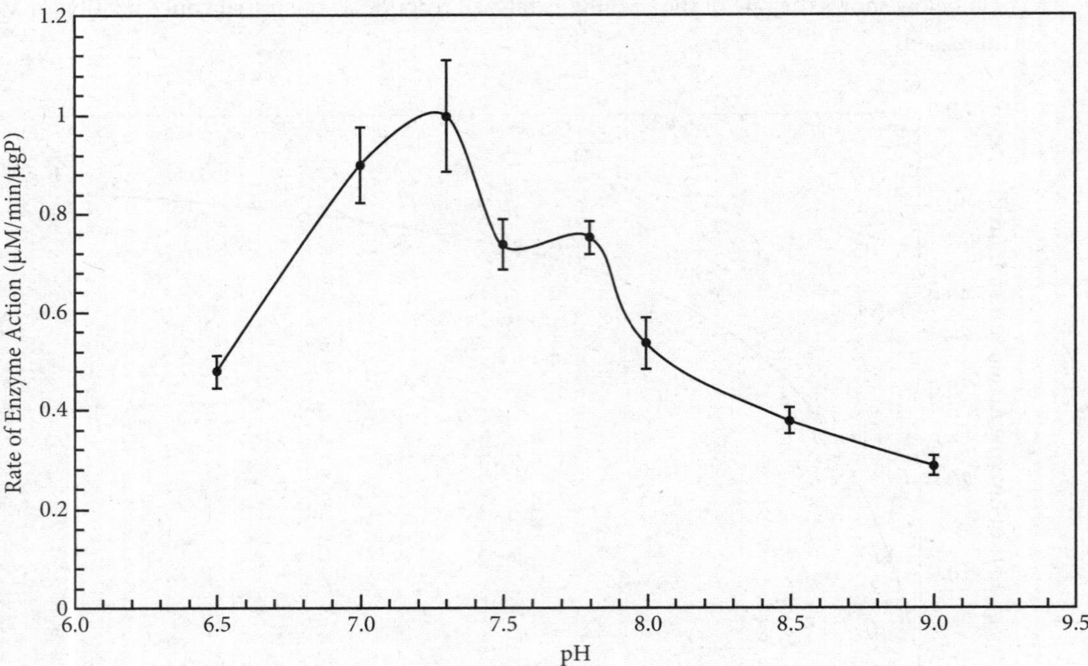

Adapted from David E. Garcia and Jay D. Keasling, "Kinetics of Phosphomevalonate Kinase from *Saccharomyces cerevisiae*," *PLoS ONE* 9, no. 1 (January 2014): e87112.

How would increasing the concentration of substrate most likely affect the rate of reaction at a pH of 7.75?

(A) The reaction rate would be lower, regardless of how the reaction rate compared with the maximum possible rate.

(B) The reaction rate would be higher unless the reaction was already at the maximum rate for that pH.

(C) Because the reaction rate at a pH of 7.75 appears to be unusually favorable, the reaction rate probably cannot increase further if substrate is added.

(D) Because the reaction rate at a pH of 7.75 is lower than the maximum rate shown at a pH of 7.25, the reaction rate will increase back to the maximum of 1.

3. Glucose-6-phosphate dehydrogenase (G6PDH) is an enzyme that can use NAD^+ or $NADP^+$ as a cofactor. The table shows the K_m and k_{cat}/K_m of a wild type and mutant version of G6PDH. The K_m represents the substrate concentration at which the reaction rate is half of the maximum rate. k_{cat}/K_m is a measure of substrate specificity that indicates the strength of preference for a particular enzyme. Data for both cofactors are shown below.

	$NADP^+$		NAD^+	
G6PDH Type	K_M (μM)	k_{cat}/K_M ($\mu M^{-1}s^{-1}$)	K_M (μM)	k_{cat}/K_M ($\mu M^{-1}s^{-1}$)
Wild type	7.5 ± 0.8	23.2 ± 2.4	5090 ± 400	$0.06 \pm 4 \times 10^{-3}$
Mutant	17696 ± 1453	$0.01 \pm 9 \times 10^{-4}$	11736 ± 804	$0.01 \pm 7 \times 10^{-4}$

Adapted from M. Fuentealba et al., "Determinants of Cofactor Specificity for the Glucose-6-Phosphate Dehydrogenase from *Escherichia coli*: Simulation, Kinetics and Evolutionary Studies," *PLoS ONE* 11, no. 3 (March 2016): e0152403.

Based on the data in the table, which of the following statements is true?

(A) The wild type enzyme works less efficiently with NAD^+ than the mutant enzyme. With this cofactor, it reaches the maximum velocity at a much higher substrate concentration than the mutant enzyme and also has higher substrate specificity.

(B) The wild type enzyme works more efficiently with NAD^+ than the mutant enzyme. With this cofactor, it reaches the maximum velocity at a much higher substrate concentration than the mutant enzyme and also has lower substrate specificity.

(C) The wild type enzyme works less efficiently with $NADP^+$ than the mutant enzyme. With this cofactor, it reaches the maximum velocity at a much lower substrate concentration and also has lower substrate specificity.

(D) The wild type enzyme works more efficiently with $NADP^+$ than the mutant enzyme. With this cofactor, it reaches the maximum velocity at a much lower substrate concentration than the mutant enzyme and also has higher substrate specificity.

4. Protein kinase C (PKC) can be inhibited by multiple inhibitors. Some are competitive inhibitors, but others are not. Which of the findings below would suggest competitive inhibition?

(A) The reaction reaches the same maximum velocity with or without the inhibitor, although at different substrate concentrations.

(B) The inhibitor is found to bind to the enzyme relatively far from its active site.

(C) Increasing the concentration of the substrate without increasing the concentration of the enzyme has no effect on the reaction rate.

(D) Raising the temperature increases the reaction rate until an optimal temperature is reached, at which point the reaction rate decreases again.

5. Researchers studied the activity of mevalonate kinase, an enzyme that may be important for biofuel production. The graph below shows the rate of a reaction as the concentration of ATP is increased while substrate concentration remains constant. The open circles represent data points collected at 30°C and the closed circles represent data points collected at 37°C.

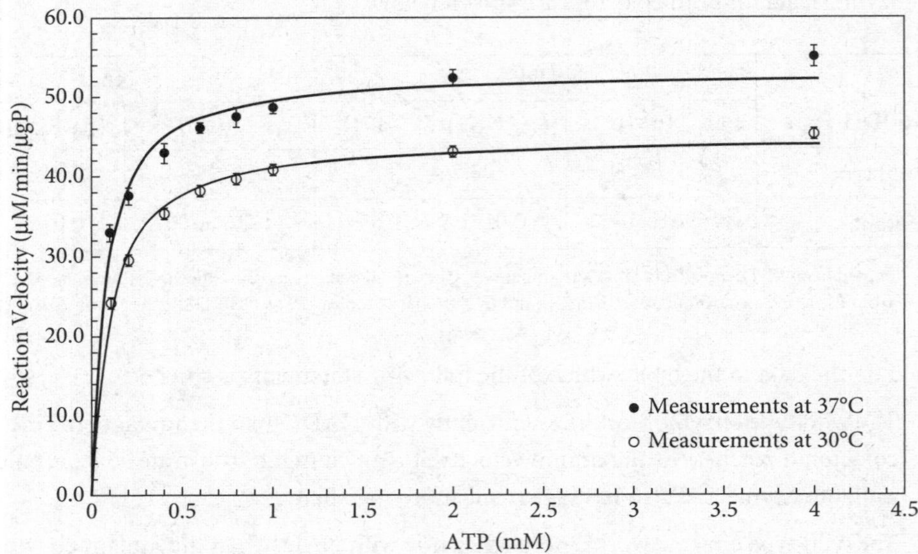

Adapted from David E. Garcia and Jay D. Keasling, "Kinetics of Phosphomevalonate Kinase from *Saccharomyces cerevisiae*," *PLoS ONE* 9, no. 1 (January 2014): e87112.

How is the reaction between enzyme and substrate modified by the different temperatures?

(A) At 30°C, the substrate molecules have less kinetic energy than at 37°C. This reduces the rate of random collisions and therefore the rate of the reaction.

(B) At 30°C, the substrate molecules have more kinetic energy than at 37°C. The excess kinetic energy makes it harder for the substrate to bind to the enzyme.

(C) At 37°C, the enzymes have formed the correct shape to maximize the reaction rate. At 30°C, the enzymes have denatured due to excess heat and the substrates have more difficulty binding to the active site.

(D) At 37°C, the concentration of the substrates is higher than at 30°C. This allows the reaction to proceed more rapidly.

6. Which of the following diagrams best represents an enzymatic reaction involving a cofactor?

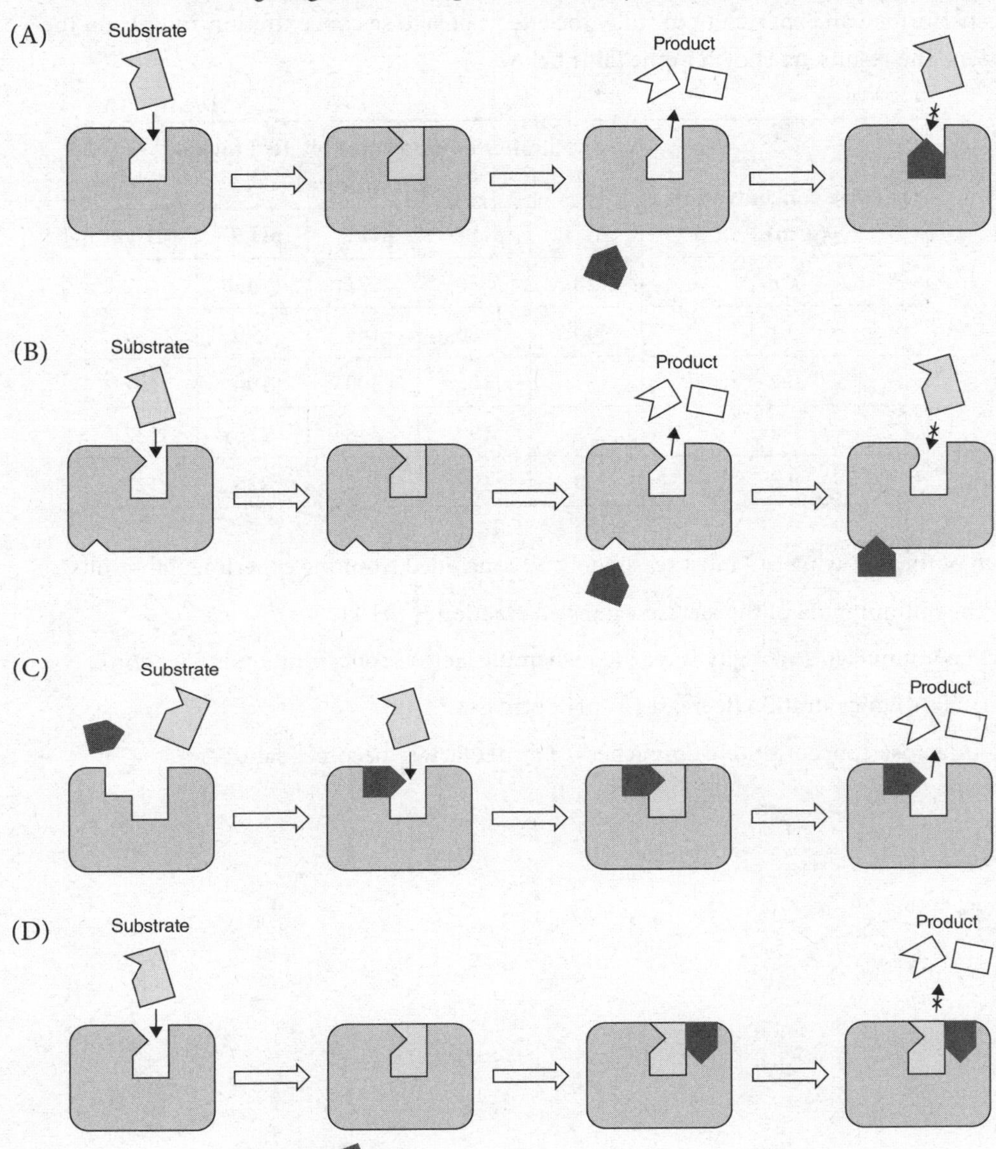

7. Lactase catalyzes the breakdown of lactose into monosaccharides (galactose and glucose). Students performed an experiment to study the effects of lactose concentration and pH on the reaction. The results are shown in the table below.

Lactose concentration (g/mL)	**Number of molecules ($\times 10^6$) formed per minute**				
	pH 3	**pH 5**	**pH 7**	**pH 9**	**pH 11**
0.05	20	36	72	46	32
0.1	40	80	145	90	50
0.2	81	165	300	190	107
0.4	98	199	348	225	122
0.8	98	199	348	225	122

Which of the following can most reasonably be concluded from the experimental results?

(A) The optimum pH of the lactase-catalyzed reaction is pH 11.

(B) Maximum lactase activity is reached when the lactose concentration is 0.05 g/mL.

(C) Lactase concentration decreases as pH decreases.

(D) As lactose concentration approaches 0.4 g/mL, lactase becomes saturated.

8. Amylase catalyzes the breakdown of starch into maltose. Students used the iodine test to investigate the hydrolysis of starch at varying temperatures over time. Adding iodine to starch results in a dark blue-black complex. The results of the reactions are shown in the figure below.

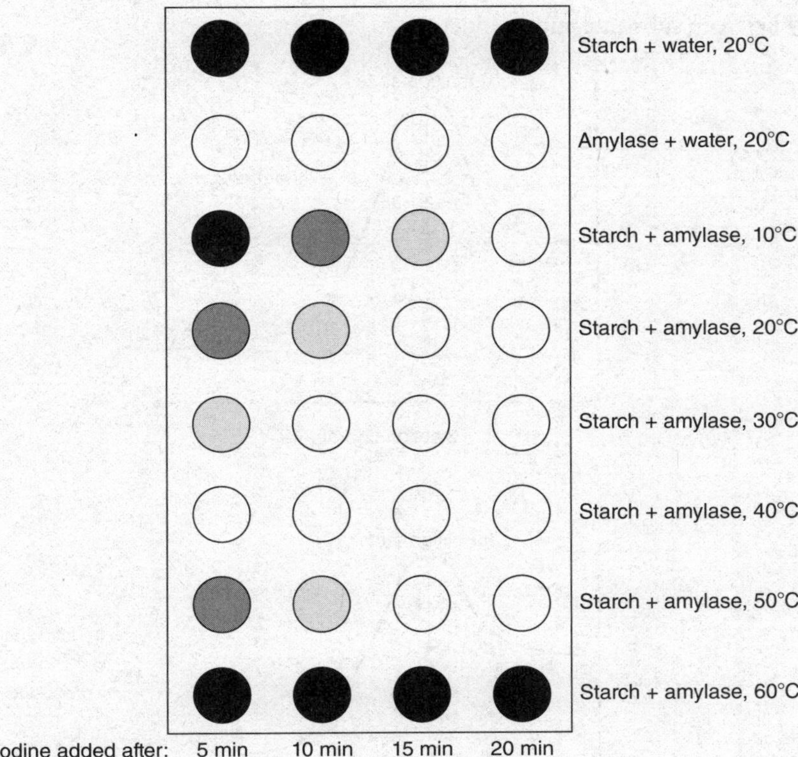

Based on this data, what is the optimum temperature range of amylase?

(A) $21°C \leq t \leq 30°C$

(B) $31°C \leq t \leq 40°C$

(C) $41°C \leq t \leq 50°C$

(D) $51°C \leq t \leq 60°C$

9. Enzymes are biological catalysts that speed up reactions. A substrate binds to an enzyme at its active site (where the chemical reaction occurs) and forms an enzyme-substrate complex. After products are formed, the enzyme returns to its original state. The graphs below show the energy of reactions with and without an enzyme, where E_a is the activation energy and ΔG is the change in free energy between substrate and product.

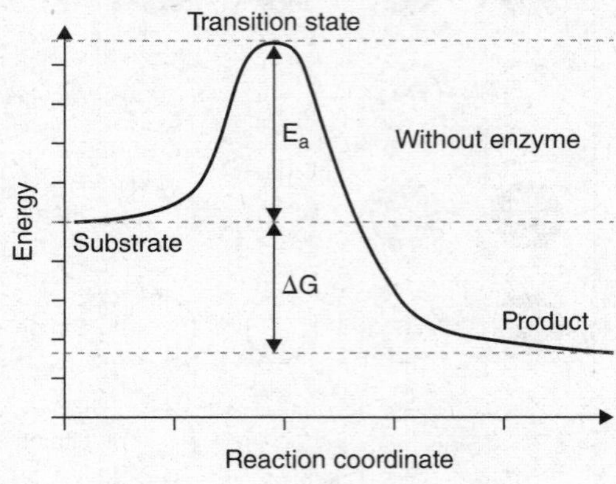

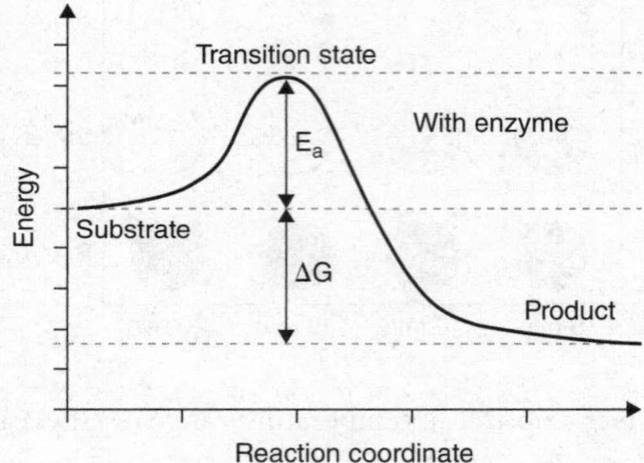

Based on the graphs, which of the following accurately describes how enzymes affect the energy of a reaction?

(A) Enzymes do not change the net energy released in the reaction.

(B) Enzymes speed up reactions by increasing the activation energy.

(C) The energy input required to reach the transition state without an enzyme increases the free energy change of the reaction.

(D) The enzyme-substrate complex facilitates the formation of the transition state to lower the free energy change of the reaction.

10. An Eadie-Hofstee diagram plots the reaction rate (v) of an enzyme as a function of the ratio between rate and substrate concentration ($v/[S]$). The y-intercept is V_{max} (the maximum rate of reaction) and the absolute value of the slope is K_m (the affinity for substrate by an active enzyme). A high K_m indicates low substrate affinity. The figure below shows a plot of competitive, noncompetitive, and no inhibition.

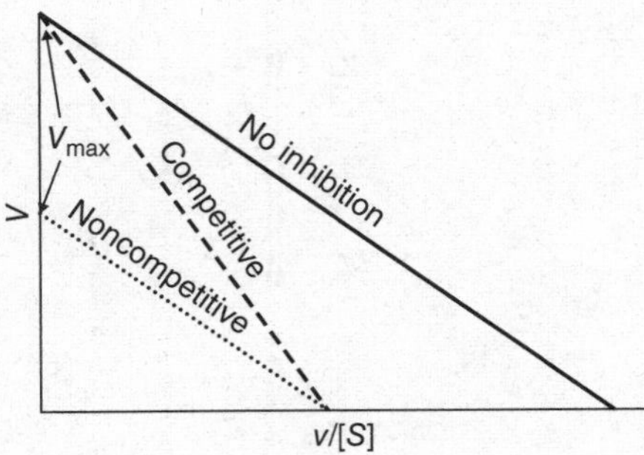

Competitive inhibition prevents the substrate from binding the enzyme, whereas noncompetitive inhibition inactivates the enzyme. Based on the figure, which of the following correctly describes the characteristics of enzyme inhibition?

(A) Noncompetitive inhibition reduces V_{max} and the amount of substrate needed to reach half V_{max}.

(B) Competitive inhibition increases K_m, so more substrate is needed to reach V_{max}.

(C) Noncompetitive inhibition has a lower V_{max} and a higher K_m than competitive inhibition.

(D) In competitive inhibition, a small amount of substrate is needed to saturate the enzyme.

Answers to this quiz can be found at the end of this chapter.

Answer Key

Test What You Already Know

1. D
2. B
3. B
4. C
5. B

Test What You Learned

1. C
2. B
3. D
4. A
5. A
6. C
7. D
8. B
9. A
10. B

Test What You Already Know score: ___________

Test What You Learned score: ___________

Use this section to evaluate your progress. After working through the pre-quiz, check off the boxes in the "Pre" column to indicate which Learning Objectives you feel confident about. Then, after completing the chapter, including the post-quiz, do the same to the boxes in the "Post" column. Keep working on unchecked Objectives until you're confident about them all!

Pre	Post	
❑	❑	**9.1** Explain how enzymes and substrates interact
❑	❑	**9.2** Explain how cofactors or coenzymes affect enzyme function
❑	❑	**9.3** Contrast competitive and noncompetitive inhibition
❑	❑	**9.4** Explain how concentration affects enzyme function
❑	❑	**9.5** Investigate how enzyme reactivity is affected by concentration

FOR MORE PRACTICE

Complete more practice online at kaptest.com. Haven't registered your book yet? Go to kaptest.com/booksonline to begin.

ANSWERS AND EXPLANATIONS

Test What You Already Know

1. D Learning Objective: 9.1

Enzymes can only catalyze one reaction at a time and it takes a certain amount of time for the reaction to occur. Unless the enzyme concentration is increased, the enzymes become saturated at high concentrations of substrate, and the rate stops increasing with increasing substrate concentration. **(D)** is thus correct. (A) is incorrect because the graph provides no information about temperature. (B) is incorrect because the varying concentrations of AR are represented by the different curves in the graph, not by the variation along the *x*-axis, as for H_2O_2. (C) is incorrect because there is no data provided in the graph that suggests that high H_2O_2 concentrations prevent the active site from changing shape as needed—the reaction rate does not decrease at higher concentrations.

2. B Learning Objective: 9.2

The lower the K_M value, the lower the concentration of substrate necessary to reach maximum velocity. Similarly, the lower the k_{cat} value, the more quickly the enzyme acts. Thus, to be the more efficient cofactor, both values should be lower, as is the case for $NADP^+$. Therefore, **(B)** is correct. (A) is incorrect because it wrongly states that the turnover rate is higher for $NADP^+$. (C) and (D) are incorrect because they wrongly name NAD+ as the more efficient cofactor.

3. B Learning Objective: 9.3

Because the *y*-intercept represents the maximum velocity and there are different values for each line, the maximum velocity changes depending on how much of the inhibitor is present. Because increasing the substrate concentration can restore a competitively inhibited reaction to its original maximum velocity, this must instead be a noncompetitive inhibitor. **(B)** is thus correct. (A) is incorrect because the *x*-intercept represents a substrate concentration, not the maximum velocity. (C) and (D) are incorrect because they mischaracterize the graph (the curves have different *y*-intercepts but the same *x*-intercept).

4. C Learning Objective: 9.4

When the reactants are added, the researchers know the exact concentrations of each. This makes it easy to examine initial reaction rates as a function of concentration. However, as the reaction proceeds, it will be more difficult to determine the exact concentration at any given point. **(C)** is thus correct. (A) is incorrect because the reaction rate is affected by multiple variables, so it does not follow the simple pattern suggested. (B) is incorrect because there is no information provided about whether the reaction generates heat. (D) is incorrect because the researchers were not acting to save time but to work with information that had already been ascertained.

5. B Learning Objective: 9.5

According to the table, each increase in concentration of 11.1 μM leads to an increase in V_{max} of about 1.5 μM/minute. Thus, V_{max} for the concentration given in the question stem should be around 11.2 μM/minute, **(B)**. (A) is too high and would more likely correspond with a concentration of 88.9 μM if you were to continue extrapolating. (C) is the V_{max} given for the concentration of 66.7 μM in the table and there is no evidence to suggest that the V_{max} won't continue to increase with increasing concentration, so this answer is too low. (D) is even lower and would correspond with a concentration around 60 μM.

Test What You Learned

1. C Learning Objective: 9.4

Because the reaction rate is already close to its maximum at 6.0 mM, the rate would not change much if the substrate concentration was increased further. At most, the rate would increase slightly. **(C)** is thus correct. (A) is incorrect because the reaction rate would not increase significantly, but only slightly. (B) is incorrect because the reaction rate would not decrease. (D) is incorrect because the change is predictable.

2. B Learning Objective: 9.5

The reaction rate is predicted to increase due to the increase in substrate concentration. This is true unless it is already at the maximum for that pH (which is not the same as the maximum at any pH because each enzyme has an optimal pH at which it functions best). Therefore **(B)** is correct. (A) is incorrect because the rate should increase with more substrate (unless already at maximum). (C) is incorrect because the reaction rate would increase unless it is already at the maximum value for that pH. The fact that the rate seems unusually favorable at 7.75 is irrelevant and could just be random variation. (D) is incorrect because there is no guarantee that the rate will increase to the value obtained for a pH of 7.25.

3. D Learning Objective: 9.2

The lowest K_M value in the table occurs for the wild type enzyme with the $NADP^+$ cofactor, which also has the highest value for k_{cat}/K_M. This indicates that the wild type enzyme with $NADP^+$ is far more efficient than any of the alternatives. **(D)** is thus correct. (A) is incorrect because the wild type enzyme is more efficient with NAD^+ and reaches maximum velocity at a lower concentration than the mutant. (B) is incorrect because the wild type enzyme with NAD^+ reaches maximum velocity at a lower concentration and has higher substrate specificity than the mutant. (C) is incorrect because the wild type enzyme is more efficient with NADP+ and has higher substrate specificity than the mutant.

4. A Learning Objective: 9.3

Because the substrate and the inhibitor compete for the active site in competitive inhibition, it is possible to increase the substrate concentration enough to reach the same maximum velocity with the inhibitor as at a lower substrate concentration without the inhibitor. Hence, **(A)** is correct. (B) is incorrect because a competitive inhibitor binds to the active site. (C) is incorrect because this is more indicative of a noncompetitive inhibitor. (D) is incorrect because temperature effects give no indication of whether the inhibitor is competitive.

5. A Learning Objective: 9.1

A certain amount of energy (the activation energy) is needed for the reaction to take place. At higher temperatures, there is more kinetic energy and the increased rate of random collisions between molecules increases the probability that they will collide in the right way to react. Thus, **(A)** correctly characterizes this relationship. (B) is incorrect because there is more kinetic energy at higher temperatures, not at lower temperatures, and because increased kinetic energy actually raises the likelihood that the substrate will bind to the enzyme. (C) is incorrect because 30°C is actually a higher temperature than 37°C, so it makes no sense to say that the protein "denatured due to excess heat." (D) is incorrect because substrate concentration remains constant, according to the question stem.

6. C Learning Objective: 9.2

Cofactors, which help catalyze reactions, are nonprotein (inorganic) species that bind to enzymes. They may change the enzyme's shape to bind the substrate or they may stabilize the enzyme's active conformation. **(C)** appropriately illustrates a cofactor binding and making a better fitting active site for substrate binding, allowing for product formation. (A), (B), and (D) are incorrect because they represent competitive inhibition, noncompetitive inhibition, and uncompetitive inhibition, respectively.

7. D Learning Objective: 9.4

According to the table, as lactose (substrate) concentration increases, lactase (enzyme) activity (represented by molecule formation) increases and then plateaus for each pH level. Since there is no further increase in the number of molecules produced at lactose concentrations greater than 0.4 g/mL, it can be concluded that lactase becomes saturated as lactose concentration approaches 0.4 g/mL. Thus, the correct answer is **(D)**. (A) is incorrect because the optimum pH occurs when lactase is most active (when the number of molecules formed is the highest), which is pH 7. (B) is incorrect because maximum lactase activity is reached at a higher lactose concentration (around 0.4 g/mL). (C) is incorrect because lactase concentration remains constant as pH changes; lactase activity, however, changes with varying pH levels.

8. B Learning Objective: 9.5

At the optimum temperature, an enzyme's catalytic activity is the greatest. At 40°C, no starch is present (clear/white) after 5 min. This suggests all the starch was hydrolyzed by amylase. At 30°C, some starch remains (light gray) after 5 min, indicating the enzyme is not as efficient at 30°C. At 50°C, the darker-colored spot at 5 min indicates amylase's catalytic activity is lower than at 30°C and 40°C. Thus, the optimum temperature of amylase will fall between 31 and 40 degrees Celsius, inclusive. **(B)** is correct.

9. A Learning Objective: 9.1

Enzymes affect the activation energy but not the free energy of the substrate and product. The net energy released in the reaction equals the free energy of the product minus the free energy of the substrate. Thus, the correct answer is **(A)**. (B) is incorrect because enzymes speed up reactions by decreasing the activation energy. (C) is incorrect because the energy input required to reach the transition state with or without an enzyme is released when the product is formed, so the free energy change of the reaction does not change. (D) is incorrect because, although the enzyme-substrate complex facilitates the formation of the transition state, the free energy change of the reaction does not increase or decrease.

10. B Learning Objective: 9.3

Noncompetitive inhibition reduces V_{max} because it reduces the concentration of active enzyme. It, however, does not change K_m, which is the amount of substrate needed to reach half V_{max} and measures the affinity of an active enzyme for its substrate. Competitive inhibition increases K_m by interfering with the binding of the substrate, so more substrate is needed to outcompete the inhibitor to reach V_{max}, which is unchanged. The steeper slope in the graph also demonstrates that K_m is higher in competitive inhibition. Thus, the correct answer is **(B)**. (A) is incorrect because, although noncompetitive inhibition reduces V_{max}, it does not change K_m. (C) is incorrect because, though noncompetitive inhibition has a lower V_{max} than competitive inhibition, competitive inhibition has a higher K_m. (D) is incorrect because, in competitive inhibition, a large amount of substrate is needed to outcompete the inhibitor and saturate the enzyme.

CHAPTER 10

Metabolism

LEARNING OBJECTIVES

In this chapter, you will review how to:

- **10.1** Describe structures required for photosynthesis
- **10.2** Explain the light-dependent reactions of photosynthesis
- **10.3** Explain energy pathways for glycolysis and fermentation
- **10.4** Explain the function of the electron transport chain
- **10.5** Compare input and output molecules and energy
- **10.6** Investigate effects on the rate of photosynthesis
- **10.7** Investigate effects on the rate of cellular respiration

TEST WHAT YOU ALREADY KNOW

1. In chloroplasts, electron transport chain (ETC) reactions establish an electrochemical gradient of protons across the thylakoid membrane. Synthesized ATP and NADPH are then used to produce carbohydrates in the chloroplast stroma. Which of the following figures most accurately illustrates ETC reactions and ATP synthesis in chloroplasts?

(A)
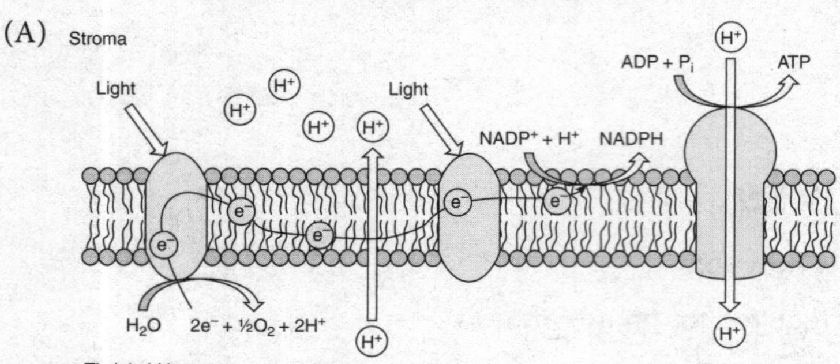

(B)
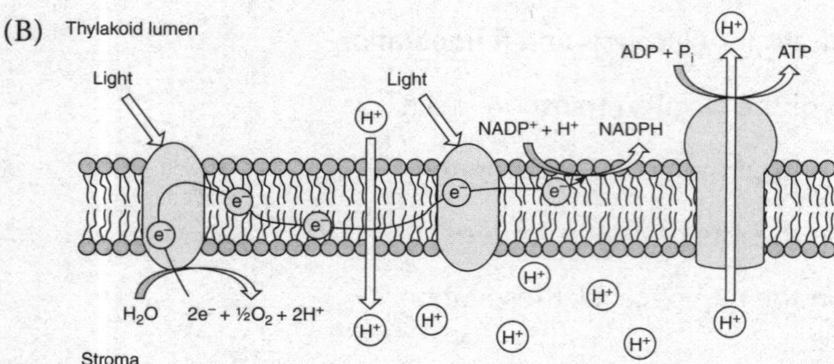

(C)
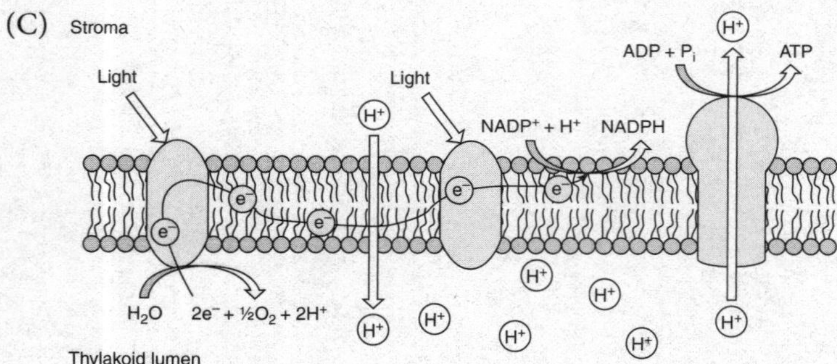

(D)
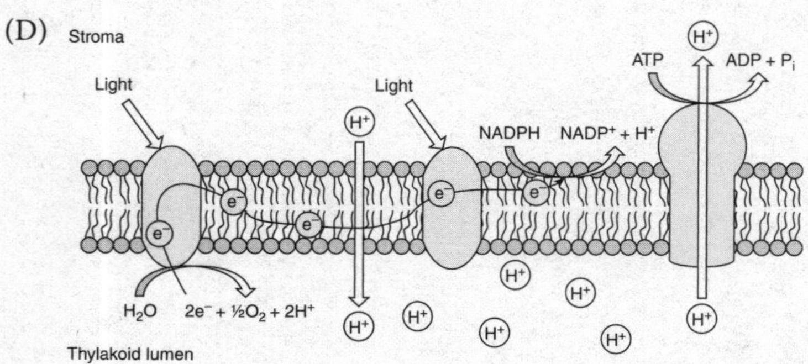

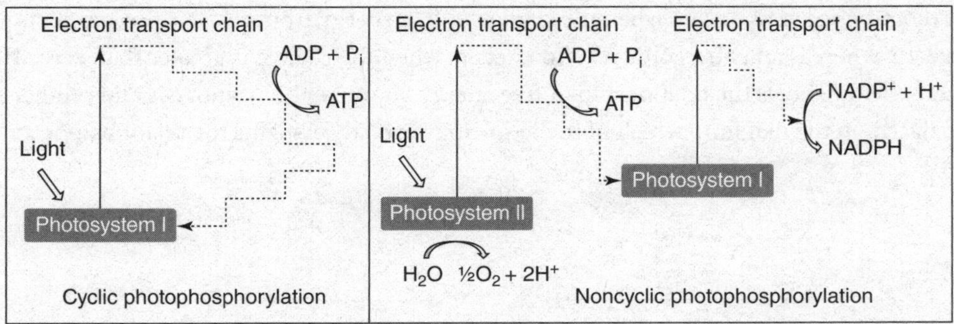

2. The diagram above illustrates cyclic and noncyclic photophosphorylation. High concentrations of NADPH in the chloroplast of a plant cell will cause a shift from noncyclic photophosphorylation to cyclic photophosphorylation. Which of the following best describes a likely result of this shift?
   (A) ATP levels will increase to meet cell energy demands.
   (B) ATP levels will become depleted as oxygen is produced.
   (C) NADPH and ATP will be used in the Krebs cycle to synthesize carbohydrates.
   (D) The reaction center in photosystem I will produce increased levels of oxygen and NADPH.

3. Human red blood cells lack mitochondria. Which of the following correctly explains the primary pathway that red blood cells use to produce energy?
   (A) Red blood cells generate ATP and NADH via aerobic respiration.
   (B) Red blood cells generate ATP and NADH via anaerobic fermentation.
   (C) Red blood cells metabolize glucose via glycolysis followed by carbon dioxide and ethanol production to produce ATP.
   (D) Red blood cells produce ATP via glycolysis followed by lactic acid production.

4. Thermogenin, an uncoupling protein found in brown adipose tissue, increases the permeability of the inner mitochondrial membrane. Which of the following accurately describes how the uncoupler affects the electron transport chain?
   (A) Thermogenin decreases the proton gradient and thus decreases ATP production.
   (B) Thermogenin decreases the proton gradient and thus increases ATP production.
   (C) Thermogenin increases the proton gradient and thus decreases ATP production.
   (D) Thermogenin increases the proton gradient and thus increases ATP production.

5. An exergonic reaction occurs when the change in free energy from reactants to products is less than zero, whereas an endergonic reaction occurs when the change is greater than zero. The diagram below shows the relationship of free energy between photosynthesis and cellular respiration. Based on the diagram, which of the following correctly explains the relationship?

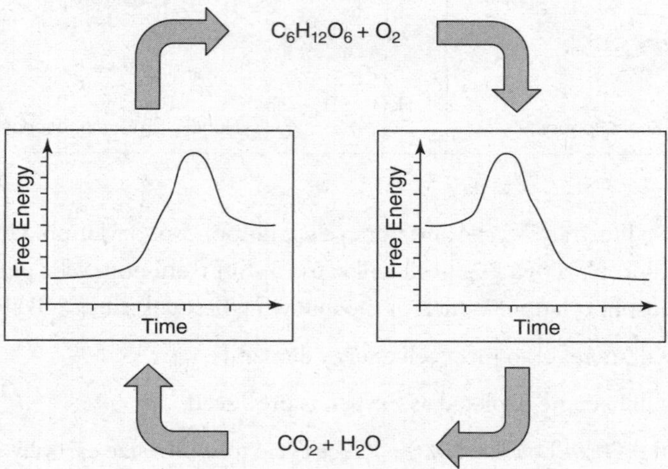

(A) Photosynthesis is exergonic, producing ATP and CO_2; cellular respiration is endergonic, consuming glucose and H_2O.

(B) Photosynthesis is exergonic, absorbing light energy to make ATP; cellular respiration is endergonic, releasing heat and CO_2.

(C) Cellular respiration is exergonic, consuming ATP to make glucose; photosynthesis is endergonic, absorbing light energy and releasing heat.

(D) Cellular respiration is exergonic, making ATP and releasing heat; photosynthesis is endergonic, converting light energy to chemical energy and glucose.

6. Dichlorophenolindophenol (DPIP), an electron acceptor that changes from blue to clear when reduced, can be used to visually determine the rate of photosynthesis. In an experiment, a solution of chloroplasts containing DPIP was divided among 4 tubes. The samples were then exposed to light (1,500 lumens) and/or heat (a temperature of 85°C), and light transmittance was measured over time using a spectrophotometer. Higher transmittance is correlated to lighter color. The results of the experiment are provided below.

	Transmittance (%)			
Time (min)	**No Light and No Heat**	**Light Only**	**Heat Only**	**Both Light and Heat**
0	23.2	20.8	20.4	20.4
5	25.3	37.8	22.9	21.1
10	26.5	48.7	20.6	21.8
15	24.2	63.4	21.6	22.7
20	25.7	77.5	22.1	23.0

Which of the following can most reasonably be concluded from the experimental results?

(A) The onset of photosynthesis is visible when DPIP is oxidized and changes from clear to blue.

(B) Chloroplasts exposed to heat had the highest rate of photosynthesis.

(C) Photosynthesis is stimulated when chloroplasts are exposed to light only.

(D) The solution in all four tubes was clear at time 0.

7. In a respiration experiment, transgenic mice (those possessing a particular genetic mutation) were compared to wild type (normal) mice. The ADP levels in the transgenic mice were found to be higher than those in the wild type. Which of the following is the most plausible explanation for this finding?

(A) The rate of oxygen consumption in mitochondria was higher in transgenic mice.

(B) The transgenic mice have impaired mitochondria, reducing their ability to convert ADP to ATP.

(C) The transgenic mice are unable to utilize ATP in the cytoplasm for metabolism.

(D) The rate of carbon dioxide consumption in cells was lower in transgenic mice.

Answers to this quiz can be found at the end of this chapter.

PHOTOSYNTHESIS

High-Yield

10.1 Describe structures required for photosynthesis

10.2 Explain the light-dependent reactions of photosynthesis

Cellular **metabolism** is the sum total of all chemical reactions that take place in a cell. These reactions can be generally categorized as either *anabolic* or *catabolic*. Anabolic processes are energy-requiring, involving the biosynthesis of complex organic compounds from simpler molecules. Catabolic processes release energy as they break down complex organic compounds into smaller molecules. The metabolic reactions of cells are coupled so that energy released from catabolic reactions can be harnessed to fuel anabolic reactions.

To survive, all organisms need energy. ATP is an energy intermediary used to drive biosynthesis and other processes. ATP is generated in mitochondria using the chemical energy of glucose and other nutrients. The energy foundation of almost all ecosystems is **photosynthesis**. Plants are **autotrophs**, or self-feeders, that generate their own chemical energy from the energy of the Sun through photosynthesis. The chemical energy that plants get from the Sun is used to produce glucose. This glucose can then be burned in plant mitochondria to make ATP, which is used to drive all of the energy-requiring processes in the plant, including the production of proteins, lipids, carbohydrates, and nucleic acids. Similarly, animals can eat plants to extract the energy for their own metabolic needs. In this way, photosynthesis is the energy foundation of most living systems.

> ✔ **AP Expert Note**
>
> **Not all autotrophs use photosynthesis. Some, such as bacteria that live near hydrothermal vents at the ocean floor, use chemosynthesis, a process that extracts energy through the oxidation of inorganic compounds.**

Photosynthesis Overview

Whereas cellular respiration takes place in all living organisms, only certain organisms contain the necessary pigments to perform photosynthesis, a process of capturing the energy of light and storing it in the chemical bonds of carbohydrates for later use. Plant cells have both **chloroplasts** and mitochondria to perform photosynthesis and cellular respiration, respectively. Photosynthesis is actually the reverse reaction of cellular respiration.

$$6\,CO_2 + 12\,H_2O + \text{light energy} \rightarrow C_6H_{12}O_6 + 6\,O_2 + 6\,H_2O$$

Perhaps the most surprising aspect of photosynthesis is the ability of photosynthetic proteins to split water molecules (H_2O). Once the water molecules have been split, the oxygen atoms are immediately released as O_2 and the hydrogen (H) atoms donate their electrons, which are used to form ATP and NADPH. The H atoms combine with the C and O atoms from CO_2 to form carbohydrates and more water. Sugars are used for energy storage, and O_2 is a waste product that other organisms use for respiration.

> ✔ **AP Expert Note**
>
> **Photosynthetic organisms are primary producers, providing food that supplies the rest of the food web. This "food" starts out in the form of glucose.**

Photosynthesis takes place in two stages: the **light** (or light-dependent) **reactions** and the **dark** (or Calvin or light-independent) **reactions**. In the light reactions, light energy is harnessed to produce chemical energy in the form of ATP and NADPH in a process called **photophosphorylation**. The dark reactions complete **carbon fixation**, the process by which CO_2 from the environment is incorporated into sugars with the help of energy released from the **oxidation** of ATP and the NADPH. In short, light reactions produce energy and dark reactions make sugars.

Photosynthesis will be explored further in the first lab section of this chapter. Comparisons of C3, C4, and CAM plants may show up within the data provided for a question. Most plants are C3. This means that the initial products of C fixation are two three-carbon molecules (phosphoglycerate, or PGA), synthesized through the intermediate enzyme **rubisco**. In C4 plants, CO_2 is initially fixed into a four-carbon molecule (oxaloacetate, also found in the Krebs cycle) by the intermediate enzyme phosphoenolpyruvic acid (PEP) in a mesophyll cell. This four-carbon molecule later releases a CO_2 molecule when it enters a bundle sheath cell. The enzyme PEP is much more likely to bind to CO_2 because it has a higher affinity for CO_2 than rubisco. C4 plants have a physiological advantage in hot, arid environments where they often have to limit the opening of stomata during the day.

✔ **AP Expert Note**

You do not need to memorize the differences between C3, C4, and CAM plants for Test Day.

Plants that go through C4 photosynthesis are grasses, which include semi-arid to arid crops like corn and sorghum. Many succulent plants, such as cacti, use an alternative method of limiting water loss in arid environments and are called CAM (crassulacean acid metabolism) plants because they collect CO_2 at night when it is cooler. CO_2 is then stored in the form of organic acids. C3, C4, and CAM plants all carry out the dark reactions in the Calvin cycle. However, C4 plants complete a carbon fixation step in separate *parts* of the plant, and CAM plants complete a carbon fixation step at separate *times*.

Photophosphorylation is driven by light energy absorbed by pigments (such as **chlorophylls**) in chloroplasts. White light is composed of many different wavelengths. Plants in particular have developed ways to use light of more than one wavelength. Chloroplasts can only use light energy if the energy is absorbed. Visible color is caused by reflected light. Plants reflect green light, so green light is not very useful in photosynthesis.

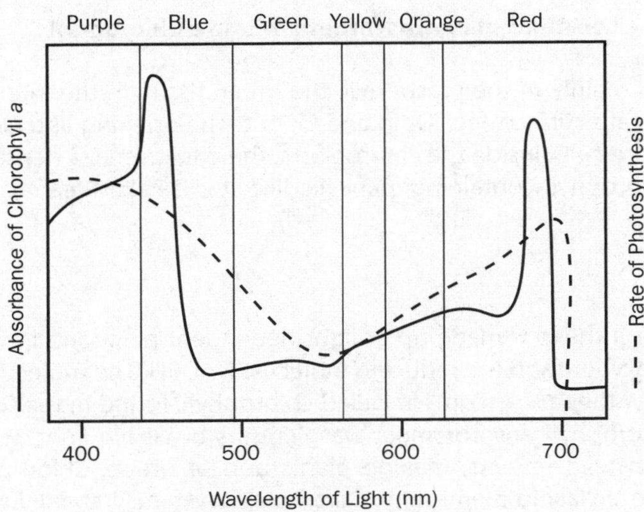

The previous image shows action spectra for the rate of photosynthesis (dotted line) and the absorbance of chlorophyll *a*. Note that the two action spectra do not correlate identically, indicating that the rate of photosynthesis depends on the presence of other photopigments as well. Chlorophyll *a* is medium green, chlorophyll *b* is yellow-green, and the carotenoids range in color from yellow to orange. Chlorophyll *a* is the only photopigment that participates in light reactions, but chlorophyll *b* and the carotenoids indirectly supplement photosynthesis by providing energy to chlorophyll *a*. (The carotenoids absorb light the chlorophyll cannot and transfer the energy to the chlorophyll.) The point to keep in mind is that many different photopigments absorb light energy from different wavelengths during photosynthesis. In order for plants to be healthy, they need blue and red light, not green.

✔ AP Expert Note

You may feel overwhelmed by the amount of information on photosynthesis, but don't panic. Only the major points are required for Test Day.

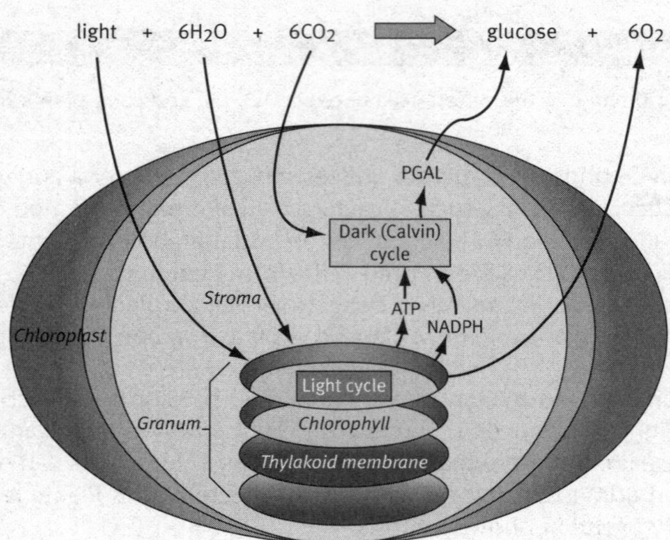

Locations of Photosynthesis within a Chloroplast

Chloroplasts are found mainly in the mesophyll, the green tissue in the interior of the plant. The stomata, pores in the leaf's surface, let CO_2 in and O_2 out. The opening and closing of the stomata is controlled by the guard cells. Inside the chloroplasts, there is stroma (a dense fluid) and thylakoid sacs (arranged into chlorophyll-containing, pancake-like stacks called grana).

Light Reactions

The first part of photosynthesis is made up of light-dependent reactions, in which light energy is used to generate ATP, oxygen, and the reducing molecule NADPH. The molecule that captures light energy to start photosynthesis is a pigment called chlorophyll, found in the thylakoid membranes of the chloroplast. Chlorophyll absorbs most wavelengths of visible light, with the exception of green light, which it instead reflects, making plants appear green. Chlorophyll is used by two complex systems in the thylakoid membrane, called photosystems I and II. Each photosystem is a complex assembly of protein and pigments in the membrane. When photons strike chlorophyll, electrons are excited and transferred through the photosystems to a reaction center. When electrons reach the reaction center, the reaction center gives up excited electrons that enter an electron transport chain, where they are used to generate chemical energy as either reduced NADPH or ATP.

Two different processes occur in the photosystems, *cyclic photophosphorylation* and *noncyclic photophosphorylation*. Both are used to generate ATP but in different ways. The ATP, in turn, is used to generate glucose in the dark reactions. Cyclic photophosphorylation occurs in photosystem I to produce ATP. In the cyclic method, electrons move from the reaction center, through an electron transport chain, then back to the same reaction center again (see figure below). The reaction center in photosystem I includes a chlorophyll called P700 because its maximal light absorbance occurs at 700 nm. This process does not produce oxygen and does not produce NADPH.

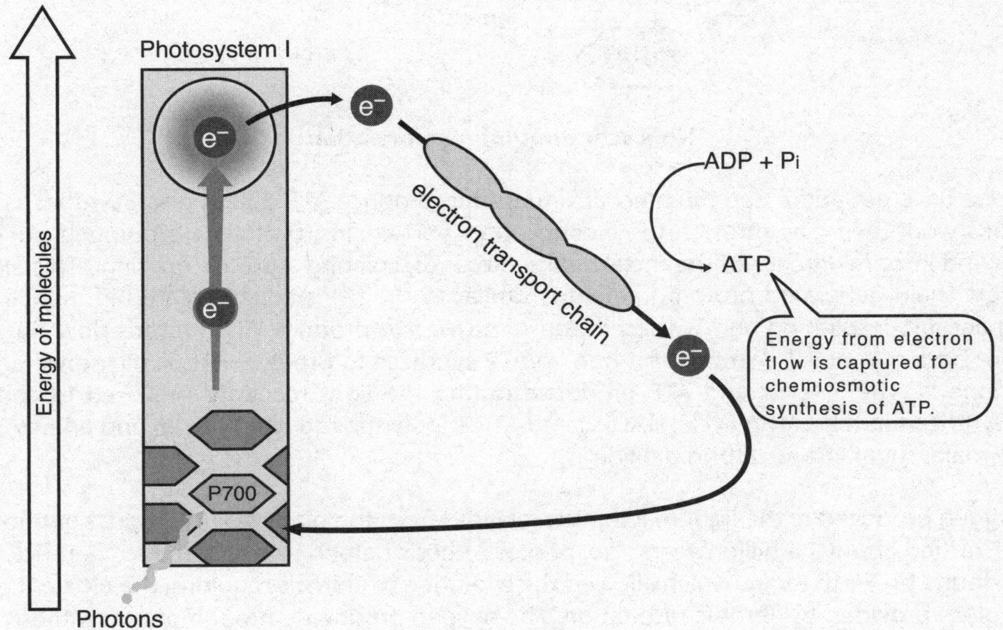

Cyclic Photophosphorylation

Noncyclic photophosphorylation starts in photosystem II (see figure on the following page). In noncyclic photophosphorylation, chlorophyll pigment absorbs light and passes excited electrons to a reaction center, a process equivalent to cyclic photophosphorylation. The photosystem II reaction center contains a P680 chlorophyll, distinct from photosystem I. From the photosystem II reaction center, the electrons are passed to an electron transport chain. In this case, however, the electrons are not returned to the reaction center at the end of the electron transport chain but are passed to photosystem I. Photosystem II replaces the electrons it lost by getting them from water, producing oxygen in the process. In this case, the electrons that enter photosystem I are used to produce NADPH.

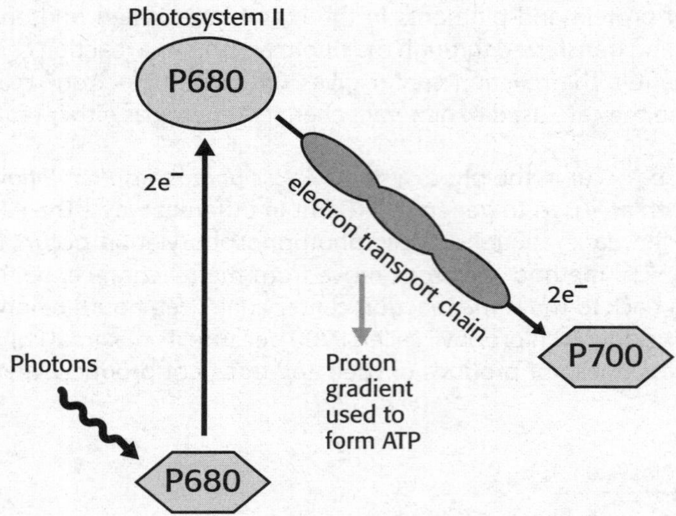

Noncyclic Photophosphorylation

So far, we have not addressed the mechanism used to produce ATP during photosynthesis. As the electrons work their way through the electron transport chains, protons are pumped out of the stroma and into the interior of the thylakoid membranes, creating a proton gradient. This electron transport chain-generated proton gradient is similar to the pH gradient created in mitochondria during aerobic respiration and is used in the same way to produce ATP. Protons flow down this gradient back out into the stroma through an ATP synthase to produce ATP, similar once again to mitochondria. The NADPH and ATP produced during the light reactions are used to complete photosynthesis in the Calvin cycle; NADPH and ATP supply the reducing power and energy necessary to make sugars from carbon dioxide.

The oxygen produced in the light reactions is released from the plant as a byproduct of photosynthesis. Starting about 1.5 billion years ago, photosynthesis helped to create the oxygen-rich atmosphere found on Earth today, which allowed the evolution of animals requiring the efficient energy metabolism provided by aerobic respiration. The oxygen produced through photosynthesis today maintains Earth's oxygen and is a key to the continued functioning of the biosphere.

Calvin Cycle

In the second phase of photosynthesis, the energy captured in the light reactions as ATP and NADPH is used to drive carbohydrate synthesis. This cycle, often called the **Calvin cycle** but also known as the Calvin-Benson cycle, fixes CO_2 into carbohydrates, reducing the fixed carbon to carbohydrates through the addition of electrons. The NADPH provides the reducing power for the reduction of CO_2 to carbohydrates, and air provides the carbon dioxide. CO_2 first combines with, or is fixed to, ribulose bisphosphate (RuBP), a five-carbon sugar with two phosphate groups. The enzyme that catalyzes this reaction is called rubisco and is the most abundant enzyme on Earth. The resulting six-carbon compound is promptly split, resulting in the formation of two molecules of 3-phosphoglycerate, a three-carbon compound. The 3-phosphoglycerate is then phosphorylated by ATP and reduced by NADPH, which leads to the formation of phosphoglyceraldehyde (PGAL). This can then be utilized as a starting point for the synthesis of glucose, starch, proteins, and fats.

✔ **AP Expert Note**

The steps of the Calvin cycle and the structure of the molecules involved are not required knowledge for the AP Biology exam.

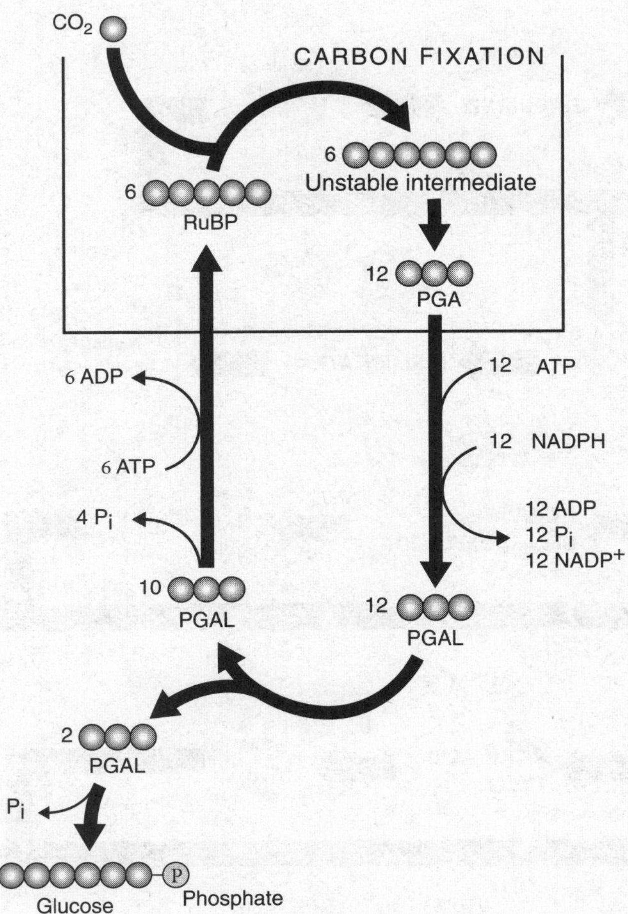

Calvin Cycle

Processes

ANAEROBIC RESPIRATION

High-Yield

10.3 Explain energy pathways for glycolysis and fermentation

Overview of Cellular Respiration

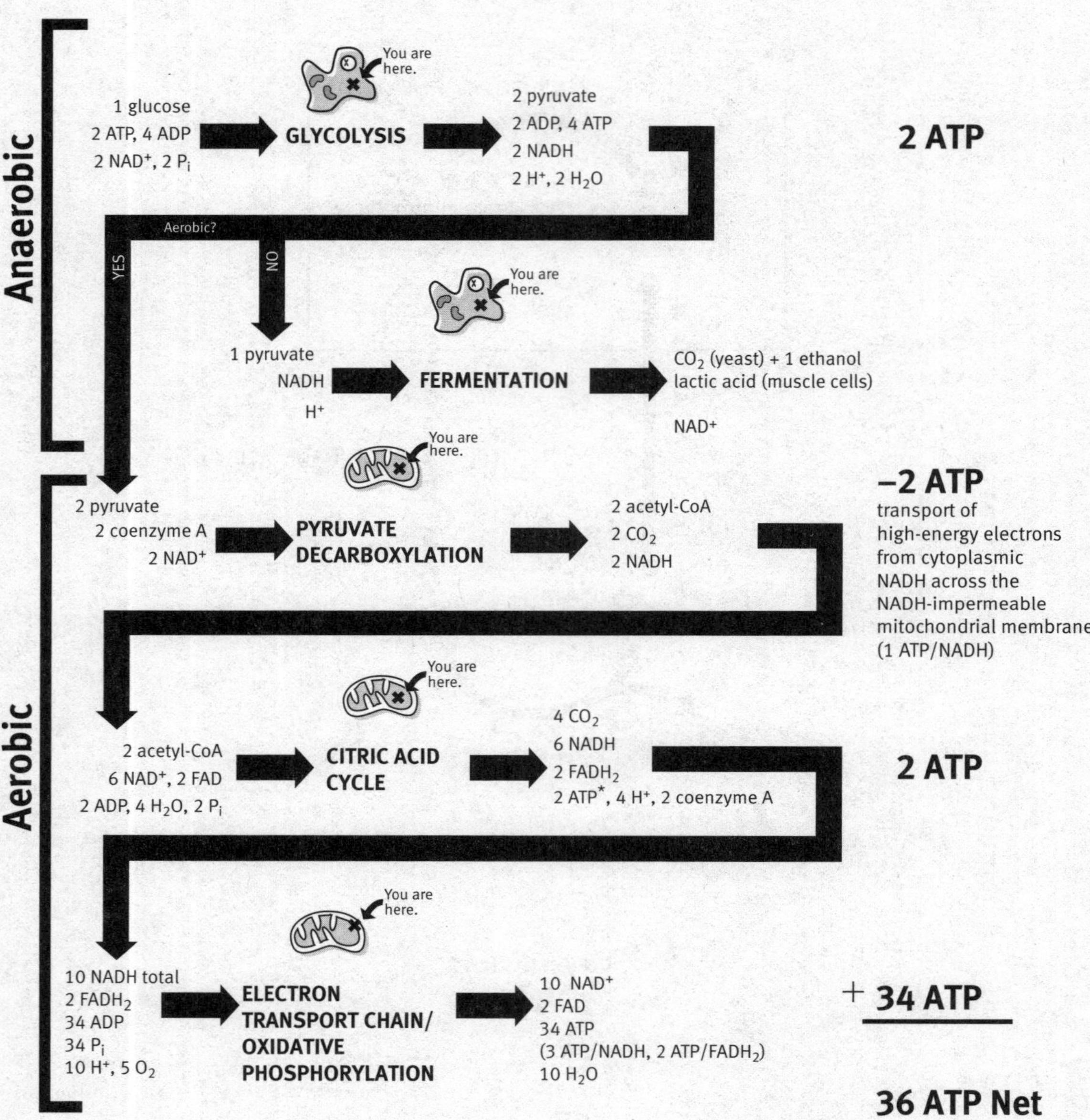

*The citric acid cycle yields a direct product of 2 GTP. The 2 GTP subsequently donate their phosphate to 2 ADP to form 2 ATP and regenerate the original 2 GDP.

General Overview of Cellular Respiration

Cellular **respiration** is the most efficient catabolic pathway used by organisms to harvest the energy stored in glucose. Whereas glycolysis yields only 2 ATP per molecule of glucose, cellular respiration can yield 36–38 ATP. Cellular respiration is an **aerobic** process; oxygen acts as the final acceptor of electrons that are passed from carrier to carrier during the final stage of glucose oxidation. The metabolic reactions of cellular respiration occur in the eukaryotic **mitochondria** and are catalyzed by reaction-specific enzymes.

Cellular respiration can be divided into five stages: glycolysis, fermentation, pyruvate decarboxylation, the citric acid cycle, and the electron transport chain.

Glycolysis

The first stage of glucose catabolism is **glycolysis**. Glycolysis is a series of reactions that lead to the oxidative breakdown of glucose into two molecules of pyruvate (the ionized form of pyruvic acid), the production of ATP, and the reduction of NAD^+ into NADH. All of these reactions occur in the cytoplasm and are mediated by specific enzymes. The glycolytic pathway is as follows:

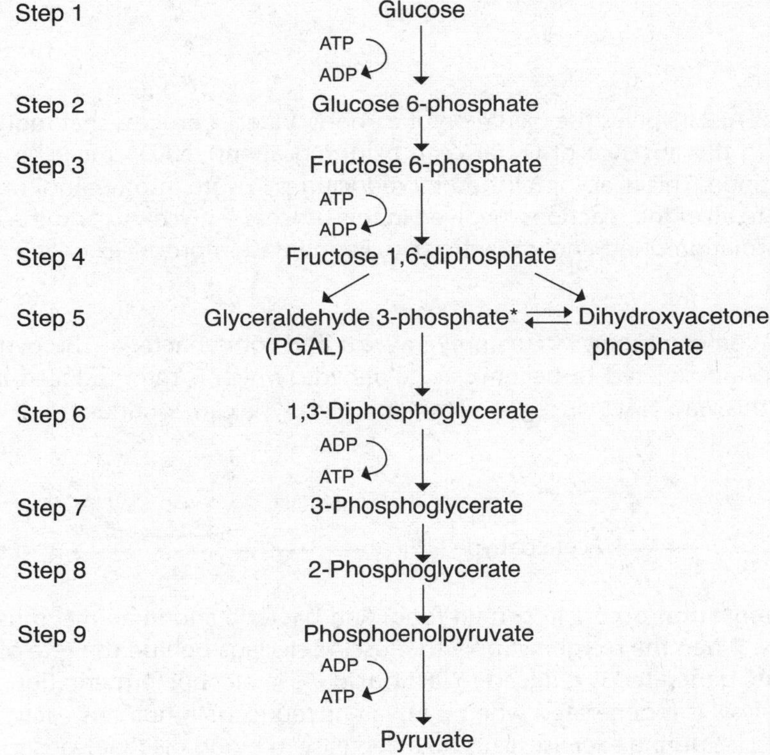

Glycolysis

> ✔ **AP Expert Note**
>
> **You do not need to know all these steps or the intermediates.**

From one molecule of glucose (a six-carbon molecule), two molecules of pyruvate (a three-carbon molecule) are obtained. During this sequence of reactions, 2 ATP are used (in steps 1 and 3) and 4 ATP are generated (2 in step 6 and 2 in step 9). Thus, there is a net production of 2 ATP per glucose molecule. This type of phosphorylation is called substrate level phosphorylation because ATP synthesis is directly coupled with the degradation of glucose without the participation of an intermediate molecule, such as NAD^+. One NADH is produced per PGAL, for a total of 2 NADH per glucose.

The net reaction for glycolysis is:

$$\text{Glucose} + 2\,\text{ADP} + 2\,P_i + 2\,NAD^+ \rightarrow 2\,\text{Pyruvate} + 2\,\text{ATP} + 2\,\text{NADH} + 2\,H^+ + 2\,H_2O$$

This series of reactions occurs in both prokaryotic and eukaryotic cells. However, at this stage, much of the initial energy stored in the glucose molecule has not been released and is still present in the chemical bonds of pyruvate. Depending on the capabilities of the organism, pyruvate degradation can proceed in one of two directions. Under **anaerobic** conditions (in the absence of oxygen), pyruvate is reduced during the process of fermentation. Under aerobic conditions (in the presence of oxygen), pyruvate is further oxidized during cellular respiration in the mitochondria.

Fermentation

The production of ATP involves the oxidation of carbohydrates, a process that requires NAD^+ as an oxidizing agent. In the absence of O_2 (also an oxidizing agent), NAD^+ must be regenerated for glycolysis to continue. This is accomplished by reducing pyruvate into ethanol or lactic acid. Fermentation refers to all of the reactions involved in this process—glycolysis and the additional steps leading to the formation of ethanol or lactic acid. Fermentation produces only 2 ATP per glucose molecule.

Alcohol fermentation commonly occurs only in yeast and some bacteria. The pyruvate produced in glycolysis is decarboxylated to become acetaldehyde, which is then reduced by the NADH to yield ethanol. In this way, NAD^+ is regenerated and glycolysis can continue.

$$\text{Pyruvate (3C)} \xrightarrow{\;\nearrow CO_2\;} \text{Acetaldehyde (2C)} \xrightarrow{\text{NADH} \;\curvearrowright\; NAD^+ + H^+} \text{Ethanol (2C)}$$

Lactic acid fermentation occurs in certain fungi and bacteria and in animal muscle cells during strenuous activity. When the oxygen supply to muscle cells lags behind the rate of glucose catabolism, the pyruvate generated is reduced to lactic acid. As in alcohol fermentation, the NAD^+ used in step 5 of glycolysis is regenerated when pyruvate is reduced. In humans, lactic acid may accumulate in the muscles during exercise, causing a decrease in blood pH that leads to muscle fatigue. Once the oxygen supply has been replenished, the lactic acid is oxidized back to pyruvate and enters cellular respiration. The amount of oxygen needed for this conversion is known as the oxygen debt.

AEROBIC RESPIRATION

High-Yield

10.4 Explain the function of the electron transport chain

10.5 Compare input and output molecules and energy

Pyruvate Decarboxylation

The pyruvate formed during glycolysis is transported from the cytoplasm into the mitochondrial matrix, where it is decarboxylated; that is, it loses a CO_2, and the acetyl group that remains is transferred to coenzyme A to form acetyl-CoA. In the process, NAD^+ is reduced to NADH.

$$\text{Pyruvate (3C)} + \text{Coenzyme A} \xrightarrow{\text{NAD}^+ \rightarrow \text{NADH} + \text{H}^+} \text{Acetyl-CoA (2C)}$$

Citric Acid Cycle

The citric acid cycle is also known as the **Krebs cycle** or the tricarboxylic acid cycle (TCA cycle). The cycle begins when the two-carbon acetyl group from acetyl-CoA combines with oxaloacetate, a four-carbon molecule, to form the six-carbon citrate. Through a complicated series of reactions, 2 CO_2 are released, and oxaloacetate is regenerated for use in another turn of the cycle.

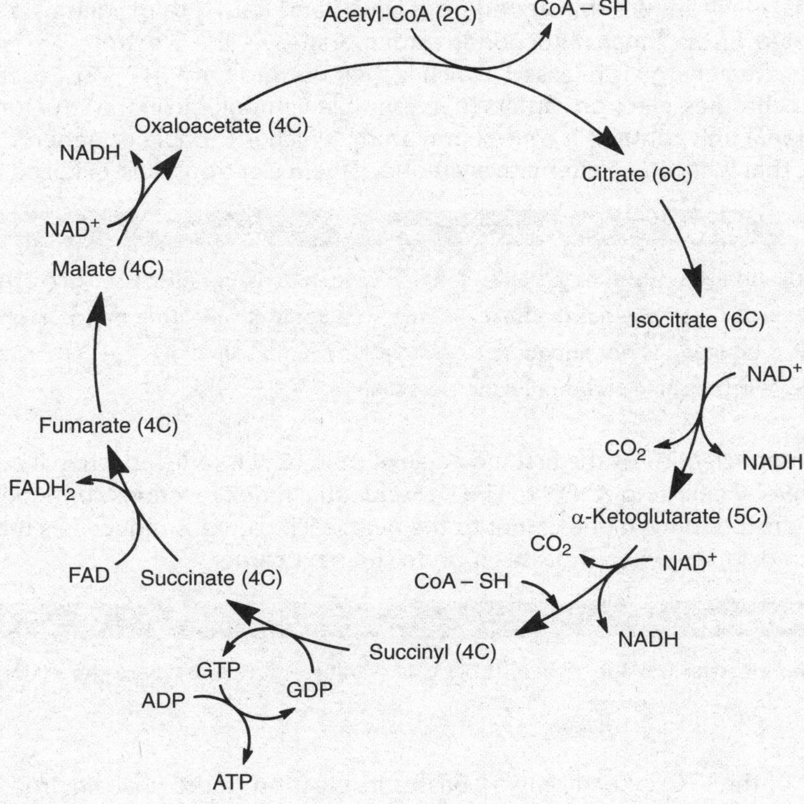

The Citric Acid Cycle

Processes

For each turn of the citric acid cycle, 1 ATP is produced by substrate level phosphorylation via a GTP intermediate. In addition, electrons are transferred to NAD^+ and FAD, generating NADH and $FADH_2$, respectively. These coenzymes then transport the electrons to the electron transport chain, where more ATP is produced via oxidative phosphorylation. To be clear, the most important function of the citric acid cycle is the production of a large number of NADH and $FADH_2$, which ultimately contribute to the production of a large number of ATP molecules.

✔ AP Expert Note

The exam will not test you on the details of pyruvate decarboxylation or the steps and intermediates of the citric acid cycle, so just be aware of the role each process plays in the larger process of aerobic respiration.

The net reaction of the citric acid cycle per glucose molecule is:

$$2 \text{ Acetyl-CoA} + 6 \text{ NAD}^+ + 2 \text{ FAD} + 2 \text{ GDP} + 2 \text{ P}_i + 4 \text{ H}_2\text{O} \rightarrow 4 \text{ CO}_2 + 6 \text{ NADH} + 2 \text{ FADH}_2 + 2 \text{ ATP} + 4 \text{ H}^+ + 2 \text{ CoA}$$

Electron Transport Chain

The **electron transport chain** (ETC) is a complex carrier mechanism located on the inside of the inner mitochondrial membrane. During oxidative phosphorylation, ATP is produced when high-energy-potential electrons are transferred from NADH and $FADH_2$ to oxygen by a series of carrier molecules located in the inner mitochondrial membrane. As the electrons are transferred from carrier to carrier, free energy is released, which is then used to form ATP. Most of the molecules of the ETC are cytochromes, electron carriers that resemble hemoglobin in the structure of their active site. The functional unit contains a central iron atom, which is capable of undergoing a reversible redox reaction; that is, it can be alternatively reduced (gain electrons) and oxidized (lose electrons).

✔ AP Expert Note

Everything the human body does to deliver inhaled oxygen to tissues (discussed in chapter 11) comes down to the role oxygen plays as the final electron acceptor in the electron transport chain. Without oxygen, ATP production is not adequate to sustain human life. Similarly, the CO_2 generated in the citric acid cycle is the same carbon dioxide we exhale.

FMN (flavin mononucleotide) is the first molecule of the ETC. It is reduced when it accepts electrons from NADH, thereby oxidizing NADH to NAD^+. Sequential redox reactions continue to occur as the electrons are transferred from one carrier to the next; each carrier is reduced as it accepts an electron and is then oxidized when it passes it on to the next carrier.

✔ AP Expert Note

Note that the electron transport chain (ETC) is like an assembly line where the majority of ATP is generated.

The last carrier of the ETC, cytochrome a_3, passes its electron to the final electron acceptor, O_2. In addition to the electrons, O_2 picks up a pair of hydrogen ions from the surrounding medium, forming water.

✔ AP Expert Note

The names of the specific electron carriers in the electron transport chain are not required knowledge for the AP Biology exam.

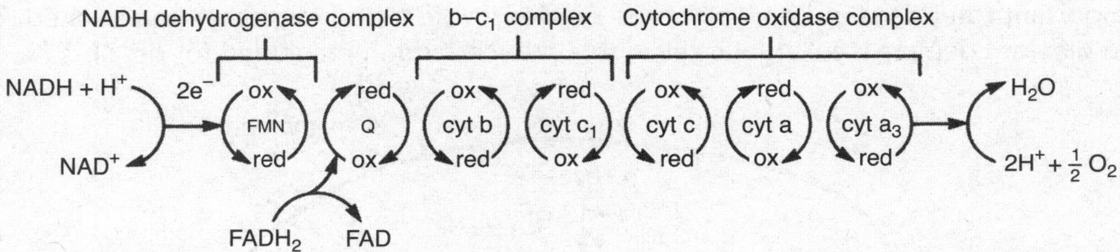

Electron Transport Chain

Without oxygen, the ETC becomes backlogged with electrons. As a result, NAD^+ cannot be regenerated and glycolysis cannot continue unless lactic acid fermentation occurs. Likewise, ATP synthesis comes to a halt if respiratory poisons such as cyanide or dinitrophenol enter the cell. Cyanide blocks the transfer of electrons from cytochrome a_3 to O_2. Dinitrophenol uncouples the electron transport chain from the proton gradient established across the inner mitochondrial membrane.

ATP Generation and the Proton Pump

The electron carriers are categorized into three large protein complexes (NADH dehydrogenase, the cytochrome b–c_1 complex, and cytochrome oxidase), as well as the carrier molecule Q. There are energy losses as the electrons are transferred from one complex to the next; this energy is then used to synthesize 1 ATP per complex. Thus, an electron passing through the entire ETC supplies enough energy to generate 3 ATP. NADH delivers its electrons to the NADH dehydrogenase complex so that for each NADH, 3 ATP are produced. However, $FADH_2$ bypasses the NADH dehydrogenase complex and delivers its electrons directly to carrier Q (ubiquinone), which lies between the NADH dehydrogenase and cytochrome b–c_1 complexes. Therefore, for each $FADH_2$, there are only two energy drops, and only 2 ATP are produced.

$$3 \times 6 \text{ NADH} \rightarrow 18 \text{ ATP}$$

$$2 \times 2 \text{ FADH}_2 \rightarrow 4 \text{ ATP}$$

$$1 \times 2 \text{ GTP (ATP)} \rightarrow 2 \text{ ATP}$$

The operating mechanism in this type of ATP production involves coupling the oxidation of NADH and $FADH_2$ to the phosphorylation of ADP. The coupling agent for these two processes is a proton gradient across the inner mitochondrial membrane, maintained by the ETC. As NADH and $FADH_2$ pass their electrons to the ETC, hydrogen ions (H^+) are pumped out of the matrix, across the inner mitochondrial membrane, and into the **intermembrane space** at each of the three protein complexes. The continuous translocation of H^+ creates a positively charged, acidic environment in the intermembrane space. This electrochemical gradient generates a proton-motive force, which drives H^+ back across the inner membrane and into the matrix. However, to pass through the membrane (which is impermeable to ions), the H^+ must flow through specialized channels provided by enzyme complexes called ATP synthetases. As the H^+ pass through the ATP synthetases, enough energy is released to allow for the phosphorylation of ADP to ATP. The coupling of the oxidation of NADH and $FADH_2$ with the phosphorylation of ADP is called **oxidative phosphorylation**.

Review of Glucose Catabolism

It is important to understand how all of the previously described events are interrelated. The following diagram depicts a eukaryotic cell with a mitochondrion magnified for detail.

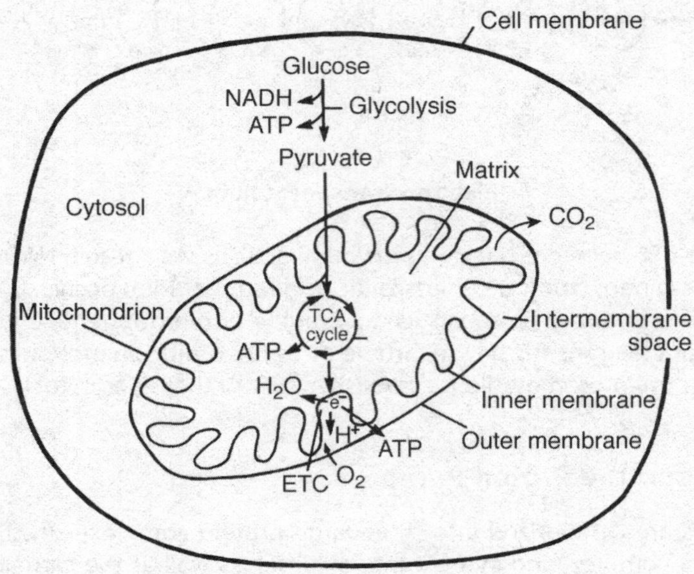

Locations of Glucose Catabolism

To calculate the net amount of ATP produced per molecule of glucose, we need to tally the number of ATP produced by substrate level phosphorylation and the number of ATP produced by oxidative phosphorylation.

Substrate Level Phosphorylation

Degradation of one glucose molecule yields a net of 2 ATP from glycolysis and 1 ATP for each turn of the citric acid cycle with two turns per glucose. Thus, a total of 4 ATP are produced by substrate level phosphorylation.

Oxidative Phosphorylation

Two pyruvate decarboxylations yield 1 NADH each for a total of 2 NADH per glucose. Each turn of the citric acid cycle yields 3 NADH and 1 $FADH_2$, for a total of 6 NADH and 2 $FADH_2$ per glucose molecule. Each $FADH_2$ generates 2 ATP, as previously discussed. Each NADH generates 3 ATP except for the 2 NADH that were reduced during glycolysis; these NADH cannot cross the inner mitochondrial membrane and must transfer their electrons to an intermediate carrier molecule, which delivers the electrons to the second carrier protein complex, Q. Therefore, these NADH generate only 2 ATP per glucose. Thus, the 2 NADH of glycolysis yield 4 ATP, the other 8 NADH yield 24 ATP, and the 2 $FADH_2$ produce 4 ATP, for a total of 32 ATP by oxidative phosphorylation.

The total amount of ATP produced during eukaryotic glucose catabolism is 4 via substrate level phosphorylation plus 32 via oxidative phosphorylation, for a total of 36 ATP. (For prokaryotes the yield is 38 ATP because the 2 NADH of glycolysis don't have any mitochondrial membranes to cross and therefore don't lose energy.) See the following table for a summary of eukaryotic ATP production.

A QUICK REFERENCE TO ENERGY PRODUCTION IN CELLULAR RESPIRATION

Glycolysis	
2 ATP invested (steps 1 and 3)	−2 ATP
4 ATP generated (steps 6 and 9)	+4 ATP (substrate)
2 NADH × 2 ATP/NADH (step 5)	+4 ATP (oxidative)
Pyruvate Decarboxylation	
2 NADH × 3 ATP/NADH	+6 ATP (oxidative)
Citric Acid Cycle	
6 NADH × 3 ATP/NADH	+18 ATP (oxidative)
2 $FADH_2 \times 2$ ATP/$FADH_2$	+4 ATP (oxidative)
2 GTP × 1 ATP/GTP	+2 ATP (substrate)
	Total: + 36 ATP

AP BIOLOGY LAB 5: PHOTOSYNTHESIS INVESTIGATION

10.6 Investigate effects on the rate of photosynthesis

In this investigation, you will learn how to measure the rate of photosynthesis by determining oxygen production. To review, photosynthesis is the process by which autotrophs capture free energy (in the form of sunlight) to build carbohydrates. The process is summarized by the following reaction:

$$2\,H_2O + CO_2 + \text{light} \rightarrow \text{carbohydrate } (CH_2O) + O_2 + H_2O$$

To determine the rate of photosynthesis by a plant cell, you can measure the production of O_2 or the consumption of CO_2. The process is not that simple, though, because photosynthesis is coupled with aerobic respiration, in which the oxygen produced is simultaneously consumed. Because measuring the consumption of carbon dioxide typically requires expensive equipment and complex procedures, you will use the floating disk procedure to measure the production of oxygen.

In this procedure, you use a vacuum to remove all air and then add a bicarbonate solution to plant (leaf) disk samples. These leaves sink until placed in sufficient light, when photosynthesis produces enough oxygen bubbles to change the buoyancy of the disk, causing it to float to the surface. Many different factors can affect the rate of photosynthesis in the real world (i.e., intensity of light, color of light, leaf size, type of plant), but the results of this experiment can also be influenced by different procedural factors (i.e., depth of solution, method of cutting disks, size of leaf disks). You will effectively be measuring net photosynthesis. The standard measurement to use after determining how long it takes each disk to float to the surface is ET_{50}, the estimated time for 50 percent of the disks to rise. This measurement will help you to aggregate your data for the second half of the investigation.

After learning how to use the floating disk procedure, you will choose one factor that affects the rate of photosynthesis and then develop and conduct an investigation of that variable. When you compare the ET_{50} for different levels of your chosen variable, you should observe that ET_{50} goes down as rate of photosynthesis goes up. This creates a nontraditional graph that is not the best display of your data (below left). Alternatively, you can use $1/ET_{50}$, which will show increasing rates of photosynthesis and, therefore, a graph with a positive slope (below right).

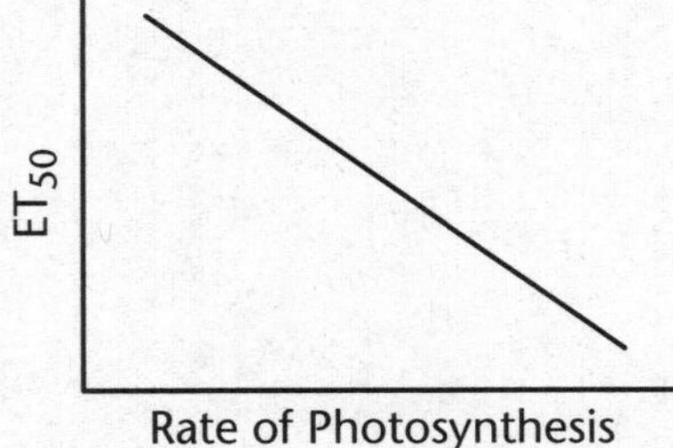

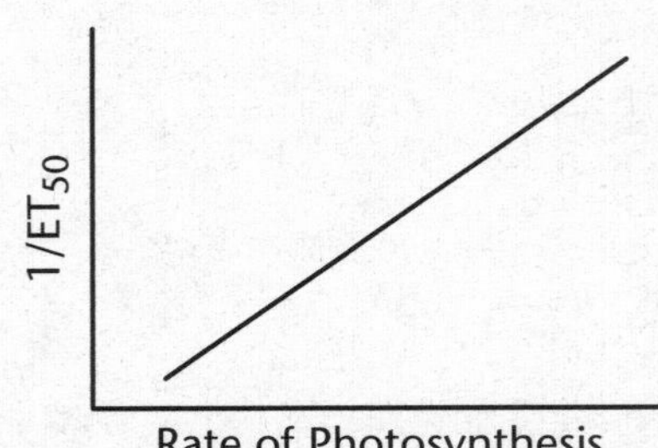

AP BIOLOGY LAB 6: CELLULAR RESPIRATION INVESTIGATION

High-Yield

10.7 Investigate effects on the rate of cellular respiration

This investigation is much more about the experimental method than about cellular respiration. In this investigation, you learn how to set up and use a respirometer to measure the change in volume of a gas, which can be assumed to be O_2, from the germinating seeds placed in water. The effect of increasing temperature on the rate of gas volume change (O_2 utilization) is also measured during the lab. One of the problems with this lab is that gas expands when it is heated. Sometimes the volume increases so much it blows the dye out of the end of the respirometer tube. You will also have the opportunity to ask your own questions and conduct your own investigations about cellular respiration.

This experiment is an excellent example of using control specimens to isolate experimental variables. Put quite simply, the dependent variable being measured is the change in gas volume in the respirometer. The hypothesis is that the change in gas volume is being caused by the utilization of O_2 by the germinating seeds. Other factors can contribute to a change in gas volume (i.e., temperature and pressure), so these variables need to be isolated from the variables you are interested in (i.e., the rate of respiration of the seeds). To isolate variables you are interested in from other variables, glass beads should be subjected to the same experimental treatment (change in temperature) that the target specimens, the live seeds, are subjected to. The glass bead sample is the control group. Any measurable change in the dependent variable in the control group must be removed from the dependent variable of the target group.

Let's say that when you heated your seeds to 35°C, the volume of gas in the respirometer of the target group (live seeds) decreased by 0.3 mL, but the volume of gas in the respirometer of the control group (glass beads) increased by 0.1 mL. Something occurred in the respirometer of the control group that caused an increase in gas volume. Perhaps the change in volume in the respirometer was created by the expansion of heated gas, or expansion was caused by CO_2 not being absorbed by the soda lime or other CO_2 absorbent. Either way, the dependent variable of temperature must be taken away from the target group because a change in volume should have occurred in the respirometer of the target group as well as in the respirometer of the control group. As a result, it can be assumed that an additional 0.1 mL (for a total of 0.4 mL) of O_2 was likely used by the germinating seeds. In the experiment, 0.4 mL of gas were not measured because 0.1 mL of gas was obscured by the expansion of gas from some unknown factor, likely increased temperature. An accurate measure of the target group can never be obtained if the control group isn't included in the experiment.

Think about how cellular respiration is a conserved evolutionary process or how different ecological habitats have modified the capture and use of free energy by different organisms. By thinking outside the box, you will be better able to apply your knowledge across the entire AP Biology exam. This is particularly helpful when addressing free-response questions, which often ask you to make connections between content and process.

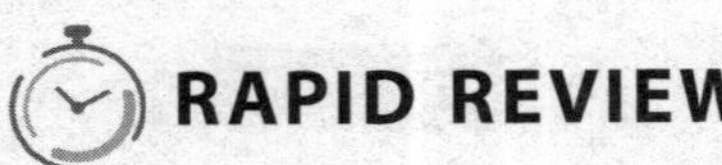

RAPID REVIEW

If you take away only 6 things from this chapter:

1. Photosynthesis is the energy foundation for almost all living systems. In addition, photosynthesis provides almost all of the oxygen present in the Earth's atmosphere.
2. All photosynthetic organisms use chloroplasts and mitochondria to perform photosynthesis and cellular respiration. Photosynthesis and cellular respiration are essentially reverse operations of each other.
3. Photosynthesis has two main parts—the light cycle and the dark cycle (the latter is usually called the Calvin or Calvin-Benson cycle). Light reactions produce energy and dark reactions make sugars. The light reactions occur in the interior of the thylakoid, while the Calvin-Benson cycle occurs in the stroma.
4. Cellular respiration is an efficient catabolic pathway and yields ATP. Cellular respiration is an aerobic process. The metabolic reactions of respiration occur in the eukaryotic mitochondria and are catalyzed by reaction-specific enzymes.
5. Cellular respiration can be divided into several stages: glycolysis, pyruvate decarboxylation, the citric acid cycle, and the electron transport chain.
6. Cellular respiration is a complex process that requires many different products and specialized molecules. Focus on the requirements and overall net production for the major steps.

TEST WHAT YOU LEARNED

1. Which of the following figures correctly illustrates the light-dependent and light-independent reactions of photosynthesis?

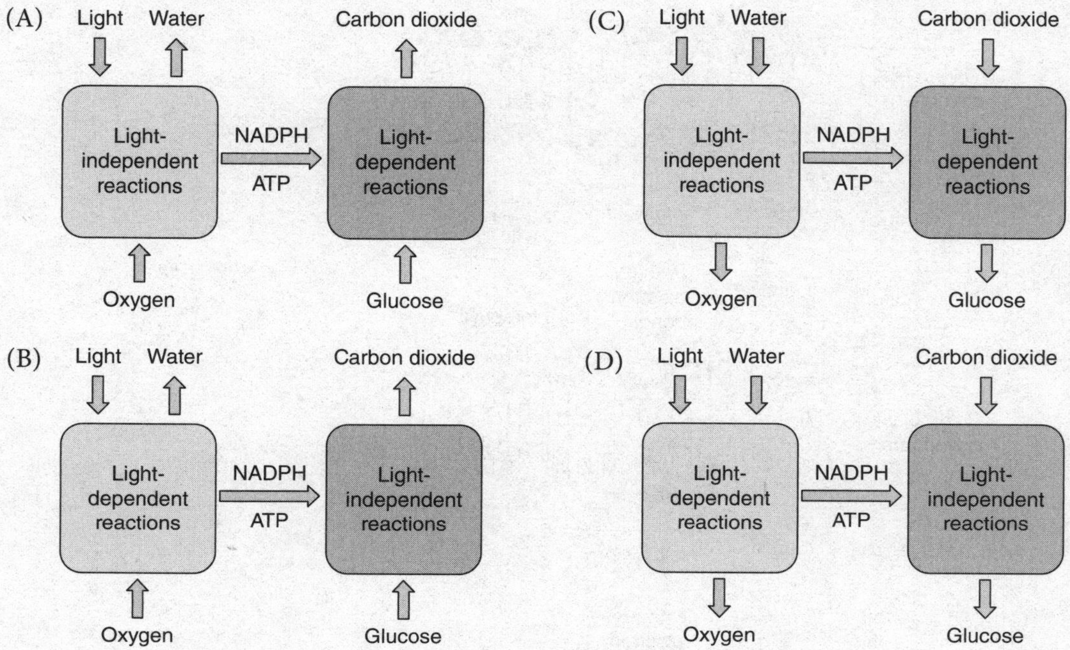

2. Autotrophs capture free energy to produce organic compounds for food while heterotrophs metabolize carbon compounds produced by other organisms as sources of free energy. Which of the following best demonstrates the function of CO_2 in autotrophic and heterotrophic metabolism?

   (A) Autotrophs use CO_2 to make glucose in the Calvin cycle, and heterotrophs produce CO_2 as a waste product in the Krebs cycle.

   (B) Autotrophs use CO_2 to make glucose in the Krebs cycle, and heterotrophs produce CO_2 as a waste product in the Calvin cycle.

   (C) Autotrophs produce CO_2 as a waste product in the Krebs cycle, and heterotrophs use CO_2 to make glucose in the Calvin cycle.

   (D) Autotrophs produce CO_2 as a waste product in the Calvin cycle, and heterotrophs use CO_2 to make glucose in the Krebs cycle.

3. Chlorophyll absorbs light energy and stores it as chemical energy by catalyzing redox reactions. Which of the following figures most accurately illustrates the structures required for photosynthesis?

(A)

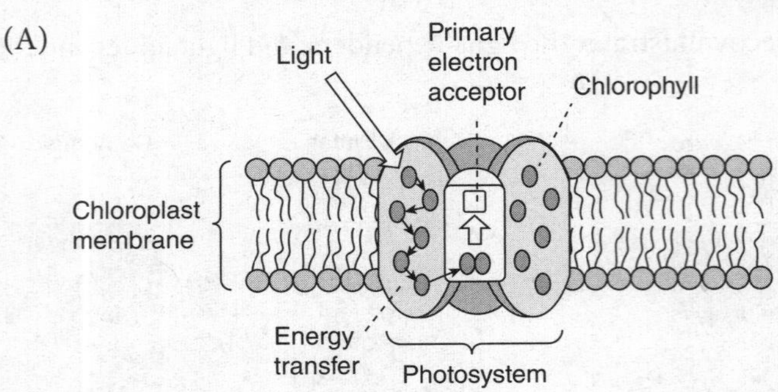

(B)

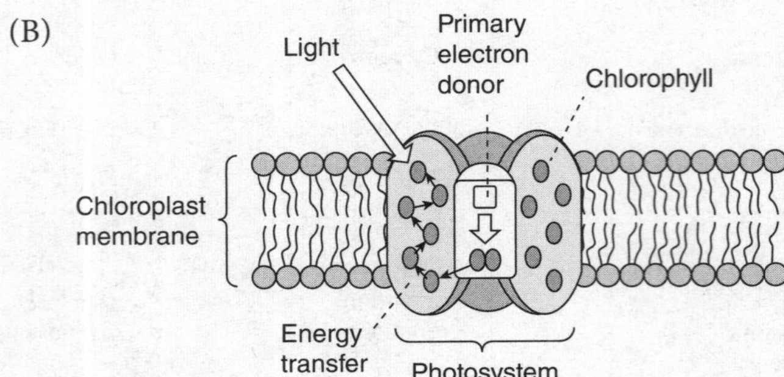

(C)

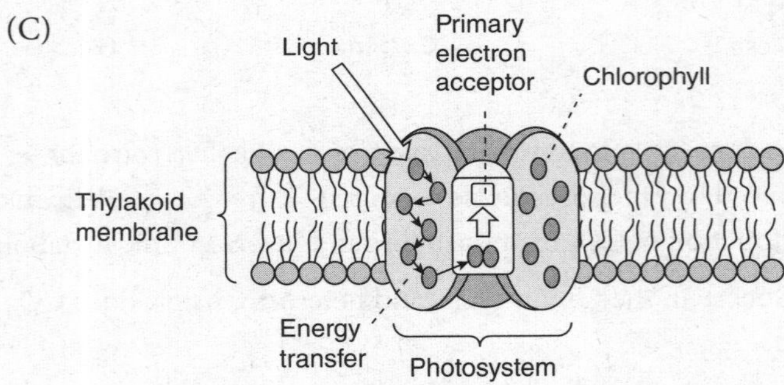

(D)

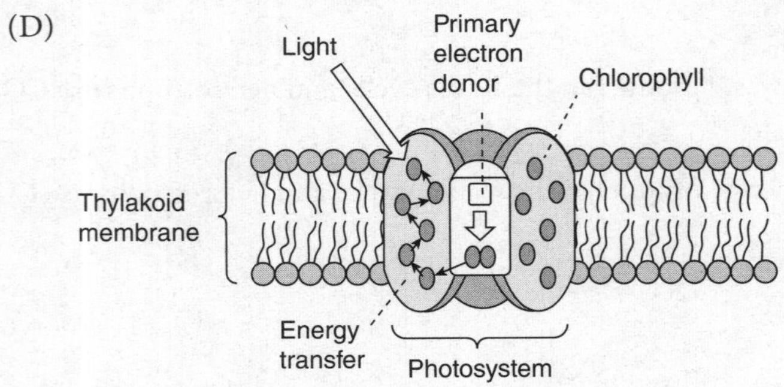

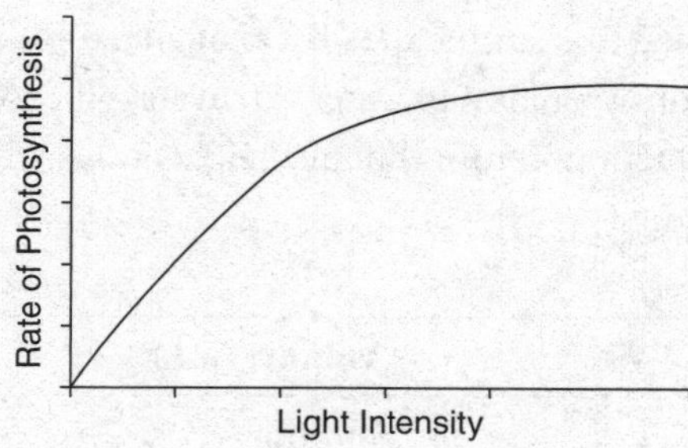

4. The graph above shows the rate of photosynthesis as a function of light intensity. The rate of photosynthesis as a function of carbon dioxide concentration exhibits similar behavior. Which of the following best explains the relationship between carbon dioxide concentration and the rate of photosynthesis?
   (A) At low concentrations, an increase in carbon dioxide has no effect on the rate of photosynthesis.
   (B) At low concentrations, carbon dioxide is a limiting factor of photosynthesis.
   (C) At high concentrations, carbon dioxide concentration is directly related to the rate of photosynthesis.
   (D) At high concentrations, carbon dioxide is a limiting factor of photosynthesis.

5. According to the Warburg effect, cancer cells have been observed to undergo aerobic glycolysis. Unlike normal cells, which carry out aerobic respiration when oxygen is available, cancer cells metabolize glucose to lactic acid via anaerobic fermentation despite the presence of oxygen. Cancer cells produce energy at glycolytic rates up to 200 times higher than those of normal cells. Which of the following best explains why cancer cells consume glucose at accelerated rates?
   (A) Aerobic glycolysis generates up to 36 ATP per molecule of glucose.
   (B) Converting glucose to lactate is less efficient than mitochondrial oxidative phosphorylation.
   (C) Lactic acid fermentation consumes more ATP than mitochondrial oxidative phosphorylation.
   (D) Oxygen is readily available in cancer cells to be consumed via glycolysis.

6. An experiment was performed to examine the effects of anaerobic respiration in yeast. Yeast was suspended in various solutions (containing water, pyruvate, glucose, sodium fluoride, and magnesium sulfate), and CO_2 production in mm (bubble height) was measured after incubation. The following data were obtained.

		Volume (mL)					CO_2
Tube	**Yeast**	**Water**	**Glucose**	**Sodium Fluoride**	**Pyruvate**	**Magnesium Sulfate**	**Produced (mm)**
1	yes	12.0	0.0	0.0	0.0	0.0	0.52
2	no	8.0	4.0	0.0	0.0	0.0	0.00
3	yes	8.0	4.0	0.0	0.0	0.0	1.49
4	yes	7.4	4.0	0.6	0.0	0.0	1.27
5	yes	5.0	4.0	3.0	0.0	0.0	0.80
6	yes	2.0	4.0	3.0	3.0	0.0	1.00
7	yes	2.0	4.0	0.0	0.0	6.0	1.98

According to the data, yeast in solution with both sodium fluoride and pyruvate produced more CO_2 than yeast in solution with sodium fluoride and no pyruvate. Which of the following statements provides the most reasonable explanation for this difference?

(A) The yeast oxidized the pyruvate, converting it to CO_2 and lactic acid.

(B) Pyruvate is a product of glycolysis, which is reduced to ethanol by yeast, so adding pyruvate would increase CO_2 production.

(C) Magnesium sulfate activates enzymes in fermentation in yeast, increasing CO_2 production.

(D) Sodium fluoride inhibits anaerobic respiration and promotes aerobic respiration, thus decreasing CO_2 production.

7. Since mitochondria are absent, the electron transport chain (ETC) in bacteria is located in the plasma membrane and is generally shorter than the mitochondrial transport chain in eukaryotes. Which of the following best describes the ETC in bacteria under aerobic conditions?
   (A) Oxygen donates an electron to initiate the bacterial ETC.
   (B) The bacterial ETC generates ATP via alcoholic fermentation.
   (C) Electrons are pumped from one side of the membrane to the other along the bacterial ETC.
   (D) The bacterial ETC has a lower ATP-to-oxygen ratio than the mitochondrial ETC.

8. Which of the following correctly identifies the products (per molecule of pyruvate) that would be generated if a person were to consume a diet that consisted only of pyruvate, rather than glucose?
   (A) If the individual were only undergoing aerobic respiration, 4 molecules of NADH and 1 molecule of $FADH_2$, in addition to 1 molecule of ATP would be generated in the Krebs cycle. The NADH and $FADH_2$ would then go through the electron transport chain, where they would be used to generate additional ATP.
   (B) If the individual were only undergoing aerobic respiration, 4 molecules of NAD^+ and 1 molecule of FAD, in addition to 1 molecule of ATP would be generated in the Krebs cycle. The NAD^+ and FAD would then go through the electron transport chain, where they would be used to generate additional ATP.
   (C) If the individual were only undergoing aerobic respiration, 7 molecules of NAD^+ and 2 molecule of FAD, in addition to 2 molecules of ATP would be generated through glycolysis and the Krebs cycle. The NAD^+ and FAD would then go through the electron transport chain, where they would be used to generate additional ATP.
   (D) If the individual were only undergoing aerobic respiration, 7 molecules of NADH and 2 molecule of $FADH_2$, in addition to 2 molecules of ATP would be generated through glycolysis and the Krebs cycle. The NADH and $FADH_2$ would then go through the electron transport chain, where they would be used to generate additional ATP.

9. A student sets up an experiment to measure the rate of photosynthesis by using a syringe to create a vacuum to remove trapped air and infiltrate the interior of plant disk samples with a solution containing bicarbonate ions that serve as a carbon source for photosynthesis. The infiltrated leaves sink in the bicarbonate solution, as shown below.

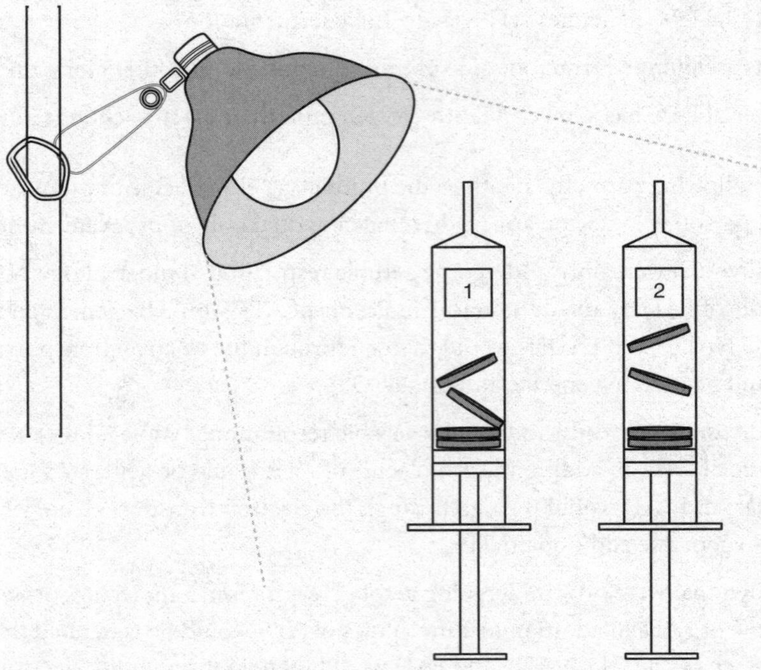

When placed in sufficient light, photosynthesis occurs, causing the plant disks to rise. The student leaves the room after 20 minutes, accidentally shutting off the light in the process. Which of the following graphs most accurately depicts the data he would likely collect through the experiment?

(A)

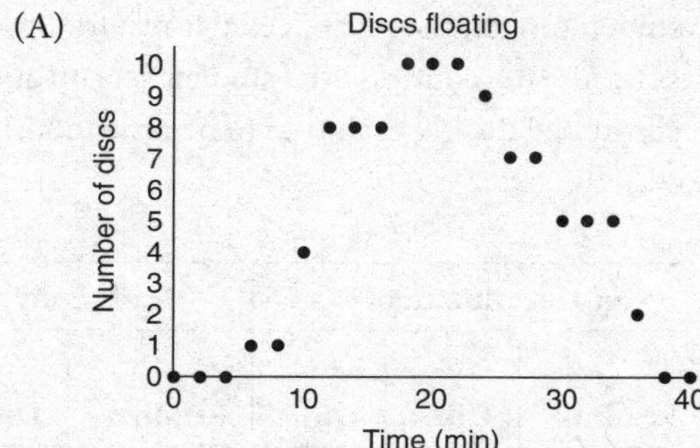
Discs floating
Number of discs
0
1
2
3
4
5
6
7
8
9
10
0
10
20
30
40
Time (min)

(B)

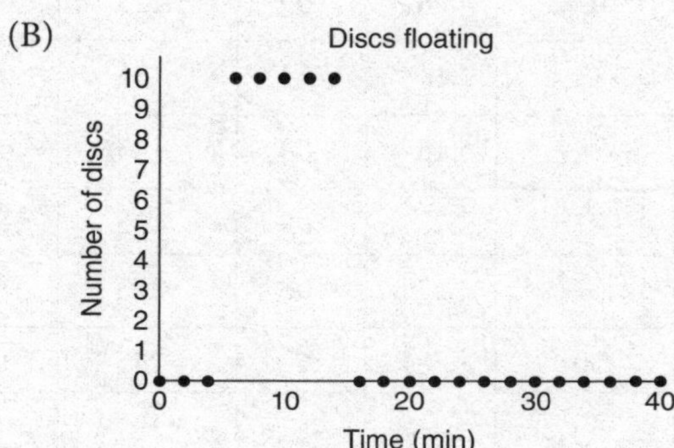
Discs floating
Number of discs
0
1
2
3
4
5
6
7
8
9
10
0
10
20
30
40
Time (min)

(C)

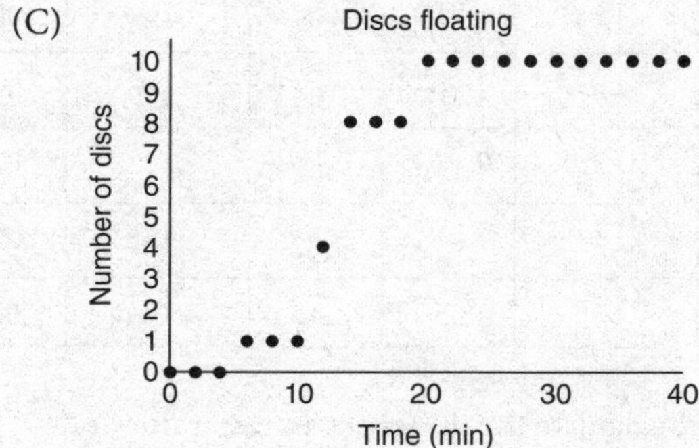
Discs floating
Number of discs
0
1
2
3
4
5
6
7
8
9
10
0
10
20
30
40
Time (min)

(D)

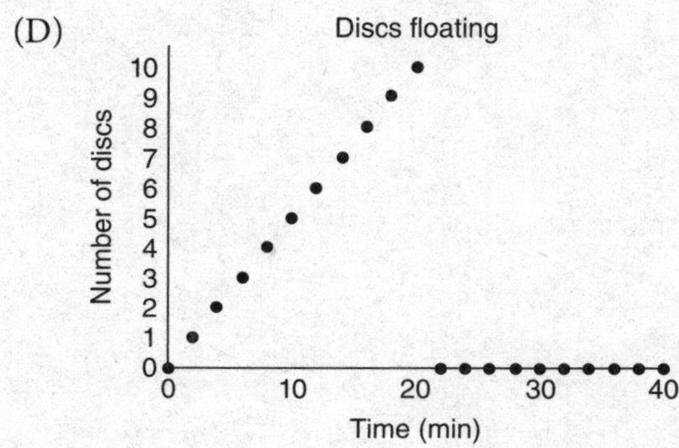
Discs floating
Number of discs
0
1
2
3
4
5
6
7
8
9
10
0
10
20
30
40
Time (min)

10. A student is attempting to observe the effects of temperature on the rate of cellular respiration for germinating peas, non-germinating peas, and glass beads (the control). The student sets up an experiment that allows for the volume of oxygen consumed during each trial to be measured. The data collected from the experiment is shown below.

		Germinating Peas			Non-Germinating Peas			Beads Only	
Temp (°C)	Time (min)	Reading	Diff.	Corr. Diff.	Reading	Diff.	Corr. Diff.	Reading	Diff.
25	0	.04			.07			.065	
25	5	.06	.02	−.03	.07	.08	.03	.115	.05
25	10	.09	.05	−.02	.15	.11	.04	.135	.07
25	15	.11	.07	−.01	.18	.16	.08	.145	.08
25	20	.15	.11	0	.23	.2	.09	.175	.11
25	25	.17	.13	−.02	.27	.23	.08	.215	.15
25	30	.22	.18	0	.30	.28	.1	.245	.18
10	0	.15			.35			.1	
10	5	.195	.045	−.015	.12	.01	−.05	.16	.06
10	10	.23	.08	.02	.13	.01	−.05	.16	.06
10	15	.25	.1	.065	.11	−.01	−.045	.135	.035
10	20	.31	.16	.14	.12	0	−.02	.12	.02
10	25	.47	.32	.285	.135	.015	−.02	.135	.035
10	30	.47	.32	.3	.17	.05	.03	.12	.02

Using the slopes of the corrected difference lines, calculate the difference in respiratory rate between germinating and non-germinating peas at 10°C seventeen minutes after the experiment began. Give your answer to the nearest hundredth.

(A) 0.01

(B) 0.02

(C) 0.14

(D) 133.33

Processes

Questions 11–12

The diagram below depicts the portion of the inner mitochondrial membrane with an ATP synthase embedded in it.

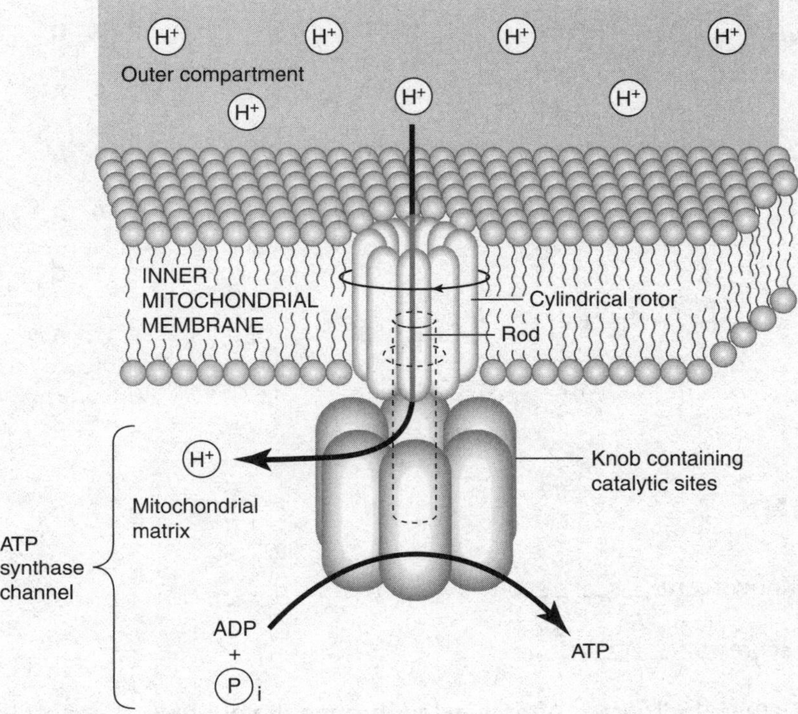

11. Which of the following is the immediate source of energy that drives the production of ATP from ADP and phosphate?
    (A) sunlight
    (B) the breakdown of glucose into carbon dioxide
    (C) hydrogen ions flowing down their concentration gradient
    (D) the pull of gravity on hydrogen ions

12. Assume that all reactions depicted in the diagram are reversible. What would happen if large amounts of ATP were supplied to the stroma side?
    (A) ATP would be hydrolyzed to ADP, causing the ATP synthase to pump hydrogen ions against their concentration gradient.
    (B) ATP would be hydrolyzed to ADP as the hydrogen ions flow down their concentration gradient.
    (C) Hydrogen ions would return through the membrane without making ATP.
    (D) ATP would continue to be synthesized from ADP as the hydrogen ions flow up their concentration gradient.

Answer Key

Test What You Already Know

1. C
2. A
3. D
4. A
5. D
6. C
7. B

Test What You Learned

1. D
2. A
3. C
4. B
5. B
6. B
7. D
8. A
9. A
10. A
11. C
12. A

REFLECTION

Test What You Already Know score: ___________

Test What You Learned score: ___________

Use this section to evaluate your progress. After working through the pre-quiz, check off the boxes in the "Pre" column to indicate which Learning Objectives you feel confident about. Then, after completing the chapter, including the post-quiz, do the same to the boxes in the "Post" column. Keep working on unchecked Objectives until you're confident about them all!

Pre	Post	
❑	❑	**13.1** Describe structures required for photosynthesis
❑	❑	**13.2** Explain the light-dependent reactions of photosynthesis
❑	❑	**13.3** Explain energy pathways for glycolysis and fermentation
❑	❑	**13.4** Explain the function of the electron transport chain
❑	❑	**13.5** Compare input and output molecules and energy
❑	❑	**13.6** Investigate effects on the rate of photosynthesis
❑	❑	**13.7** Investigate effects on the rate of cellular respiration

FOR MORE PRACTICE

Complete more practice online at kaptest.com. Haven't registered your book yet? Go to kaptest.com/booksonline to begin.

ANSWERS AND EXPLANATIONS

Test What You Already Know

1. C Learning Objective: 13.1

In chloroplasts, as electrons move through the electron transport chain, protons are pumped from the stroma into the thylakoid lumen, creating a proton gradient and producing NADPH. Protons flow down this gradient back out into the stroma through an ATP synthase to produce ATP. **(C)** correctly illustrates this process. (A) is incorrect because the flow of protons is reversed. (B) is incorrect because the locations of the stroma and thylakoid lumen are switched. (D) is incorrect because ATP and NADPH are being expended instead of produced.

2. A Learning Objective: 13.2

The concentration of NADPH increases in the chloroplast when the plant cell is not producing sufficient ATP to meet cell energy demands. In cyclic photophosphorylation, electrons move from the reaction center in photosystem I (P700), through an electron transport chain, and then back to the same reaction center to produce ATP. This process does not produce oxygen or NADPH. Thus, shifting from noncyclic photophosphorylation to cyclic photophosphorylation will only increase ATP levels. ATP is then used in the Calvin cycle to make sugars to meet cell energy demands. Therefore, the correct answer is **(A)**. (B) is incorrect because the opposite is true: ATP levels will increase and oxygen is not produced. (C) is incorrect because carbohydrates are synthesized in the Calvin cycle. (D) is incorrect because oxygen and NADPH are produced in noncyclic photophosphorylation.

3. D Learning Objective: 13.3

Lacking mitochondria, red blood cells predominantly produce energy by anaerobic respiration. Thus, red blood cells produce ATP via glycolysis followed by lactic acid fermentation, **(D)**. (A) is incorrect because aerobic cellular respiration occurs in mitochondria, which red blood cells lack. (B) is incorrect because the products of anaerobic fermentation are ATP and NAD^+ (glycolysis produces ATP and NADH, and fermentation regenerates NAD^+). (C) is incorrect because human red blood cells undergo lactic acid fermentation, not alcohol fermentation.

4. A Learning Objective: 13.4

By increasing the permeability of the inner mitochondrial membrane, thermogenin uncouples the electron transport chain from oxidative phosphorylation. Electrons are transferred to O_2 to generate H_2O, but protons that have been pumped out of the mitochondrial matrix into the intermembrane space return to the mitochondrial matrix due to the "leaky" membrane. Thus the proton gradient, which is needed to drive ATP production, is decreased, and energy is released as heat instead of used for ATP production. Thus, the correct answer is **(A)**. This mechanism is used by hibernating animals to stay warm. (B) is incorrect because decreasing the proton gradient would decrease ATP production. (C) and (D) are incorrect because thermogenin decreases the proton gradient.

5. D Learning Objective: 13.5

During cellular respiration, cells break down glucose ($C_6H_{12}O_6$) to make ATP. The products (CO_2 and H_2O) have less energy than the reactants, so the process is exergonic. During photosynthesis, plants use sunlight to make glucose from carbon dioxide and water. The products have more energy than the reactants, so the process is endergonic. **(D)** appropriately matches each process to its change in energy. (A) and (B) are incorrect because photosynthesis is not exergonic and cellular respiration is not endergonic. (C) is incorrect because cellular respiration consumes glucose to make ATP.

6. C Learning Objective: 13.6

DPIP acts as a substitute for $NADP^+$ in photosynthesis. During photosynthesis, DPIP is reduced and becomes colorless, which results in an increase in light transmittance. According to the data, an increase in light transmittance over time is found for the tube exposed to light only, making **(C)** correct. (A) is incorrect because DPIP changes from blue to clear when reduced during photosynthesis. (B) is incorrect because the tube exposed to light only had the highest rate of photosynthesis (highest percent transmittance). (D) is incorrect because the solution in all four tubes was blue at time 0.

7. B Learning Objective: 13.7

Aerobic phases of cellular respiration in eukaryotes occur in mitochondria, which synthesize ATP from ADP. Higher than normal levels of ADP suggest that the transgenic mice are unable to convert ADP to ATP as efficiently because the mutation has impaired their mitochondria, **(B)**. (A) is incorrect because a higher rate of oxygen consumption in mitochondria would decrease ADP levels (electrons are transferred to oxygen when ADP is phosphorylated to make ATP). (C) is incorrect because ADP levels would be greater when ATP is utilized ($ATP \rightarrow ADP + P_i$). (D) is incorrect because carbon dioxide is released, not consumed, in cellular respiration.

Test What You Learned

1. D Learning Objective: 10.2

In the light-dependent reactions of photosynthesis, light energy splits water, releasing oxygen, and NADPH and ATP are produced. In the light-independent reactions of photosynthesis, carbon dioxide is reduced by NADPH and ATP to make glucose. Thus, the correct answer is **(D)**. (A) and (B) are incorrect because photosynthesis consumes water and carbon dioxide and produces oxygen and glucose ($6CO_2 + 6H_2O \rightarrow C_6H_{12}O_6 + 6O_2$). (C) is incorrect because the light-dependent and light-independent reactions are switched: light-dependent reactions produce NADPH and ATP for light-independent reactions.

2. A Learning Objective: 10.5

Autotrophs convert free energy in sunlight to ATP and NADPH to produce carbohydrates from carbon dioxide in the Calvin cycle. Heterotrophs metabolize carbohydrates by hydrolysis, releasing carbon dioxide in the Krebs cycle while synthesizing ATP, NADH, and $FADH_2$. Thus, the correct answer is **(A)**. (B) is incorrect because the Calvin cycle and Krebs cycle are switched: autotrophs use CO_2 to make glucose in the Krebs cycle, and heterotrophs produce CO_2 as a waste product in the Calvin cycle. (C) and (D) are incorrect because heterotrophs do not use CO_2 to make glucose.

3. C Learning Objective: 10.1

Photosynthesis converts light energy to chemical energy. First, chlorophyll molecules arranged in photosystems embedded within thylakoid membranes of chloroplasts trap energy. The absorbed light energy of the photon is then transferred to an electron, and the excited electron is transferred to a primary electron acceptor. The chlorophyll is oxidized (loses an electron), and the electron is transferred to the electron transport chain. **(C)** correctly illustrates this process. (A) and (B) are incorrect because the light reactions occur in the thylakoid membrane, not the chloroplast membrane. (D) is incorrect because the direction of the energy transfer arrows is reversed and because the molecule should be a primary electron acceptor, not a donor.

4. B Learning Objective: 10.6

Since the rate of photosynthesis as a function of carbon dioxide concentration behaves similarly to the rate of photosynthesis as a function of light intensity, the graph can be used to represent the relationship between carbon dioxide concentration and the rate of photosynthesis. According to the graph, at low light intensity, the rate of photosynthesis increases as light intensity increases, whereas at high light intensity, the rate of photosynthesis plateaus (an increase in light intensity does not increase the rate of photosynthesis). Thus, light intensity is a limiting factor of photosynthesis at low light intensity. Carbon dioxide has the same effect, so at low carbon dioxide concentrations, carbon dioxide is a limiting factor of photosynthesis. Therefore, **(B)** is correct. (A) is incorrect because, at low concentrations, carbon dioxide and the rate of photosynthesis are directly related. (C) and (D) are incorrect because, at high concentrations, carbon dioxide does not affect the rate of photosynthesis.

5. B Learning Objective: 10.3

Metabolism of glucose to lactate generates only 2 ATP per molecule of glucose, whereas oxidative phosphorylation generates up to 36 ATP per molecule of glucose. Thus, a high rate of glucose uptake is required to meet the cancer cell's energy needs using this less efficient method, making **(B)** correct. (A) is incorrect because the cancer cells convert glucose to lactic acid (normally an anaerobic process) via aerobic glycolysis (glycolysis in the presence of oxygen), which only generates 2 ATP per molecule of glucose. (C) is incorrect because neither lactic acid fermentation nor mitochondrial oxidative phosphorylation consumes ATP. (D) is incorrect because although oxygen is present, it is not consumed in aerobic glycolysis.

6. B Learning Objective: 10.7

During anaerobic respiration in yeast (alcoholic fermentation), glycolysis breaks down glucose into pyruvate, which is then converted to carbon dioxide and ethanol. Thus, increasing the amount of pyruvate would, in turn, increase the amount of carbon dioxide and ethanol produced. **(B)** is the appropriate match. (A) is incorrect because pyruvate is oxidized under aerobic conditions and lactic acid is not a product of yeast fermentation. Although (C) is factually true, it does not explain why the solution with sodium fluoride and pyruvate increased respiration, and therefore is incorrect. (D) is incorrect because, according to the data, sodium fluoride does not promote aerobic respiration and also because aerobic respiration would increase CO_2 production.

7. D Learning Objective: 10.4

The electron transport chain produces a proton gradient that drives ATP production. During aerobic respiration, oxygen is the final electron acceptor in the ETC and is reduced to form water, marking the completion of the ETC. Since bacterial ETCs are usually shorter than mitochondrial ETCs, the proton gradient (used to drive ATP synthesis) that is generated by bacterial ETC will be less than that generated by mitochondrial ETC. Thus, bacterial ETC will produce less ATP per oxygen molecule. The correct answer is **(D)**. (A) is incorrect because oxygen is not an electron donor but the final electron acceptor. (B) is incorrect because alcoholic fermentation does not occur under aerobic conditions. (C) is incorrect because protons, not electrons, are pumped from one side of the membrane to the other to form a proton gradient.

8. A Learning Objective: 10.5

The person would not obtain energy from glycolysis because the pyruvate would go directly into pyruvate decarboxylation and then the Krebs cycle. Therefore, for each molecule of pyruvate consumed, 1 NADH would be generated through the oxidation reaction converting pyruvate to acetyl-CoA. One round of the Krebs cycle would then produce 3 molecules of NADH, 1 molecule of $FADH_2$, and 1 molecule of ATP. The NADH and $FADH_2$ would then enter the electron transport chain. where they would be used to generate more ATP. This best matches **(A)**. (B) is incorrect because NADH and $FADH_2$ are made during the Krebs cycle. NAD^+ and FAD are regenerated through the electron transport chain to allow the Krebs cycle to run continuously. (C) and (D) are incorrect because the pyruvate would not go through glycolysis since the purpose of glycolysis is to generate pyruvate from glucose.

9. A Learning Objective: 10.6

As photosynthesis occurs, the dense bicarbonate is converted to oxygen, which affects the buoyancy of the plant disc, allowing it to rise. However, when the student accidentally shuts off the light, the plants stop undergoing photosynthesis and instead begin to undergo cellular respiration. Plant respiration consumes the oxygen bubbles that had been generated during photosynthesis, causing the disks to sink again as a result. This is best shown in **(A)**. (B) and (D) can be eliminated because the disks would not all sink immediately after the light was turned off. (C) can be eliminated because, after the light is turned off, the plants would naturally begin to undergo cellular respiration, which would definitely affect their buoyancy.

10. A Learning Objective: 10.7

Before finding the difference between respiratory rates, the rates themselves must be calculated. As stated in the question stem, these rates can be calculated by finding the slopes of each experimental group at seventeen minutes. The corrected values are used rather than the original readings to account for any minute changes that may have affected the readings. The corrections are based on the glass beads since glass beads do not engage in cellular respiration. These calculations are shown below.

Germinating peas:

Time (min)	Reading	Diff.	Corr. Diff.
15	0.25	0.1	0.065
20	0.31	0.16	0.14

$$\text{Slope} = \frac{\text{corr. diff. 2} - \text{corr. diff. 1}}{\text{time 2} - \text{time 1}} = \frac{0.14 - 0.065}{20 - 15} = \frac{0.075}{5} = 0.015$$

Non-germinating peas:

Time (min)	Reading	Diff.	Corr. Diff.
15	0.11	-0.01	-0.045
20	0.12	0	-0.02

$$\text{Slope} = \frac{\text{corr. diff. 2} - \text{corr. diff. 1}}{\text{time 2} - \text{time 1}} = \frac{-0.02 - (-0.045)}{20 - 15} = \frac{0.025}{5} = 0.005$$

Thus, the difference in respiratory rate between the germinating and non-germinating peas is 0.015 – 0.005, which equals 0.01. **(A)** is correct.

11. C Learning Objective: 10.4

The diagram shows hydrogen ions flowing through the ATP synthase, down their concentration gradient, **(C)**. Although sunlight and the breakdown of glucose into carbon dioxide might provide energy to create the hydrogen ion gradient, they are not shown in the diagram and are not the immediate source of energy. Therefore, (A) and (B) are incorrect. In a mitochondrion, the membrane folds around in three dimensions. Thus, it would be equally correct to draw the diagram in the opposite orientation so that the hydrogen ions appear to be flowing up. Gravity has little to no effect at this scale, so (D) is incorrect.

12. A Learning Objective: 10.4

With a high enough concentration of ATP, breakdown of ATP would be favored. This would cause the pumping of hydrogen ions against their concentration gradient, matching **(A)**. (B) is incorrect because the energy released from ATP hydrolysis would drive hydrogen ions *up* their concentration gradient. (C) is incorrect because hydrogen ions cannot pass through the enzyme without spinning the rotor and causing a reaction. (D) is incorrect as both ATP synthesis and the movement of hydrogen ions up their concentration are energy requiring processes and could not occur together.

CHAPTER 11

The Cell Cycle

LEARNING OBJECTIVES

In this chapter, you will review how to:

11.1 Describe the three phases of interphase

11.2 Explain the process and function of mitosis

11.3 Explain the process and function of meiosis

11.4 Explain how meiosis promotes genetic diversity and evolution

11.5 Investigate genetic probabilities in cell division

TEST WHAT YOU ALREADY KNOW

1. Below is a diagram of the cell cycle, in which the length of each labeled arc is roughly proportional to the time that a cell spends in that corresponding stage of the cell cycle.

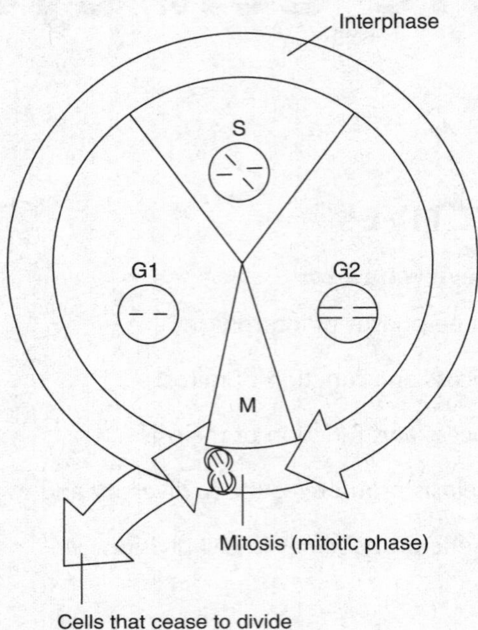

Which of the following most accurately describes a reason for the relative lengths of these stages?

(A) During the S phase, the cell synthesizes all the proteins necessary to carry out the various functions of the cell, so the S phase must be shorter than the G phases so that newly synthesized proteins have time to carry out their functions.

(B) Most of the time of the cell cycle is dedicated to the G phases so that the cell can grow, absorb nutrients, and synthesize the various biomolecules and organelles necessary for the survival and effective functioning of daughter cells.

(C) The M phase is shorter than the S phase because if the M phase took longer, the newly-synthesized chromosomes would become so unstable that one of the two daughter cells would inherit an incomplete set of chromosomes.

(D) The two G phases must be exactly the same length to ensure that the cell cycle repeats at perfectly even intervals, as the cell cannot control the cell cycle except by dividing at very consistent intervals of time.

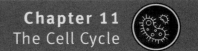

2. A cell culture is treated with a drug that effectively prevents DNA synthesis, but which affects no other process in the cell. How would the administration of this drug affect mitosis in a culture of otherwise healthy adult epithelial cells?

   (A) It would not affect the cells because each cell already has two copies of the whole genome, which is enough to allow for division.

   (B) It would not affect the cells because cells cease to divide once an organism reaches its full size at adulthood.

   (C) It would affect the cells because they would still be able to divide once but would end up with only half the ordinary amount of DNA.

   (D) It would affect the cells because they would not progress beyond the synthesis phase of the cell cycle and would, therefore, fail to divide.

3. The diagram below illustrates the effects of nondisjunction during meiosis I and II.

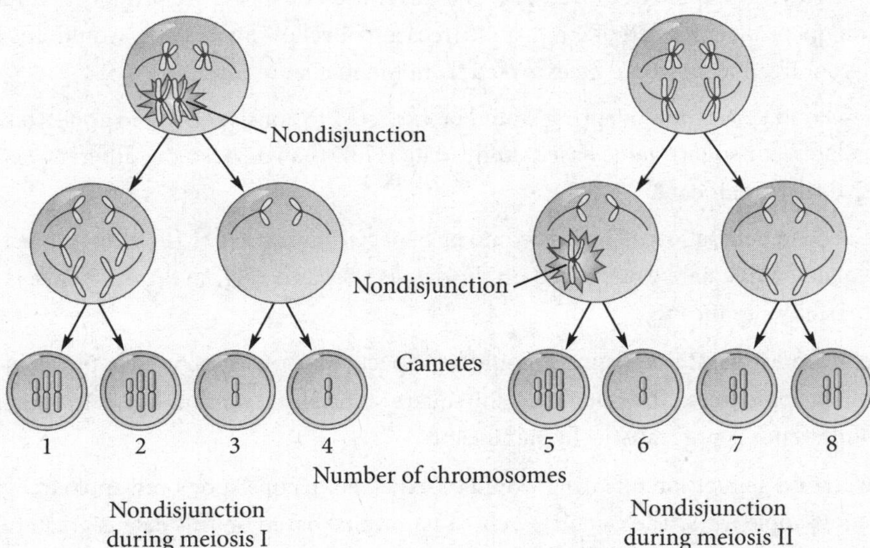

   What would happen if gamete 2 undergoes fertilization by combining with a normal gamete from the same species?

   (A) Gamete 2 has an extra chromosome. If it undergoes fertilization, there would be three copies of the longer chromosome in the offspring.

   (B) Gamete 2 has an extra chromosome. If it undergoes fertilization, there would be two copies of the longer chromosome in the offspring.

   (C) Gamete 2 has one fewer chromosome. If it undergoes fertilization, there would be one copy of the longer chromosome in the offspring.

   (D) Gamete 2 has one fewer chromosome. If it undergoes fertilization, there would be two copies of the longer chromosome in the offspring.

4. Which of the following best characterizes the evolutionary purpose of the chromosomal "crossing over" that occurs during meiotic division?

   (A) The structural integrity of chromosomes is optimized by fitting together the most stable pieces of different chromosomes.

   (B) The organism is able to swap out damaged or mutated portions of chromosomes, thereby improving the health and fitness of the resultant gametes.

   (C) Organisms can create superior offspring by combining only the best genes of each chromosome and passing on only those copies with good genes.

   (D) The random shuffling of genetic material allows for greater genetic diversity among offspring, thereby increasing the general fitness of a population.

5. A homozygous red apple (*RR*) tree was crossed with a homozygous yellow apple (*yy*) tree resulting in first generation offspring that produced only red apples. Next, a first generation offspring is crossed with a yellow apple tree and 10 second generation offspring are produced. If the second generation offspring consisted of 6 red apple trees and 4 yellow apple trees, would this be statistically different from what would be expected from Mendelian ratios?

   (A) The second generation offspring would be expected to consist of 10 red apple trees. The calculated chi-square value based on the data is less than 3.84, so the difference is not statistically significant.

   (B) The second generation offspring would be expected to consist of 10 yellow apple trees. The calculated chi-square value based on the data is less than 3.84, so the difference is not statistically significant.

   (C) The second generation offspring would be expected to consist of 5 red apple trees and 5 yellow apple trees. The calculated chi-square value based on this data is less than 3.84, so the difference is not statistically significant.

   (D) The second generation offspring would be expected to consist of 5 red apple trees and 5 yellow apple trees. The calculated chi-square value based on this data is greater than 3.84, so the difference is statistically significant.

Answers to this quiz can be found at the end of this chapter.

INTERPHASE

11.1 Describe the three phases of interphase

The cell cycle is the life cycle of the cell, including reproduction. In **somatic cells**, the cell cycle consists of the three stages of **interphase**, followed by mitosis, as shown in the following figure.

S = DNA replicates
G2 = Gap 2 (cell gets ready to divide)
M = Mitosis (cell division)
G1 = Gap 1 (cell grows)

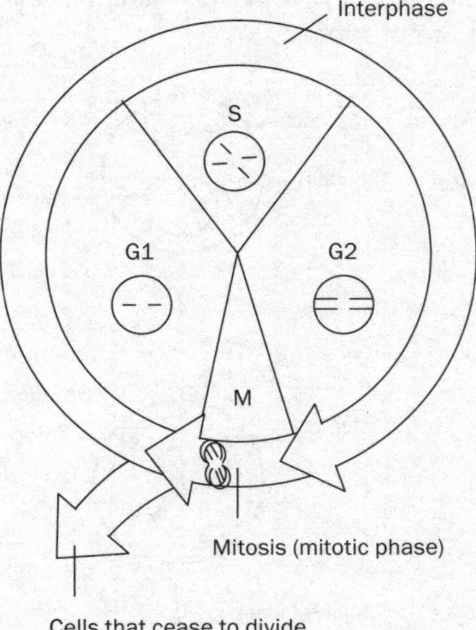

The Cell Cycle

Most somatic cells spend about 90 percent of the cell cycle in interphase. Interphase begins with G_1, the first gap phase, in which the cell grows and synthesizes RNA, proteins, and organelles. After sufficient growth has occurred, the cell passes the G_1 checkpoint and proceeds to the next phase. During the S phase, the synthesis phase, DNA replication occurs. Before replication, each chromosome in the cell consists of a single **chromatid** but, after replication, each consists of two identical sister chromatids, attached at the centromere. Next comes G_2, the second gap phase, in which the cell continues to grow and synthesize more proteins. After the cell grows enough to pass the G_2 checkpoint, it will then proceed on to the phases of mitosis.

✔ AP Expert Note

G_0

Some somatic cells cease dividing after they become specialized, entering a resting phase known as G_0. Specific cellular conditions can cause these resting cells to reenter the cell cycle and begin dividing again.

MITOSIS

High-Yield

11.2 Explain the process and function of mitosis

Mitosis (also known as the M phase) is the process through which a cell replicates and divides. The following diagram illustrates the four phases of mitosis (**prophase**, **metaphase**, **anaphase**, and **telophase**) and situates them within the larger cell cycle, which also includes interphase and cytokinesis, the division of cytoplasm in animal cells that follows the division of the nucleus when two distinct cells are produced.

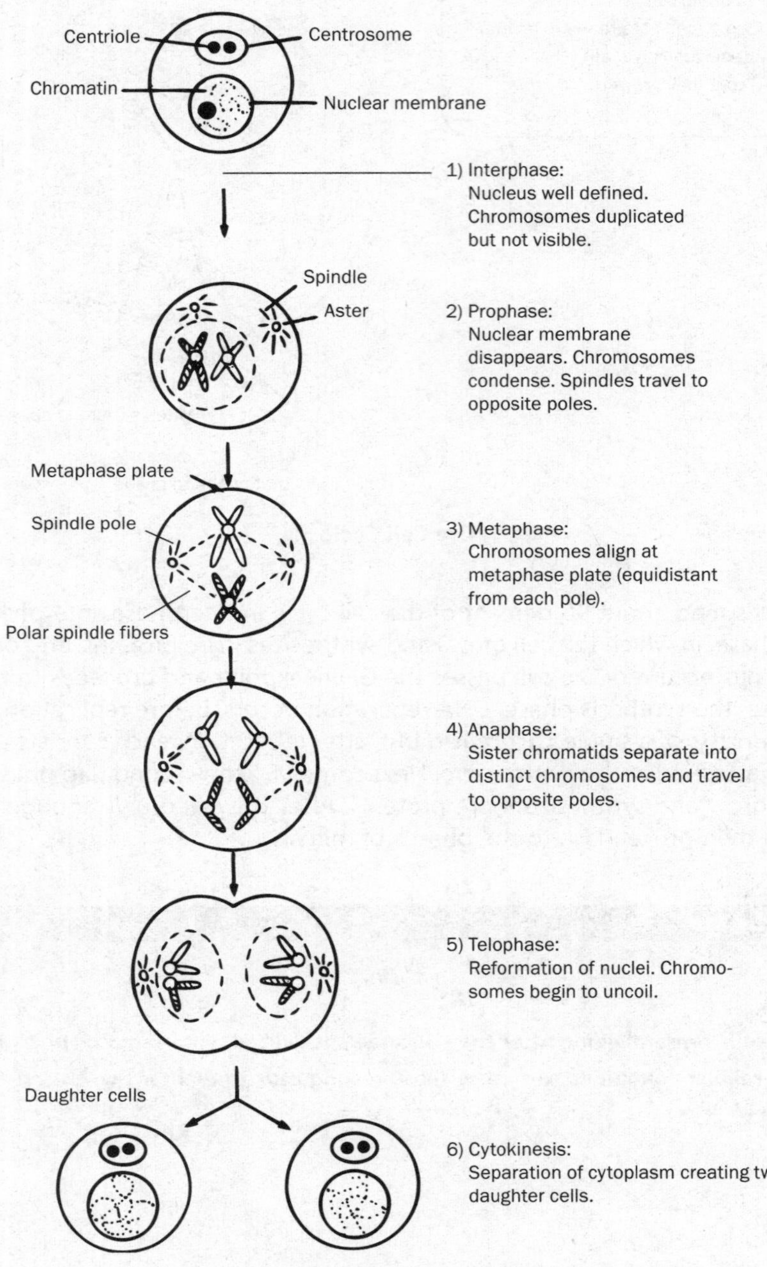

The Place of Mitosis in the Cell Cycle

✔ **AP Expert Note**

You are not required to know the names of the phases of mitosis or meiosis for the AP exam. Instead, focus on the overall processes (replication, alignment, and separation) and the theme of heredity.

Cell division is primarily controlled through genetics. The DNA in a cell controls whether a cell divides at all and at what rate. Certain hormones and other chemicals need to be present for a cell to divide. Cell division can be controlled in tissue cultures by inhibiting protein synthesis or affecting nutrient availability. Cell division is reduced by crowding through contact inhibition.

Phases of Mitosis

Among the key aspects of mitosis are the following:

- Chromosomes shorten and thicken in the nucleus, and the nuclear membrane dissolves.
- The mitotic spindle of microtubules is formed.
- The contractile ring of actin develops around the center of the cell.

For mitosis to work, a single pair of centrioles will copy themselves during the S phase, and the two pairs will move to opposite poles of the cell. These pairs of centrioles form the foundation for centrosomes, microtubule organizing centers that will shoot linked tubulin proteins across the cell as mitosis begins. Keep in mind, however, that the centrioles themselves are not necessary for microtubules to form from the centrosome areas of the cell. In fact, plant cells have centrosomes without centrioles. The thing to remember is that the centrosome regions form the two poles (like north and south) on opposite ends of the cells, between which microtubule spindle fibers will form.

Despite the conventional division into distinct stages, mitosis is a continuous process that does not stop between each phase. Four of these stages comprise mitosis, and the final stage—**cytokinesis**—completes the process of cell division as the cell pinches in two.

Prophase

In prophase, **chromatin** shows up under the microscope as well-defined chromosomes. These chromosomes are an X shape, two sister chromatids connected by a centromere, a specific DNA sequence. The mitotic **spindle** begins to form and elongate from the centrosome regions during prophase.

Metaphase

During metaphase, kinetochore microtubules push equally from opposite poles so that chromosomes are aligned in the middle of the cell. This center area where the alignment occurs is called the metaphase plate.

Anaphase

In anaphase, paired sister chromatids separate as kinetochore microtubules shorten rapidly. The polar microtubules lengthen as the kinetochore microtubules shorten, thereby pushing the poles of the cell farther apart.

Telophase

During telophase, separated sister chromatids group at opposite ends of the cell, near the centrosome region, having been pulled there by the receding microtubules. A new nuclear envelope forms around each group of separated chromosomes. At this point, mitosis has ended.

Cytokinesis

In cytokinesis, the contractile ring of actin protein fibers shortens at the center of the cell. A cleavage furrow, or indentation, is created as the ring contracts. As one side of the cell contacts the other, the membrane pinches off and two cells exist where one did before.

MEIOSIS

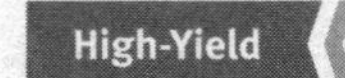

11.3 Explain the process and function of meiosis

11.4 Explain how meiosis promotes genetic diversity and evolution

Meiosis is the process that sexually reproducing organisms use to generate **gametes**, **haploid** cells that combine (usually with gametes from another member of the same species) to produce offspring with a full **diploid** set of chromosomes. The overarching purpose of meiosis is to facilitate sexual reproduction, which increases the genetic diversity of a population. This allows for greater variation within a species, which can help that species adapt to changing circumstances and evolve.

> ✔ **AP Expert Note**
>
> **Ploidy**
>
> **Ploidy refers to the number of sets of chromosomes within the cells of an organism. An organism or cell that has two sets of chromosomes is called diploid (designated as 2*N*). A diploid organism has one set of chromosomes from each parent, for a total of two sets. When a cell or organism has only one set of chromosomes, it is called haploid (1*N* or *N*). Note that diploid organisms will produce haploid gametes, but their somatic cells will be diploid. In some cases, cells have several sets of chromosomes and are called polyploid (3*N*, 4*N*, etc.). There are a few organisms that exist in a natural state with haploid or polyploid cells, but most organisms are diploid.**

Meiosis occurs in two steps, *meiosis I* and *meiosis II*. Each step includes a round of division and is divided into four phases. Meiosis II is essentially mitosis with half the number of chromosomes, but meiosis I contains some key differences from both meiosis II and mitosis.

Meiosis I

The four phases of meiosis I are described below.

Prophase I

The chromatin condenses into chromosomes, the spindle apparatus forms, and the **nucleoli** and **nuclear membrane** disappear. Homologous chromosomes (chromosomes that code for the same traits, one inherited from each parent) come together and intertwine in a process called **synapsis**.

Processes

Because at this stage each chromosome consists of two sister chromatids, each synaptic pair of homologous chromosomes contains four chromatids and is therefore often called a **tetrad**.

Sometimes chromatids of homologous chromosomes break at corresponding points and exchange equivalent pieces of DNA; this process is called **crossing over**. Note that crossing over occurs between homologous chromosomes and not between sister chromatids of the same chromosome. (The latter are identical, so crossing over would not produce any change.) Those chromatids involved are left with an altered but structurally complete set of genes. The chromosomes remain joined at points, called chiasmata, where the crossing over occurred. Such genetic recombination can unlink linked genes, thereby increasing the variety of genetic combinations that can be produced via gametogenesis. Recombination among chromosomes results in increased genetic diversity within a species. Note that sister chromatids are no longer identical after recombination has occurred.

Metaphase I

Homologous pairs (tetrads) align at the equatorial plane, and each pair attaches to a separate spindle fiber by its kinetochore.

Anaphase I

The homologous pairs separate and are pulled to opposite poles of the cell. This process is called disjunction, and it accounts for a fundamental Mendelian law: independent assortment. During disjunction, each chromosome of paternal origin separates (or disjoins) from its homologue of maternal origin and either chromosome can end up in either daughter cell. Thus, the distribution of homologous chromosomes to the two intermediate daughter cells is random with respect to parental origin. Each daughter cell will have a unique pool of alleles (genes coding for alternative forms of a given trait; e.g., yellow flowers or purple flowers) from a random mixture of maternal and paternal origin. In other words, each daughter cell will almost certainly have some chromosomes of paternal origin and some of maternal origin, rather than all paternal or all maternal.

Telophase I

A nuclear membrane forms around each new nucleus. At this point, each chromosome still consists of sister chromatids joined at the centromere. The cell divides (by cytokinesis) into two daughter cells, each of which receives a nucleus containing the haploid number of chromosomes. Between cell divisions, there may be a short rest period, or interkinesis, during which the chromosomes partially uncoil.

Meiosis II

This second division is very similar to mitosis, except that meiosis II is not preceded by chromosomal replication.

Prophase II

The centrioles migrate to opposite poles, and the spindle apparatus forms.

Metaphase II

The chromosomes line up along the equatorial plane. The centromeres divide, separating the chromosomes into two sister chromatids.

Anaphase II

The sister chromatids are pulled to opposite poles by the spindle fibers.

Telophase II

A nuclear membrane forms around each new (haploid) nucleus. Cytokinesis follows and two daughter cells are formed. Thus, by the completion of meiosis II, four haploid daughter cells are produced per gametocyte. (In human females, only one of these becomes a functional gamete.)

Comparison with Mitosis

In addition to having different functions (mitosis allows for the replication of somatic cells, while meiosis enables sexual reproduction), there are a number of notable differences between mitosis and meiosis. These differences are summarized in the following table and figure.

MITOSIS VS. MEIOSIS						
Process	**Number of Chromosomes in Parent Cell**	**Prophase/ Prophase I**	**Metaphase/ Metaphase I**	**Anaphase/ Anaphase I**	**Number of Daughter Cells**	**Number of Chromosomes in Daughter Cells**
Mitosis	2*N*	Replicated chromosomes come into view as sister chromatids	Individual chromosomes align at the metaphase plate	Centromeres separate and sister chromatids travel to opposite poles	2	2*N*
Meiosis	2*N*	Chromosomes form tetrads by synapsis, crossing over at chiasmata	Pairs of homologous chromosomes align at the metaphase plate	Synapsis ends and homologous chromosomes travel to opposite poles; sister chromatids travel to the same pole	4	*N*

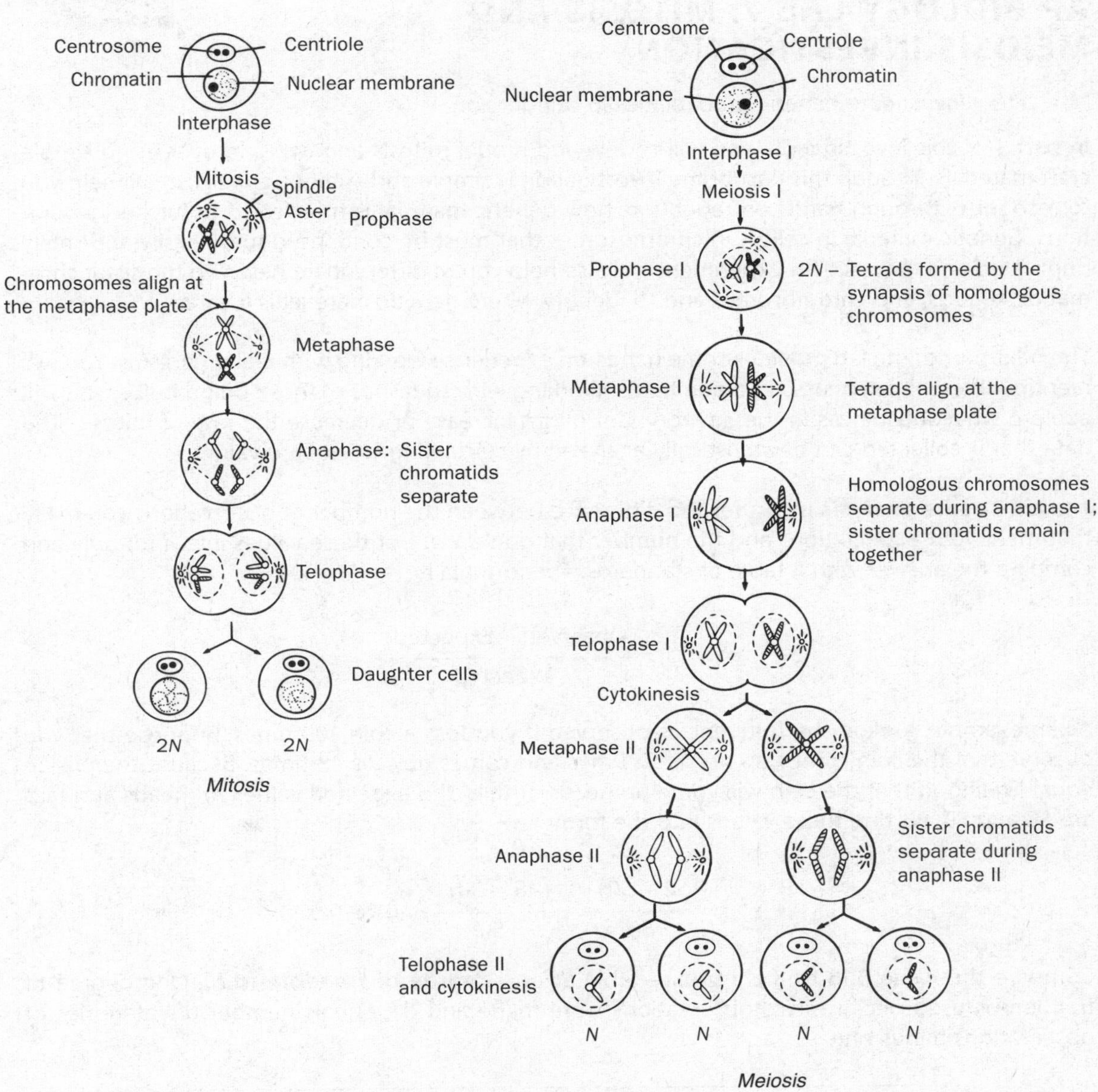

Comparison of Mitosis and Meiosis

AP BIOLOGY LAB 7: MITOSIS AND MEIOSIS INVESTIGATION

11.5 Investigate genetic probabilities in cell division

In part 1 of this investigation, you will review and model mitosis and crossing over using simple craft materials. Though this part of the investigation is simple and straightforward, it will help you to recognize, through hands-on repetition, how genetic material is transferred to further generations. Genetic material in cells is in chromosomes that must be condensed to be easily and safely duplicated and moved. This experiment will also help you to differentiate between the sister chromatids, tetrads, and chromosomes and to identify where genetic material is located.

The next part of the lab provides some hands-on experience working with real organisms. You will identify different cell phases in onion bulbs. Working with squashes of those onion bulbs, you will explore what substances in the environment might increase or decrease the rate of mitosis. The data that is collected can be statistically analyzed by calculating chi-square values.

The **chi-square analysis** measures the difference between the number of observations you make that meet your expectations and the number that don't. You put these values into a formula and compare the answer with a table of standards. The formula is:

$$\chi^2 = \sum \frac{(\text{Observed} - \text{Expected})^2}{\text{Expected}}$$

As an example, look at the frequencies observed if you toss a coin 100 times. Suppose that you observe that the coin comes up heads 52 times and comes up tails 48 times. Because there is an equal likelihood that the coin will come up heads or tails, the expected values for heads and tails are 50 and 50. Putting these values into the formula:

$$\chi^2 = \frac{(52-50)^2}{50} + \frac{(48-50)^2}{50} = 0.16$$

Compare the value 0.16 to a chi-square table at one **degree of freedom (d.f.)**. One degree of freedom is used because two observations were made and d.f. equals number of categories of observations minus one.

Probability (*p*)	**Degrees of Freedom (d.f.)**				
	1	2	3	4	5
0.05	3.84	5.99	7.82	9.49	11.1

At $p = 0.05$ in the chi-square table, the value obtained from the previous equation would have to be at least 3.8 for your observations to be statistically different from what is expected. The value calculated was only 0.16, so it can be concluded that although your observations weren't exactly $\frac{50}{50}$, they were not statistically significantly different from the expected outcome. There are many considerations and assumptions to make when performing statistical analyses, which would be covered in greater depth in a class on statistical analysis. For the chi-square analysis, there are three key points to remember:

- The formula for the chi-square statistic:

$$\chi^2 = \sum \frac{(\text{Observed} - \text{Expected})^2}{\text{Expected}}$$

- The formula for degrees of freedom: d.f. = # of categories of observations − 1
- Your calculated χ^2 must be greater than the corresponding p value in the table for the outcome to be significantly different than what you expected.

✔ AP Expert Note

The formula for chi-square and the degrees of freedom table will be provided to you in the equations sheet on Test Day, so you do not need to memorize them. You are most likely to see chi-square tested in grid-in questions, but it could also appear in other question types.

Part 3 of the lab asks you to explore the differences between normal cells and cancer cells. Remember that cancer typically results from a mutation in a gene or protein that controls the cell cycle. Comparing pictures of chromosomes from normal and HeLa cells, you will be able to count the number of chromosomes found in each type of cell and identify differences in appearance. The main conclusion to draw from this part of the investigation is that in normal cells, division is blocked if there is damage to or mutations in DNA. In contrast, this ability to block division is lacking in cancer cells, and they divide uncontrollably.

Finally, you will review meiosis and nondisjunction events by modeling. Based on this review, you will measure crossover frequencies and genetic outcomes in a fungus. This part of the investigation will also give you solid experience with the microscope by preparing slides and counting fungal asci (singular: ascus) under the microscope. Though you will not need to use a microscope or describe how to use a microscope on the AP Biology exam, these experiences are helpful for answering short and long free-response questions, which often ask you to describe experimental setups and procedures.

RAPID REVIEW

If you take away only 3 things from this chapter:

1. Interphase consists of three stages: G_1 (first gap phase), S (synthesis, in which DNA replication occurs), and G_2 (second gap phase). Interphase is followed in somatic cells by mitosis.
2. Mitosis is the process in which a cell produces two identical daughter cells. Its stages include prophase, metaphase, anaphase, and telophase. Cytokinesis occurs immediately following mitosis and refers to the splitting of the cell into two new cells.
3. Meiosis refers to the process by which sexually reproducing organisms produce sex cells (gametes) with half the chromosomes (haploid) of the rest of the organism's cells (which are diploid). It consists of two rounds of division, meiosis I and meiosis II, and it results in the creation of four gametes. In sexual reproduction, the male and female gametes join to create a new organism with the normal diploid number of chromosomes.

TEST WHAT YOU LEARNED

1. The effects of a particular insecticide on vegetable roots were studied. Two batches of parsnip roots were immersed in water, and one batch was treated with 0.5 M solution of insecticide while the other was untreated. After 5 days, samples were collected and prepared from both batches and the tips were observed under a microscope. The total number of cells and those undergoing mitosis were counted. The data is shown in the table below.

Parsnip Tips with Insecticide			Parsnip Tips without Insecticide		
Phase	**Cell Count**	**% Cells in Phase**	**Phase**	**Cell Count**	**% Cells in Phase**
Interphase	575	88%	Interphase	330	75%
Mitosis	78	12%	Mitosis	110	25%
Total Cell Count	653		**Total Cell Count**	440	

Does the insecticide have a statistically significant effect on the mitotic division of cells in parsnip root tips?

(A) The calculated chi-square value based on this data is greater than 3.84. Therefore, the 0.5 M solution of insecticide does not have a significant negative effect on the mitotic division of cells in parsnip root tips.

(B) The calculated chi-square value based on this data is less than 3.84. Therefore, the 0.5 M solution of insecticide does not have a significant negative effect on the mitotic division of cells in parsnip root tips.

(C) The calculated chi-square value based on this data is greater than 3.84. Therefore, the 0.5 M solution of insecticide has a significant negative effect on the mitotic division of cells in parsnip root tips.

(D) The calculated chi-square value based on this data is less than 3.84. Therefore, the 0.5 M solution of insecticide has a significant negative effect on the mitotic division of cells in parsnip root tips.

2. Which is the most likely cause of an error during meiosis that results in daughter cells with an incorrect number of chromosomes?

(A) Displacement of sister chromatids to opposite poles of the dividing cell

(B) Recombination of genetic material during the earlier stages of meiosis

(C) Activation of the securin proteins that separate chromatids during anaphase II

(D) Nondisjunction of the genetic material at any point during meiosis

3. Certain cell types may become senescent, meaning that they no longer divide. A simple example of this phenomenon can be found in skin cells, which originate from live progenitor cells beneath the surface layers of skin. Progenitor cells will undergo a simple mitotic division in which one daughter cell remains anchored to the basilar membrane, while the other daughter cell travels toward the surface of the skin and ceases to undergo further mitotic divisions. How do the genetic contents of these two daughter cells compare?

   (A) During any normal mitotic division, all the genetic information is copied and passed along to both daughter cells equally.

   (B) Modifications are made to the genetic information passed on to the mobile cell, which renders it unable to divide further.

   (C) The anchored daughter cell retains all the genetic information so that it can continue to create new cells, while the mobile cell retains none and later dies.

   (D) Neither cell retains the full set of genetic information after division, leading to the gradual aging of the skin.

4. Before a cell can undergo cellular division, it must progress through an interphase stage in which the cell matures and produces proteins needed for division. Which of the following best describes the interphase stage?

   (A) The cell grows, replicates its DNA, and prepares to divide during interphase. If a genetic mutation is introduced during interphase and not repaired, the daughter cells will inherit the mutation.

   (B) The cell grows, replicates its DNA, and prepares to divide during interphase. None of the genetic mutations introduced during interphase are inherited by the daughter cells.

   (C) The cell replicates its DNA and prepares to divide during interphase, but does not grow. None of the genetic mutations introduced during interphase are inherited by the daughter cells.

   (D) The cell replicates its DNA and prepares to divide during interphase, but does not grow. If a genetic mutation is introduced during interphase and not repaired, the daughter cells will inherit the mutation.

5. A couple with black hair has three children with black hair and one child with red hair. How did this most likely occur?

   (A) A child with red hair is possible through genetic drift, in which two chromosomes exchange genetic information, increasing the diversity of the individual's genome.

   (B) Each member of the couple was heterozygous, carrying one dominant gene for black hair and one recessive gene for red hair.

   (C) Having a child with red hair is possible if genes on the father's Y chromosome mutated and the mutated Y chromosome was passed on to the red-haired child.

   (D) Giving birth to a child with red hair can occur through gene flow, in which new genes enter a population as a result of migration.

Questions 6–7

The diagram below shows crossing over, a process that happens to homologous chromosomes during prophase I of meiosis.

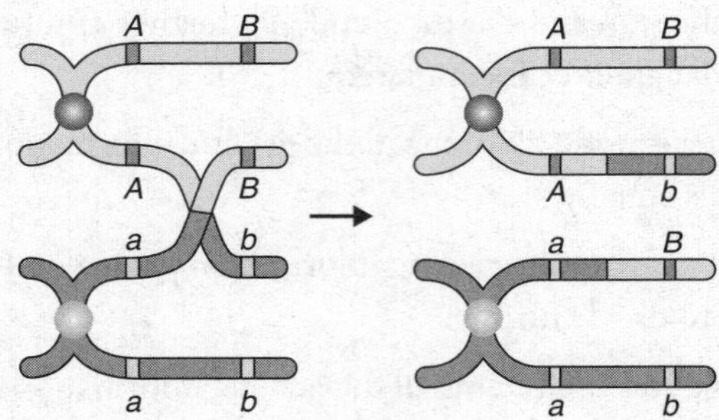

6. Which of the following best describes the importance of the process shown?
   (A) The process contributes to genetic diversity by recombining alleles in new patterns.
   (B) The process contributes to evolution by causing random point mutations.
   (C) The process contributes to genetic drift by preventing heterozygosity.
   (D) The process contributes to embryonic development by ensuring that all cells have the same DNA.

7. Which of the following is true of the segments labeled *B* and *b*, which represent different alleles?
   (A) They likely differ in their entire sequence.
   (B) Segment *B* is made of DNA, while segment *b* is composed of RNA.
   (C) They likely differ in only one or a few nucleotides.
   (D) They likely have the exact same sequence.

Questions 8–9

Fungi are capable of reproducing both sexually and asexually. Both types of reproduction lead to the production of spores, which are then dissipated away from the parent organism via the wind or by latching onto a passing animal. Fungi produce a large number of spores to increase the likelihood that one of them will land in an environment that will support its growth and survival.

Fungi reproduce asexually via fragmentation, budding, or spore production. Fragmentation occurs when a fungal mycelium splits into pieces, allowing each piece to grow into its own independent mycelium. During budding, a large bulge forms on one side of the cell, the nucleus divides, and the bud detaches from the parent cell. Spore production is the most common form of asexual reproduction. The resulting spores are genetically identical to the parent that produces them.

Fungal sexual reproduction begins with the production of two mating types. If both of the mating types are present in the same mycelium, it is termed homothallic, or self-fertile, reproduction. If the mating types are present in two different mycelia, it is termed heterothallic reproduction. There are a lot of variations in fungal sexual reproduction, but all of them involve three stages. It begins with plasmogamy, during which two haploid cells fuse, allowing two haploid nuclei to coexist in one cell. During the next step, karyogamy, the haploid nuclei fuse to form a diploid zygote nucleus. Finally, gametes of different mating types are generated. These spores are then spread through the environment.

8. A common phenomenon that occurs during fungal sexual reproduction, whether homothallic or heterothallic, is the pairing of homologous chromosomes and the exchanging of different segments of their DNA. Which of the following best describes an advantage of this process of "crossing over"?

   (A) It increases the number of chromosomes in the offspring, providing increased genetic diversity in the species.

   (B) It allows for the best genes from each mating type to be combined into one chromosome, providing increased fitness in the offspring.

   (C) It enables the mating types to replace regions of DNA that have been damaged in some way with each others' normal DNA.

   (D) It involves the random recombination of DNA , providing increased genetic diversity among offspring.

9. A certain type of fungus maintains a commensalistic relationship with a species of drought-tolerant oak tree known to grow in northern Nebraska. Recent changes in logging practices in the area have led many local loggers to cut down many of these oak trees. In response, the fungi begin to reproduce rapidly, in an attempt to ensure their continued survival. Which of the following provides the most likely method of reproduction that the fungi uses in these circumstances?

   (A) Budding, because it allows the fungi to reproduce quickly in response to their rapidly changing environment

   (B) Sexual spore production, because it increases the likelihood of variations that may help the fungi adapt to the changing environment

   (C) Fragmentation, because it does not require the exertion of a large amount of energy, making it the best option during this stressful situation

   (D) Asexual spore production, because there is no need to waste time finding a compatible mate

10. A student prepares two samples of fungi in culture. She then adds lectin, a protein that binds to certain carbohydrates, to one of the samples, leaving the other one untreated as a control. This specific lectin is known to induce mitosis by binding to a membrane-bound antigen on the cell surface. This causes an intracellular signal cascade that causes calcium release from the endoplasmic reticulum (ER) and, ultimately, cell replication.

After the samples are stored for two days, the number of cells from each group in interphase and mitosis are observed and shown in the table below.

	Interphase	**Mitosis**	**Total**
Control	296	50	346
Treated	322	176	498
Total	618	226	844

The values that were expected can be calculated as given in the following table.

	Interphase	**Mitosis**
Control	[(Total Control) × (Total Interphase)] ÷ Overall Total	[(Total Control) × (Total Mitosis)] ÷ Overall Total
Treated	[(Total Treated) × (Total Interphase)] ÷ Overall Total	[(Total Treated) × (Total Mitosis)] ÷ Overall Total

Calculate the chi-squared value based on the data provided. Give your answer to the nearest whole number.

(A) 1

(B) 4

(C) 46

(D) 59

11. On average, most human cells become senescent, or stop dividing, after 52 divisions. This is known as the Hayflick limit. However, one characteristic that makes cancer cells abnormal is that they never reach this limit. Which of the following correctly explains this phenomenon?

(A) During the 51st division of a normal cell, the DNA that is passed on to the daughter cells is reversibly modified to prevent it from dividing further; however, one of the mutations occurring in cancerous cells reverses this modification, allowing the cancer cell to continue dividing.

(B) In normal cells, mitosis alternates with interphase in the cell cycle, allowing it to enter into a stage where it no longer divides; however, one of the mutations occurring in cancerous cells makes them unable to enter the interphase portion of the cell cycle, so they are therefore forced to continuously divide in the mitotic stage.

(C) During mitosis in a cell that has already divided 51 times, the parent cell passes on all of its genetic information to only one of the daughter cells, so that the other can no longer divide; however, one of the mutations occurring in cancerous cells counteracts this, allowing both daughter cells to receive the parent's DNA.

(D) As normal cells pass through the cell cycle, the protective ends of their DNA degrade slightly with each division so that by the time they reach their 52nd division, the ends are completely gone; however, one of the mutations occurring in cancerous cells generates a protein that counteracts the chromosomal degradation, allowing the cells to continue dividing.

12. Many viral infections, such as infections by the human papillomavirus (HPV), increase the likelihood of the host developing cancer at the site of infection. Which of the following best characterizes the mechanism by which such infections give rise to cancerous growths?

(A) Viruses may produce toxic proteins that attract the immune system to the virally infected cells, thereby distracting the immune system from detecting cancerous cells.

(B) Viruses disrupt the cell's ability to synthesize its own DNA and RNA, thereby preventing the cell from growing and replicating.

(C) Viruses can increase the rate at which cells grow and divide, thereby leading to uncontrolled cell growth and proliferation.

(D) Viruses can inhibit appropriate cell growth by consuming the cell's nutrients, thereby reducing the cell's ability to defend itself from carcinogenic chemicals.

Answer Key

Test What You Already Know

1. B
2. D
3. A
4. D
5. C

Test What You Learned

1. C
2. D
3. A
4. A
5. B
6. A
7. C
8. D
9. B
10. C
11. D
12. C

REFLECTION

Test What You Already Know score: ___________

Test What You Learned score: ___________

Use this section to evaluate your progress. After working through the pre-quiz, check off the boxes in the "Pre" column to indicate which Learning Objectives you feel confident about. Then, after completing the chapter, including the post-quiz, do the same to the boxes in the "Post" column. Keep working on unchecked Objectives until you're confident about them all!

Pre	Post	
❑	❑	**11.1** Describe the three phases of interphase
❑	❑	**11.2** Explain the process and function of mitosis
❑	❑	**11.3** Explain the process and function of meiosis
❑	❑	**11.4** Explain how meiosis promotes genetic diversity and evolution
❑	❑	**11.5** Investigate genetic probabilities in cell division

FOR MORE PRACTICE

Complete more practice online at kaptest.com. Haven't registered your book yet? Go to kaptest.com/booksonline to begin.

ANSWERS AND EXPLANATIONS

Test What You Already Know

1. B Learning Objective: 11.1

There are many potential reasons that could account for the relative lengths of these stages, so this question can be effectively answered by eliminating flawed answer choices that contain incorrect descriptions of events during the cell cycle. (A) is incorrect because it suggests that protein synthesis occurs only during the S phase, without addressing that the S (Synthesis) phase is so named because DNA is synthesized during that phase. In fact, protein synthesis occurs throughout interphase, especially during the G_1 and G_2 phases (the first and second Gap phases) while the cell is growing and synthesizing new proteins in preparation for the S phase and the M phase (Mitosis). (C) is incorrect because the newly synthesized chromosomes made during the S phase must maintain their stability throughout G_2, which is already much longer than either the S or M phases. Furthermore, if the chromosomes did become unstable, then both daughter cells would be affected because the new chromosomes contain both old and newly synthesized strands of DNA. (D) assumes that because the two phases are drawn to the same proportion, they are exactly the same, even though the question stem states that the lengths in the figure are only "roughly proportional" to the relative durations of these stages. (D) also ignores the various mechanisms that cells possess to regulate the cell cycle, including checkpoints, cyclin-dependent kinases, and other cell signaling factors. **(B)** correctly describes the purpose of G_1 and G_2 (first and second Gap phases) with an appropriate description of the activities taking place within a cell during that stage.

2. D Learning Objective: 11.2

DNA synthesis is a necessary step in the cell cycle of a mitotically dividing cell, and epithelial cells are a common example of mitotically dividing cells (only a few types of cells do not routinely divide, such as neurons and cardiac muscle cells). Checkpoints in the cell cycle prevent mitosis in the event that certain events do not happen as planned. In order to move from the synthesis phase into the second growth phase before mitosis, DNA synthesis must be completed. However, if DNA synthesis were blocked by a drug, chromosomes could not be copied and the cell would remain unable to complete the synthesis phase and progress toward mitosis, thereby preventing the cell from dividing. Thus, **(D)** is correct. (A) and (B) are incorrect because the cells would be affected. (C) is incorrect because the cells would not even be able to divide once, as noted above.

3. A Learning Objective: 11.3

The cell depicted contains two pairs of chromosomes, so each of its gametes would ordinarily contain only two chromosomes total. Gamete 2 actually contains three chromosomes (one extra), including two copies of the longer chromosome. When gamete 2 combines with a standard gamete during fertilization, it will add one more copy of the longer chromosome, for a total of three copies. **(A)** is thus correct. (B) is incorrect because fertilization will lead to three copies of the longer chromosome, not two. (C) and (D) are incorrect because gamete 2 has one extra chromosome, not one fewer chromosome.

4. D Learning Objective: 11.4

Like other features of meiotic division and fertilization, the primary purpose of sexual reproduction is to increase the genetic fitness of a population by increasing the extent of genetic diversity within the population as a whole. **(D)** most directly references this purpose, making it correct. Though there are error-checking mechanisms in place to ensure the integrity of chromosomes, crossing over is not one of those mechanisms, so (A) and (B) are both incorrect. Although (C) provides an explanation in terms of the genetic fitness of progeny, the "best" genes are selected by natural selection at an organismal level and not at a molecular level due to crossing over.

5. C Learning Objective: 11.5

The first generation offspring should all have the genotype *Ry* (because they receive an *R* allele from the homozygous red parent and a *y* allele from the homozygous yellow parent). Thus, when a *Ry* (red) apple tree is crossed with a *yy* (yellow) apple tree, half of the offspring should be *Ry* (red) and the other half should be *yy* (yellow). Out of 10 offspring, 5 would be red and 5 yellow. This eliminates (A) and (B). To determine whether this is statistically significant, calculate chi-square as follows:

phenotype	o	e	o–e	$(o–e)^2$	$(o–e)^2/e$
red	6	5	1	1	0.2
yellow	4	5	−1	1	0.2
					$\chi^2 = 0.4$

Because 0.4 is less than 3.84, the threshold for $p = 0.05$ with 1 degree of freedom, the result is not statistically significant, making **(C)** correct. (D) is incorrect because it misrepresents the chi-square value.

Test What You Learned

1. C Learning Objective: 11.5

The results without insecticide suggest that three-quarters of the cells are ordinarily expected to be in interphase and one-quarter in mitosis. With a sample of 653 total cells, that means that about 490 should be in interphase and 163 in mitosis. This allows the following chi-square calculation:

phase	o	e	o–e	$(o–e)^2$	$(o–e)^2/e$
interphase	575	490	85	7225	14.745
mitosis	78	163	−85	7225	44.325
					$\chi^2 = 59.07$

Because 59.07 is greater than 3.84 (the significance threshold for $p = 0.05$ with 1 degree of freedom), the effect of the insecticide is statistically significant. **(C)** is thus correct. (A) is incorrect because it suggests the effect is not significant. (B) and (D) are incorrect because they wrongly suggest that the chi-square value is less than 3.84.

2. D Learning Objective: 11.3

Nondisjunction is an error in the separation of chromosomes, which can occur during meiosis (in either meiosis I or meiosis II) and which leads to genetic abnormalities. It produces gametes with either one missing or one extra chromosome. **(D)** is thus correct. (A) is incorrect because chromosomes normally separate to opposite poles of the cell during meiosis. Recombination involves the exchange of material between chromosomes but does not itself alter the number of chromosomes in gametes, so (B) is also incorrect. (C) is incorrect because chromatids are supposed to separate during anaphase II; only the failure to separate properly could cause a discrepancy in chromosome number.

3. A Learning Objective: 11.2

Except in a very small number of cell types (e.g., erythrocytes and platelets), mitosis results in complete copies of genetic information being passed along to both daughter cells equally, so **(A)** is correct. Most differences in the structure and function of daughter cells are attributable primarily to the different mRNA and protein contents between the two cells, and not to the presence, absence, or direct modification of genetic material. Thus, (B), (C), and (D) are incorrect.

4. A Learning Objective: 11.1

During the 3 stages of interphase (G_1, S, and G_2), there is pronounced growth, DNA replication, and crucial preparation for cellular division. DNA replication errors during this process, if not detected and fixed, would result in mutations being passing on to daughter cells. **(A)** is thus correct. (B) is incorrect because unrepaired mutations would be passed on. (C) and (D) are incorrect because cells do in fact grow during interphase.

5. B Learning Objective: 11.4

The most likely explanation for an offspring having a genetically influenced trait that neither of its parents possesses is that the parents were carriers (heterozygous) for the trait. **(B)** is thus correct. (A) is incorrect because it misidentifies crossing over as "genetic drift" and because crossing over only results in the exchange of genetic information between chromosomes, so it could not bring into being a new allele for red hair. (C) is incorrect because the Y chromosome predominantly contains genes that influence primary and secondary sexual characteristics in males, so it is unlikely to influence hair color. (D) is incorrect because gene flow can only alter the allelic frequencies of a population as a whole but cannot change the genotypes of existing members of a population, such as the parents described in the question stem.

6. A Learning Objective: 11.4

Crossing over shuffles alleles into new patterns, producing genetic diversity **(A)**. It does not create random point mutations, so (B) is incorrect. (C) is incorrect because crossing over does not prevent heterozygosity, which is a measure of allele diversity in a population. (D) is incorrect because crossing-over ensures that nuclei are variable, not that they have the same DNA.

Processes

7. C Learning Objective: 11.4

Most differing alleles are the result of single nucleotide changes, **(C)**. (A) is incorrect because only small changes are necessary to change the function of a protein. (B) is incorrect because information is stored on chromosomes as DNA sequences. (D) is incorrect because the exact same DNA sequence could not result in different alleles.

8. D Learning Objective: 11.3

Crossing over increases the genetic diversity, and therefore the fitness, of a population by allowing homologous chromosomes to exchange segments of DNA, sometimes even serving to unlink linked genes. Thus, **(D)** is correct. (B) and (C) can be eliminated because the parent organisms have no control over which regions of their chromosomes are exchanged. (A) is incorrect because altering the number of chromosomes in an offspring is usually a fatal mutation. Additionally, at the end of meiosis, organisms have gone from diploid to haploid, allowing them to become diploid once again after zygote formation.

9. B Learning Objective: 11.4

Due to the change in environment, the fungi are most likely to utilize sexual reproduction because it gives them the highest chance of adapting to their new surroundings. Thus, **(B)** is correct. (A), (C), and (D) all represent forms of asexual reproduction, which would not provide vary much genetic variation. Therefore, if their commensalistic partner is disappearing, it is unlikely that these forms of reproduction would help them evolve either to live independently or to find a new organism to coexist with. (C) and (D) are also incorrect because they list benefits of asexual reproduction that are most useful during periods of relative stability.

10. C Learning Objective: 11.5

The calculations yield the following values:

Group	*o*	*e*	*o–e*	$(o-e)^2/e$
Control Interphase	296	253	43	7.31
Control Mitosis	50	93	−43	19.88
Treated Interphase	322	365	−43	5.07
Treated Mitosis	176	133	43	13.90

Summing up the values in the final column, as per the chi-square formula, yields a value of 46.16, which rounds down to 46. **(C)** is correct.

11. D Learning Objective: 11.2

Mitosis is the process during which a cell replicates and divides. Within the cell cycle of somatic cells, it occurs after interphase, during which the cell grows, DNA replicates, and the cell becomes ready to divide. While a somatic cell contains the resources necessary to divide an unlimited amount of times during its lifetime, it is limited by its own DNA. Each time DNA is replicated, tiny bits of the protective regions on the end, known as telomeres, become degraded. After approximately 52 divisions, these regions are completely degraded, triggering the cell to enter apoptosis. However, cancer cells are able to divide continuously and without a limit because they make a protein known as telomerase that regenerates the protective ends of DNA. This best matches **(D)**. (A) is incorrect because DNA is not reversibly modified during replication; it is permanently degraded. (B) is incorrect because, like all cells, cancer cells have to enter interphase to allow their DNA to replicate, prior to dividing. (C) is incorrect because normally, except in a few specific types of cells, a complete copy of genetic information is passed on to both daughter cells. In the cells where this does occur, it is unlikely that cancer would alter this process.

12. C Learning Objective: 11.1

Cancer is defined principally as uncontrolled cell growth due to dysregulation of the cell cycle, so the correct answer will most directly address this mechanism. **(C)** does exactly this, so it is correct. (A) assumes too much about the potential limitations of the immune system, and neglects an explanation of why cancers tend to originate at the site of viral infection, as opposed to some other part of the body. (B) references disruption of the cell cycle, but a cell that cannot grow or divide could not grow to create a tumor. Although viruses do divert the host cell's nutrients, this would not directly account for why cells at the site of infection would become cancerous by growing and multiplying faster than other cells, rendering (D) incorrect.

CHAPTER 12

Homeostasis

LEARNING OBJECTIVES

In this chapter, you will review how to:

12.1 Describe energy flow in biological systems

12.2 Explain how the laws of thermodynamics apply to living systems

12.3 Describe how changes in energy affect living systems

12.4 Describe coupled reactions important for life

12.5 Contrast positive and negative feedback mechanisms

12.6 Describe thermoregulation processes

TEST WHAT YOU ALREADY KNOW

1. The following diagrams represent biotic and abiotic components of the arctic tundra. Which diagram represents the most accurate flow of energy between the components of the ecosystem?

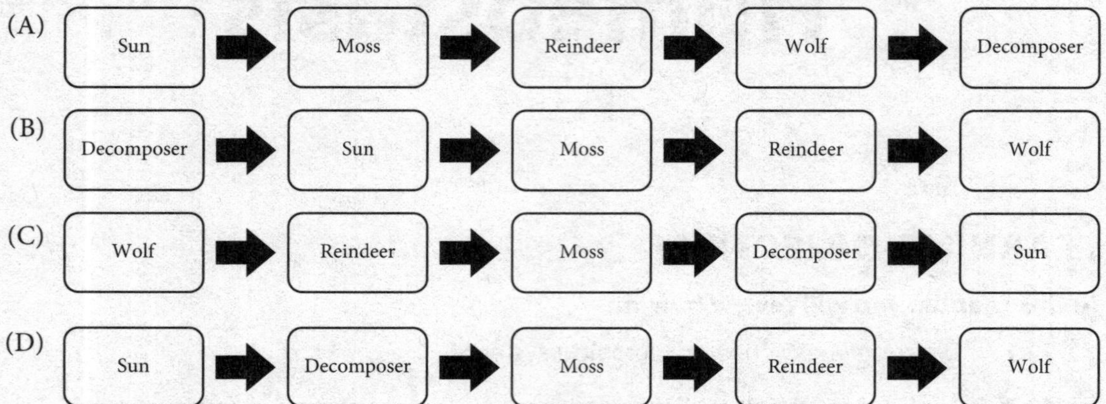

2. A small pond is populated by algae, paramecia, and amoebae. Sunlight is absorbed by the microscopic algae. Paramecia eat the algae. Amoebae feed on the paramecia and algae. Which of the following statements is correct about the pond, based on the laws of thermodynamics?
   (A) Algae transform all of the solar energy into chemical energy in the form of sugars.
   (B) Solar energy is only converted into chemical energy in this ecosystem.
   (C) Photosynthesis is an endergonic process because algae have more energy at the end of the process.
   (D) Photosynthesis is an exergonic process because algae expend the energy from photosynthesis to perform cellular work.

3. Yeast cells obtain energy by metabolizing sugars produced by other organisms. In an experiment, yeast cells are mixed with barley seeds, which supply a source of starch for fermentation. Fermentation is measured by estimating the production of CO_2 in a respirometer, a device that measures changes in gas volume. Which of the following experimental conditions would most likely yield the highest level of fermentation?
   (A) Mixing barley seeds and yeast in water
   (B) Scratching barley seeds to release starch before adding yeast
   (C) Presoaking barley seeds in water to activate enzymes that break down starch
   (D) Germinating barley seeds until seedlings emerge before mixing them with the yeast

4. The following diagram illustrates a classic experiment in chemiosmosis in which isolated thylakoid membranes are incubated in the dark at pH 4 and, after equilibration, transferred to a buffer at pH 8 containing ADP and P_i. ATP was synthesized and accumulated in the medium.

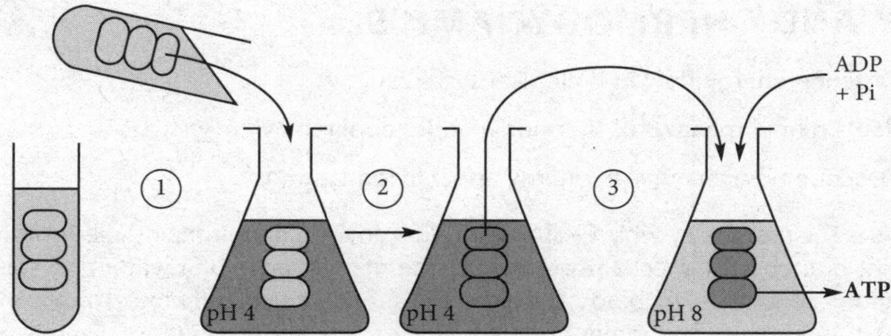

Which of the following hypotheses about the production of ATP in the thylakoid membrane is best supported by the experiment?

(A) ATP formation in the thylakoid membrane is driven by the oxidation of NADPH.

(B) ATP synthesis depends on the light reactions of photosynthesis.

(C) ATP synthesis does not require an intact thylakoid membrane.

(D) ATP synthesis depends on an H^+ gradient across the thylakoid membrane.

5. Platelets circulate in the blood in an inactive state. Damage in a blood vessel exposes and releases molecules not normally found in blood, which activates platelets. The activated platelets adhere to each other and to the blood vessel. This sets off an enzymatic cascade that activates additional platelets. Plugging of the hole in the blood vessel finally shuts off the response of the platelets. Which of the following statements best describes the response of the platelets?

(A) The response of the activated platelets is a negative feedback mechanism because it stops the loss of blood.

(B) Activation of platelets is a positive feedback loop because it acts as an amplifier of the response.

(C) Platelets do not respond to positive feedback because there would be no way to shut off activation once the damage is repaired.

(D) The response of the activated platelets promotes an increase in the total number of platelets.

6. The temperature setpoint of the body is determined by the hypothalamus. When core body temperature rises above the setpoint, responses such as sweating bring about cooling. When the temperature drops below it, warming mechanisms, such as narrowing of blood vessels and shivering, increase body temperature. A child develops a temperature of 40°C (104°F) because an ear infection has reset her hypothalamus to a higher core body temperature. Which of the following symptoms would the child most likely experience while the infection persists?

(A) Profuse sweating

(B) Chills and shivering

(C) Feeling overheated

(D) Thirst and decreased urination

Answers to this quiz can be found at the end of this chapter.

ENERGY AND THERMODYNAMICS

High-Yield

12.1 Describe energy flow in biological systems

12.2 Explain how the laws of thermodynamics apply to living systems

12.3 Describe how changes in energy affect living systems

Homeostasis is the process by which a stable internal environment is maintained within an organism. Important homeostatic mechanisms include the maintenance of a water and solute balance (osmoregulation), regulation of blood glucose levels, and the maintenance of a constant body temperature. In mammals, the primary homeostatic organs are the kidneys, liver, large intestine, and skin.

The maintenance of homeostasis within an organism uses up a considerable amount of energy. Thus, before considering specific homeostatic processes, it is essential to understand a few points on energy flow and thermodynamics.

Energy Flow

The ultimate energy source for living organisms is the Sun. Autotrophic organisms, such as green plants, convert sunlight into energy stored in the bonds of organic compounds (chiefly glucose) during the anabolic process of photosynthesis. Autotrophs do not need an exogenous supply of organic compounds. Heterotrophic organisms obtain their energy catabolically, via the breakdown of organic nutrients that must be ingested. Note in the following energy flow diagram that some energy is dissipated as heat at every stage.

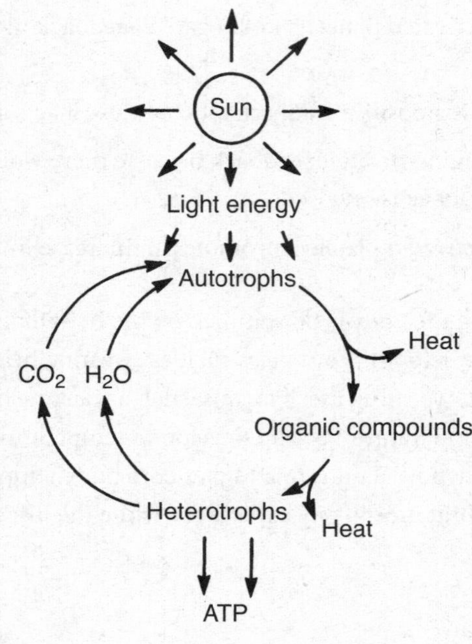

Energy Flow

Thermodynamics

Thermodynamics deals with heat, energy, and work and can be discussed in terms of temperature, internal energy, entropy, and pressure. For the AP Biology exam, you should be familiar with the three laws of thermodynamics.

The First Law of Thermodynamics

The first law maintains that the increase in internal energy of a closed system is equal to the difference of the heat supplied to the system and the work done by the system. This is a variation of the law of the conservation of energy that states that energy cannot be created or destroyed, but can only change form. The classic example of this is how kinetic energy can be converted to potential energy or (within the body) chemical energy can be converted to kinetic energy.

The Second Law of Thermodynamics

The second law states that the entropy of a closed system tends to increase. In other words, energy spreads out over time rather than spontaneously consolidating. Eventually, differences in temperature, pressure, and chemical potential tend to even out in a physical system that is isolated from the outside world. Consider, for example, how "normal" each of the following seems: hot coffee cools down, iron rusts, and balloons deflate. The thermal energy in the hot coffee is spreading out to the cooler air that surrounds it. The chemical energy in the bonds of elemental iron and oxygen is released and dispersed as a result of the formation of the more stable, lower-energy bonds of iron oxide (rust). The energy of the pressurized air is released to the surrounding atmosphere as the balloon deflates. These examples have a common denominator: in each of them, energy of some form is going from being localized or concentrated to being spread out or dispersed.

The Third Law of Thermodynamics

According to the third law, the entropy of a system approaches a constant value as the temperature approaches zero. Therefore, at absolute zero (the coldest possible temperature), entropy reaches its minimum value. At absolute zero, nothing can be colder and no heat energy remains in a system.

Bioenergetics

Biological thermodynamics, also known as bioenergetics, is the study of energy transformation in biology. This involves looking at energy transformations and transductions in and between living things, including their major functions down to the cellular level, and understanding the function of the chemical processes underlying these transductions.

Changes in free energy availability affect the ability of organisms to maintain organization, grow, and reproduce. Organisms that acquire excess free energy will either use the energy for growth or store the energy, while organisms that acquire insufficient free energy will lose mass and ultimately die. Since the energetic costs of reproduction are large, depending on energy availability, different organisms use various reproductive strategies. For instance, plants and animals may reproduce seasonally during the most energetically favorable part of the year (determined by ambient temperature and food availability). A biennial plant takes two years to complete its life cycle. In the first year, the plant germinates and grows, and then becomes dormant in the colder months. During the subsequent year, the plant produces flowers and fruits. Animals may undergo reproductive diapause, a period during which growth or development is suspended and metabolism is decreased in response to adverse environmental conditions such as temperature extremes or reduced food and water availability. Changes in energy also cause disruptions at the population level by affecting population size and number.

✔ AP Expert Note

The most important energy transformations in organisms are found in the processes of photosynthesis and cellular respiration, discussed in chapter 10 of this book. At a larger scale, energy transformations can also be studied at the ecosystem or biome level—see chapter 19 for more on this ecological perspective.

COUPLED REACTIONS AND CHEMIOSMOSIS

12.4 Describe coupled reactions important for life

A **coupled reaction** is one in which transport across a membrane is coupled with a chemical reaction. Of all the things to remember here, there is one that is particularly important: **chemiosmosis** is used by cells to generate ATP by moving H^+ ions across a membrane, down a concentration gradient, as shown in the figure on the following page. Special membrane proteins called **ATPases** create proton channels to convert ADP to ATP when a proton passes through. Aerobic respiration and photosynthesis utilize chemiosmosis, whereas glycolysis and other forms of ATP creation do not.

The mitochondrion moves H^+ into the **intermembrane space** via the **electron transport chain**, which creates a proton gradient across the inner membrane. The energy to do this comes from the breakdown of food. The chloroplast moves H^+ into the **thylakoid space** in a way very similar to the mitochondrion, but it drives the oxidative phosphorylation process with light energy. This process is called **photophosphorylation**.

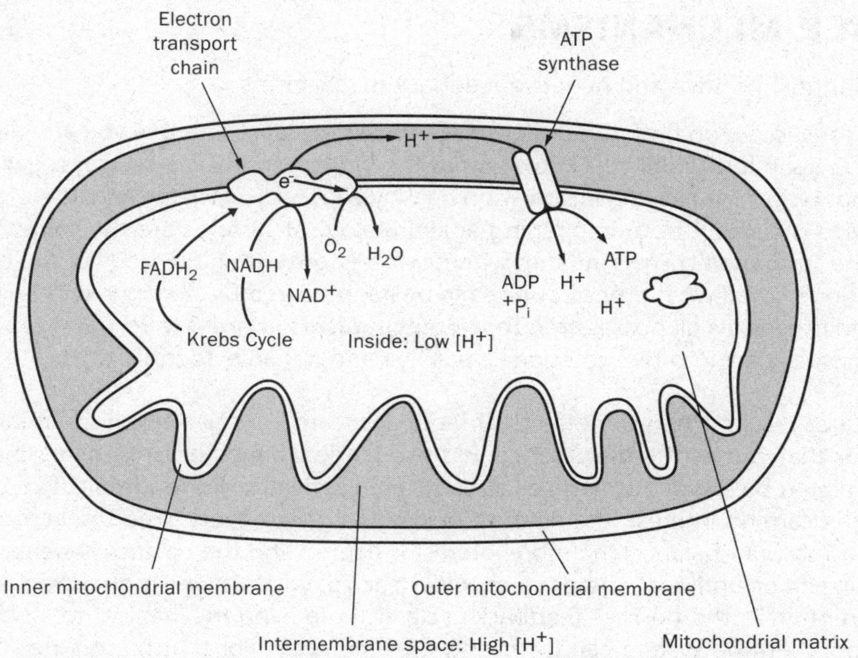

Chemiosmosis in a Mitochondrion

> ✔ **AP Expert Note**
>
> **The processes that drive electron transport and chemiosmosis are similar in both photosynthesis and cellular respiration.**

The structures that perform chemiosmosis are examples of form specialized for function. The invaginations of the inner membrane (cristae) inside the mitochondrion provide an increased surface area. Even the proteins in the membrane are aligned in a way that spans the membrane width, and their juxtaposition provides the proper architecture to allow chemical reactions to take place. The most important thing to remember about fine protein structures is that their location and form are what allow chemiosmosis to happen.

Catabolism

Catabolism is the breaking down of complex substances into simple substances, making energy available in the process. In **anaerobic catabolism**, there is no electron transport chain or oxygen (O_2) available to carry out a reaction. Anaerobic reactions require almost as much energy to carry out as they yield. From an evolutionary perspective, it is much more advantageous to have the ability to utilize the oxidative properties of oxygen, as reactions in **aerobic catabolism** do, because more energy can be generated. For more on the aerobic and anaerobic catabolism of glucose, see chapter 10.

Keep in mind that not all energy comes from sugar in the form of glucose. Organisms use food in the form of carbohydrates (starch), protein, and fat. Carbohydrates ultimately break down or are transformed into glucose. The amino acids from proteins ultimately enter the citric acid cycle as pyruvate and acetyl-CoA. Fats are broken down into glycerols that can be converted to pyruvate and fatty acids that are converted to acetyl-CoA. All of the pyruvate and acetyl-CoA from the breakdown of food enter the citric acid cycle and the electron transport chain.

FEEDBACK MECHANISMS

12.5 Contrast positive and negative feedback mechanisms

Input is vital for success on both the organismal and cellular level. Imagine you are playing a video game and you keep losing. All you know is that the video game suddenly says "game over," but you never know why. You have no information on why you lost your game and, therefore, you have no idea how to prevent it from happening again. Biological systems are the same. They require input from the surrounding environment, whether it be from within the body or the external environment, to understand what processes need to be adapted in order to survive. This input is called feedback, and organisms all possess **feedback mechanisms** in order to respond to stimuli. These mechanisms usually fall into two categories: positive and negative feedback.

Positive feedback results when the effects of feedback from a system result in an increase in the original factor that causes the disturbance. Positive feedback mechanisms increase or accelerate the output created by a stimulus that has already been activated. For example, if you have a bank account that is earning interest, the amount of interest grows every time the account increases. The higher the account balance, the more interest is earned and the balance increases even more. This can go on and on until some other mechanism (such as withdrawing money from your account) causes a disruption in the positive feedback cycle. If there is no mechanism to stop the positive feedback loop, then the system can lose control of the cycle. For that reason, positive feedback systems are considered unstable. To use an ecological example, positive feedback can be seen with climate change. An initially small perturbation in the environment can positively feed back onto itself, growing and growing until the problem yields huge effects on the climate.

Negative feedback is the opposite of positive feedback. As more feedback is received, it causes the processes that brought about the initial change to slow down or stop altogether. Self-regulating systems tend to function by using negative feedback. It allows for stability within a system because it reduces the effects of fluctuations. Negative feedback loops allow a system to have the necessary amount of correction at the most important time. One of the simplest examples of negative feedback is one of the human body's methods of thermoregulation. An increase in core body temperature will stimulate the body to produce sweat. As the body's sweat production increases, it causes a drop in body temperature. This decrease in body temperature will turn off the mechanism for sweat production.

THERMOREGULATION

12.6 Describe thermoregulation processes

Thermoregulation is one aspect of homeostasis that is commonly featured in the AP Biology exam. The hypothalamus is the brain region that acts to control the body temperature of an organism that is able to set its internal temperature. Mammals fall into this category. Hormones such as **adrenaline** and the thyroid hormones can increase metabolic rate and, subsequently, heat production. Muscles can generate heat by contracting rapidly (shivering). Heat loss is regulated through the contraction or relaxation of precapillary sphincters.

Alternative mechanisms are used by some mammals to regulate body temperature. Panting is a cooling mechanism that results in the evaporation of water from the respiratory passages. Sweating is also important in increasing evaporative heat loss so that the body cools. Fur is used to trap heat, and hibernation during the winter conserves energy by decreasing heart rate, breathing, and metabolism. Animals able to regulate their internal temperature even in the face of a changing external temperature are called endotherms, or homeotherms. Mammals and birds are capable of this type of regulation, yet reptiles, amphibians, and most other animals are not and are known to be cold-blooded, or ectotherms.

There are two laws you should know that relate heat and body size:

1. Bigger bodies produce less body heat per pound per hour.
2. Bigger bodies lose less body heat per pound per hour.

Metabolic heat production drops in a very specific manner as body size increases. Compared with an elephant, which might weigh 10,000 pounds, a small mammal weighing only 1 pound produces about 10 times more heat per pound than the elephant does. Yet of the two animals, the elephant stays warmer because it has much less overall surface area compared to internal volume than does the small mammal. In other words, the small mammal gives off much more heat to its surroundings.

RAPID REVIEW

If you take away only 5 things from this chapter:

1. Homeostasis is the process by which a stable internal environment is maintained within an organism. Our primary homeostatic organs are the kidneys, liver, large intestine, and skin.
2. There are three laws of thermodynamics. Together they discuss the conservation of energy, entropy, and absolute zero.
3. Bioenergetics, or biological thermodynamics, studies how chemical energy is broken down and converted to usable energy within the biological system. This can be at the ecosystem, organismal, or cellular level.
4. Chemiosmosis produces energy from the movement of H^+ ions across a membrane down a concentration gradient in both photosynthesis and respiration. Catabolism is the breaking down of complex substances into simple substances, making energy available in the process.
5. Thermoregulation refers to the physiological processes that come together to maintain a stable body temperature in warm-blooded animals.

TEST WHAT YOU LEARNED

1. Consider the chemical reactions X and Y. The progress of each reaction is plotted as a function of the free energy of the compounds involved in the reaction.

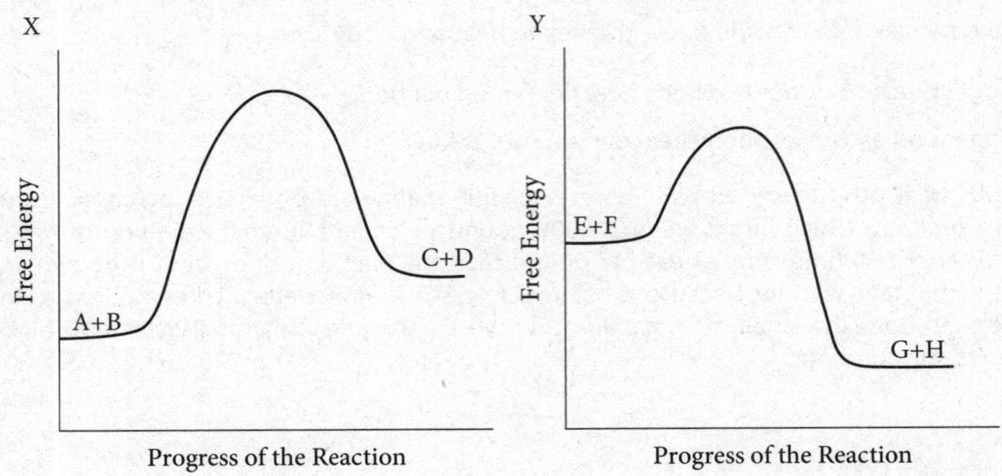

Based on the free energy plots, a biochemist considers the following hypotheses on whether the reactions would take place in a living organism. Which of them is best supported by the graphs?

(A) Reaction X is endergonic and cannot take place in a living cell.

(B) Reaction Y would proceed on its own without the input of energy.

(C) Reaction X would proceed if an enzyme lowered the free energy of C+D.

(D) Reaction X would proceed if it is coupled to reaction Y.

2. The second law of thermodynamics states that the entropy of a closed system must always increase. This means that such systems always move to a more disordered state. However, cells constantly synthesize macromolecules that are complex and highly ordered. Which of the following statements best explains the connection between the second law of thermodynamics and the biosynthesis of macromolecules in a living cell?

(A) Releasing water molecules in the biosynthesis of macromolecules increases entropy.

(B) The entropy of the surroundings increases as the cell synthesizes macromolecules.

(C) The entropy in the cell increases as it synthesizes macromolecules.

(D) The second law of thermodynamics does not apply to biological systems.

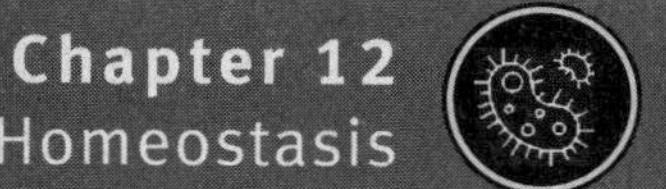

3. The Cretaceous-Paleogene boundary coincides with a mass extinction that led to the disappearance of more than three-fourths of all plant and animal species living on Earth. One popular hypothesis proposes that a large meteorite collided with the Earth, sending large clouds of debris into the atmosphere that blocked sunlight. How would such a catastrophic event most likely have changed the levels of energy in the biosphere?
   (A) The clouds blocking the Sun caused a long period of cold temperatures, which stopped energy flow.
   (B) The asteroid impact caused major earthquakes and tsunamis that wiped out ecosystems.
   (C) Darkness made photosynthesis nearly impossible, removing the foundation of the food web.
   (D) Most life was extinguished upon meteorite impact or soon thereafter, stopping the flow of energy.

4. A bicyclist eats cereal for breakfast before a race. The flow of energy for this scenario can be represented by a diagram that shows energy flowing from the Sun to the grain to the bicyclist's muscles. In which of the following shapes should the diagram be made to best represent the flow of energy up these three levels?
   (A) A rectangle, because energy is neither created nor destroyed as it flows up the levels
   (B) A circle, because energy cycles through the levels
   (C) An upright pyramid, because energy is lost at each level as heat
   (D) An inverted pyramid, because the biker's muscles accumulated energy from all of the lower levels

5. Zoologists estimated the variation in the body sizes of woodrats over the last 20,000 years by dating pellets with ^{14}C. They plotted their results as the relative increase or decrease in body size relative to present size. Temperatures are below average during glacial periods and above average during interglacial periods.

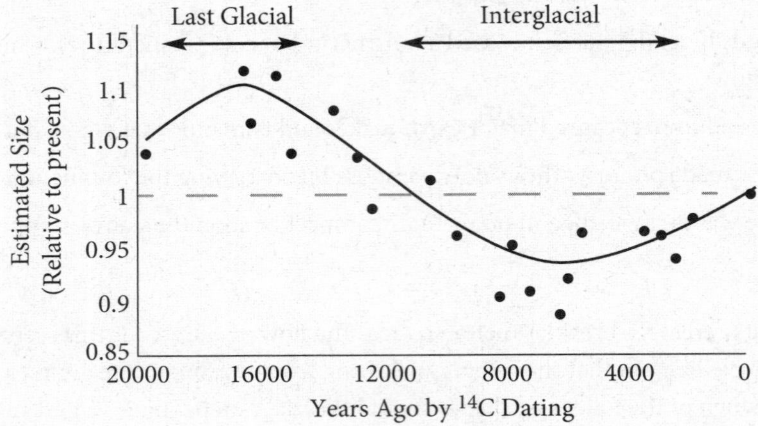

Adapted from Felisa A. Smith, Julio L. Betancourt, and James H. Brown, "Evolution of Body Size in the Woodrat Over the Past 25,000 Years of Climate Change," *Science* 270, no. 5244 (December 1995): 2012–2014.

Which of the following statements is best supported by the data?

(A) Body size is independent of ambient temperature.

(B) Body size decreases when ambient temperature increases.

(C) Body size increases when ambient temperature increases.

(D) Body size increases as a function of evolution.

6. The hypothalamus secretes thyrotropin-releasing hormone (TRH), which causes the anterior pituitary to release thyroid-stimulating hormone (TSH) in the bloodstream. The thyroid gland responds to TSH by secreting the hormone thyroxine, which contains iodine. When iodine is missing in a diet, thyroxine is not secreted by the gland. Individuals whose diet is deficient in iodine may develop a goiter, an enlarged thyroid gland. Which of the following hypotheses best explains the formation of a goiter?

(A) TSH is regulated by a negative feedback loop and keeps stimulating the thyroid when thyroxine production is low.

(B) Without enough iodine present, cells in the gland keep dividing and the organ becomes enlarged.

(C) Thyroxine precursors accumulate in the gland and begin to form new compounds that cause it to become enlarged.

(D) If the dietary content of iodine is too low, the gland enlarges to maximize uptake from the bloodstream.

7. Following ingestion of a meal, an increase in blood glucose levels stimulates insulin secretion by pancreatic beta cells. Insulin activates glucose utilization and storage via positive feedback. Decreased blood glucose levels, in turn, stop insulin secretion via negative feedback.

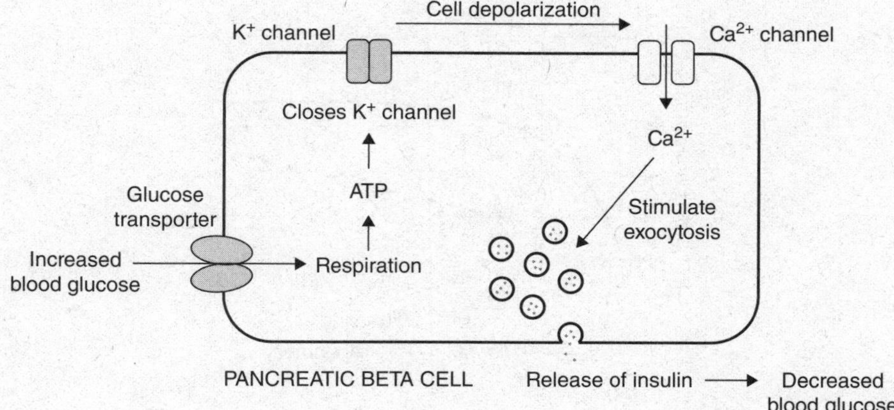

Patients with type 2 diabetes have high glucose levels because they do not produce enough insulin to overcome insulin resistance (decreased ability to respond to insulin). Based on the diagram, which of the following describes the treatment that is most likely to be effective for type 2 diabetes?

(A) A drug that stimulates insulin secretion by blocking ATP-dependent potassium channels

(B) A drug that closes the voltage-gated calcium channels to inhibit exocytosis of insulin

(C) A drug that opens the ligand-gated potassium channel to promote insulin secretion

(D) A drug that inhibits glucose uptake and usage to increase glucose levels

8. Researchers directed a narrow beam heat lamp at the thorax of tethered sphinx moths to simulate heat generation from flight. The results are shown in the graphs below.

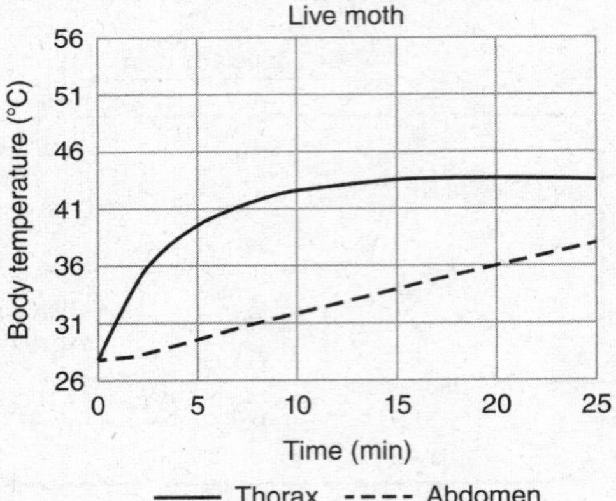

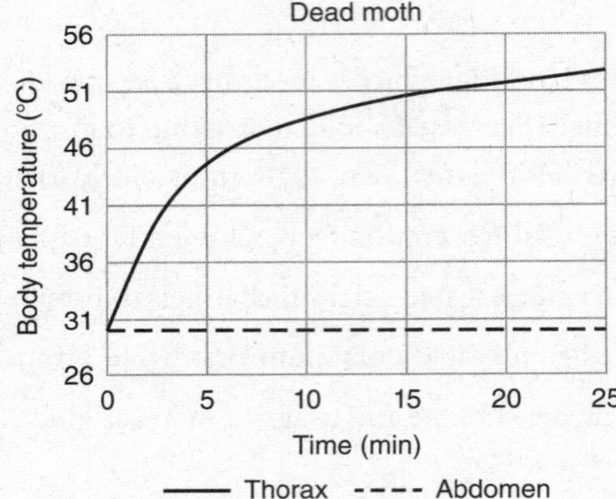

In a second experiment, the researchers investigated the effect of circulation between the thorax and abdomen of free-flying moths on body temperature. For half of the moth population, the researchers prevented the circulation between the thorax and abdomen. Which of the following is a likely result of the second experiment?

(A) In free-flying moths with circulation unblocked, the thorax overheated and the temperature of the abdomen did not change.

(B) In free-flying moths with circulation unblocked, the abdomen heated up and the temperature of the thorax plateaued.

(C) In free-flying moths with circulation blocked, the thorax and abdomen both overheated.

(D) In free-flying moths with circulation blocked, the temperature of the abdomen decreased, while the temperature in the thorax increased.

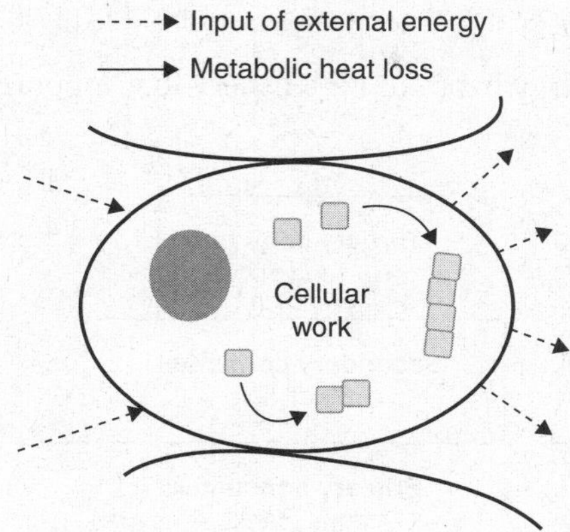

9. The second law of thermodynamics states that the entropy of a closed system tends to increase. The diagram above shows a living cell. A living cell is an open system that does not violate the second law of thermodynamics. This is most likely explained by which of the following statements?

   (A) All the external energy obtained by the cell is used to form ordered structures.

   (B) Cellular work that combines smaller components into larger structures increases the entropy, while metabolic heat loss decreases the entropy of the environment.

   (C) To maintain its high degree of structural organization, the cell releases energy as heat to increase entropy.

   (D) A living cell is at equilibrium with its environment, so the total energy is conserved.

10. The diagram below shows energy and nutrient flow through an ecosystem. Values indicate energy available as food and are in kilojoules.

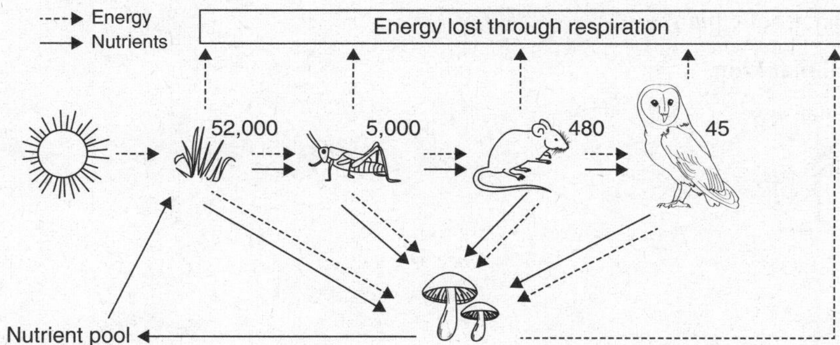

Calculate the percentage of energy available to the primary consumer that is passed to the tertiary consumer. Express your answer to the nearest 0.1%.

   (A) 0.1

   (B) 0.9

   (C) 9.4

   (D) 99.1

Processes

Questions 11–14

The diagram below shows the energy pyramid for a fishery that supports tuna fish as the top-level carnivore.

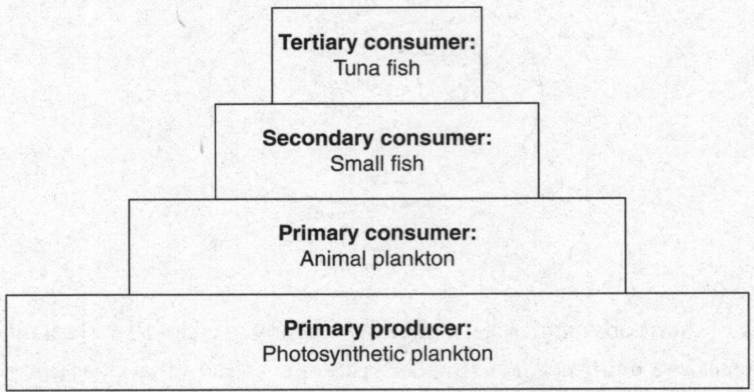

11. Assuming that there are 2,000 J of energy available in the form of small fish, which of the following best matches modern theories of energy flow in ecosystems?

    (A) There are 10,000 J of energy available in photosynthetic plankton.

    (B) There are 2 J of energy in tuna fish.

    (C) There are 2,000 J of energy in animal plankton.

    (D) There are 200 J of energy in tuna fish.

12. Humans can choose to feed at different trophic levels. Eating which of the following would support the maximum number of people?

    (A) photosynthetic plankton

    (B) animal plankton

    (C) small fish

    (D) tuna fish

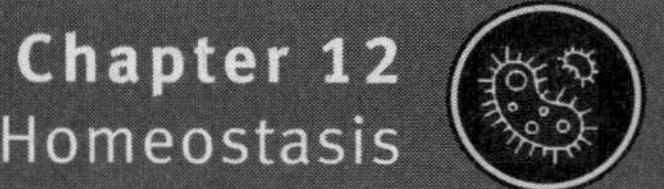

13. Which of the following activities could increase the number and/or size of tuna fish, and therefore the amount of energy, that could be harvested?
    (A) removing mercury from the environment
    (B) harvesting 10% of the small fish
    (C) harvesting 10% of the animal plankton
    (D) fertilizing patches of ocean to increase the number of photosynthetic plankton

14. Scientists measured mercury levels in tissue samples of organisms from different trophic levels in this ecosystem. Which of the following is true?
    (A) Tuna fish have the lowest concentrations of mercury because there are fewer tuna fish than other organisms.
    (B) Small fish have higher concentrations of mercury than photosynthetic plankton because the toxin concentration gets magnified at each trophic level.
    (C) Animal plankton have the highest concentrations of mercury because most of them are classified as invertebrates.
    (D) Photosynthetic plankton have the highest concentrations of mercury because they actively take up mercury as protection against animal plankton.

Answer Key

Test What You Already Know

1. A
2. C
3. C
4. D
5. B
6. B

Test What You Learned

1. D
2. B
3. C
4. C
5. B
6. A
7. A
8. B
9. C
10. B
11. D
12. A
13. D
14. B

REFLECTION

Test What You Already Know score: ___________

Test What You Learned score: ___________

Use this section to evaluate your progress. After working through the pre-quiz, check off the boxes in the "Pre" column to indicate which Learning Objectives you feel confident about. Then, after completing the chapter, including the post-quiz, do the same to the boxes in the "Post" column. Keep working on unchecked Objectives until you're confident about them all!

Pre	Post	
❑	❑	**12.1** Describe energy flow in biological systems
❑	❑	**12.2** Explain how the laws of thermodynamics apply to living systems
❑	❑	**12.3** Describe how changes in energy affect living systems
❑	❑	**12.4** Describe coupled reactions important for life
❑	❑	**12.5** Contrast positive and negative feedback mechanisms
❑	❑	**12.6** Describe thermoregulation processes

FOR MORE PRACTICE

Complete more practice online at kaptest.com. Haven't registered your book yet? Go to kaptest.com/booksonline to begin.

ANSWERS AND EXPLANATIONS

Test What You Already Know

1. A Learning Objective: 12.1

Energy enters the ecosystem from the Sun. Moss captures some of this energy through photosynthesis. The primary consumer, the reindeer, feeds on moss. The wolf, the secondary consumer, obtains energy from the reindeer. The wolf will eventually die and be consumed by decomposers. **(A)** is thus correct. (B), (C), and (D) are incorrect because they list the components in the wrong order.

2. C Learning Objective: 12.2

Photosynthesis is an endergonic process, in which the products have more free energy than the reactants. **(C)** is thus correct. (A) is incorrect because not all of the solar energy is transformed; some of the energy is lost as heat. (B) is incorrect because the solar energy is also converted into mechanical energy in the movements of paramecia and amoebae. (D) is incorrect because photosynthesis is not exergonic.

3. C Learning Objective: 12.3

Activating enzymes that break down starch into simple sugars would enhance fermentation by the yeast. Therefore, **(C)** is correct. (A) and (B) are incorrect because the yeast are only capable of fermenting simple sugars, not starch, so neither experimental treatment would lead to much fermentation. (D) is incorrect because germinating seedlings would only deplete starch reserves, leaving even fewer carbohydrates available for fermentation.

4. D Learning Objective: 12.4

The conditions of the experiment create an artificial H^+ gradient across the thylakoid membranes. The flow of H^+ is what allows ATP synthase in the membranes to produce ATP. Thus, the experiment supports the hypothesis in **(D)**. (A) is incorrect because no NADPH was added to the flasks. (B) is incorrect because the experiment was conducted in darkness. (C) is incorrect because the membranes were not damaged in the experiment.

5. B Learning Objective: 12.5

The platelet behavior described in the question stem is a classic example of positive feedback, in which one response (the activation of the platelets) leads to more of the same effect. **(B)** is thus correct. (A) is incorrect because this is positive feedback, not negative feedback. (C) is incorrect because there is a separate negative feedback loop (described in the question stem) that stops platelet activation. (D) is incorrect because platelet activation promotes activation of inactive platelets, but does not cause more platelets to be produced.

6. B Learning Objective: 12.6

Chemicals released during the infection reset the hypothalamus to a higher temperature setpoint. Physiological responses such as shivering raise the core temperature to match the new setpoint. Thus, even when the child has a temperature higher than normal, she will experience symptoms like chills and shivering until her body reaches the new setpoint of 40°C. **(B)** is thus correct. (A) is incorrect because sweating only occurs when physiological temperature is above the setpoint. (C) is incorrect because the child will feel cold, not hot, until her body reaches the adjusted setpoint. (D) is incorrect because these are responses to dehydration, such as might happen when someone sweats profusely, which is not expected in this set of circumstances.

Test What You Learned

1. D Learning Objective: 12.4

Endergonic reactions happen all the time in cells because they are coupled with exergonic reactions. Reaction X, which is endergonic, could proceed in a cell if coupled with reaction Y, which is exergonic. **(D)** is thus correct. (A) is incorrect because endergonic reactions can take place in living cells. (B) is incorrect because reaction Y still needs to reach its activation energy; some energy input would be required. (C) is incorrect because enzymes can only lower activation energy; they cannot change the free energy of products or reactants.

2. B Learning Objective: 12.2

The cell and its surroundings form a closed system. The total entropy of the system increases even as the entropy of the cell decreases. Therefore, **(B)** is correct. (A) is incorrect because the release of water molecules does not increase entropy. (C) is incorrect because entropy actually decreases when complex molecules are synthesized. (D) is incorrect because the laws of thermodynamics apply to all physical systems, including organisms.

3. C Learning Objective: 12.3

With few exceptions, all energy enters the biosphere as light from the Sun through the process of photosynthesis. After the asteroid impact, debris blocking the Sun prevented photosynthesis, taking out the foundation of the food web, **(C)**. (A) is incorrect because energy flow does not stop during periods of cold temperatures, but only diminishes. (B) is incorrect because these events do not have an impact on the energy levels in the biosphere. (D) is incorrect because energy flow would not stop if most organisms were killed, but would just be reduced.

4. C Learning Objective: 12.1

Energy flows from the Sun to a producer, then to a consumer, then to the breakdown of energy-rich carbohydrates by catabolism, and finally to ATP supporting sliding filaments in the muscles. Energy is lost at each step as heat dissipated to the surrounding environment. The traditional way of representing this loss of energy is a pyramid, as in **(C)**. (A) is incorrect because some energy is dissipated into the environment as heat with each step. (B) is incorrect because energy does not circulate in this scenario. (D) is incorrect because energy does not accumulate at each level, but diminishes since some is lost as heat with each step.

5. B Learning Objective: 12.6

Body size was larger when temperatures were colder and smaller when temperatures were warmer. Thus, body size decreases when temperature increases, as in **(B)**. (A) is incorrect because body size is inversely correlated with ambient temperature. (C) is incorrect because it states the opposite of what occurs. (D) is incorrect because body size both increased and decreased over the course of woodrat evolution.

6. A Learning Objective: 12.5

Thyroid hormone production is regulated by a negative feedback loop. If insufficient iodine is available to produce thyroxine, the anterior pituitary will not be inhibited and will continue producing TSH. This is what leads to the formation of the goiter. **(A)** is thus correct. (B) is incorrect because the lack of iodine does not stimulate cell division. (C) is incorrect because the thyroxine precursors do not form new compounds in the gland. (D) is incorrect because the gland does not become enlarged in order to increase iodine filtration.

7. A Learning Objective: 12.5

Increased blood glucose activates the metabolism of glucose, which leads to the generation of ATP. The binding of ATP to ligand-gated potassium channels closes the potassium channels. This causes depolarization of the cell. The membrane potential opens voltage-gated calcium channels, allowing calcium to enter the cell and trigger exocytosis of secretory vesicles containing insulin. A patient with type 2 diabetes needs more insulin to overcome insulin resistance. A drug that closes potassium channels or opens calcium channels would increase insulin secretion. Thus, the correct answer is **(A)**. (B) and (D) are incorrect because inhibiting insulin secretion and increasing glucose levels would not overcome insulin resistance. (C) is incorrect because closing, not opening, the potassium channel would promote insulin secretion.

8. B Learning Objective: 12.6

The first experiment shows that for the live moths, the thorax prevents the moths from overheating by transferring the heat to the abdomen: the temperature of the thorax plateaus while the temperature of the abdomen increases. For the dead moths, the thorax overheats while the temperature of the abdomen remains at the ambient temperature. In the second experiment, metabolic heat from the contraction of flight muscles would increase the body temperature in the thorax of the free-flying moths. The free-flying moths with circulation intact would exhibit body temperatures like that of the live moth, while the moths with circulation blocked would exhibit body temperatures like that of the dead moth. The correct answer is thus **(B)**. (A) is incorrect because, with circulation intact, the thorax would not overheat and the temperature of the abdomen would increase. (C) is incorrect because only the thorax would become overheated when the circulation is blocked. (D) is incorrect because the temperature in the abdomen of the free-flying moths with circulation blocked would not decrease, but would stay at the ambient temperature.

9. C Learning Objective: 12.2

A living cell is in a state of non-equilibrium in which its high degree of structural organization (combining smaller components into larger more ordered structures) is maintained by releasing energy obtained from the environment as heat. The release of energy as heat increases entropy, so the second law of thermodynamics is not violated. **(C)** appropriately summarizes this. (A) is incorrect because part of the external energy obtained is released as heat. (B) is incorrect because combining smaller components into larger structures decreases entropy and metabolic heat loss increases the entropy of the surroundings. (D) is incorrect because a living cell is in a state of non-equilibrium.

10. B Learning Objective: 12.1

The energy available to the primary consumer (grasshopper) is 5,000. The percentage of energy that is passed to the tertiary consumer (owl) is $45/5000 \times 100\%$, which equals 0.9%. **(B)** is correct. Note: grass is a producer, the mouse is the secondary consumer, and the fungi are decomposers.

11. D Learning Objective: 12.3

Approximately 10% of the energy at one trophic level is transferred to the next, and 200 J is 10% of 2,000 J, making **(D)** correct. (A) underestimates the amount of photosynthetic plankton. There would need to be 200,000 J of energy in photosynthetic plankton to support 2,000 J of energy in small fish. (B) underestimates the amount of energy in tuna fish. (C) underestimates the amount of energy in animal plankton. There would have to be 20,000 J of energy in animal plankton to support 2,000 J of energy in small fish.

12. A Learning Objective: 12.3

Feeding at the lowest possible trophic level maximizes the amount of energy available to people. At each higher trophic level, energy is lost due to the organisms' requirements to maintain cells, grow, and reproduce. Thus, **(A)** is correct. For example, if there were 200,000 J of energy in photosynthetic plankton, it could either support 20,000 J of energy to people or to animal plankton. That 20,000 J of animal plankton in turn could either support 2,000 J of energy to people or to small fish. That 2000 J of small fish could either support 200 J of energy to people or to tuna fish. That 200 J of tuna fish could support 20 J of energy to people. As a result, (B), (C), and (D) are not correct.

13. D Learning Objective: 12.4

Increasing the number of photosynthetic plankton would lead to an increase in energy available at every successive trophic level, making **(D)** correct. Removing mercury, as (A) suggests, would make the tuna healthier, but that action would not necessarily affect the energy available. Harvesting any organisms from the lower trophic levels would make less energy available to support tuna fish. Therefore, (B) and (C) would lead to a decrease in the number of tuna fish that could be harvested.

14. B Learning Objective: 12.6

Pollutants, such as mercury, are biologically magnified at higher trophic levels because each higher trophic level accumulates the mercury that was absorbed at each lower trophic level. Tuna fish, therefore, have accumulated mercury that the small fish accumulated from the animal plankton that was accumulated from the photosynthetic plankton. **(B)** correctly demonstrates this relationship. The tuna fish have the highest concentrations of mercury, so (A) is incorrect. Animal plankton have accumulated mercury from only a single trophic level below them. Therefore, animal plankton do not have the highest concentrations of mercury, making (C) incorrect. Photosynthetic plankton have the lowest concentrations of mercury since they accumulate mercury only from the ocean and not from any trophic level below them. Therefore, (D) is incorrect.

PART 5

Transformations of Life

CHAPTER 13

Molecular Genetics

LEARNING OBJECTIVES

In this chapter, you will review how to:

- **13.1** Differentiate between the structures of DNA and RNA
- **13.2** Explain the function of DNA and RNA
- **13.3** Describe the process of DNA replication
- **13.4** Describe the process and product of transcription
- **13.5** Describe the steps of translation
- **13.6** Describe post-translational modifications to polypeptides
- **13.7** Investigate analysis of DNA

TEST WHAT YOU ALREADY KNOW

1. A scientist feeds fluorescently labeled nucleotides to cells and measures the fluorescence emission to quantify the amount of RNA synthesized. Which of the following fluorescently labeled compounds would give the most reliable measure of the amount of newly synthesized RNA?

    (A) Deoxyribose adenine phosphate

    (B) Deoxyribose uracil phosphate

    (C) Ribose thymine phosphate

    (D) Ribose uracil phosphate

2. Lactose fermenting strains of *Streptococcus pneumoniae* are designated as *lac*$^+$, whereas strains deficient in lactose fermentation are designated as *lac*$^-$. Extracts from *lac*$^+$ cells are incubated separately with three different enzymes: extract A with protease K (which digests proteins), extract B with DNases (which digests DNA), and extract C with RNase H (which digests RNA). After treatment with enzymes, each extract is incubated with live *lac*$^-$ cells and grown on a medium containing an indicator. Blue colonies indicate that lactose is being fermented. Colorless colonies indicate that the cells cannot ferment lactose.

    Which of the following would most likely be observed after growing the cultures?

    (A) Only colorless colonies appear from cells incubated with extract A.

    (B) Only colorless colonies appear from cells incubated with extract B.

    (C) Only colorless colonies appear from cells incubated with extract C.

    (D) Blue colonies appear from cells grown with all three extracts.

3. A molecular biologist is studying a mutant strain of *E. coli* that has a defect occurring during DNA replication. In order to understand the cells' mutation, she labels the cells with radioactive thymidine to monitor the synthesis of DNA. After several hours of incubation, she recovers both long and short strands of labeled DNA molecules. The results are not modified by treatment of the sample with mung bean nuclease, an enzyme that digests single-stranded DNA.

    These results best support which of the following hypotheses?

    (A) The *E. coli* are deficient in gyrases, which unwind DNA molecules.

    (B) The *E. coli* are deficient in DNA polymerase I, which fills in gaps in the DNA molecules.

    (C) The *E. coli* are deficient in DNA polymerase III, which synthesizes two new strands simultaneously.

    (D) The *E. coli* are deficient in ligases, which catalyze the formation of phosphodiester bonds.

4. Researchers discovered that two proteins with different primary structures, calcitonin (a parathyroid hormone) and the neurotransmitter peptide CGRP (expressed in neurons), can be mapped to the same gene. Two distinct mRNA molecules are transcribed from the gene in question. The two mRNA sequences differ in length and sequence, with the exception of identical sequences found in the region preceding the protein-encoding regions. Which of the following hypotheses would best explain these observations?

   (A) The sequences of the two mRNA molecules have different mutations.

   (B) The proteins are encoded by the sense and antisense strands of DNA.

   (C) Alternative splicing takes place in different tissues.

   (D) The proteins undergo extensive post-translational modifications.

5. A part of an mRNA molecule with the following sequence is being read by a ribosome:

   5′-UCG-GCA-CAU-UUA-UAU-GUU-3′

CODON TABLE

	SECOND POSITION								
	U		C		A		G		
FIRST POSITION	code	amino acid	code	amino acid	code	amino acid	code	amino acid	THIRD POSITION
U	UUU	phe	UCU	ser	UAU	tyr	UGU	cys	U
	UUC		UCC		UAC		UGC		C
	UUA	leu	UCA		UAA	STOP	UGA	STOP	A
	UUG		UCG		UAG	STOP	UGG	trp	G
C	CUU	leu	CCU	pro	CAU	his	CGU	arg	U
	CUC		CCC		CAC		CGC		C
	CUA		CCA		CAA	gln	CGA		A
	CUG		CCG		CAG		CGG		G
A	AUU	ile	ACU	thr	AAU	asn	AGU	ser	U
	AUC		ACC		AAC		AGC		C
	AUA		ACA		AAA	lys	AGA	arg	A
	AUG	met	ACG		AAG		AGG		G
G	GUU	val	GCU	ala	GAU	asp	GGU	gly	U
	GUC		GCC		GAC		GGC		C
	GUA		GCA		GAA	glu	GGA		A
	GUG		GCG		GAG		GGG		G

Using the codon table above, determine the amino acid sequence encoded.

   (A) ser-ala-his-leu-tyr-val

   (B) ser-ala-gln-leu-(stop signal)

   (C) ser-arg-val-asn-ile-gln

   (D) leu-tyr-ile-tyr-thr-ala

Transformations

6. Cellular proliferation in cancer can be regulated by signaling receptors belonging to the receptor tyrosine kinase (RTK) family. Binding of a ligand to RTKs triggers phosphorylation of tyrosine residues in the cytoplasmic domain of the protein. One approach to cancer treatment has been the use of small molecule inhibitors that prevent phosphorylation of RTKs.

   Which of the following statements correctly represents how the small molecule inhibitors are regulating RTKs?

   (A) The RTKs are regulated at the transcription level.

   (B) The RTKs are regulated at the RNA processing level.

   (C) The RTKs are regulated at the translation level.

   (D) The RTKs are regulated at the post-translational level.

7. The diagram below shows a DNA plasmid with a total length of 2,000 base pairs. The tick marks indicate cleavage sites for restriction enzymes (enzyme EcoRI and enzyme HaeIII).

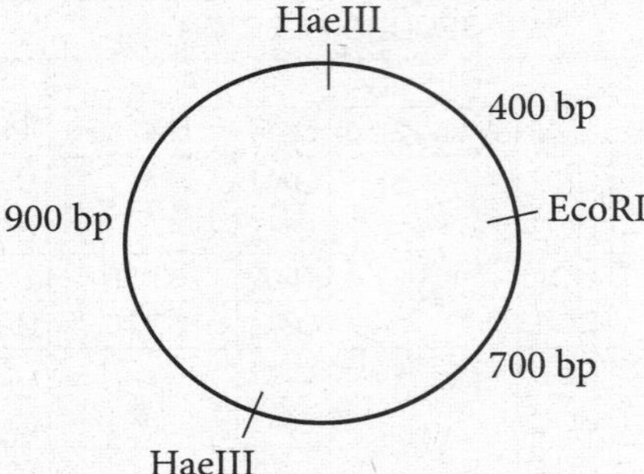

   The plasmid was digested with the enzyme HaeIII to completion and the fragments were separated by gel electrophoresis. Which of the following results would most likely be observed?

   (A) Two fragments measuring 400 bp and 900 bp

   (B) Two fragments measuring 700 bp and 900 bp

   (C) Two fragments measuring 900 bp and 1,100 bp

   (D) Three fragments measuring 400 bp, 700 bp, and 900 bp

Answers to this quiz can be found at the end of this chapter.

NUCLEIC ACIDS

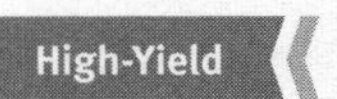

13.1 Differentiate between the structures of DNA and RNA

13.2 Explain the function of DNA and RNA

✔ AP Expert Note

Themes in Molecular Genetics

There are three key themes of molecular genetics to remember:

1. **Be familiar with the structures of DNA and RNA and how these building blocks provide both a simple system for duplication and a wealth of genetic diversity that codes for variation in all existing and extinct organisms.**
2. **Know that ultimately DNA is a code for the translation of proteins, that these proteins impart function to all of the biological processes that make an organism alive, and that control of expression of the genes for these proteins largely controls bioactivity.**
3. **As a complement to your understanding of evolution, understand DNA's role as the root of genetic change in the form of mutations.**

DNA and RNA Structures

You have undoubtedly learned that **DNA** and **RNA** molecules are long strands of nucleotides linked together by their sugar-phosphate backbone between the 5′ and 3′ carbons of deoxyribose or ribose. DNA has a **deoxyribose** sugar backbone while RNA has a **ribose** backbone. RNA also contains the nitrogenous base uracil instead of thymine. Both strands of complementary DNA serve as templates for duplication during mitosis and meiosis, whereas only one strand serves as a template for transcription. The simple sugar-phosphate backbone allows for exact copies of itself to be reproduced during propagation.

There are only four different **nitrogenous bases** that make up the **nucleotides** in DNA: **adenine**, **cytosine**, **guanine**, and **thymine**. These bases pair along complementary strands of DNA. Guanine pairs with cytosine, and thymine pairs with adenine. In RNA, thymine is replaced with **uracil**, which pairs with adenine. Therefore, four different nucleotides form a code for protein synthesis in all of the organisms that have ever existed on Earth. It is amazing that the combination of nitrogenous bases in the DNA of all organisms on Earth leads to such genetic variety.

Nucleotides are the basic building blocks of DNA and RNA. Millions of nucleotides make up the DNA in each cell. However, because there is so much DNA, our DNA is packaged into smaller tightly wound structures, **chromosomes**. Because each nucleotide has a nitrogen base (A, C, G, T) and it takes three bases to code for one amino acid, the average gene that codes for 300 amino acids is approximately 900 nucleotides long. Each chromosome may have hundreds to thousands of genes. Chromosomes are made up of millions of nucleotides wound tight and held together by histone proteins.

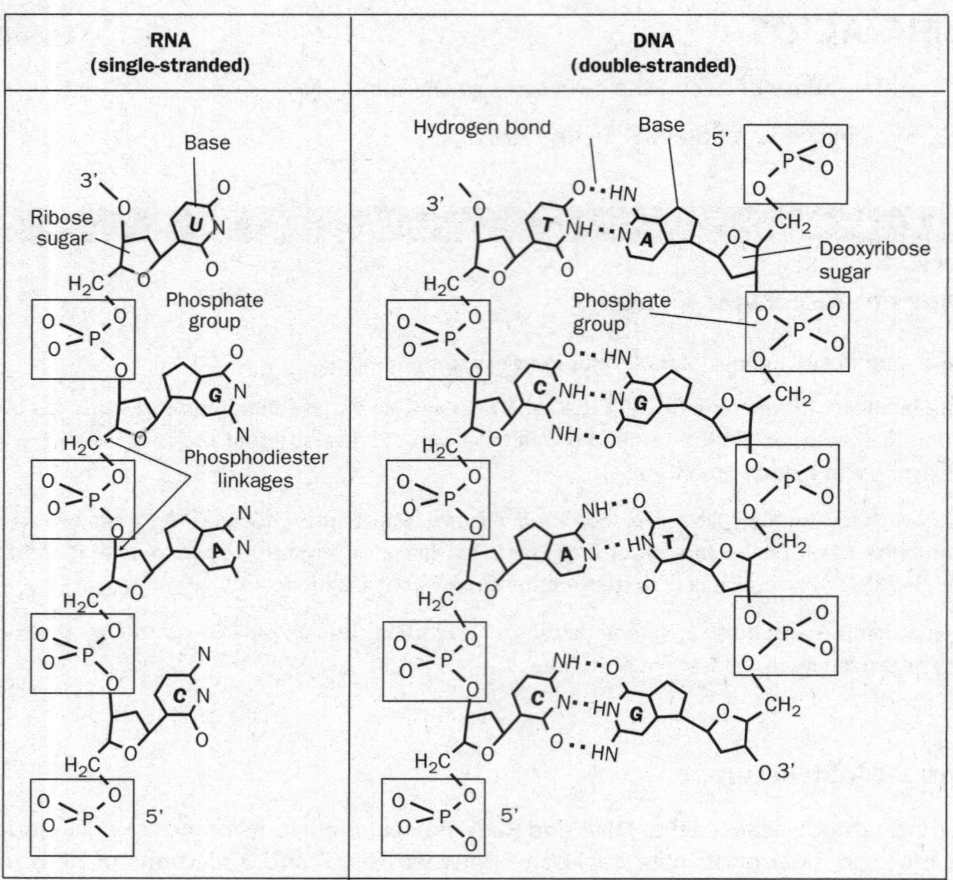

Structural Comparison of DNA and RNA

✔AP Expert Note

The molecular structures of the different nucleotides are outside the scope of the AP Biology exam.

DNA and RNA Functions

The primary function of DNA is to serve as heritable genetic material, containing the recipes for all of the proteins an organism can produce. RNA, in contrast, has a variety of forms with differing functions:

- *mRNA*: messenger RNA; delivers genetic instructions from nucleus to ribosomes
- *tRNA*: transfer RNA; brings amino acids to ribosomes for translation
- *rRNA*: ribosomal RNA; structural component of ribosomes
- *hnRNA*: heterogeneous nuclear RNA; synthesized from a DNA template by transcription—sometimes called pre-mRNA
- *RNAi*: RNA interference; when small segments of RNA bind to specific mRNA to block gene expression

DNA REPLICATION

High-Yield

13.3 Describe the process of DNA replication

It is important to understand that the strands of the DNA double helix are oriented in an anti-parallel manner to each other—with one strand 5′ to 3′ and the other side 3′ to 5′. The 3′ to 5′ designations refer to the carbons that make up the sugar in each nucleotide, and the nucleotides of each strand are connected by hydrogen bonds to a nucleotide in the complementary strand. DNA polymerase II adds from the 5′ end of the incoming nucleotide to the 3′ position of the ribose in the last nucleotide on the DNA strand being synthesized. However, DNA polymerase cannot bind until an RNA primer, a short strand of complementary RNA nucleotides, has been attached to a separated strand by an enzyme called primase.

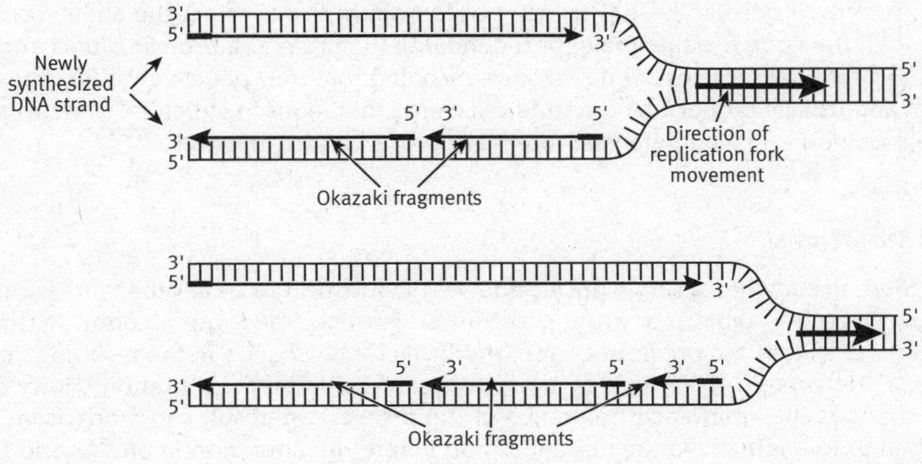

DNA Replication

DNA helicase unzips the DNA to ready it for replication. DNA replication is semiconservative, meaning that half of the original DNA is conserved in each daughter molecule. This means that the new DNA uses the existing DNA strands as a template. When DNA splits and copies, although the two resulting DNA strands that are created are identical to each other, one strand is made up of the "old" DNA strand while the other daughter strand is made of newly synthesized DNA. Okazaki fragments are formed because DNA polymerase can only read the older DNA in a 3′ to 5′ direction and synthesize a new strand in a 5′ to 3′ direction. It works away from the origin of replication on one side and then jumps back to follow the unzipping DNA, but can only continue to work in a direction away from the replication fork, thus leaving fragments of DNA that have to be linked later by DNA ligase.

✔ AP Expert Note

Replication always adds new nucleotides in the 5′ to 3′ direction—no exceptions! To help you remember, use the mnemonic device: ***You read up on a topic and write down your notes.*** **Polymerase reads up (3′ to 5′) and writes down (5′ to 3′).**

Mutations

The goal of a cell is usually to maintain the same code in its DNA, base pair by base pair. Sometimes changes occur spontaneously during replication or are due to environmental factors such as irradiation. These changes in the DNA code are called **mutations**. Mutations can involve a change in only one base or several bases. Mutations are also placed in one of two categories, base-pair substitutions or insertions and deletions. **Base-pair substitutions** occur when one base pair is incorrectly reproduced and exchanged with a different base pair. An **insertion** is when any number of extra base pairs are added to the code, and a **deletion** is when any number are removed from the code.

The effect of mutations depends upon where they occur in the code. Mutations of introns often have no effect on an individual because they are not translated into proteins. Likewise, base-pair substitutions in third position base pairs for a codon often do not affect the amino acid it codes for (remember, the code is degenerate, or redundant); therefore, the protein is unchanged. Most mutations in structural proteins are deleterious, meaning that they negatively affect the nature of the protein and usually produce a nonviable cell. Some mutations in structural proteins are viable and, if the mutation is in a gamete, are potentially passed on to offspring.

Types of Mutations

Point mutations occur when a single nucleotide base is substituted by another. If the substitution occurs in a noncoding region, or if the substitution is transcribed into a codon that codes for the same amino acid as the previous codon, there will be no change in the resulting amino acid sequence of the protein. This type of point mutation is a "silent" mutation. However, if the mutation changes the amino acid sequence of the protein, the result can range from an insignificant change to a lethal change depending on where the alteration in amino acid sequence takes place.

Frameshift mutations involve a change in the reading frame of an mRNA. Because ribosomes and tRNAs "read" the mRNA in sections of three bases (codons), if a base is inserted or deleted due to faulty transcription or a mutation in the actual DNA, the reading of the resulting mRNA will shift, and this is called a frameshift mutation. Base insertions and deletions, particularly toward the start of the protein's amino acid sequence, can render the remaining structure nonfunctional as almost every amino acid along the sequence gets changed.

Nonsense mutations produce a premature termination of the polypeptide chain by changing one of the codons to a stop codon. Beta-thalassemia is a hereditary disease in which red blood cells are produced with little or no functional hemoglobin for oxygen carrying. The different forms of this disease can be produced by a variety of mutations, including point mutations, frameshift mutations, and nonsense mutations.

The following diagram is useful for understanding these different types of mutations.

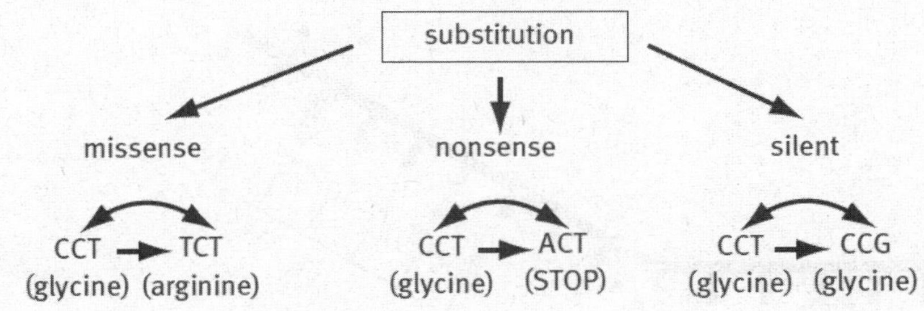

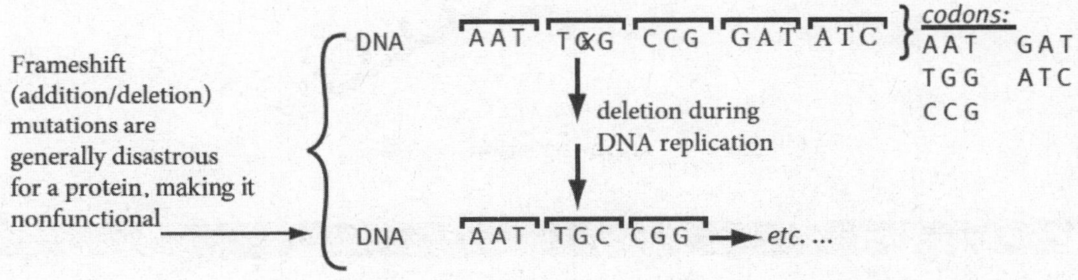

Varieties of Mutations

TRANSCRIPTION

13.4 Describe the process and product of transcription

Transcription is the process of "mirroring" a sequence of nucleotides on an original strand of DNA with a strand of **complementary bases** of mRNA (substituting uracil for thymine). Occurring in the nucleus, transcription starts at a special region called a promoter and ends at another special region called a terminator. The mRNA then undergoes a sequence of post-transcriptional modifications. For reasons not fully understood, the original DNA contains many regions that are not coded for in protein synthesis. These regions are called **introns**, and they are excised before the mRNA enters the cytoplasm for protein translation. **Exons** are the regions that code for protein synthesis in mRNA and remain after the introns have been excised. Transcription factors and regulatory proteins, which play an important role in transcription control, are discussed in chapter 14.

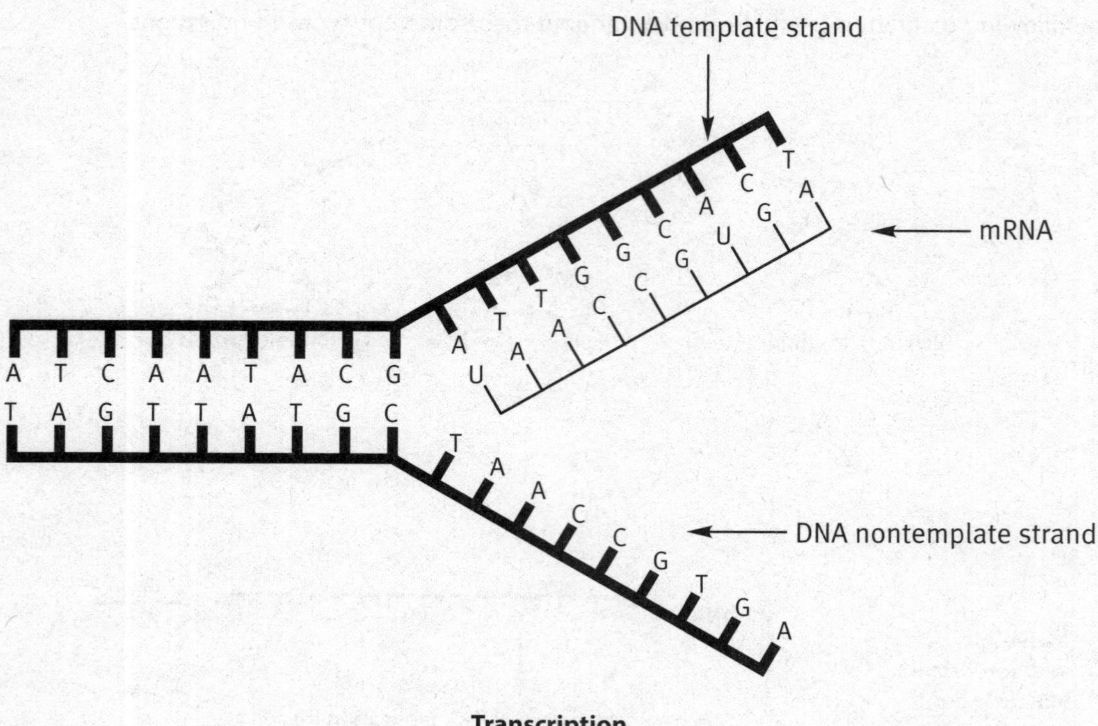

Transcription

> ✔ **AP Expert Note**
>
> **The mnemonic device *INtrons are IN the way, EXons are EXpressed* can be used to remember the different regions in DNA.**

Stages of Transcription

Initiation

The first stage of transcription is *initiation*. During this stage, RNA polymerase binds to the **promoter** located upstream from the genes to be "read." The bound RNA polymerase causes DNA to unwind, exposing a single-strand that serves as the template for transcription.

One of the most common elements in eukaryotic promoters is the **TATA box**. Although not present in "housekeeping genes" and other developmental genes, such as the homeotic genes, TATA boxes are A-T-rich regions of DNA that are involved in positioning the start of transcription. The reason for this is that regions of DNA rich in adenine and thymine tend to separate more easily than those rich in Cs and Gs. Adenine and thymine form only two hydrogen bonds across the double helix, while C and G form three. Separation of DNA at the TATA boxes in the promoter regions of genes allows DNA to unzip at those regions for RNA polymerase to access the DNA template. Other types of promoters known as *internal promoters* can be found within the introns of genes. These promoters occur especially in genes that encode rRNA and tRNA molecules.

Elongation

In the *elongation* stage, RNA polymerase transcribes the DNA template strand to produce a complementary antiparallel RNA strand. Reading from the 3′ end of the template strand, RNA polymerase builds an hnRNA (pre-mRNA) chain in the 5′ to 3′ direction.

Termination

The *termination* stage occurs when RNA polymerase encounters a **terminator** sequence located downstream from the promoter. The terminator signals RNA polymerase to detach from the template strand and release the hnRNA chain for post-transcriptional processes. Introns are removed and a 5′ methylguanosine cap and a 3′ poly-A tail are added to complete the mRNA. The mRNA then exits via the nuclear pores and goes to the ribosomes in the cytoplasm for translation.

TRANSLATION

13.5 Describe the steps of translation

13.6 Describe post-translational modifications to polypeptides

In a eukaryotic organism, DNA is found in the **nucleus** of the cell. **Protein synthesis** takes place in the cytoplasm, so information from the coding regions of DNA in the nucleus must be moved to the cytoplasm. Information is moved to the cytoplasm through **mRNA (messenger RNA)**.

Once in the cytoplasm, mRNA acts as the template for protein **translation**. A short series of three bases, called a triplet or **codon**, codes for a **tRNA (transfer RNA)** that carries a specific amino acid. There are 64 (4^3) possible triplets, but there are only 20 amino acids. Even allowing for the start signal codon AUG as a site to begin protein translation and several codons acting as stop signals, there is considerable redundancy (also known as degeneracy) in the **genetic code**.

The **ribosome** performs protein synthesis. Once an mRNA enters the cytoplasm, several ribosomes attach to it, creating multiple lengths of the protein simultaneously. Elongation continues until a stop codon is reached and the ribosome disengages the mRNA. The newly synthesized polypeptide usually undergoes transformation (i.e., removal of terminal amino acids) before achieving its active state.

Transformations

> ✔ **AP Expert Note**
>
> **You do not need to memorize the genetic code, codons, or the structure of the 20 amino acids for the AP Biology exam.**

Stages of Translation

Initiation

The first stage of translation is *initiation*. During this stage, the mRNA attaches itself to the ribosome. The first codon on the mRNA is always AUG (the start sequence), which codes for the amino acid methionine. tRNA brings the first amino acid and places it in its proper place.

Elongation

Subsequent amino acids are then brought to the ribosome and joined together by peptide bonds in the *elongation* stage.

Termination

The *termination* stage is always triggered by one of three stop codons: UAA, UGA, or UAG. They stop protein synthesis and release the protein from the ribosome.

> ✔ **AP Expert Note**
>
> **To remember the start codon (AUG), think of the month that is the start of the school year for many students: AUGust. A mnemonic to help remember the three stop codons (UAA, UGA, UAG) is: U Are Annoying, U Go Away, U Are Gone.**

Post-Translational Modifications

After translation, synthesized polypeptides or proteins may undergo post-translational modifications, which generally include covalent enzymatic alterations to amino acids or the C-terminus or N-terminus. Post-translational modifications influence a protein's structure and specify a protein's function. Examples of post-translational modifications are phosphorylation, ubiquitination, and glycosylation. Phosphorylation is involved in various cellular processes, such as the cell cycle, apoptosis, differentiation, and enzyme activity. By adding a phosphate group (via kinase) and/or removing a phosphate group (via phosphatase), phosphorylation and its reverse, dephosphorylation, can activate or inactivate a protein. Similarly, ubiquitination adds ubiquitin, a small regulatory protein, which acts as a signal that turns transcription levels on, off, up, or down. Glycosylation—the attaching of a carbohydrate molecule—promotes protein folding, improves protein stability, and is involved in immune recognition.

AP BIOLOGY LAB 9: BIOTECHNOLOGY: RESTRICTION ENZYME ANALYSIS OF DNA INVESTIGATION

13.7 Investigate analysis of DNA

This experiment involves splicing DNA using restriction enzymes and visualizing these fragments using gel electrophoresis. A gel is a matrix composed of a polymer that acts like filter paper. Think of the gel as a thick, tangled jungle. DNA fragments are negatively charged, so they will move away from the negative pole of an applied current and move toward a positive pole. As the DNA fragments move away from the negative terminal of the charge applied, they get tangled in the jungle. The smaller the fragments of DNA, the faster they can move through the tangle. The movement of the fragments can be adjusted by changing the density of the gel (the tangles in the jungle) or by changing the quantity of charge (the pushing/pulling power) applied to the gel.

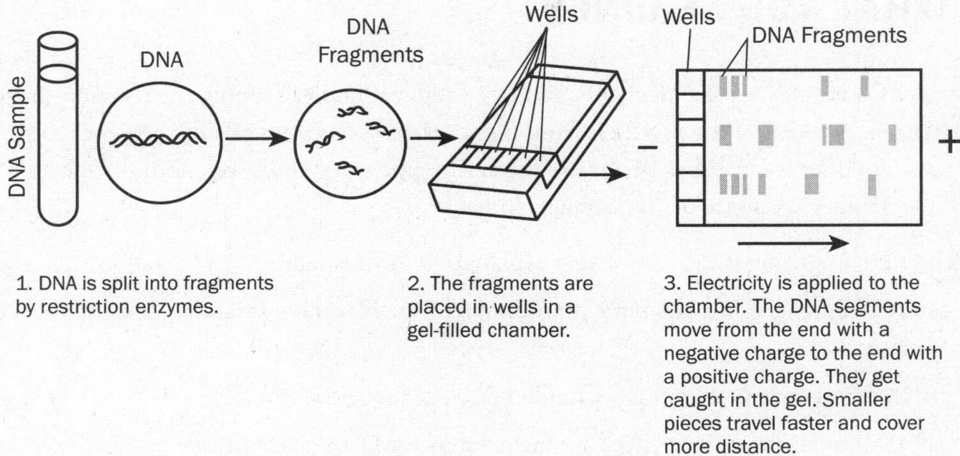

DNA Sample
DNA
DNA Fragments
Wells
Wells
DNA Fragments
–
+
1. DNA is split into fragments by restriction enzymes.
2. The fragments are placed in wells in a gel-filled chamber.
3. Electricity is applied to the chamber. The DNA segments move from the end with a negative charge to the end with a positive charge. They get caught in the gel. Smaller pieces travel faster and cover more distance.

Diagram of Gel Electrophoresis

Restriction enzymes are manufactured by molecular genetics laboratories to have a high specificity for a series of base pairs on a strand of DNA or RNA. When the enzyme finds and binds to these base pairs, it snips the DNA. The more this series of base pairs appears on the DNA or RNA strand, the more cuts the restriction enzyme makes in the strand. After the strand is digested by the restriction enzyme, the fragments with different lengths can be separated using gel electrophoresis. This technique is commonly used to determine genetic differences between organisms, as long as the appropriate region of DNA can be isolated and the right combination of restriction enzymes can be determined. The most important thing to remember from this lab is that smaller fragments move farther on the gel. They move farther *because* they are smaller. All of the fragments move for the same reason: they are charged molecules moving away from an applied charge of the same kind (positive or negative).

RAPID REVIEW

If you take away only 6 things from this chapter:

1. Nitrogenous base pairs make up DNA and RNA: adenine pairs with thymine (DNA only) or uracil (RNA only) and cytosine pairs with guanine.
2. DNA replication is a semiconservative process, in which one of the antiparallel strands of DNA is preserved and the other strand is newly synthesized. New nucleotides are added in a 5′ to 3′ direction.
3. In transcription, the DNA strands separate and mRNA copies one side. The mRNA takes the information to the ribosome, where protein synthesis occurs.
4. In translation, tRNA carries amino acids to the mRNA and assembles them into proteins based on the mRNA code. Proteins are often modified after translation, giving them their final structure.
5. Mutations are the source of genetic change. Types include base-pair substitutions, which affect one amino acid, as well as insertions and deletions, which shift the genetic code and affect many amino acids.
6. Scientists can modify an organism's DNA by adding new genes.

TEST WHAT YOU LEARNED

1. A biochemist is studying the mechanism of DNA replication by stopping the reaction shortly after its initiation. Analysis of the products from the reaction on a gel reveals the presence of single-stranded and double-stranded DNA. The shorter fragments of DNA were found to be attached to a complementary sequence of 10 ribonucleotides.

   Which of the following is the most likely explanation for the data?

   (A) Short fragments of mRNA bind at random to available single-stranded DNA sequences in the cell.

   (B) tRNAs bind to available single-stranded DNA sequences in the cell.

   (C) Short ribonucleotide sequences are the primers for DNA polymerases.

   (D) Short sequences of RNA attached to DNA represent an artifact of the isolation procedure.

2. Hormone response elements are short sequences of DNA to which a hormone-hormone receptor complex can bind. In a mammalian cell, a mutation inserts an estrogen-response element upstream of the transcription initiation site of the gene encoding the protein tubulin. Tubulin is a component of the cytoskeleton. Which of the following results would most likely occur as a result of this mutation?

   (A) The expression of tubulin would become estrogen responsive.

   (B) Baseline expression of tubulin would increase.

   (C) The amino acid sequence of tubulin would be changed.

   (D) Transcription of the tubulin gene could not undergo termination.

3. A researcher wishes to synthesize insulin *in vitro* and decides to use a cytoplasmic wheat germ extract that contains ribosomes, tRNAs, amino acids, enzymes, and translation factors. She isolates several types of macromolecules from pancreatic cells and incubates them with the wheat germ extract.

   Based on this information, the insulin will only be produced when

   (A) DNA from pancreatic cells is added to the mixture

   (B) ribosomes from pancreatic cells are added to the mixture

   (C) mRNA from pancreatic cells is added to the mixture

   (D) both DNA and mRNA from pancreatic cells are added to the mixture

4. Several point mutations in a DNA sequence changed the mRNA transcript encoded by a gene as shown below.

   Original sequence:

   5′-UCG-GCA-CAU-UUA-UAU-GUU-3′

   New sequence:

   5′-AGC-GCA-CAU-UUG-UAA-GUU-3′

CODON TABLE

FIRST POSITION		SECOND POSITION								THIRD POSITION
		U		C		A		G		
		code	amino acid	code	amino acid	code	amino acid	code	amino acid	
U		UUU	phe	UCU	ser	UAU	tyr	UGU	cys	U
		UUC		UCC		UAC		UGC		C
		UUA	leu	UCA		UAA	STOP	UGA	STOP	A
		UUG		UCG		UAG	STOP	UGG	trp	G
C		CUU	leu	CCU	pro	CAU	his	CGU	arg	U
		CUC		CCC		CAC		CGC		C
		CUA		CCA		CAA	gln	CGA		A
		CUG		CCG		CAG		CGG		G
A		AUU	ile	ACU	thr	AAU	asn	AGU	ser	U
		AUC		ACC		AAC		AGC		C
		AUA		ACA		AAA	lys	AGA	arg	A
		AUG	met	ACG		AAG		AGG		G
G		GUU	val	GCU	ala	GAU	asp	GGU	gly	U
		GUC		GCC		GAC		GGC		C
		GUA		GCA		GAA	glu	GGA		A
		GUG		GCG		GAG		GGG		G

Using the codon table provided, which of the following statements is accurate?

(A) The changes in the sequence caused a substitution.

(B) The changes in the sequence generated a shorter peptide.

(C) The changes in the sequence generated a longer peptide.

(D) The changes in the sequence did not change the peptide.

5. The mRNA encoding the insulin polypeptide is translated on ribosomes attached to the rough endoplasmic reticulum (RER). Studies show that the protein encoded by the mRNA contains almost twice as many amino acids as the active insulin. Insulin is secreted into the bloodstream, along with an inactive peptide, at a 1:1 molar ratio.

   Which of the following best explains the observed results?

   (A) The precursor of insulin contains a signal sequence and an internal region that are excised after translation.

   (B) Insulin contains hydrophobic core residues that allow the molecule to be embedded in the RER membrane.

   (C) Insulin is translated as a dimer and digested into monomers.

   (D) Insulin is degraded in the RER into amino acids that are later reassembled prior to secretion.

6. A study shows that an isolated nucleic acid polymer contains 32% adenine, 15% guanine, and 25% cytosine. Which of the following conclusions is most probable based on this data?

   (A) The percentage of uracil is 32%.

   (B) The percentage of thymine is 32%.

   (C) The nucleic acid is a double-stranded DNA molecule.

   (D) The nucleic acid is a single-stranded RNA molecule.

7. The diagram below shows a DNA plasmid with a total length of 2,000 base pairs. The tick marks indicate cleavage sites for restriction enzymes (enzyme EcoRI and enzyme HaeIII).

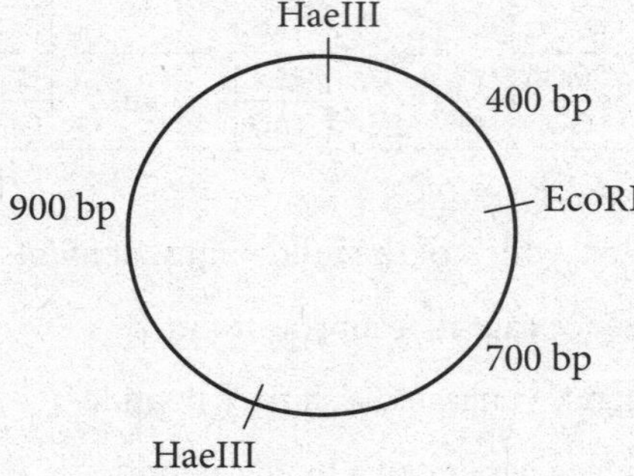

   The plasmid was digested with the enzyme EcoRI to completion and separated by gel electrophoresis. Which of the following results would most likely be observed?

   (A) One 1,600 bp fragment

   (B) One 2,000 bp fragment

   (C) Two fragments of 400 bp and 1,600 bp

   (D) Two fragments of 700 bp and 1,300 bp

Questions 8–9

Cancer is characterized by uncontrolled cell division. Many different mutations can contribute to cancer. Several of them affect the MAP kinase pathway, as shown in the figure below.

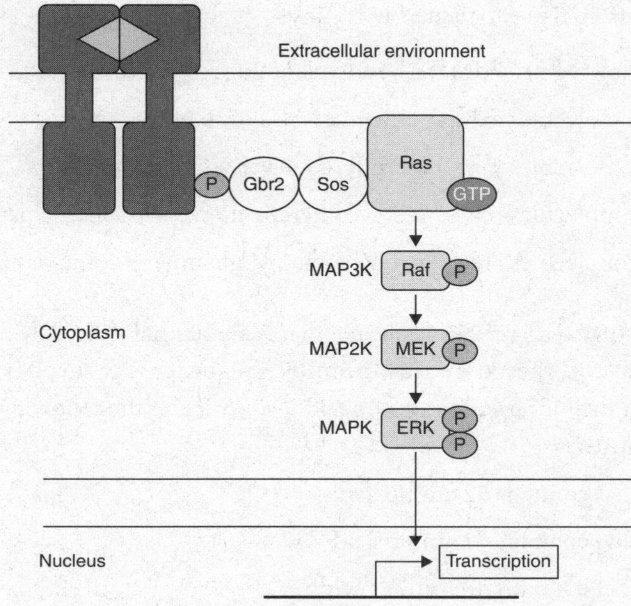

8. Which of the following would produce the greatest effect on transcription?
   (A) doubling the number of receptors
   (B) doubling the amount of Raf
   (C) doubling the rate at which Raf phosphorylates MEK
   (D) doubling the rate at which MEK phosphorylates ERK

9. The *Ras* gene is mutated in many forms of cancer. Given that this pathway leads to the transcription of genes involved in mitosis, which of the following would most likely lead to cancerous cell proliferation?
   (A) mutation of Ras so that the protein it encodes constantly phosphorylates Raf
   (B) mutation of Ras so that the protein it encodes no longer phosphorylates Raf
   (C) mutation of Ras so that the protein it encodes phosphorylates MAPK instead of MAP3K
   (D) mutation of Ras so that the protein it encodes cannot interact with Sos

10. Researchers isolate mRNA molecules from several muscle types of the same animal. These mRNA are transcribed from the gene encoding the protein myoglobin. The mRNA molecules are purified by binding them to beads coated with poly(T). The binding is accomplished using a polymer of thymidine residues (polydT), which binds to poly-adenosine tails. Gel electrophoresis shows that the lengths of the mRNAs vary from tissue to tissue, but is consistent among muscle cells of the same type.

    Which of the following provides the best explanation for the results obtained by the researchers?

    (A) The mRNA molecules exhibit different levels of stability.

    (B) The mRNA molecules may have different levels of polyadenylation.

    (C) The mRNA molecules may undergo different alternative splicing patterns.

    (D) The mRNA molecules are capped by a methyl guanine in some tissues but not in others.

11. Scientists have isolated an mRNA molecule that is a potential candidate for encoding a small peptide growth factor. They know that the mRNA sequence that encodes this peptide is 156 nucleotides long, including start and stop codons. Which of the following peptides could be encoded by the mRNA?

    (A) A peptide that contains 52 amino acids

    (B) A peptide that contains 51 amino acids

    (C) A peptide that contains 462 amino acids

    (D) A peptide that contains 468 amino acids

12. A researcher is interested in knowing whether a particular virus has a DNA or RNA genome and analyzes the structure of the individual virions. Which of the following should she expect to observe if the virus has an RNA genome?

    (A) Single-stranded nucleic acids

    (B) Nucleic acids containing adenine

    (C) Nucleic acids containing phosphate

    (D) Nucleic acids containing ribose

13. A biochemist studies protein post-translational modification and protein trafficking in the cell. She is identifying a modification that allows intrinsic proteins to form stable associations with the cell plasma membrane. Which of the following post-translational modifications is most likely to occur on these intrinsic proteins, and why?

    (A) Tyrosine residues will be phosphorylated to facilitate protein insertion in the membrane.

    (B) Polysaccharide groups will be added to the protein to increase its polarity.

    (C) Lipid groups will be added to the protein to increase its hydrophobicity.

    (D) Disulfide bridges will be formed among the proteins to stabilize interactions with the membrane.

The answer key to this quiz is located on the next page.

Transformations

Answer Key

Test What You Already Know

1. D
2. B
3. D
4. C
5. A
6. D
7. C

Test What You Learned

1. C
2. A
3. C
4. B
5. A
6. D
7. B
8. A
9. A
10. C
11. B
12. D
13. C

REFLECTION

Test What You Already Know score: __________

Test What You Learned score: __________

Use this section to evaluate your progress. After working through the pre-quiz, check off the boxes in the "Pre" column to indicate which Learning Objectives you feel confident about. Then, after completing the chapter, including the post-quiz, do the same to the boxes in the "Post" column. Keep working on unchecked Objectives until you're confident about them all!

Pre	Post	
❑	❑	**13.1** Differentiate between the structures of DNA and RNA
❑	❑	**13.2** Explain the function of DNA and RNA
❑	❑	**13.3** Describe the process of DNA replication
❑	❑	**13.4** Describe the process and product of transcription
❑	❑	**13.5** Describe the steps of translation
❑	❑	**13.6** Describe post-translational modifications to polypeptides
❑	❑	**13.7** Investigate analysis of DNA

FOR MORE PRACTICE

Complete more practice online at kaptest.com. Haven't registered your book yet? Go to kaptest.com/booksonline to begin.

ANSWERS AND EXPLANATIONS

Test What You Already Know

1. D Learning Objective: 13.1

RNA contains phosphate, the sugar ribose, and the nitrogenous bases adenine, cytosine, guanine, and uracil. Only **(D)** includes the appropriate sugar and an appropriate RNA base that is distinct from the bases of DNA. (A) and (B) are incorrect because they contain the sugar deoxyribose, a component of DNA. (C) is incorrect because RNA does not contain thymine, a base found only in DNA.

2. B Learning Objective: 13.2

Because the *lac*⁻ cells contain everything necessary to construct enzymes that can ferment lactose except for the DNA with the *lac*⁺ genes, the only cells that would be adversely affected would be those incubated with extract B, which loses those genes due to the action of the DNases. Because those cells would be unable to ferment lactose, they would appear as colorless. **(B)** is thus correct. (A) is incorrect because the *lac*⁻ cells contain all the proteins and amino acids necessary to construct lactose-fermenting enzymes, provided they have the proper genes (which they can get from extract A), so they would appear blue. (C) is incorrect because the *lac*⁻ cells contain all the RNA molecules and nucleotides necessary to transcribe and translate the lactose-fermenting enzymes, provided they have the proper genes (which they can get from extract C), so they would also appear blue. (D) is incorrect because the cells incubated with extract B will appear colorless, as explained above.

3. D Learning Objective: 13.3

Short strands of DNA would ordinarily not be found in the cellular environment, so something must be preventing these short strands from being joined together into longer strands. Joining nucleotides together requires the creation of phosphodiester bonds, so if enzymes that created these bonds (ligases) were deficient in the cell, it could explain the existence of these short strands. **(D)** is thus correct. (A) is incorrect because DNA must be unwound in order to be replicated, and it is clear that some replication is occurring since the newly labeled nucleotides are finding their way into DNA strands. (B) is incorrect because there were no single-stranded DNA molecules (as indicated by the unchanged results after adding mung bean nuclease), so DNA polymerase I must have filled in the gaps on the new strands. (C) is incorrect because new strands were in fact synthesized.

4. C Learning Objective: 13.4

In order for the same gene to produce two distinct mRNA transcripts, different exons from the gene must be spliced together for each protein. Thus, alternative splicing in different tissues explains the existence of these proteins that come from the same gene, making **(C)** correct. (A) is incorrect because mutations occur largely at random, so it would be highly improbable to have two distinct proteins consistently produced from the same gene as a result of separate mRNA mutations. (B) is incorrect because the mRNA sequences are identical at some points, which would not be possible if both the sense and antisense strands were being used. (D) is incorrect because the proteins have differing amino acid sequences, which would not result simply from post-translational modifications. Moreover, (D) would not explain why the mRNA sequences are themselves different.

5. A Learning Objective: 13.5

Messenger RNA is read in the 5′ to 3′ direction, so to convert the codons to amino acids, simply move from left to right and match each one to the table. This yields a sequence of ser-ala-his-leu-tyr-val, as in **(A)**. (B) is incorrect because it mistakenly interprets the third and fifth codons in the sequence. (C) is incorrect because it provides the sequence if an antisense mRNA were translated (UCG becomes AGC, GCA becomes CGU, etc.), but mRNA is read directly by the ribosome without the creation of an antisense strand. (D) is incorrect because it provides the sequence if the mRNA strand were read from 3′ to 5′, but this is the opposite of the direction of reading in translation.

6. D Learning Objective: 13.6

Because phosphorylation directly modifies existing proteins, it must be regulated after translation occurs, making it a post-translational modification. Therefore, **(D)** is correct. (A), (B), and (C) are incorrect because these are all processes that occur before a protein has been constructed.

7. C Learning Objective: 13.7

The enzyme would break the plasmid at the points labeled "HaeIII" in the diagram. This would create two fragments, one with a length of 900 base pairs and the other with a length of $400 + 700 = 1{,}100$ base pairs. **(C)** is thus correct. (A) and (B) are incorrect because they fail to account for the total length of one of the strands. (D) is incorrect because it presents the results if both HaeIII and EcoRI were used.

Test What You Learned

1. C Learning Objective: 13.3

In order for DNA polymerases to begin their work, small RNA primers must first attach to the single strands of DNA after they have been separated by helicase. **(C)** is thus correct. (A) is incorrect because mRNA does not bind to DNA at random to form sequences. (B) is incorrect because tRNA does not bind to DNA, but rather to mRNA at ribosomes. (D) is incorrect because there is nothing in the description of the isolation procedure that suggests the researcher is adding ribonucleotides.

2. A Learning Objective: 13.4

The insertion of an estrogen response element upstream of the initiation site would make the expression of tubulin become estrogen responsive, making **(A)** correct. (B) is incorrect because the baseline expression of tubulin would remain unchanged; the only effect would be to make expression estrogen responsive. (C) is incorrect because the mutation does not occur in the region that encodes for tubulin, but rather upstream of that region. (D) is incorrect because termination takes place downstream of the initiation site, not upstream.

3. C Learning Objective: 13.2

Given that the extract contains all the cellular machinery necessary for translation of insulin except for mRNA, the insulin will only be produced when pancreatic mRNA (some of which will code for insulin) is added to the mixture. **(C)** is thus correct. (A) is incorrect because the gene for insulin cannot be transcribed and translated without mRNA. (B) is incorrect because the mixture already contains ribosomes and because mRNA is needed. (D) is incorrect because the DNA is not necessary; the mRNA transcripts are sufficient, given that all of the machinery for translation is already present in the mixture.

4. B Learning Objective: 13.5

The conversion of the fourth codon from UAU to UAA, changes it from a tyr to a stop signal. This is a nonsense mutation that will cause the sequence to be prematurely terminated, generating a shorter peptide sequence, **(B)**. (A) is incorrect because the amino acids in the sequence remain unchanged; the only difference is that the fourth codon becomes a stop signal. (C) is incorrect because the protein is made shorter, not longer. (D) is incorrect because a change in fact occurred: the peptide sequence was made shorter.

5. A Learning Objective: 13.6

The most likely reason for an mRNA sequence to be far longer than an active protein is some type of post-translational modification that excises a segment of the protein. **(A)** thus provides the best explanation. (B) is incorrect because insulin is secreted into the bloodstream, as noted in the question stem, not embedded in a membrane. (C) is incorrect because the question stem suggests that the excised region is an inactive peptide; a dimer would be split into two active proteins. (D) is incorrect because it would be redundant and a waste of energy to split up the protein into amino acids and then reassemble it.

6. D Learning Objective: 13.1

To find which conclusion is most probable, use process of elimination to rule out impossible or unlikely conclusions. The percentages given in the question stem add up to 72%, so the remaining base (whether uracil or thymine) would have to constitute 28% of the nucleotides. This eliminates (A) and (B). If the molecule were double-stranded, it would be expected that there would be equal amounts of cytosine and guanine, as well as equal amounts of adenine and either thymine or uracil. Because the percentages are unequal, (C) can also be ruled out. Thus, the most probable conclusion is that this a single-stranded RNA molecule, **(D)**.

7. B Learning Objective: 13.7

The EcoRI enzyme would only cleave the plasmid at one location, which would not separate the plasmid into multiple fragments, but would leave a single linear strand of DNA 2,000 base pairs long. **(B)** is thus correct. (A) is incorrect because it is too short. (C) and (D) are incorrect because the plasmid is cleaved in only one location, not two, as would be required to have two fragments.

8. A Learning Objective: 13.3

The signal transduction pathway amplifies the signal at every step. Therefore, the earlier in the pathway the doubling is made, the more of an effect it will have. **(A)** targets the earliest step in the process, receptor-ligand binding. (B), (C), and (D) are all later in the pathway and therefore would have less of an effect.

9. A Learning Objective: 13.3

Constantly phosphorylating Raf would permanently turn on the production of mitosis-promoting genes. This would lead to unregulated cell division, a hallmark trait of cancer. Thus, **(A)** is correct. If Raf phosphorylation were turned off, mitosis would happen less frequently, so (B) is incorrect. Directly phosphorylating MAPK would decrease the frequency of mitosis because there would be no amplification effect of the cascade, so (C) is incorrect. (D) is incorrect because if Ras could no longer interact with Sos, there would be no activation of mitosis by this pathway.

10. C Learning Objective: 13.4

The difference in mRNA lengths among various tissues points to different splicing patterns being used in different tissues. This is also evident by the similar-sized fragments found in similar tissue, like among muscle cells of the same type. Thus, the correct answer is **(C)**. (A) is incorrect because mRNA molecules exhibiting different levels of stability would degrade into lengths that vary from experiment to experiment. (B) is incorrect because all mRNAs were isolated by binding to poly(T) and, therefore, must have had poly(A) tails. (D) is incorrect because the absence of a methyl-guanosine cap would not make a noticeable change in the size of the molecules.

11. B Learning Objective: 13.5

Amino acids are encoded by codons, which are sequences of three nucleotides. An mRNA sequence that is 156 nucleotides long has $156 \div 3$ or 52 codons. The start codon encodes the amino acid methionine, and the last codon is the stop codon, which is not translated. Thus, the peptide would contain 51 amino acids. The correct answer is **(B)**. (A) is incorrect because although the mRNA sequence contains 52 codons, the last codon does not encode an amino acid. (C) and (D) are incorrect because three nucleotides encode one amino acid. The number of nucleotides is divided by 3, not multiplied by 3.

12. D Learning Objective: 13.1

Nucleic acids are composed of nucleotides made of a pentose sugar, a phosphate group, and a nitrogenous base. RNA contains the pentose sugar ribose, whereas DNA contains deoxyribose. Both RNA and DNA contain the nitrogenous bases adenine, guanine, and cytosine, but differ in that RNA contains uracil and DNA has thymine. Thus, the correct answer is **(D)**. Typically, DNA is present in double-stranded form, and RNA is present in single-stranded form. However, some viruses have a double-stranded RNA genome, so (A) is incorrect. (B) is incorrect because adenine is present in both DNA and RNA. (C) is incorrect because phosphate is found in all nucleic acids.

13. C Learning Objective: 13.6

The cell plasma membrane is composed of a phospholipid bilayer, so intrinsic proteins would be stabilized by lipid modification. The addition of lipid groups would increase its hydrophobicity and facilitate the insertion of intrinsic proteins in the lipid environment of the membrane. Therefore, the correct answer is **(C)**. (A) is incorrect because phosphorylation does not increase hydrophobicity since phosphate groups are charged. (B) is incorrect because intrinsic proteins do not need to increase polarity to form stable interactions with lipids in the membrane. (D) is incorrect because disulfide bridges do not increase the solubility of proteins in the membrane.

CHAPTER 14

Development

LEARNING OBJECTIVES

In this chapter, you will review how to:

- **14.1** Describe the process of fertilization and genetic transfer
- **14.2** Describe positive and negative gene control mechanisms
- **14.3** Explain how gene regulation results in gene expression
- **14.4** Explain how gene expression leads to cell specialization

TEST WHAT YOU ALREADY KNOW

1. *Lymnaea stagnalis* snails can have left-coiling (sinistral) or right-coiling (dextral) shells, shown in the figure as A and B, respectively.

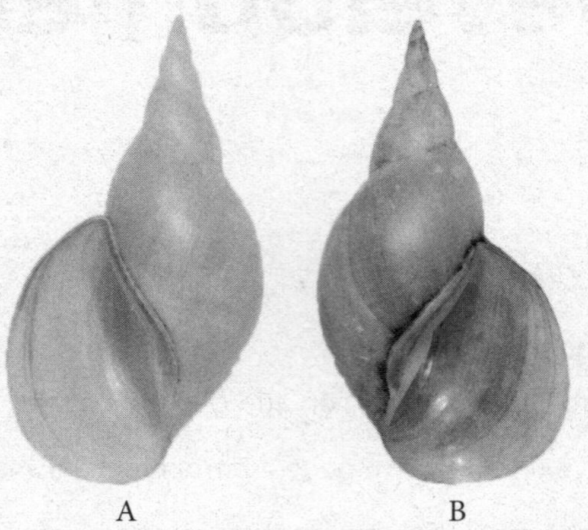

Lymnaea stagnalis shells can be left-handed/sinistral (A) or right-handed/dextral (B).

Adapted from Edmund Gittenberger, Thomas D. Hamann, and Takahiro Asami, "Chiral Speciation in Terrestrial Pulmonate Snails," *PLoS ONE 7*, no. 4 (April 2012): e34005.

Although researchers know that shell development is heavily influenced by genetics, they are interested in the effects of the environment on shell development as well. To study this, researchers could raise snails in different laboratory environments. Which of the following possibilities would be least likely from crosses of true-breeding, male, left-coiling snails and true-breeding, female, right-coiling snails?

(A) After the offspring were randomly divided into groups to be raised in two different laboratory environments, those in one environment all resembled type A, while those in the other environment all resembled type B.

(B) After the offspring were randomly divided into groups to be raised in two different laboratory environments, all of the snails in both environments resembled type A.

(C) After the offspring were randomly divided into groups to be raised in two different laboratory environments, all of the snails in both environments resembled type B.

(D) After the offspring were randomly divided into groups to be raised in four different laboratory environments, there were similar amounts of each shell type but slightly more of type A.

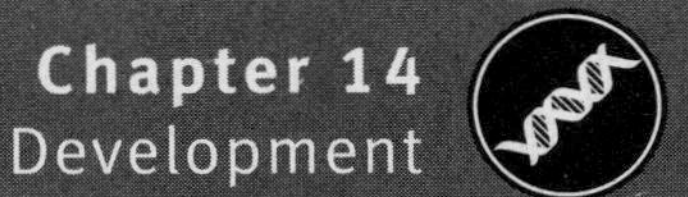

2. If a particular gene product is needed during early development, the gene must be turned on at the appropriate time. The default setting is off and then the binding of an activator turns on the gene when it is needed. Which statement gives the best description of this type of gene control?
   (A) It is constitutive expression because the gene is expressed throughout development.
   (B) It is positive control because the gene is turned on by the presence of an activator once it is needed.
   (C) It is negative control because the gene is turned on and is capable of being turned off.
   (D) It is an operon because operons are always controlled by repressors.

3. The table below compares gene expression in larval honey bee workers and queens. The honey bee genes were compared with genes of known function in *Drosophila melanogaster* in order to predict their functions within honey bees. The light bars represent the percent of genes that are differentially expressed in queens. The dark bars represent values for workers.

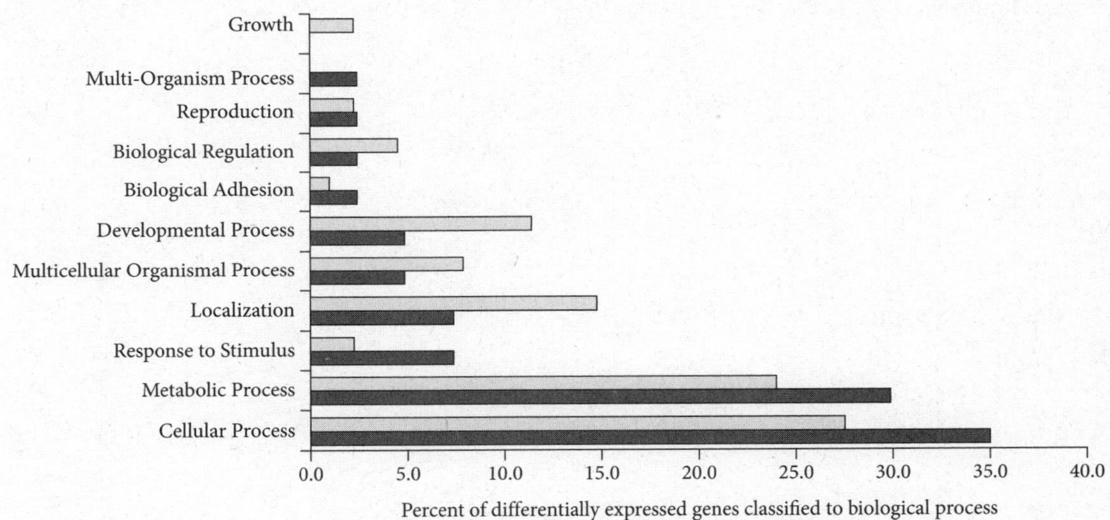

Adapted from Ana Durvalina Bomtorin et al., "Hox Gene Expression Leads to Differential Hind Leg Development between Honeybee Castes," *PLoS ONE 7*, no. 7 (July 2012): e40111.

Based on this data, what can most reasonably be concluded about the difference between queens and workers?
   (A) Workers may need to express a greater proportion of regulatory genes than queens as they are exposed to so many complex demands while outside of the hive.
   (B) Queens may need a lower proportion of genes expressed during development than workers as they have a relatively simple lifestyle that does not require leaving the hive regularly.
   (C) Workers may need to express a higher proportion of genes related to sensory organs to allow them to respond to a variety of stimuli while foraging.
   (D) Queens may need a wider range of metabolic functions compared with workers because of their important role in the colony.

4. Honeybees have two female castes, workers and queens, which are phenotypically very different. Researchers have discovered that feeding bee larvae a diet rich in a protein called royalactin increases the amount of juvenile hormone that they produce. Higher hormone levels result in the development of queen bee characteristics. However, they have also discovered that feeding royalactin to fruit flies causes similar phenotypic changes. What is the most reasonable explanation for this finding?

   (A) Royalactin affects gene expression by regulating different genes in the fruit fly and the honeybee.

   (B) Royalactin affects gene expression by regulating similar genes in the fruit fly and the honeybee.

   (C) Royalactin acts directly as a hormone by binding to different receptors in the fruit fly and the honeybee.

   (D) Royalactin acts directly as a hormone by binding to similar receptors in the fruit fly and the honeybee.

Answers to this quiz can be found at the end of this chapter.

FERTILIZATION AND EARLY DEVELOPMENT

14.1 Describe the process of fertilization and genetic transfer

Sexual reproduction of diploid (2*N*) organisms involves the recombination and transmission of genetic material from parents to offspring via vertical gene transfer and **fertilization**. (Note: horizontal gene transfer is the transmission of genetic material from a donor to a recipient organism that is not its offspring.) For sexual reproduction to occur, cells undergo **gametogenesis**—the process that forms **egg (ovum)** and **sperm (spermatozoon)** cells or gametes via meiosis. Recall that meiosis, which was discussed in chapter 11, is the process that forms haploid cells (containing one set of chromosomes, 1*N*). During fertilization, a sperm cell fertilizes the egg cell to produce a single cell called a **zygote**. (Sexual reproduction can also occur when a hermaphroditic organism fertilizes its own eggs to form offspring.) The fusion of one sperm with one egg, both of which are haploid, restores the diploid number of chromosomes. Since each zygote is genetically different, fertilization increases genetic variation and contributes to the survival of a species.

Fertilization is the first step in embryogenesis—the process by which an embryo forms and develops. After fertilization, the zygote undergoes **cleavage** (cell division without an increase in mass) and cellular differentiation to form a multicellular organism.

Fertilization can occur within the body or outside of the body. Internal fertilization typically occurs in the female reproductive organs of mammals and birds, while external fertilization usually occurs in aquatic environments and can be seen in frogs, sea urchins, and most fish. In plants, fertilization is preceded by **pollination**, the transferring of pollen to the female reproductive organs of a plant. Additionally, in angiosperms (flowering plants), double fertilization occurs—one sperm cell fertilizes the egg cell, and the other fuses to form the endosperm, which provides nutrients to the developing embryo in the seed.

The details of embryonic development are not tested on the AP Biology exam.

GENE REGULATION AND CELL SPECIALIZATION

14.2 Describe positive and negative gene control mechanisms

14.3 Explain how gene regulation results in gene expression

14.4 Explain how gene expression leads to cell specialization

In multicellular organisms, the developmental pathways from zygote to fully formed organism (adult) involve regulatory systems and differential **gene expression** governed by internal and external signals. Though each eukaryotic cell contains the same genome, each cell does not express the same set of genes. Some genes may be continually expressed, but most gene expression is regulated to maximize energy usage and increase metabolic fitness. Otherwise, cells would not be able to obtain enough resources to stay alive.

Gene Regulation and Expression

Regulatory systems in both prokaryotes and eukaryotes consist of genes that encode proteins (**repressors** and **activators**), which interact with DNA sequences to control gene expression via negative and positive control mechanisms. In **negative control**, genes are expressed until they are switched off by a repressor; in **positive control**, genes are expressed when they are turned on by an activator. Thus, repressors decrease expression and activators increase expression. **Inducers**, which bind to repressors and activators, increase gene expression by disabling repressors and enabling activators.

Prokaryotic cells generally control gene expression at the **transcriptional level**, while eukaryotic cells regulate gene expression not only at the transcriptional level but also at the post-transcriptional, translational, and post-translational levels. **Transcription factors** include hundreds or thousands of proteins that exert transcriptional control over the genome. Each protein has a highly conserved DNA-binding domain, which allows it to attach to specific DNA sequences or regulatory proteins. Together with regulatory proteins, transcription factors can act as repressors or activators.

An example of a regulatory DNA sequence is an **enhancer** sequence—a noncoding region of DNA located upstream or downstream of a promoter. As discussed in chapter 13, DNA is transcribed when RNA polymerase binds to a promoter sequence. The binding of transcription factors to enhancers enhances the transcription of an associated gene. Experimental removal of enhancers has been shown to cause drastic decreases in gene transcription.

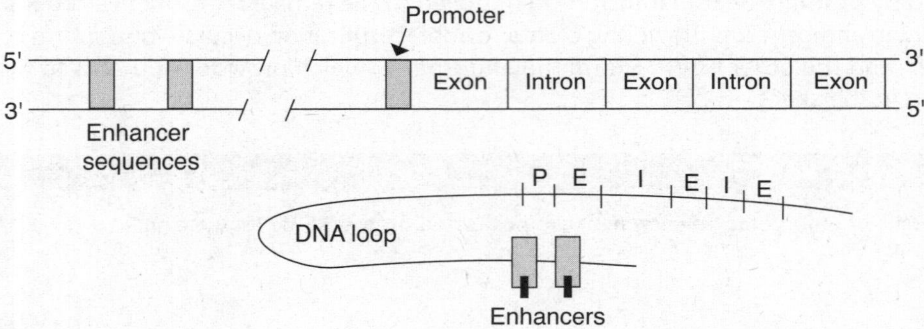

One model of regulated gene expression that illustrates inducible (positive) and repressible (negative) systems is the *lac* (lactose) **operon** in *E. coli*, which is required to transport and metabolize lactose for carbon and energy. An operon is a series of genes (including a promoter, an operator, and a terminator) that synchronize to perform a biological function. The following diagram shows that when lactose is not present in the growth medium, a protein repressor binds to the operator (regulatory site) to inhibit the transcription of the operon. When lactose (the inducer) is present, it binds to the repressor. The release of the repressor from the operator causes the transcription of the operon by RNA polymerase to commence. The transcribed mRNA codes for the proteins that metabolize lactose, which eventually leads to the removal of lactose from the medium. When lactose is gone, the repressor binds to the operator again and the operon is back to square one. This is a classic example of a negative feedback loop. After all, there is no need for the proteins that metabolize lactose unless the substrate is available. This is also an example of positive control because the expression of the genes is induced by the presence of the substrate.

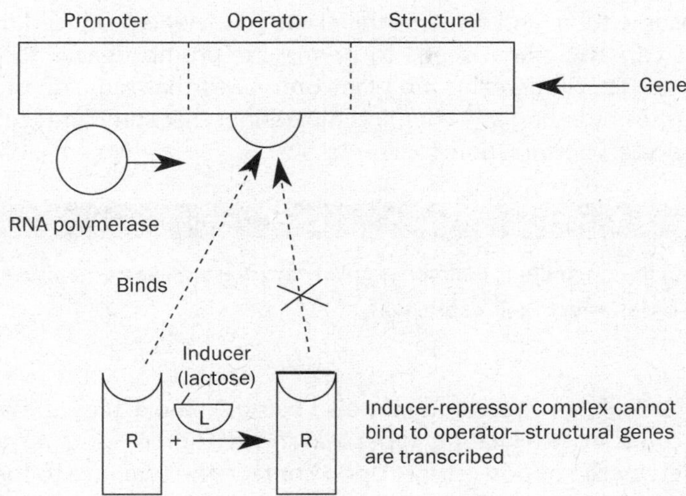

The *Lac* Operon in *E. coli*

Another important concept to appreciate from this model of gene expression is that the promoter and operator of an operon are "unaware" of what happens at another part of the DNA strand, unless the DNA performs a specific activity on them. For example, if the gene to produce human insulin were placed within a bacterium's operon that produces enzymes for metabolizing glucose, the insulin gene would also be transcribed when glucose signals the transcription of the glucose metabolizing genes. Scientists can then harvest the insulin from the bacterium. This type of genetic manipulation has revolutionized the discipline of molecular genetics.

Cell Differentiation and Specialization

Every cell in multicellular organisms carries out very specific and different functions in different structures. How does a cell in the eye of an eagle express the proper genes to make proteins for eye development and function, while a cell in the pectoral muscle of an eagle expresses proteins to build muscle tissue and use energy? The regulation of gene expression in eukaryotes causes cells to differentiate. Since most eukaryotic cells in a given organism contain the same set of genetic material, differential expression determines the cells' structure and function. Each cell type undergoes **differentiation** and **specialization** to produce specific proteins with particular functions according to its needs. For instance, regulation determines whether cells become lens cells that are sensitive to light, muscle cells that contract, or nerve cells that transmit information. Another example is the expression of the SRY (sex-determining region Y) gene on the Y chromosome, which initiates male sexual development. Errors or changes in gene regulation during development may lead to severe and detrimental consequences.

Both **intercellular** (external) and **intracellular** (internal) signal transmissions regulate gene expression. Signals may activate transcription and tell the developing organism which genes to express and when to express them. Additionally, external signals may trigger intracellular signaling pathways or signal transduction cascades. For example, in eukaryotes, cAMP is a small molecule that is involved in amplifying the message in intracellular signaling pathways. In bacteria, cAMP regulates metabolic gene expression. When glucose concentration is low, cAMP accumulates and transcription factors are activated to turn on the lac operon. In plant cells, ethylene triggers fruit-ripening. Special receptors bind to ethylene, which triggers a cascade that turns off anti-ripening genes. Ripened fruits invite animals to consume them and disperse their seeds. Ethylene produced during certain developmental conditions can also signal seeds to germinate, prompt leaves to change colors, and trigger flower petals to die. Gibberellins are plant hormones that regulate growth and influence developmental processes by acting as a chemical messenger that stimulates cell elongation, breaking and budding, and seed germination.

✔ **AP Expert Note**

The important thing to remember is that genes are always present in an organism's DNA, but certain environmental variables affect their expression.

Molecules in the extracellular environment may also be signal molecules released from other cells. Haploid yeast cells come in two mating types—a and α (alpha)—and respond to the mating pheromones produced by the opposite type. Upon contact, the two fuse to form a diploid zygote. Signals in embryos of multicellular organisms typically regulate expression of certain genes in targeted cells via **induction**. In cell-to-cell interactions, cytokines, molecules secreted by cells, signal other cells to begin cell replication and division. During pattern formation, cells of the developing embryo communicate their relative position. The concentration of morphogens tells cells the proximity of the cell releasing the signal and determines the spatial organization of tissues and organs. Cells exposed to higher concentrations of morphogen will develop differently than cells exposed to lower concentrations of morphogen. Hox genes (homeotic genes), which are also involved in pattern formation, control the development of appendages like antennae and legs. When one of the Hox genes is mutated, the wrong body part forms.

RAPID REVIEW

If you take away only 5 things from this chapter:

1. Fertilization, the first step in embryonic development, is the fusion of one sperm with one egg (gametes) to produce a zygote that becomes a fully formed organism.
2. In gene regulation, negative control occurs when a repressor switches genes off by inhibiting transcription, while positive control occurs when an activator switches genes on by promoting transcription. Inducers are molecules that bind to repressors and/or activators to increase gene expression.
3. Operons generally occur in bacterial genomes and are sets of genes that perform a biological function like metabolizing lactose. When the substrate is absent, a repressor binds to the operator to inhibit transcription of the operon, and when present, the substrate binds to the repressor and induces transcription.
4. Though each contains the same set of genetic material, cells in multicellular organisms have specific and different functions. The regulation of gene expression causes cells to undergo differentiation and specialization.
5. Intercellular and intracellular signals regulate gene expression. For example, in embryos of multicellular organisms, the concentration of morphogens determines spatial organization of tissues and organs.

TEST WHAT YOU LEARNED

1. Queen, worker, and drone honeybees have dramatically different morphologies and roles in the hive, even though all these individuals possess a similar genome and live in the same hive. A researcher finds that bee morphology is driven by an external chemical signal. Based on this finding, which of the following hypotheses would be the most appropriate to test next?
   (A) Differences in incubation temperature during development affect which genes are expressed.
   (B) Differences in diet during larval development affect which genes are expressed.
   (C) The number of mutations in the coding DNA regions of queens versus workers affect which genes are expressed.
   (D) The differences in the ways that queens, workers, and drones use their hindlimbs affect which genes are expressed.

2. *Lymnaea stagnalis* snails usually have right-coiling shells, and left-coiling shells are rare. Studies have shown that the allele for a right-coiling shell is dominant to the allele for a left-coiling shell. A cross between a true-breeding left-coiling female snail and a homozygous right-coiling male snail produces entirely left-coiling offspring. However, a cross between a right-coiling female and a homozygous-left coiling male produces entirely right-coiling offspring. Based on these findings, which of the following explanations is most probable?
   (A) The male snail does not contribute any genetic material during fertilization in this species, meaning that the species must be parthenogenetic.
   (B) The direction of shell coiling is controlled solely by environmental factors, so the alleles contributed by the mother and father do not affect the shell morphology of the offspring.
   (C) During the earliest stages of development, only mRNA from the male's sperm is used for gene expression to determine the direction of shell coiling.
   (D) During the earliest stages of development, mRNA from the mother's genes affects the direction of shell coiling.

3. Researchers have found that a gene called Tbx5 is important in vertebrate forelimb development. Activation of this gene can occur along the entire region of the body where limbs develop. However, a repressive complex including Hoxc9 prevents *Tbx5* expression in areas of the body posterior to the forelimb region. The figure below shows the length of a developing embryo with the head (cephalic) region at the top and the tail (caudal) region at the bottom. It shows a neural tube (NT), which will become the spinal cord; somites; and lateral plate mesoderm (LPM), which will give rise to limbs and other structures.

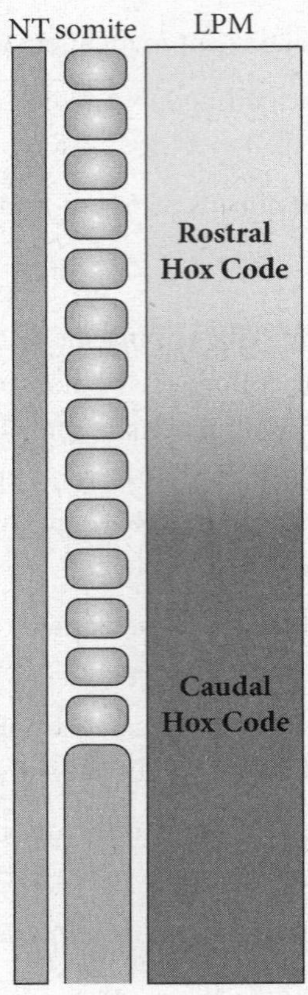

Adapted from Satoko Nishimoto et al., "A Combination of Activation and Repression by a Colinear Hox Code Controls Forelimb-Restricted Expression of Tbx5 and Reveals Hox Protein Specificity," *PLoS Genet 10*, no. 3 (March 2014): e1004245.

Based on the information provided, which of the following is the most reasonable conclusion?

(A) The Hoxc9 regulatory complex will exert negative control in the caudal region, so the forelimb will develop in the rostral region.

(B) The Hoxc9 regulatory complex will exert negative control in the rostral region, so the forelimb will develop in the rostral region.

(C) The Hoxc9 regulatory complex will exert positive control in the caudal region, so the forelimb will develop in the caudal region.

(D) The Hoxc9 regulatory complex will exert positive control in the rostral region, so the forelimb will develop in the caudal region.

4. The figure shows the expression of four different genes associated with hindlimb development in honeybee workers and queens. The developmental stages shown are those most critical for hindlimb development, which is believed to begin around stage L4.

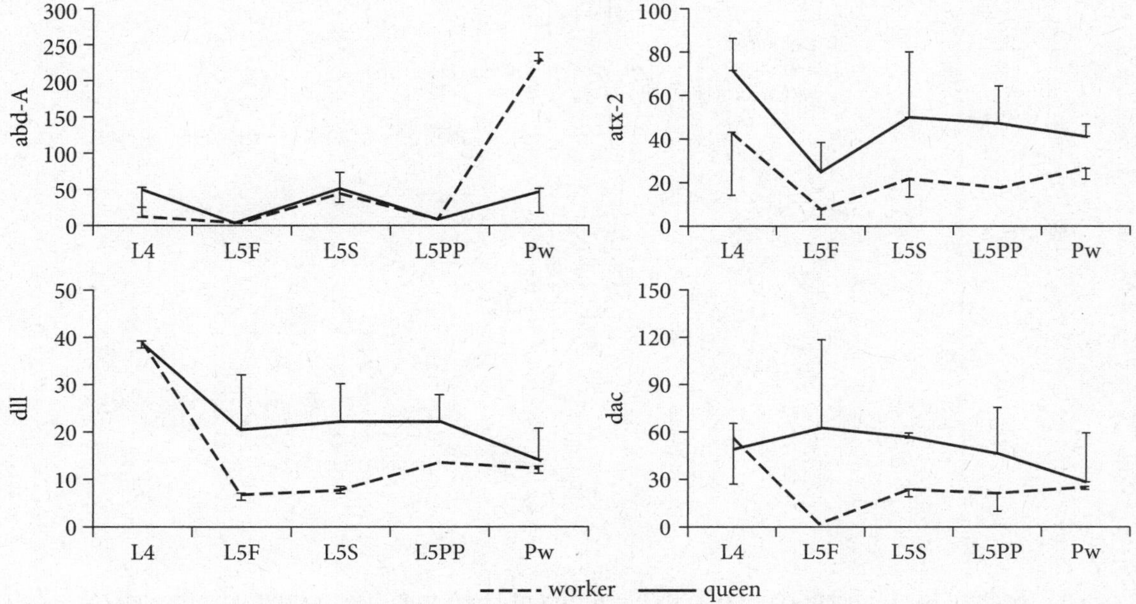

Expression of genes (identified on the *y*-axes) as development progresses from a larval stage (L4) to the white-eyed pupa stage (Pw). The *x*-axis represents stage of development and the *y*-axis represents relative transcript levels (a measure of gene expression).

Adapted from Ana Durvalina Bomtorin et al., "Hox Gene Expression Leads to Differential Hind Leg Development between Honeybee Castes," *PLoS ONE* 7, no. 7 (July 2012): e40111.

Which gene appears to be most important for the difference in hindlimb development in workers compared with queens, and why?

(A) The atx-2 gene is probably the most important for worker hindlimb specialization because it is expressed the most in workers at the L4 stage.

(B) The dac gene is probably the most important for worker hindlimb specialization because it is expressed at a relatively constant level throughout the stages shown.

(C) The abd-A gene is probably the most important for worker hindlimb specialization because it increases more dramatically in workers than in queens at the white-eyed pupa stage.

(D) The dll gene is probably the most important for worker hindlimb specialization because its expression dramatically drops after the L4 stage in workers but drops less in queens.

5. All of the cells shown in the diagram below were produced by mitosis and cell division from the same individual. Which of the following best explains how cells with these diverse shapes and functions resulted?

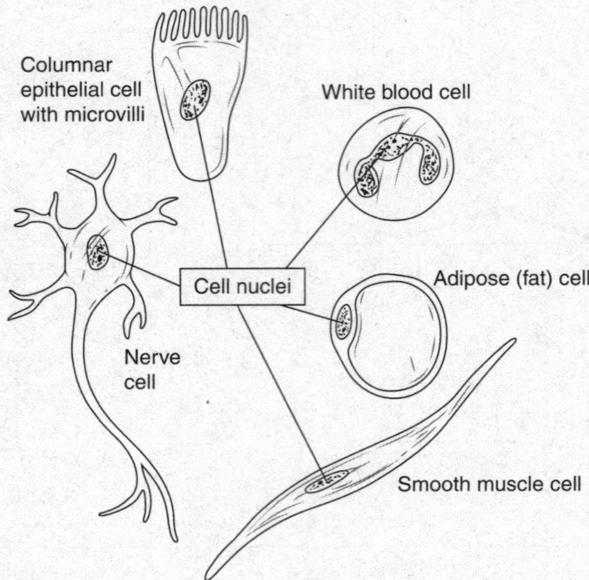

(A) Each of the different types of cells has different DNA that directs the shape the cell takes.

(B) Each of the different types of cells expresses a different set of transcription factors. These turn on the specific genes necessary in the cell that has them.

(C) These shapes are random. Cells take the shape of the space they have.

(D) Cells evolve as the organism grows so that the correct types are left behind.

6. The lactose operon in the bacterium *E. coli* is shown in the figure below.

Promoter	Operator	Lac Z	Lac Y	Lac A

This operon, also called the *lac* operon, is an example of negatively-controlled gene expression. When lactose is not available, a repressor protein binds to the operator region and reduces the transcription of the structural genes *lac*Z, *lac*Y, and *lac*A to very low levels. When lactose is present in the medium, allolactose, a molecule generated by lactose metabolism, binds the repressor and releases it from the operator region.

In a mutant strain of *E. coli*, a mutation in the operator region prevents the binding of the repressor protein. Researchers insert a wild type operator gene into the mutant cells using a plasmid. Which of the following results will they most likely observe?

(A) The bacterial cells cannot metabolize lactose.

(B) The expression of the *lac* operon can only be induced by lactose.

(C) The bacterial cells will continually express the *lac* operon.

(D) The bacterial cells synthesize lactose.

7. Gibberellins are a family of hormones that promote germination in seeds. A number of gibberellin-dependent transcription factors mediate the effect of the hormone on the seeds. A student is investigating the mechanism of gibberellin action on germination. He sprayed seeds with the hormone gibberellin and noted changes in the concentration of proteins and mRNA in the seeds. He found a decrease in proteins that attach to gibberellin-dependent transcription factors. This drop in protein concentration preceded an increase in the transcription of gibberellin-dependent mRNAs. Which of the following provides the best interpretation of the student's finding?

(A) Gibberellins remove inhibitory proteins that prevent transcription of gibberellin-dependent proteins.

(B) Pre-existing proteins must be degraded to provide amino acids for newly synthesized gibberellin-dependent proteins.

(C) Gibberellin-dependent transcription factors are not essential for the synthesis of gibberellin-dependent proteins.

(D) Gibberellins must bind to DNA for transcription to take place.

8. A medical researcher working with an infertile couple measures the number of sperm cells present in three equally-sized samples of semen before artificial insemination. She finds that the first sample has 125,000,000 cells/mL; the second sample has 95,000,000 cells/mL; and the third sample has 110,000,000 cells/mL. She combines all three samples and estimates that 1,100 sperm cells per initial mL of semen will reach the fallopian tubes. Based on her estimate, what percentage of the sperm cells will reach the fallopian tubes where fertilization occurs?

(A) 0.001% (B) 0.1% (C) 1% (D) 10%

9. Hsp70 heat shock proteins are a family of chaperones, which are proteins that guide the correct folding of other, newly formed proteins. Expression of Hsp70 increases after cells are subjected to a heat shock, allowing cells to better tolerate heat stress. Heat shock proteins are encoded by DNA sequences that include temperature sensitive RNA sequences, known as RNA thermometers. The 5′-UTR region of these RNA thermometers fold in a conformation that prevents ribosomes from binding when temperatures are low.

   Wild type *E. coli* strains synthesize Hsp70 shock proteins after exposure to heat. A microbiologist isolates a mutant strain of *E. coli* that thrives at 22°C, but quickly dies when transferred to a temperature of 40°C. To study the heat response in the mutant strain, the microbiologist transforms mutant cells with the following plasmids and records the results.

Plasmid	5′-UTR	Coding Sequence of Hsp70	Growth at 40°C
A	Mutant	Mutant	−
B	Wild type	Wild type	+
C	Mutant	Wild type	−
D	Wild type	Mutant	+

   Which of the following best describes the mutant *E. coli* strain?

   (A) The strain has a mutation in the RNA thermometer region.

   (B) The strain expresses a mutant Hsp70 protein that does not function as a chaperone.

   (C) The strain contains a cytoplasmic factor that inactivates Hsp70 proteins.

   (D) The strain contains a cytoplasmic factor that inactivates RNA thermometer sequences.

10. The lactose (*lac*) operon of *E. coli* is under dual control from a repressor protein that binds to the operator in the absence of available lactose and the catabolite activator protein (CAP). The repressor prevents transcription of a polycistronic (single) mRNA containing the coding information for structural genes *lac*Z, *lac*Y, and *lac*A. CAP binds to a sequence upstream of the promoter when glucose levels are low. Binding of CAP is required for the transcription of the structural genes.

    In an experiment, *E. coli* cells are grown on a medium supplemented with lactose as the only source of carbon. A scientist measures transcription of the structural genes associated with the *lac* operon. How many distinct strands of mRNA are transcribed from the *lac* operon under the described experimental conditions?

    (A) 1 (B) 2 (C) 3 (D) 4

11. Hox genes belong to a family of conserved genes that determine the characteristics of different segments of the body. Hox genes regulate the differentiation of stem cells during development. Different Hox genes are activated or deactivated based on where the stem cells are located in the body. For example, in mouse embryos, Hox10 genes are active in the lower back region, where there are no ribs, but inactive in the rib cage region. This different pattern of Hox10 gene expressions ensures that ribs develop in the thorax. Hox genes are also found in snakes, which possess ribs along their entire body.

    Based on this information, which of the following is most reasonable to conclude about the role of the Hox genes in snakes?

    (A) Hox10 genes may be present, but are always active in snakes.

    (B) Hox10 genes may be present, but are always inactive in snakes.

    (C) Snakes never possessed the Hox10 genes since snakes evolved before mice.

    (D) Snakes have likely lost all Hox genes because they no longer possess a body plan with different segments.

12. Scientists investigated the events surrounding the attachment of a sperm cell to an egg in a certain species of mouse. They found that receptors on the sperm bound to glycoproteins displayed on the zona pellucida of the egg. They also found that the mouse sperm was unable to fertilize hamster eggs. Which of the following is the most likely explanation for this experimental observation?

    (A) The enzymes of a mouse sperm cell cannot digest the zona pellucida of a hamster egg.

    (B) The hamster egg secretes enzymes that kill sperm cells.

    (C) The glycoprotein receptors on sperm and egg are species-specific.

    (D) The zygote produced from the mouse sperm and a hamster egg could not undergo cell division and would degenerate.

13. Molecular biologists cloned the gene for protein X from a library of mouse mRNA. They transformed a compatible cell line with two vectors, one vector containing the cloned gene and a second vector containing the enhancer region of protein X. They stimulated the cells with an inducer known to stimulate expression of protein X. Which of the following experimental results will most likely be observed?

    (A) The cells will not transcribe the mRNA because the site where RNA polymerase II attaches is missing.

    (B) No transcription will be observed because cells in tissue culture do not express foreign DNA.

    (C) The cells will transcribe the mRNA for protein X, but the mRNA will be unstable and rapidly degraded.

    (D) No or little transcription will occur because enhancer sequences must be *cis* to (on the same chromosome as) the open reading frame.

Answer Key

Test What You Already Know

1. A
2. B
3. C
4. B

Test What You Learned

1. B
2. D
3. A
4. C
5. B
6. C
7. A
8. A
9. A
10. A
11. B
12. C
13. D

REFLECTION

Test What You Already Know score: __________

Test What You Learned score: __________

Use this section to evaluate your progress. After working through the pre-quiz, check off the boxes in the "Pre" column to indicate which Learning Objectives you feel confident about. Then, after completing the chapter, including the post-quiz, do the same to the boxes in the "Post" column. Keep working on unchecked Objectives until you're confident about them all!

Pre	Post	
❑	❑	**14.1** Describe the process of fertilization and genetic transfer
❑	❑	**14.2** Describe positive and negative gene control mechanisms
❑	❑	**14.3** Explain how gene regulation results in gene expression
❑	❑	**14.4** Explain how gene expression leads to cell specialization

FOR MORE PRACTICE

Complete more practice online at kaptest.com. Haven't registered your book yet? Go to kaptest.com/booksonline to begin.

ANSWERS AND EXPLANATIONS

Test What You Already Know

1. A Learning Objective: 14.1

According to the question stem, "shell development is heavily influenced by genetics," so it would be unlikely for handedness of the shell to vary entirely based on environmental factors. This is precisely what is suggested in (A), in which the environment seems to be the sole determinant of handedness. Because **(A)** offers the least likely outcome, it is correct. (B) is incorrect because this result is consistent with sinistral being the dominant trait. (C) is incorrect because this result is consistent with dextral being the dominant trait. (D) is incorrect because this result is consistent with a more complex non-Mendelian inheritance pattern, with environmental factors having limited influence.

2. B Learning Objective: 14.2

According to the question stem, the gene is off by default but can be expressed when an activator binds and turns it on. This is positive control, as indicated in **(B)**. (A) is incorrect because the gene is not expressed throughout development, but is off by default (constitutive expression means that a gene is continually expressed). (C) is incorrect because the default for the gene is off and it is turned on by an activator; in negative control, a gene is on by default and is turned off with a repressor. (D) is incorrect because the gene is controlled by an activator, not a repressor; in addition, operons are not always controlled by repressors.

3. C Learning Objective: 14.3

According to the table, workers express more genes relating to responding to stimuli than do queens. This suggests that they may need greater use of their sense organs, perhaps for activities such as foraging. **(C)** is thus correct. (A) is incorrect because queens express more genes relating to biological regulation. (B) is incorrect because queens express considerably more genes relating to developmental processes. (D) is incorrect because workers express more genes relating to metabolic processes.

4. B Learning Objective: 14.4

In order for phenotypic changes to be similar, the genes affected in both fruit flies and honeybees should also be similar. **(B)** is thus correct. (A) is incorrect because the regulation of different genes is unlikely to produce similar phenotypic changes in the two species. (C) and (D) are incorrect because royalactin is described in the question stem as increasing juvenile hormone levels, so it is unlikely to act as a hormone itself.

Test What You Learned

1. B Learning Objective: 14.3

The question stem specifies that "bee morphology is driven by an external chemical signal." One possible source of this external chemical signal could be the food that is fed to larvae; larvae fed different foods could develop different body types. **(B)** is thus correct. (A) is incorrect because the question stem specifies that the differences in morphology are driven by an external chemical signal, not a temperature differential. (C) is incorrect because it is highly unlikely that random mutations could produce the consistent differences in morphology between queens and workers. (D) is incorrect because the question stem specifies an external chemical signal is responsible, not a difference in the use of body parts.

2. D Learning Objective: 14.1

In both of the examples provided in the question stem, the direction of coiling in the mother determines the direction of coiling in all offspring. This suggests that something from the mother, perhaps mRNA transcribed from her genes, is definitive for determining shell handedness. **(D)** is thus correct. (A) is incorrect because males and females are actually crossed in the experiment, so the species could not be parthenogenetic. (B) is incorrect because the coiling direction of the mother seemed to have a definitive influence on coiling direction in offspring, so environmental factors could not be the only determinant. (C) is incorrect because the shell handedness of the males had no influence on the offspring.

3. A Learning Objective: 14.2

According to the question stem, the Hoxc9 regulatory complex represses forelimb development in the posterior areas of the body, which correspond to the caudal region. Thus, the complex is exerting negative control in the caudal region, which will lead to forelimb development in the rostral region. **(A)** is thus correct. (B) is incorrect because the negative control is actually exerted in the caudal region, not the rostral region. (If negative control were exerted in the rostral region, then forelimb development would be inhibited in the rostral and allowed to progress in the caudal region instead.) (C) and (D) are incorrect because repression is a type of negative control, not positive control.

4. C Learning Objective: 14.4

Because abd-A increases much more in workers than in queens, it is likely to be important for the differences in worker hindlimbs. **(C)** is thus correct. (A) is incorrect because queens actually express atx-2 the most at the L4 stage. (B) is incorrect because dac is expressed at a relatively constant level in queens, but it appears to decrease in workers. (D) is incorrect because the difference in dll expression between queens and workers is relatively minimal.

5. B Learning Objective: 14.3

Since these cells were all produced by mitosis, they all have the same genetic makeup. The differences depend on which genes are expressed and which are not. Gene expression is controlled primarily by transcription factors. The difference between nerve cells and smooth muscle cells, for example, lies in the fact that there are different transcription factors present in nerve cells than what is present in smooth muscle cells. As a result, different genes are expressed in different cells. **(B)** correctly summarizes this concept. Since mitosis produces cells with identical nuclei, (A) is incorrect. Cell differentiation is not a random process, so (C) is incorrect. (D) is incorrect because cells develop as organisms grow; cells do not evolve.

6. C Learning Objective: 14.2

The repressor may bind to the wildtype operator on the plasmid, but the operator upstream of the structural genes is still not blocked. Thus, the bacterial cells will continually express the *lac* operon. The correct answer is **(C)**. (A) is incorrect because the binding of the repressor to the wildtype operator does not affect the expression of the *lac* operon, so the enzymes needed to metabolize lactose would be continuously produced. (B) is incorrect because the operator next to the structural genes is not blocked by a repressor, so the *lac* operon would be expressed whether lactose is present or absent. (D) is incorrect because bacterial cells do not synthesize lactose, which is a sugar found mostly in milk.

7. A Learning Objective: 14.3

The measured proteins are most likely inhibitors of the gibberellin-dependent transcription factors that prevent the transcription factors from binding to the DNA. Gibberellin induces transcription by removing the inhibitory proteins. Once the inhibitory proteins are degraded, the transcription factors can bind to DNA and transcription from DNA into gibberellin-dependent mRNAs takes place. The correct answer is **(A)**. (B) is incorrect because the degraded proteins are inhibitors of the gibberellin-dependent transcription factors not a source of amino acids. (C) is incorrect because transcription would not occur without the gibberellin-dependent transcription factors, so they are essential for the synthesis of gibberellin-dependent proteins. (D) is incorrect because gibberellins act through gibberellin-dependent transcription factors for transcription.

8. A Learning Objective: 14.1

The average number of sperms per mL is the sum of the three samples divided by 3: (125,000,000 + 95,000,000 + 110,000,000)/3 or 110,000,000. Since the number of sperms that reached the fallopian tubes is 1,100 per mL, the percentage of sperm cells that will reach the fallopian tubes is 1,100/110,000,000*100% or 0.001%. Therefore, **(A)** is correct.

9. A Learning Objective: 14.4

The 5′-UTR region of the RNA thermometer enables growth at 40oC. The data shows that a mutation in this area prevents growth at 40oC regardless of whether the coding sequence of Hps70 is mutated or not. Thus, the correct answer is **(A)**. (B) is incorrect because the Hsp70 protein from the mutant confers protection to heat stress if the 5′-UTR contains the wild type RNA thermometer. (C) is incorrect because the wild type and mutant Hsp70 proteins are functional in the mutant cells at high temperature if they are synthesized. (D) is incorrect because the RNA thermometer from a wild type cell is functional in the mutant cells.

10. A Learning Objective: 14.3

Since lactose is the only source of carbon, glucose levels are low. This signals the repressor to unbind from the operator and CAP to bind to the promoter, both of which turn on the *lac* operon. Hence, a single mRNA strand containing the structural genes for *lacZ, lacY,* and *lacA* is transcribed from the operon. The correct answer is **(A)**.

11. B Learning Objective: 14.4

Based on the information, the expression of Hox10 genes in mice inhibit rib development. Since snakes possess ribs along their entire body, Hox10 genes most likely are inactive in snakes. Thus, the correct answer is **(B)**. (A) is incorrect because if Hox10 genes worked the same way in snakes as in mice, expression of Hox10 genes would cause snakes to lack ribs throughout their bodies. (C) is incorrect because we cannot assume, from the information provided, that snakes do not possess the Hox10 genes. (D) is incorrect because Hox genes are present, and snakes do exhibit a body plan with different segments, such as the head.

12. C Learning Objective: 14.1

The ligand-receptor interaction between sperm and egg is species specific and prevents sperm from a different species from binding to the egg. Thus, the correct answer is **(C)**. (A) is incorrect because if the sperms do not bind, the acrosomal reaction will not take place. (B) is incorrect because the hamster egg can be fertilized by hamster sperm. (D) is incorrect because no fertilization has taken place; therefore, there is no zygote.

13. D Learning Objective: 14.2

Enhancers must be located on the same DNA for the transcription complex to form efficiently and predictably. Since the enhancer region of protein X and the gene for protein X are in different vectors, no or little transcription will occur. Hence, the correct answer is **(D)**. (A) is incorrect because although, the promoter is present and polymerase II could attach to the DNA, the interaction must be stabilized by a transcription complex that requires the enhancer region to be located upstream or downstream. (B) is incorrect because cells in culture are able to transcribe foreign DNA (the structure of DNA is universal). (C) is incorrect because the cells will not transcribe mRNA if the transcription complex does not form.

CHAPTER 15

Inheritance

LEARNING OBJECTIVES

In this chapter, you will review how to:

15.1 Describe chromosome structure and function

15.2 Compare chromosomal mutations and abnormalities

15.3 Explain and predict results of Mendelian inheritance

15.4 Explain why some traits are non-Mendelian

TEST WHAT YOU ALREADY KNOW

1. To understand more about the function of chromatin compaction, researchers examined the effects of radiation exposure on condensed and decondensed chromatin. The figure below shows the results from chromatin that was exposed to different amounts of radiation in Gray (Gy) units. The left column for each dose shows all DNA, whereas the right column shows only DNA that has experienced double strand breaks (DSBs). DSBs indicate radiation damage. The rows show DNA that was condensed, decondensed, or previously decondensed and then later recondensed at the time of irradiation.

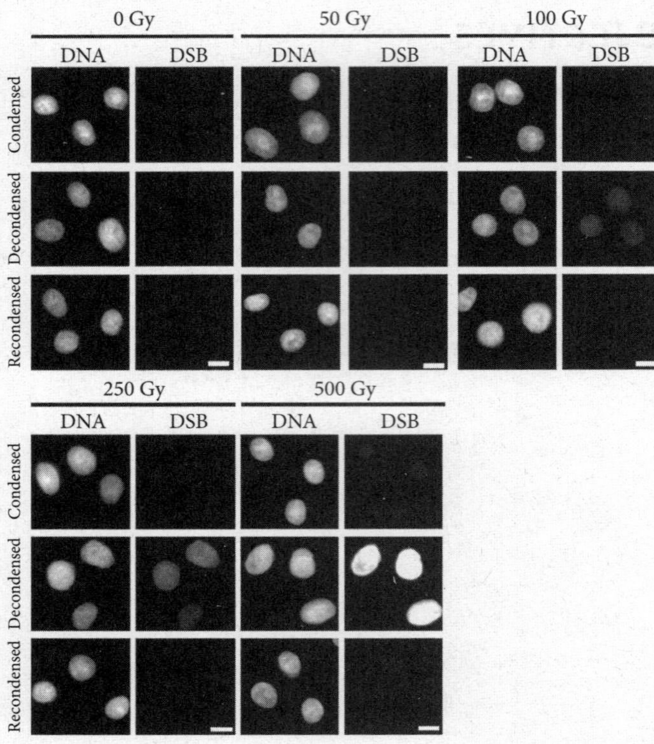

Adapted from Hideaki Takata et al., "Chromatin Compaction Protects Genomic DNA from Radiation Damage," *PLoS ONE* 8, no. 10 (October 2013): e75622.

Which of the following conclusions can most justifiably be drawn from these results?

(A) Condensed DNA increasingly experienced damage due to double strand breaks as radiation intensity increased, but decondensed DNA was protected from this damage.

(B) Decondensed DNA increasingly experienced damage due to double strand breaks as radiation intensity increased, but condensed DNA was protected from this damage.

(C) Both decondensed and recondensed DNA experienced damage due to double strand breaks in a dose-dependent manner.

(D) Decondensed DNA showed no damage in even the highest radiation treatment, suggesting that decondensation offered some type of protection from radiation damage.

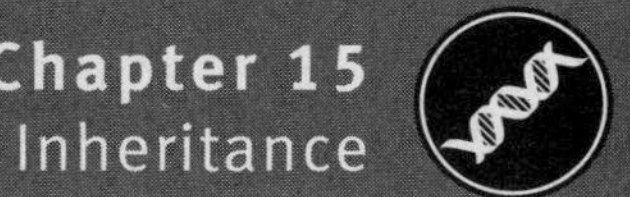

2. To understand gene regulation, researchers have carefully examined the structure and function of the *lac* operon in *E. coli*. The *lac* operon contains three genes that code for proteins associated with lactose metabolism. It also contains an operator region that binds to a repressor and turns off the expression of all three genes. If a mutation occurred in the operator region of the *lac* operon, what would most likely happen?

   (A) Because a mutation had occurred, the operon could not work properly and the genes would not be expressed.

   (B) Because a mutation had occurred, the proteins associated with lactose metabolism might not work properly.

   (C) Because the repressor would be affected by the mutation, the repressor would bind too strongly to the operator and the genes could not be expressed at all.

   (D) Because the repressor could not recognize the operator, the repressor would not bind and the genes would be expressed.

3. Purebred Birman cats have distinctive white feet (called "gloves") that develop from having two copies of a Mendelian gloving allele. What could most reasonably be concluded if two cats that did not have gloves were crossed and produced a kitten with gloving?

   (A) The gloving allele is dominant and a mutation must have occurred for the kitten to have gloving.

   (B) The gloving allele is dominant and it was passed by one parent to the kitten.

   (C) The gloving allele is recessive and both of the parents must have been carriers.

   (D) The gloving allele is recessive and a single parent must have passed it to the kitten.

4. Glucose-6-phosphate dehydrogenase (G6PD) is an X-linked gene whose mutation is associated with a reduced risk of malaria. The figure shows G6PD enzyme activity in men and women by genotype.

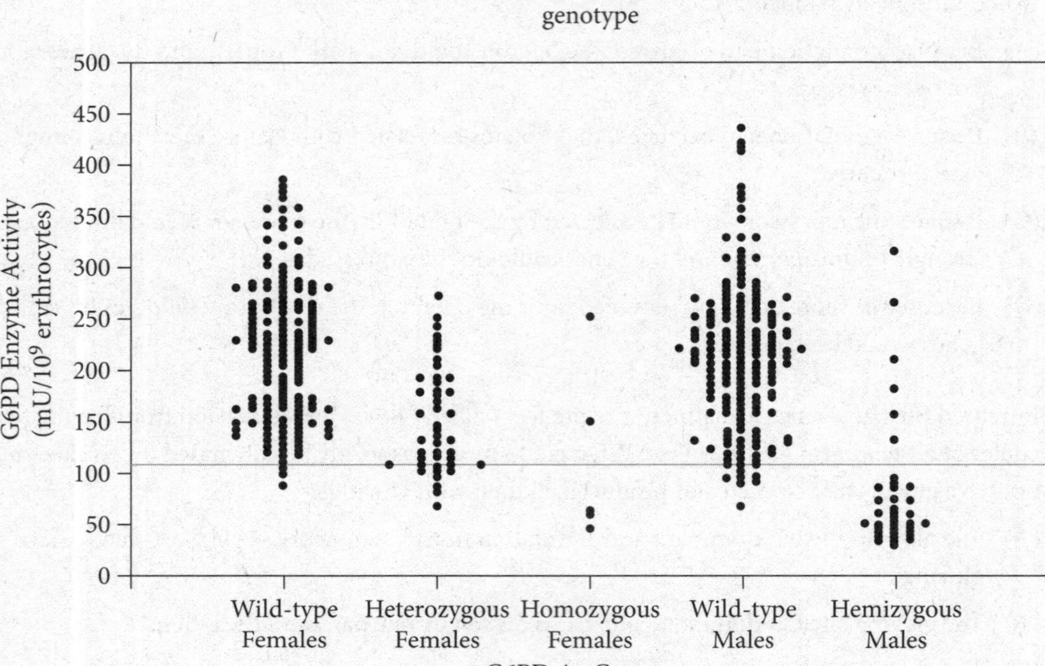

Adapted from M. K. Johnson et al., "Impact of the Method of G6PD Deficiency Assessment on Genetic Association Studies of Malaria Susceptibility," *PLoS ONE* 4, no. 9 (September 2009): e7246.

The researchers found that one copy of the mutant allele was sufficient to reduce the risk of malaria in men but homozygosity of the mutant allele was required to reduce the risk of malaria in women. What is the most likely explanation for these results?

(A) Women possess two copies of the X chromosome, so both copies must contain the recessive mutant allele to yield resistance to malaria.

(B) Gene expression is impacted by whether an individual is likely to be affected by malaria, meaning that individuals in high-risk areas are more likely to have the mutant allele.

(C) Hemizygous individuals have the greatest protection against malaria because they produce more G6PD enzyme.

(D) Women who are homozygous for the mutant allele and men who are hemizygous for the mutant allele are not able to produce any functional G6PD enzyme.

Answers to this quiz can be found at the end of this chapter.

CHROMOSOMES

High-Yield

15.1 Describe chromosome structure and function

15.2 Compare chromosomal mutations and abnormalities

Chromosomes are condensed bodies of DNA molecules that store codes for the translation of several different kinds of proteins. These proteins dictate how an organism is put together and functions. In the most simplified way of looking at the chromosomal theory of inheritance, each section of DNA that translates a different protein is called a **gene**. For example, a region of DNA may code for a protein that controls eye color, so this region of DNA is called the gene for eye color. In other words, a *gene* refers to a specific location on a chromosome.

Alleles specifically code for the traits of an organism. In the case of eye color, there might be an allele for blue eye color and an allele for green eye color. The eye color that an organism ends up with (its **phenotype**) is dictated by which alleles are placed in the gene location for eye color (its **genotype**). Sometimes an organism will have a genotype that includes an allele that is not expressed in its phenotype.

In a typical diploid (2*N*) eukaryotic organism, each cell has two copies of each chromosome. These pairs of chromosomes are a result of the combination of two haploid gametes formed by meiosis, one from the mother and one from the father. The mother and father each provided one allele for each gene, leaving the offspring with either two of the same allele (e.g., two for blue eye color or two for green eye color) or one of each allele (e.g., one allele for blue eye color and one allele for green eye color). In many genes, there is one allele that is **dominant** over the other and hides the expression of the other allele in the phenotype of the offspring. The other allele is called **recessive**. Geneticists generally use uppercase letters for dominant alleles (*B*) and lowercase letters for recessive alleles (*b*). If blue eye color is the dominant allele (indicated as *B*) and green eye color is the recessive allele (indicated as *b*), an offspring could have one of three different genotypes (*BB*, *Bb*, and *bb*) when its parents' gametes combine. The phenotype of an offspring with the genotype *BB* is blue eyes. An offspring with the genotype *Bb* will also have blue eyes because the blue eye color allele is dominant. Only offspring with the genotype *bb* will have a green eye phenotype.

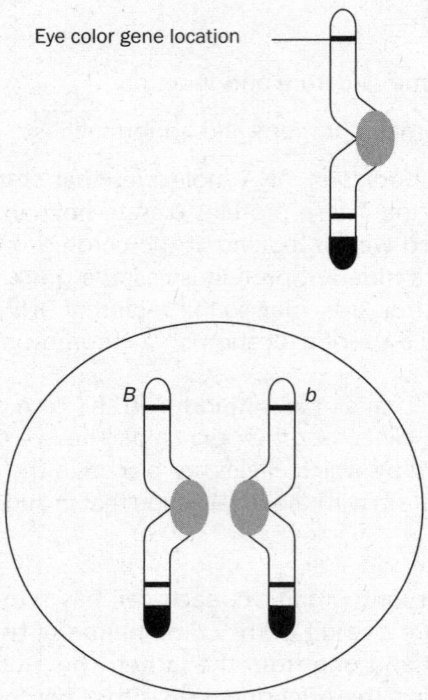

The diagram above shows a cell from an animal with one allele for blue eye color (*B*) and one allele for green eye color (*b*). This animal's genotype for eye color is *Bb,* and its phenotype is blue eyes because the blue eye color allele is dominant. The chromosomes are essentially identical in appearance, but the alleles have a slightly different DNA code.

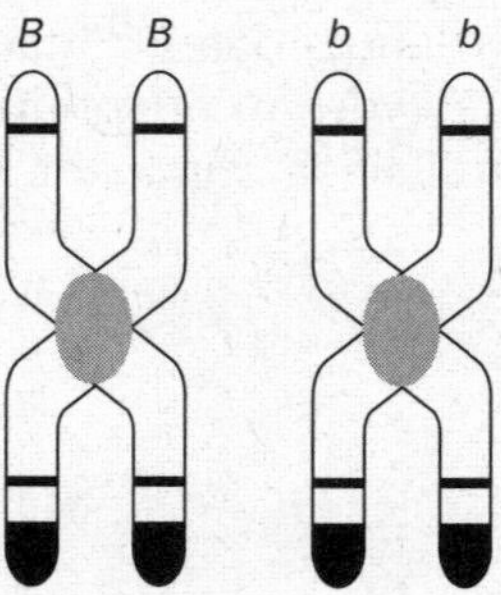

The figure above shows chromosomes after DNA replication. The DNA strands have been doubled so that there is an exact duplicate of each original chromosome linked by a centromere to sister chromatids. There are now two copies of the *B* allele and two copies of the *b* allele.

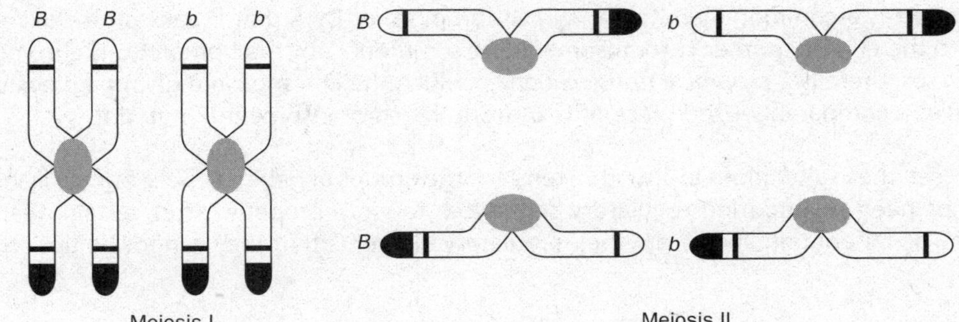

After the cell undergoes complete meiosis, the original cell with 2*N* chromosomes has divided into four daughter cells, each with 1*N* chromosomes. The daughter cells are four gametes, two with the *B* allele and two with the *b* allele. Because the original animal cell had both a *B* allele and a *b* allele, it can produce gametes with alleles for either blue or green eye color. This occurs in all individuals of the same species. An individual with the recessive green eye color (its genotype is *bb*) can only produce gametes with *b* alleles. When an organism has two of the same alleles (e.g., *BB* or *bb*), it is called **homozygous** for that gene. If both alleles are the dominant allele, the genotype is called **homozygous dominant**. If both alleles are recessive, the genotype is called **homozygous recessive**. If the organism has different alleles (e.g., *Bb*), it is called **heterozygous** for the gene.

Because the green eye color allele is recessive, it can be discerned that an individual in the population with green eyes has the genotype *bb*. An individual with blue eyes can have either a *BB* genotype or a *Bb* genotype. The genotype of a blue-eyed individual can be determined by mating the blue-eyed individual with a green-eyed individual. This is called a **test cross**. If all of the offspring have blue eyes, then the blue-eyed individual was homozygous dominant. The genotypes of all of the offspring from a mating between a *BB* genotype and a *bb* genotype can only be *Bb*.

If any of the offspring have green eyes, the blue-eyed adult must have been heterozygous. The genotypes of the offspring from a mating between a *Bb* genotype and a *bb* genotype are either *Bb* or *bb*. Mating between two heterozygous adults can produce offspring with three different genotypes: *BB*, *Bb*, and *bb*. In this case, two blue-eyed adults can produce a green-eyed offspring.

Chromosome Alterations

Several types of chromosomal breakage can occur in the course of DNA replication or at other points in the cell cycle. Some of these alterations can cause drastic changes in the ability of a cell to produce certain proteins. Many birth defects can be traced to defects in chromosome structure passed down through sperm or egg cells, in which one of the following alterations has taken place:

Deletion is when a chromosomal fragment, either from the end of a chromosome or from somewhere in the middle, is lost as the chromosome replicates.

If the fragment that detaches from a chromosome during a deletion event reattaches itself to the homologous chromosome, that chromosome will then have two sets of identical genes in a particular region (a *duplication*), while the original chromosome where the deletion occurred will be shorter than normal.

There is also the possibility that the deleted fragment could attach back into the chromosome from which it came, but in a reverse direction. This is known as *inversion*.

Translocation is a common alteration in which the piece of DNA that breaks off a chromosome attaches to the end of another chromosome, most commonly not the homologous chromosome. In some cases, there is a *reciprocal translocation*, in which the chromosome giving a piece of DNA also receives a comparably-sized piece of DNA from the chromosome it donated to.

In many cases, these alterations in structure render large groups of genes useless, especially because most genes need neighboring regulatory sequences to work properly. After certain alterations, genes may be far enough away from their regulatory elements that they cannot be transcribed.

MENDELIAN INHERITANCE

High-Yield

15.3 Explain and predict results of Mendelian inheritance

After learning the basics of genetics, it is important to review some specific information about inheritance that might appear on the exam. You need to be familiar with mathematical principles of simple probability. Relax; it's not as hard as it sounds. Even if there are more than two alleles for a gene, an individual can only have two alleles on each pair of chromosomes. Ending up with a combination of alleles is about as simple as tossing a coin.

An organism that is homozygous for a particular gene (*AA*) produces four gametes each with an *A* allele. Each of the gametes is one out of a possible four gametes, but because all of the gametes have an *A*, four out of four of the gametes have an *A*. Four out of four is a probability of one because $\frac{4}{4} = 1$. You can look at the question more intuitively by saying that because there are only *A* alleles in the parent cell, all of the daughter cells (100%, or $\frac{1}{1}$) will have the *A* allele. If an organism is heterozygous (*Aa*), it produces two gametes with an *A* allele (2 out of 4) and two gametes with an *a* allele (2 out of 4). The probability of either the *A* or *a* allele in the gametes is $\frac{2}{4} = \frac{1}{2}$. Or again, intuitively, if the parent cell has one *A* and one *a* allele, it will produce gametes that are half *A* (50%, or $\frac{1}{2}$) and half *a* (50%, or $\frac{1}{2}$).

Now consider two genes and an individual that is homozygous for both (*AABB*). Although there are two genes to consider, the organism still produces four gametes from a single parent cell. Each gamete gets an *A* allele and a *B* allele. What is the probability that a gamete will have both an *A* and *B* allele? Four out of four will have an *A* and four out of four will have a *B* so all (four out of four) will have *AB* ($\frac{4}{4} \times \frac{4}{4} = \frac{16}{16} = 1$).

If the individual is heterozygous for one of the genes (*AABb*), the probability of the *A* allele stays the same, but now only half of the gametes get a *B* allele ($\frac{2}{4} = \frac{1}{2}$) and the other two get a *b* allele. What is the probability of producing a gamete with both an *A* and *B* allele from this individual? Four out of four will have an *A* and two out of four will have a *B*, so only half (two out of four) will have *AB* ($\frac{4}{4} \times \frac{2}{4} = \frac{8}{16} = \frac{1}{2}$).

If the individual is heterozygous for both genes (*AaBb*), the probability of each allele in the gametes (*A*, *a*, *B*, or *b*) is $\frac{1}{2}$, so the probability of *AB* is $\frac{1}{2} \times \frac{1}{2} = \frac{1}{4}$. The probability of *Ab* would also be $\frac{1}{4}$, as would be the probability of *aB* and the probability of *ab*.

✔ **AP Expert Note**

The Two Laws of Mendelian Inheritance

Law of Segregation—This describes the separation of alleles in the parent genotype during the process of gametogenesis. There can be a maximum of only two different alleles in a single diploid parent; half the gametes get one allele and the other half get the other allele.

Law of Independent Assortment—This suggests that different genes sort into different gametes, independently of each other. For example, the sorting of alleles for eye color is not affected by the sorting of alleles for hair color. These laws explain the 3 : 1 and 9 : 3 : 3 : 1 ratios of phenotypes observed in monohybrid and dihybrid crosses.

Once you can figure out what gametes will be produced from which individuals, you can figure out the genotypes of offspring from a **Punnett square**. The following is a simple cross between two heterozygous individuals for one gene. A mating of this kind is called a **monohybrid cross**.

	A	*a*
A	*AA*	*Aa*
a	*aA*	*aa*

Monohybrid Cross

Typically, the gametes of the sperm are recorded along the left side of the Punnett square, and those of the egg are recorded across the top. The allele from each sperm is paired with the allele from each egg where the columns and rows meet, showing the genotypes of the offspring. The different indications of *Aa* and *aA* are used only to demonstrate the source of the alleles; they are the same genotype.

This cross results in three different genotypes (*AA*, *Aa*, and *aa*), but only two phenotypes because the dominant trait is expressed in both the *AA* and *Aa* individuals. The dominant trait shows up in the offspring in a ratio of 3 to 1.

You can put together a **dihybrid cross** (tracking two genes—*A* and *B*—rather than one) just as easily as a monohybrid cross. Just put the gametes of the male in the left column and the gametes of the female across the top.

	AB	*Ab*	*aB*	*ab*
AB	*AABB*	*AABb*	*AaBB*	*AaBb*
Ab	*AAbB*	*AAbb*	*AabB*	*Aabb*
aB	*aABB*	*aABb*	*aaBB*	*aaBb*
ab	*aABb*	*aAbb*	*aabB*	*aabb*

Dihybrid Cross

The dihybrid cross produces the famous phenotypic ratio of 9 : 3 : 3 : 1. If you consider a large number of offspring, there will be on average 9 out of every 16 that express the dominant phenotypes for both genes, 3 out of every 16 that express the dominant phenotype of one gene, 3 out of every 16 that express the dominant phenotype of the other gene, and only 1 out of every 16 that expresses the recessive phenotypes for both genes.

✔ AP Expert Note

Notice patterns in the ratios that occur with specific types of crosses. For example, crossing a heterozygous dominant (*Rr*) with a homozygous recessive (*rr*) always gives a 1 : 1 ratio of dominant to recessive phenotypes (half dominant phenotype; half recessive phenotype). This can speed up the process of figuring out phenotypic ratios and probabilities.

These are the basics of inheritance. However, genes do not operate in isolation from one another; this makes genetics more complex than Mendel's experiments and the Punnett square might suggest.

Pedigrees

Pedigrees are family trees that enable us to study the inheritance of a particular trait across many related generations. Males are typically designated as squares on the pedigree, while females are designated as circles. Those who phenotypically show a trait are shaded, while those who carry a trait but do not show it are either half-shaded or given a dot inside their circle or square.

In humans, genetic traits can be classified by whether they are transmitted on autosomal chromosomes (numbers 1–22) or on **sex chromosomes** (almost always the X, because the Y chromosome carries a limited number of genes). In addition, traits can be dominant or recessive. For the AP Biology exam, you should be familiar with the characteristics of different inheritance patterns so you can easily spot patterns. (Note that sex-linked traits, including both X-linked and Y-linked, count as non-Mendelian inheritance patterns.)

Autosomal Dominant

- Males and females are equally likely to have the trait.
- Traits do not skip generations.
- The trait is present if the corresponding gene is present.
- There is male-to-male and female-to-female transmission.

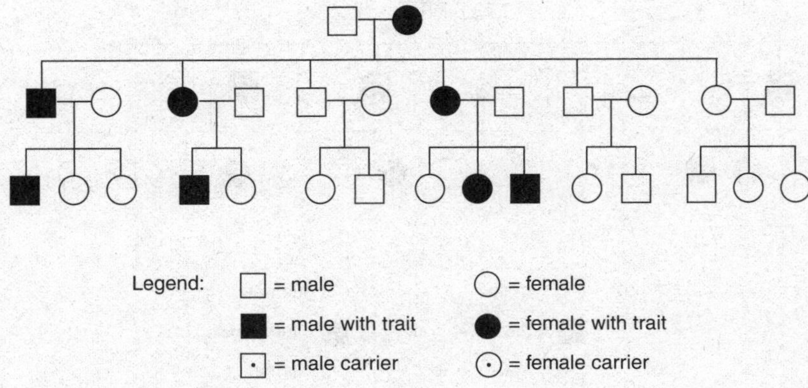

Autosomal Recessive

- Males and females are equally likely to have the trait.
- Traits often skip generations.
- Only homozygous individuals have the trait.
- Traits can appear in siblings without appearing in parents.
- If a parent has the trait, those offspring who do not have it are heterozygous carriers of the trait.

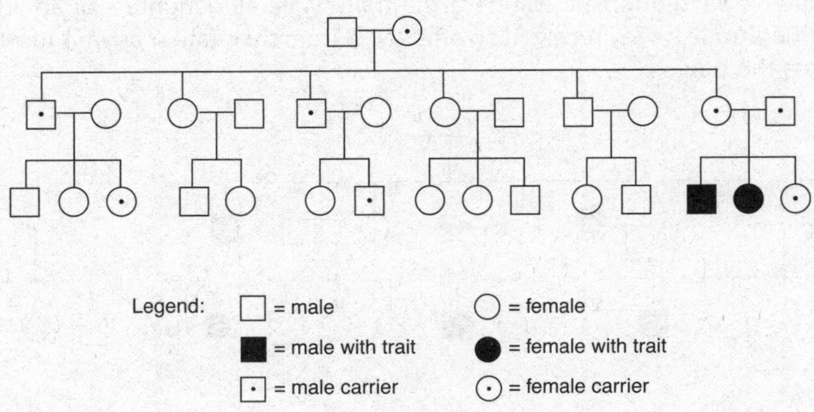

X-Linked Dominant

- All daughters of a male who has the trait will also have the trait.
- There is no male-to-male transmission.
- A female who has the trait may or may not pass on the affected X to her son or daughter (unless she has two affected Xs).

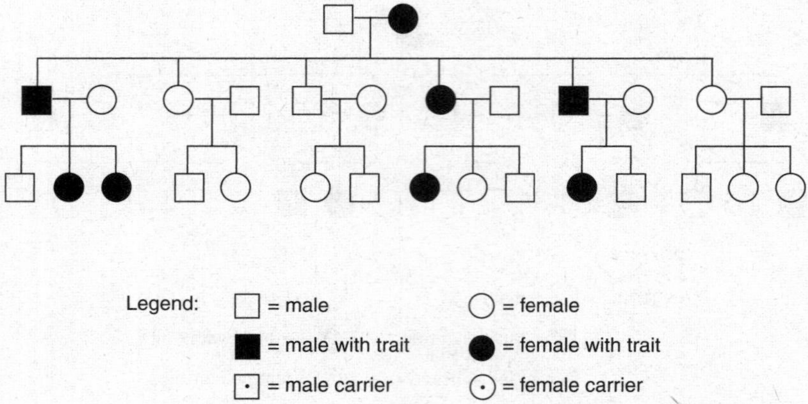

X-Linked Recessive

- Males are more likely to express the trait than females because males only need one copy for expression while females need two.
- Males cannot be carriers.
- All daughters of a male who has the trait will be carriers for the trait (unless they also receive an affected X from their mother as well, in which case they will express the trait).
- There is no male-to-male transmission.
- All sons of an affected female will have the trait, while all daughters of an affected female will be carriers (unless they received an affected X from their father as well, in which case they will express the trait).

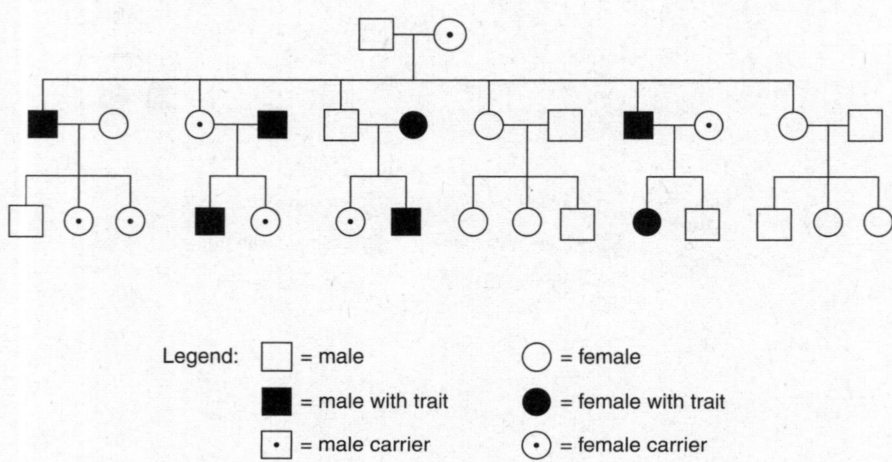

NON-MENDELIAN INHERITANCE

15.4 Explain why some traits are non-Mendelian

Mendelian genetics, however, does not apply to all genes. Inheritance patterns that do not follow **Mendelian laws** fall under non-Mendelian inheritance. For instance, genes may be linked to other genes on the same chromosome. Thus, the linked genes may not sort independently and may be inherited together. If linked genes are far apart on the chromosome, **crossing over** (exchange of genes between homologous chromosomes) during meiosis may separate them. The farther away genes are from each other, the more likely they are to be separated as a result of crossing over. The phenotypic ratios for linked genes, accordingly, will differ from those based on Mendelian inheritance.

Sex-linked traits also do not follow ratios predicted by Mendel's laws. Determined by genes on sex chromosomes (X and Y), sex-linked traits often appear more frequently in one sex (females are XX and males are XY). Since the Y chromosome in mammals and flies is small and carries few genes, X-linked recessive traits, such as hemophilia, are always expressed in males. The genotype of males for X-linked genes is called **hemizygous**, because they only have one copy of genes exclusive to the X chromosome. Sex-limited genes—though present in both sexes—are only expressed in one sex, so the same genotype can result in the expression of different phenotypes. Milk production, for example, is expressed only in female mammals, whereas pattern baldness is expressed exclusively in human males.

Extranuclear inheritance is another example of non-Mendelian inheritance since it describes the inheritance of genes outside the nucleus. Extranuclear inheritance occurs typically in mitochondria and chloroplasts, which contain genes that replicate independently of cell division. In vegetative segregation, the parent cell randomly distributes genes from mitochondria and chloroplasts to daughter cells, whereas in single-parent inheritance, offspring only receive genes from one parent. In humans, mitochondrial DNA is inherited entirely from the mother (maternal inheritance).

✔ AP Expert Note

Phenotype and Genotype: Other Factors to Consider

Many different factors beyond those encompassed in Mendel's laws can affect phenotype and genotype in offspring. You should be familiar with the following:

- **Incomplete dominance** is a form of inheritance in which *both* heterozygous alleles are expressed. This means that the offspring will display a combined phenotype that is distinct from both parent organisms. For example, a plant with red flowers and a plant with white flowers might produce offspring that have pink flowers.
- **Codominance** is an inheritance pattern that occurs when neither of the alleles is completely recessive or dominant. In the heterozygote, both alleles are expressed equally in the phenotype. For instance, the offspring of a plant with red flowers and a plant with white flowers might express both red and white petals. Another example of codominance is the ABO blood groups of humans. The *A* and *B* alleles are codominant to each other (*O* is recessive), so a person with AB blood expresses both A and B antigens on red blood cells.
- **Multiple allele inheritance** involves more than two alleles coding for a certain trait, which results in more than two phenotypes. Human blood type (ABO) is an example of multiple allele inheritance.
- *Polygenic inheritance* is a type of inheritance in which several interacting genes control a single trait. Many traits result from the additive influences of multiple genes; skin color is one common example of a polygenic trait.
- *Genetic recombination* is a molecular process by which an organism's genes are rearranged in its offspring. Through this process, two alleles can be separated and replaced by different alleles, thereby changing the genetic makeup but preserving the structure of the gene. Chromosomal crossing over is an example of a mechanism by which this process takes place.
- *Gene transfer* can be vertical or horizontal. Vertical gene transfer occurs when an organism receives genetic material (i.e., DNA) from a parent organism or from a predecessor species. Horizontal gene transfer occurs when an organism transfers genetic material to cells that are not its offspring. Examples of horizontal gene transfer in bacteria were discussed in chapter 6.

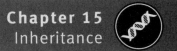

RAPID REVIEW

If you take away only 5 things from this chapter:

1. Chromosomes are the basic units of inheritance. How they reorganize and combine directly influences the genetic material present in an offspring.
2. Different versions of a gene that code for the same trait are called alleles. In classical (Mendelian) genetics, an individual receives one allele from each parent. Individuals with matching alleles are homozygous for that trait while those with different alleles are heterozygous. Usually, one version of the allele is dominant (e.g., brown eye color) and the other is recessive (e.g., blue eye color). Heterozygotes are "ruled" by the dominant allele.
3. Geneticists perform test crosses to determine the genetic makeup (genotype) of organisms displaying the dominant phenotype. A Punnett square is used to illustrate a test cross. Mendel's Law of Segregation states that an individual's alleles separate during meiosis, and either may be passed on to the offspring. Mendel's Law of Independent Assortment states that inheritance of a particular allele for one trait does not affect inheritance of other traits.
4. Sex-linked genes in humans occur on the X or Y chromosome. Males who inherit recessive X-linked genes from their mothers always express the trait in question, since men have only one X chromosome.
5. Pedigrees are family trees that enable us to study the inheritance of a particular trait across many related generations. They can also help us determine the type of transmission through generations: autosomal recessive/dominant, X-linked recessive/dominant, and so on.

TEST WHAT YOU LEARNED

1. In order to develop better melons, researchers want to use a mutagen, ethyl methanesulfonate (EMS), to induce mutations in a melon crop. The numbers and rates of mutations caused by EMS are shown in Table 1 and the rates of different mutation types are shown in Table 2.

Table 1

EMS Dose	M1 Plants	Induced Mutations	Mutation Frequency
1%	617	19	1/588 kb
2%	1473	67	1/356 kb
3%	40	3	1/146 kb

Table 2

Missense	Nonsense	Splicing	Silent
65.1%	2.4%	1.2%	31.3%

Both tables adapted from Fatima Dahmani-Mardas et al., "Engineering Melon Plants with Improved Fruit Shelf Life Using the TILLING Approach," *PLoS ONE* 5, no. 12 (December 2010): e15776.

Based on the information in the tables, how could the researchers best design their mutation protocol?

(A) Researchers should use the highest possible EMS dose to reduce the mutation rate to a manageable amount and then begin by examining splicing mutations to determine whether any produce new enzymes that improve melon quality.

(B) Researchers should use the highest possible EMS dose and then begin by examining the silent mutations to determine whether any are associated with improved melon quality.

(C) Researchers should use a relatively low EMS dose and then begin by examining the missense mutations to determine whether any are associated with improved melon quality.

(D) Researchers should use a relatively low EMS dose and then begin by examining the nonsense mutations to determine whether any produce new enzymes that improve melon quality.

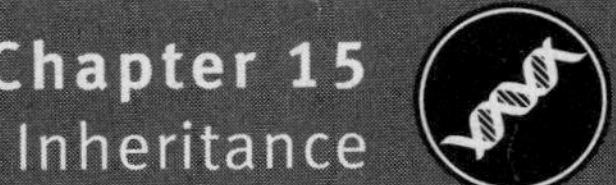

2. Human ABO blood types are inherited through a codominant pattern of inheritance. The possible blood types include A, B, AB, and O, representing different antigens on the red blood cell membrane. Individuals with type A blood have an A antigen and individuals with type B blood have a B antigen. Individuals with type O blood do not have either of these antigens on the surface of their red blood cells. What most likely happens in individuals with type AB blood?

   (A) Individuals with AB blood have a unique antigen that combines features of the A and B antigens within a single molecule, resulting in AB antigens that are present on the surface of their blood cells.

   (B) Individuals with AB blood have some cells that have A antigens on them and other cells that have B antigens on them.

   (C) Individuals with the AB blood type have a gene that codes for the A antigen and another gene that codes for the B antigen, but it is unpredictable which antigens will appear on the surface of their blood cells.

   (D) Individuals with AB blood have one gene that codes for the A antigen and another gene that codes for the B antigen, so both antigens are present on their blood cells.

3. Eukaryotic chromosomes can be tightly packed as heterochromatin or can be more loosely packed as euchromatin. What is the best prediction regarding the expression of genes in heterochromatin and euchromatin?

   (A) RNA polymerase can access the promoters in euchromatin more easily than it can access the promoters in heterochromatin, so euchromatin is transcribed more readily than heterochromatin.

   (B) RNA polymerase can access the promoters in heterochromatin more easily than it can access the promoters in euchromatin, so heterochromatin is transcribed more readily than euchromatin.

   (C) DNA polymerase can access promoters in euchromatin more easily than it can access the promoters in heterochromatin, so euchromatin is transcribed more readily than heterochromatin.

   (D) DNA polymerase can access promoters in heterochromatin more easily than it can access the promoters in euchromatin, so heterochromatin is transcribed more readily than euchromatin.

4. Coat color in cats is controlled by multiple genes. For example, cats can have a dominant autosomal allele called "agouti" for a particular coat pattern or can have a recessive allele for a solid coat color. Additionally, they can have a standard, full-color coat or a dilute, pale coat depending on whether they have the recessive dilute allele. A test cross was performed between an agouti, full-color male and a female that was homozygous recessive for both traits. Which of the following would be true regarding the results of the cross?
   (A) If all of the offspring had agouti, full-color coats, then the male is probably homozygous dominant for both traits.
   (B) If all of the offspring had agouti, full-color coats, then the male is probably heterozygous for both traits.
   (C) If all of the offspring had solid, dilute coats, then the male is probably homozygous dominant for both traits.
   (D) If half of the offspring had agouti, full-color coats and the other half had solid, dilute coats, then the male is probably homozygous dominant for both traits.

5. The pedigree diagram below shows the inheritance of a rare disorder over three generations. Circles represent females, and squares represent males. Shaded symbols indicate individuals with the disorder.

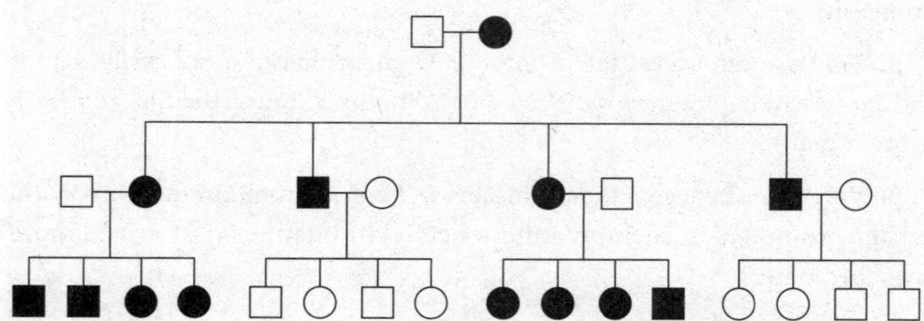

An individual with a rare disorder investigated the incidence of the disorder in his family and drew the pedigree diagram above. He contends that the gene for the disorder is on the mitochondrial chromosome, which is passed from mother to child in the cytoplasm of the egg. The sperm does not donate mitochondria. Is he correct? Why or why not?
   (A) He is correct. The pedigree supports mitochondrial inheritance. All children of affected women have the disorder. No children of nonaffected women have the disorder.
   (B) He is incorrect. The pedigree supports sex-linked recessive inheritance because the diagram shows that males are more likely to have the disorder than females.
   (C) He is incorrect. The pedigree supports autosomal dominant inheritance because the disorder shows up in every generation.
   (D) He is correct. The pedigree supports mitochondrial inheritance. Some males are carriers, which explains why some of the males do not display the trait.

6. A flower grower attempting to develop new varieties of a particular species took pure-breeding white flowers and crossed them with pure-breeding red flowers. The F1 offspring were all pink. The flower grower hypothesized that this was the result of incomplete dominance. In incomplete dominance, two alleles are expressed. The phenotype that results is intermediate between the two. In this case, she hypothesized that R is a red allele and that W is a white allele, so a heterozygote would be pink. Which of the following would refute the hypothesis that the inheritance of flower color in this case was incomplete dominance?

   (A) Self-crossing the F1 flowers and obtaining offspring that differed significantly from a 1:2:1 ratio of red:pink:white

   (B) Backcrossing the F1 flowers to the white parent and obtaining offspring that differed significantly from a 1:2:1 ratio of red:pink:white

   (C) Self-crossing the F1 flowers and obtaining offspring that differed significantly from a 1:1 ratio of red:pink

   (D) Self-crossing the F1 flowers and obtaining any red or pink flowers

7. Researchers have studied how genetic variants of *Arabidopsis thaliana* respond to infection by vascular wilt fungus. The figure below shows the fraction of susceptible, intermediate, and resistant plants in four different genetic variations. The trait does not appear to be explained by Mendelian models of inheritance.

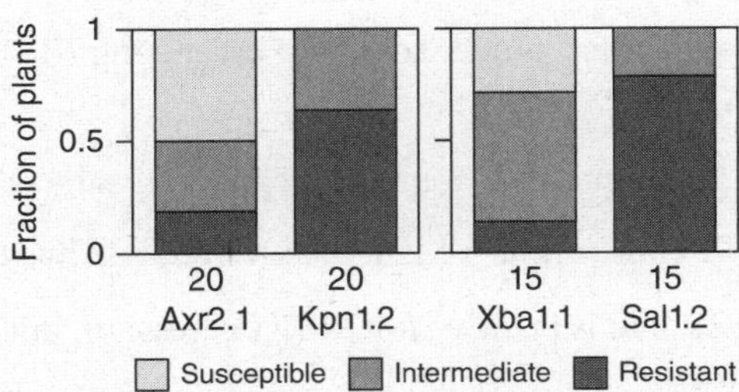

Adapted from Shen Y, Diener AC "*Arabidopsis thaliana* Resistance to Fusarium Oxysporum 2 Implicates Tyrosine-Sulfated Peptide Signaling in Susceptibility and Resistance to Root Infection." PLoS Genet 9(5)(2013): e1003525.

A researcher produces 240 seedlings from line Axr2.1. Based on this information and the figure above, how many of the seedlings would possess at least some level of resistance to vascular wilt fungus?

   (A) 20

   (B) 48

   (C) 72

   (D) 120

8. There has been considerable interest in understanding the behavior of cancer cells to improve treatments. One abnormality that has been detected in certain types of cancer is activation of X chromosomes that are normally inactive as Barr bodies. The table below shows characteristics of different cancer cell lines. "XIST" refers to "X-inactive specific transcript," which is important in X chromosome inactivation. "Xi" refers to an inactive X chromosome and "Xa" refers to an active X chromosome.

Cell Line	XIST Expression	X Chromosome Status
HMEC*	positive	normal Xs
Hs578t	negative	loss Xi + gain Xa
MDA MB 231	negative	loss Xi + gain Xa
T47D	negative	loss Xi + gain Xa
HCC 2185	negative	loss Xi + gain Xa
MCF7	positive	loss Xi + gain Xa

*HMEC represents normal human epithelial mammary cells, used as a control.
Adapted from Sirchia SM, Tabano S, Monti L, Recalcati MP, Gariboldi M, Grati FR, et al. (2009) Misbehaviour of *XIST* RNA in Breast Cancer Cells. PLoS ONE 4(5): e5559.

Based on the data in the table, must cancer cells stop producing XIST in order for an individual to have two active X chromosomes?

(A) Yes, because all of the cell lines that are XIST negative also have two active X chromosomes.
(B) No, because the MCF7 cancer line is XIST positive but still has two active X chromosomes.
(C) No, because the HMEC line is positive for XIST expression, but still has two active X chromosomes.
(D) Yes, because the T47D cancer line is XIST positive but has lost Xi.

9. Like humans, cats can have blood-clotting disorders such as hemophilia. In cats, hemophilia B is more common in male cats than in female cats. Which of the following modes of inheritance best explains this finding?

(A) The allele is recessive and found on the X chromosome.
(B) The allele is dominant and found on the X chromosome.
(C) The allele is recessive and autosomal.
(D) The allele is dominant and autosomal.

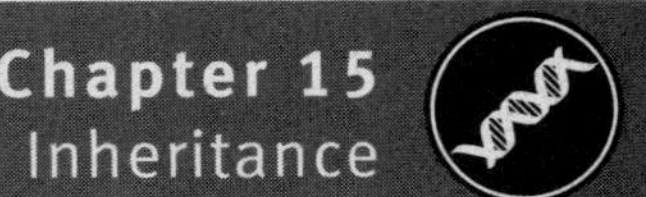

10. There is some evidence that bronchial hyperresponsiveness, common in individuals with asthma and chronic obstructive pulmonary disease (COPD), is correlated with high IgE levels. These findings were made by studying siblings who both have the disease. Which of the following conclusions is best supported by these findings?

    (A) Genes for this trait are located close together on the same chromosome.

    (B) Bronchial hyperresponsiveness is caused by elevated IgE levels.

    (C) Individuals develop asthma and COPD entirely due to environmental factors.

    (D) Genes for these traits are located on opposite ends of the same chromosome.

11. When analyzing the genome of a male child with a genetic disease, researchers discovered that he had two copies of chromosome 1 that were both inherited from the same parent. Which of the following does not provide a valid explanation of how this could have occurred?

    (A) Nondisjunction occurred during both oogenesis and spermatogenesis. An ovum lacking a copy of chromosome 1 was fertilized by a sperm with two copies of chromosome 1.

    (B) Nondisjunction occurred during both oogenesis and spermatogenesis. An ovum with two copies of chromosome 1 was fertilized by a sperm that lacked chromosome 1.

    (C) Nondisjunction occurred during oogenesis. An ovum with two copies of chromosome 1 was fertilized by a normal sperm, resulting in a diploid zygote.

    (D) Nondisjunction occurred during oogenesis. An ovum with two copies of chromosome 1 was fertilized by a sperm with one copy of chromosome 1, which was then extruded during early development.

12. The dilute coat color in cats is caused by a Mendelian autosomal recessive allele. Cats with two of these alleles have lighter fur color compared with heterozygous cats or those lacking the allele. A researcher crossed heterozygous cats and collected data from their 232 offspring. What is the predicted number of offspring with dilute fur?

    (A) 0

    (B) 58

    (C) 116

    (D) 174

13. Consider the figure below, which shows homologous recombination during meiosis. These chromosomes are from a single parent and individual genes are labeled as A and B.

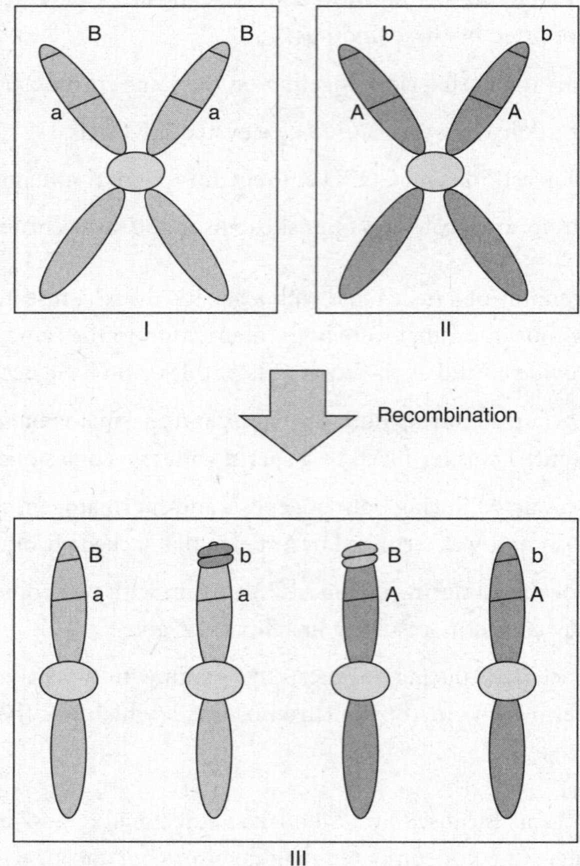

Which of the following is true regarding possible gametes that this individual could produce?

(A) The individual will always produce a mix of Ab, AB, ab, and aB gametes.

(B) The individual can only produce AB and ab gametes or Ab and aB gametes.

(C) Depending on where crossing over occurs, the individual can produce AB and ab gametes or a combination of Ab, AB, ab, and aB gametes.

(D) Depending on where crossing over occurs, the individual can produce Ab and aB gametes or a combination of Ab, AB, ab, and aB gametes.

The answer key to this quiz is located on the next page.

Answer Key

Test What You Already Know

1. B
2. D
3. C
4. A

Test What You Learned

1. C
2. D
3. A
4. A
5. A
6. A
7. D
8. B
9. A
10. A
11. C
12. B
13. D

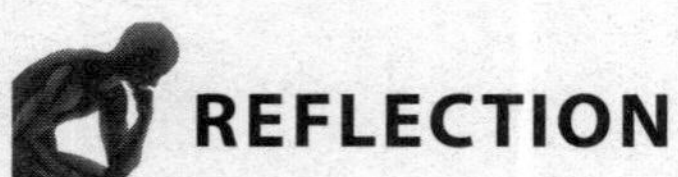

REFLECTION

Test What You Already Know score: __________

Test What You Learned score: __________

Use this section to evaluate your progress. After working through the pre-quiz, check off the boxes in the "Pre" column to indicate which Learning Objectives you feel confident about. Then, after completing the chapter, including the post-quiz, do the same to the boxes in the "Post" column. Keep working on unchecked Objectives until you're confident about them all!

Pre	Post	
☐	☐	**15.1** Describe chromosome structure and function
☐	☐	**15.2** Compare chromosomal mutations and abnormalities
☐	☐	**15.3** Explain and predict results of Mendelian inheritance
☐	☐	**15.4** Explain why some traits are non-Mendelian

FOR MORE PRACTICE

Complete more practice online at kaptest.com. Haven't registered your book yet? Go to kaptest.com/booksonline to begin.

ANSWERS AND EXPLANATIONS

Test What You Already Know

1. B Learning Objective: 15.1

The only DNA that showed double strand breaks was the decondensed DNA, and these breaks became more substantial as the radiation intensity increased. This suggests that condensation provides protection from radiation damage. **(B)** is thus correct. (A) is incorrect because it reverses the results for condensed and decondensed DNA. (C) is incorrect because only the decondensed DNA experienced damage. (D) is incorrect because decondensed DNA was the only type to show DSBs.

2. D Learning Objective: 15.2

A mutation in the operator region would most likely prevent the repressor from recognizing the operator, thereby stopping it from binding. With no repressor bound, the genes would be continuously expressed. Therefore, **(D)** is correct. (A) is incorrect because only the operator region would be affected and because this would cause the genes to be expressed. (B) is incorrect because the genes for lactose metabolism proteins would be unaffected by a mutation to the operator region. (C) is incorrect because the repressor would not itself be changed by a mutation to the operator region and because the mutation is more likely to prevent the binding of the repressor, rather than cause it to bind too tightly.

3. C Learning Objective: 15.3

The most likely explanation for a trait to appear in the offspring of parents who both lack that trait is that the gene for that trait is recessive and that both parents are heterozygous (carriers). **(C)** is thus correct. (A) is incorrect because it provides a less likely explanation of the result; non-deleterious mutations are quite rare, so it is far more probable that the parents were simply carriers of a recessive gloving gene. (B) is incorrect because a dominant allele is always expressed, so it would not be possible in this circumstance for both parents to lack gloving. (D) is incorrect because a recessive allele will not be expressed unless it is received from both parents.

4. A Learning Objective: 15.4

If men require only one copy of the mutant allele to have malaria resistance, but women must be homozygous (i.e., require two copies), then the malaria resistance allele must be X-linked recessive. **(A)** is thus correct. (B) is incorrect because the results provide no information about the geographic distribution of the allele. (C) is incorrect because males with the mutant allele have less G6PD activity, not more. (D) is incorrect because the results indicate that even homozygous females and hemizygous males display some amount of G6PD activity, so it is not true that they produce no functional G6PD.

Test What You Learned

1. C Learning Objective: 15.2

According to Table 1, even a 1% EMS dose led to a substantial number of mutations. Table 2 indicates that missense mutations are the most common. Indeed, it is more likely that missense mutations (which change the identities of one or more amino acids without otherwise altering protein structure) would be more likely to produce favorable results than mutations that altered the structure more dramatically. **(C)** is thus correct. (A) is incorrect because splicing mutations are relatively rare, as indicated in Table 2, and because a high dose is not needed to induce mutations (and too high of a dose could produce too many mutations, which is more likely to be deleterious to the plant). (B) is incorrect because silent mutations have no impact on amino acid sequences in a protein, so there is no chance that they could lead to improvements. (D) is incorrect because nonsense mutations add a premature stop codon, which truncates proteins in a way that usually renders them inoperable; thus, these mutations are less likely to result in benefits.

2. D Learning Objective: 15.4

In a codominant inheritance pattern, an organism will display both of the traits if its genome includes both of the codominant alleles. Thus, individuals with type AB blood will produce red blood cells with both A and B antigens, **(D)**. (A) is incorrect because each allele codes for an antigen separately. There is no way to combine the characteristics of both antigens into a single hybrid. (B) is incorrect because AB individuals will display both A and B antigens on all red blood cells. (C) is incorrect because it is safe to predict that all of the individual's red blood cells will contain both antigens.

3. A Learning Objective: 15.1

RNA polymerase must be able to bind to a promoter (a specific DNA sequence) in order for transcription to occur. This is only possible when the DNA is relatively loosely packed. Thus, euchromatin will be transcribed more readily than heterochromatin, making **(A)** correct. (B) is incorrect because polymerase will access euchromatin more easily than heterochromatin. (C) and (D) are incorrect because DNA polymerase is operative in DNA replication, not in transcription.

4. A Learning Objective: 15.3

Because the agouti and full-color traits are dominant, according to the question stem, then it is quite likely that the male is homozygous dominant for both traits if all of the offspring were agouti and full-color. **(A)** is thus correct. (B) is incorrect because a male that was heterozygous for both traits would likely lead to a greater mix of offspring phenotypes when mated with the specified female. (C) and (D) are incorrect because a parent that was homozygous dominant for both traits would only produce offspring that are agouti and full-color.

5. A Learning Objective: 15.4

The mother in the first generation gives the disorder to all of her children. In the second generation, only the females pass on the disorder; they pass it on to 100% of their children. This supports mitochondrial inheritance, making **(A)** correct. If (B) was correct, the females in the first generation would have gotten a dominant allele from their father and not shown the disorder. The reasoning in (C) is not sound. Dominant alleles don't always show up in every generation, and the pedigree doesn't match this even if it was true. Since mitochondria are not inherited through the father, (D) is incorrect.

6. A Learning Objective: 15.4

If this is a case of incomplete dominance, heterozygotes resulting from self-crossing the F1 flowers should produce a 1:2:1 ratio of red:pink:white, **(A)**. Any ratio that significantly differed from this would cause the flower grower to reject the hypothesis of incomplete dominance. Backcrossing the F1 flowers to the white parent would give a ratio of 1:1 pink:white. Therefore, (B) is incorrect. Since the expected ratio for self-crossing the F1 flowers is 1:2:1 of red:pink:white, (C) and (D) are incorrect.

7. D Learning Objective: 15.4

According to the figure, approximately half of the Avr2.1 plants are either intermediate or completely resistant. Therefore, approximately 50% of the 240 total plants produced should exhibit some level of resistance: $\frac{1}{2} \times 240 = 120$. Therefore, **(D)** is correct.

8. B Learning Objective: 15.2

Evaluate each answer choice. According to the data, (A) is incorrect because the MCF7 cell line, which is XIST positive, has two active X chromosomes. Thus, cancer cells do not need to stop producing XIST to have two active X chromosomes. **(B)** is the correct answer. Since the MCF7 cancer line is XIST positive but still has two active X chromosomes, there must be other factors involved in X chromosome inactivation. (C) is incorrect because the HMEC line is the control line of healthy cells and has normal X chromosomes, meaning that one is inactivated in each cell. (D) is incorrect because the T47D cancer cell line has lost Xi but is XIST negative.

9. A Learning Objective: 15.3

Hemophilia B being more common in male cats than female cats suggests X-linked recessive inheritance. X-linked recessive alleles are more common in males because they only need to inherit the recessive allele from one parent, rather than from both parents (as is true for females), in order to exhibit the trait. Therefore, the correct answer is **(A)**. (B) is incorrect because there is no reason to predict that an X-linked dominant trait would be more common in males unless other factors were involved. Similarly, (C) and (D) are incorrect because there is no reason to predict that an autosomal recessive or dominant trait would be more common in males unless other factors were involved.

10. A Learning Objective: 15.1

The finding that this disease is shared by siblings suggests the disease is caused by a genetic component. When genes are located close together on the same chromosome, they are likely to stay together during meiosis and be inherited together. As a result, these traits will typically be shared by siblings. Hence, the correct answer is **(A)**. (B) is incorrect because correlation does not necessarily reflect causation and there is insufficient information to draw this conclusion from the passage. (C) is incorrect because there is no evidence presented that the disease is entirely caused by environmental conditions and because the evidence that is given suggests a genetic component. (D) is incorrect because, when genes are located far apart on a chromosome, they are often separated by crossing over during prophase I of meiosis. This would likely prevent them from being inherited in exactly the same way in siblings.

11. C Learning Objective: 15.2

Evaluate each answer. (A) is a valid explanation because, in this example, both copies of chromosome 1 came from the sperm. This illustrates uniparental disomy. (B) is also an example of uniparental disomy because both copies of chromosome 1 came from the egg. For **(C)**, if two copies of chromosome 1 were in the egg and a normal haploid sperm fertilized the egg, then the zygote would be triploid, not diploid. **(C)** is not an accurate explanation and thus the correct answer. (D) is a valid explanation because self-correction may occur if there is an extra or missing chromosome. A single chromosome can duplicate itself or an extra chromosome can be extruded.

12. B Learning Objective: 15.3

A cross between two heterozygous individuals is expected to produce a 3:1 ratio of dominant to recessive offspring. Since cats with two recessive alleles exhibit dilute fur, ¼ of the total offspring (232) is expected to have dilute fur: $\frac{1}{4} \times 232 = 58$. Therefore, **(B)** is correct.

13. D Learning Objective: 15.1

If the genes are tightly linked and no crossing over occurs, then all of the chromosomes will show one of the parental types (i.e., Ab and aB). If there is a high level of recombination, then about half of the chromosomes will be recombinant, nonparental types (AB and ab) and about half will be parental types (Ab and aB). Thus, the correct answer is **(D)**. (A) is incorrect because, depending on the location of crossing over, the A and B genes may or may not be on different sides of the strand breakage, so a mixture of all four gametes (AB, ab, Ab, and aB) will not always be produced. (B) is incorrect because, if crossing over occurs, the individual can produce all four gametes (AB, ab, Ab, and aB). (C) is incorrect because the individual will produce Ab and aB gametes, not AB and ab gametes, if crossing over does not occur.

CHAPTER 16

Evolution

LEARNING OBJECTIVES

In this chapter, you will review how to:

- **16.1** Differentiate between types of selection affecting evolution
- **16.2** Describe the support for the common ancestry theory
- **16.3** Describe genetic drift and gene flow
- **16.4** Explain speciation and extinction
- **16.5** Explain how molecular biology and biochemistry support evolution
- **16.6** Explain how common structures and features support evolution
- **16.7** Describe how populations continue to evolve
- **16.8** Investigate the effects of artificial selection on evolution
- **16.9** Investigate allele distribution mathematical modeling

TEST WHAT YOU ALREADY KNOW

1. A population of rabbits settled in a desert environment where the soil is covered with small boulders and shrubs. Desert foxes are the major predators that the rabbits encounter. Small rabbits hide under rocks and shrubs to avoid detection. Desert foxes usually avoid large rabbits that can inflict serious wounds by kicking and biting. Ecologists survey the population size of the rabbits. Which of the following graphs is most likely to represent the distribution of body sizes in the rabbit population after several generations?

(A)

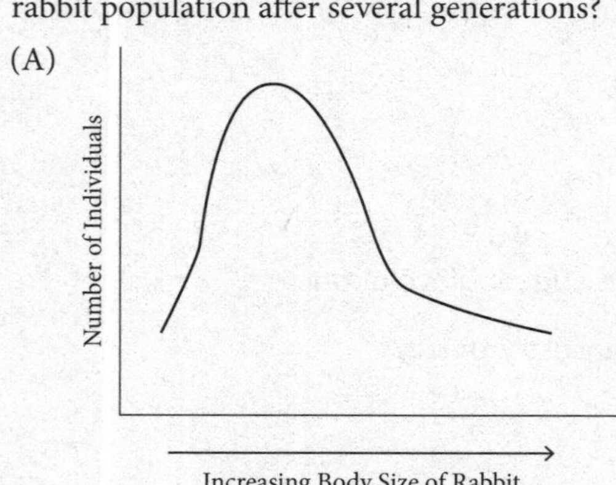

(C)

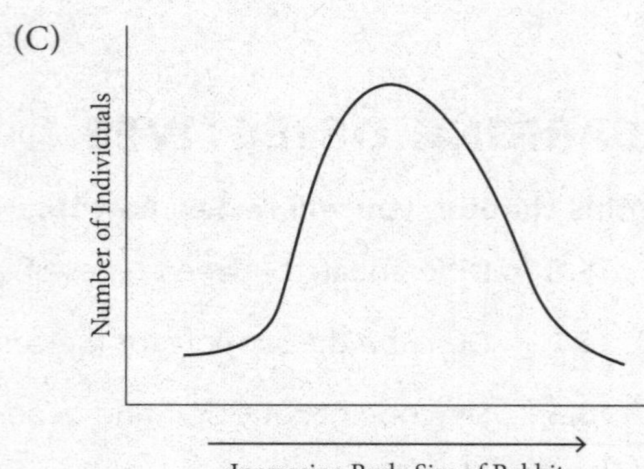

(B)

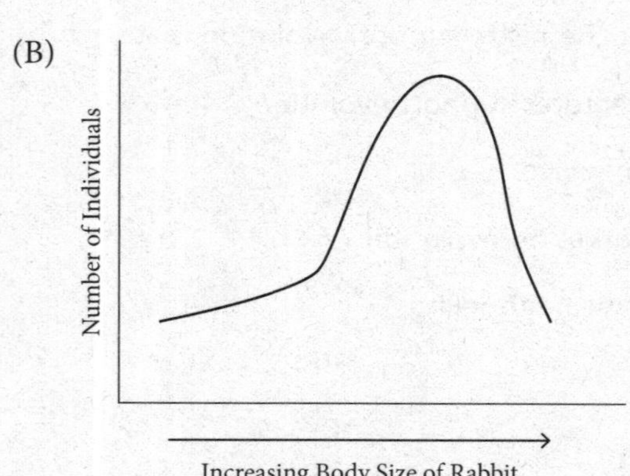

(D)

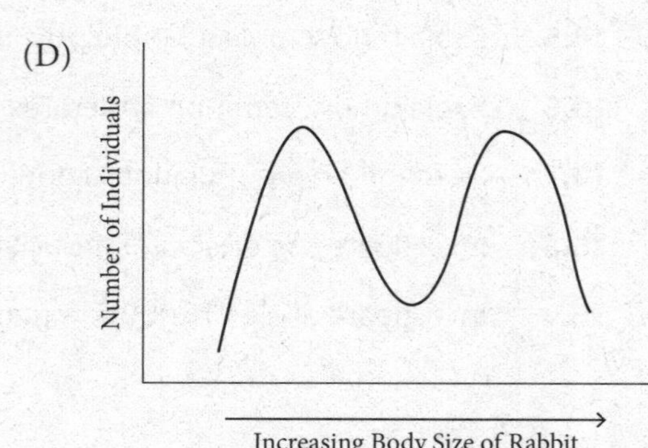

2. Research has shown that the glycolysis pathway exists in almost all organisms. Glycolysis exists among Bacteria, Archaea, and Eukarya and is catalyzed by homologous enzymes in all three domains. Which of the following hypotheses is supported by this finding?
   (A) Glycolysis is an inefficient process to produce ATP in all extant organisms.
   (B) Glycolysis is a primitive process that appeared in a common ancestor of all present-day organisms.
   (C) Glycolysis is present in many different organisms because it has appeared several times during evolution.
   (D) Glycolysis appeared only after organisms developed mitochondria to support metabolism.

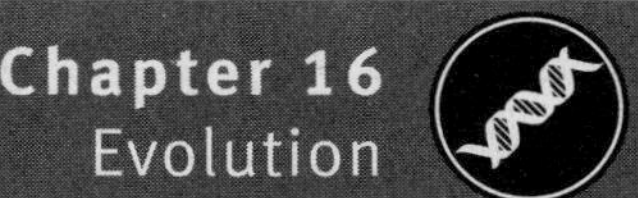

3. A small flock of finches was blown over from the mainland onto a neighboring island. On the mainland, a small beak is the most abundant trait, but a large proportion of birds that landed on the island happened to have large beaks. On the island, many types of seeds are available, giving birds of all beak sizes the same opportunity for feeding. Many years later, ecologists survey the distribution of the beak sizes in the island finches. Which of the following statements represents the most likely finding of the survey?

   (A) The size of beaks is equally distributed between large and small because the population experiences stabilizing selection.

   (B) Most birds have a large beak as a result of genetic drift because the trait was overrepresented in the small founding population.

   (C) There are more birds with small beaks because it reflects the makeup of the mainland bird population.

   (D) Most birds have a large beak because it gives a competitive advantage over other birds.

4. The soil in areas surrounding mines is often polluted with toxic heavy metals. These toxins prevent most grasses from growing. Buffalo grass is an unusual type of grass that can tolerate heavy metals in the soil. Populations in polluted areas have evolved a distinct genotype for tolerating the metal toxins. This resistant buffalo grass grows in close proximity to nonresistant buffalo grass that can only propagate in unpolluted areas, but resistant buffalo grass does not breed with nonresistant buffalo grasses.

   Based on these findings, which best describes the resistant buffalo grass populations?

   (A) Resistant buffalo grass has not undergone speciation because it is not a distinct geographical population.

   (B) Resistant buffalo grass is not a new species of grass because it has the morphological appearance of neighboring buffalo grasses.

   (C) Resistant buffalo grass has undergone speciation because it is not interbreeding freely with neighboring buffalo grasses.

   (D) Resistant buffalo grass is a new species of grass because it is planted next to mines to detoxify the ground.

5. Fungi, the kingdom that includes mold, yeast, and mushrooms, was formerly considered a part of the kingdom Plantae, which includes land plants. Published studies drew a comparison of base substitutions in actin, a major cytoskeleton protein. A summary of the data is shown in the table below.

Percent of Nucleotide Substitutions in Actin Genes

	Yeast	Soybean	Maize	Fruit Fly	Sea Urchin	Rat Skeletal Muscle	Human Cardiac Muscle
Yeast		17.7	18.7	10.9	11.6	11.6	11.8
Soybean	17.7		10.3	14.5	13.3	13.6	13.7
Maize	18.7	10.3		14.8	12.2	13.9	14.1
Fruit Fly	10.9	14.5	14.8		5.0	6.0	5.8
Sea Urchin	11.6	13.3	12.2	5.0		5.5	5.5
Rat Skeletal Muscle	11.6	13.6	13.9	6.0	5.5		0.57
Human Cardiac Muscle	11.8	13.7	14.1	5.8	5.5	0.57	

Adapted from Robin C. Hightower and Richard B. Meagher, "The Molecular Evolution of Actin," *Genetics* 114, no. 1 (September 1986): 315–332.

Based on the data, was the reclassification of Fungi as a distinct kingdom appropriate?

(A) The classification should have been revised because yeast are more closely related to maize than to humans.

(B) The classification should not have been revised because yeast are more closely related to soybeans than to animals.

(C) The classification should have been revised because fungi are more closely related to animals than to plants.

(D) The classification should not have been revised because fungi occupy a distinct evolutionary branch.

6. Snakes have no visible limbs and move by slithering on their ventral scales. The skeletons of boas and pythons show the remnants of small bones which are characterized as embryonic hind limb buds. Some scientists call them vestigial structures that indicate that snakes descended from ancient lizards that lost their limbs during the course of evolution. Which of the following findings would best support the hypothesis that snakes are descendants of lizard ancestors rather than a new class of animals?

(A) Snakes with fully developed hind limbs were discovered in the fossil record.

(B) Snakes can develop hind limbs if they exercise by climbing trees.

(C) A snake is found that is born with hind legs that they lose soon after birth.

(D) Lizards and snakes have similar scales that demonstrate their shared lineage.

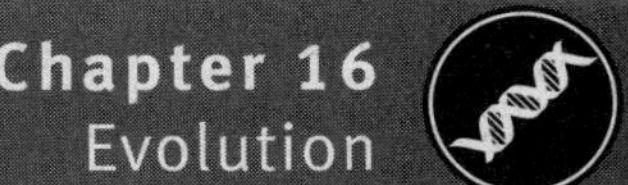

7. The common cuckoo lays its eggs in the nests of other birds so that other species will take on the costs of raising the baby cuckoos. The ashy-throated parrotbill is one species of bird that is parasitized by the cuckoo in this way. Based on this information, what selection pressures are most likely faced by cuckoo populations that parasitize parrotbills?

   (A) There is strong selection pressure on the cuckoos to produce distinctive eggs that can be quickly recognized as unique.

   (B) There is strong selection pressure on the cuckoos to produce large eggs so that their offspring can outcompete parrotbill offspring.

   (C) There is strong selection pressure on the cuckoos to produce eggs with variable coloration to increase the likelihood that some will resemble parrotbill eggs.

   (D) There is strong selection pressure on the cuckoos to produce eggs that are as similar as possible to those of the parrotbills.

8. Researchers have developed two lines of rats, named low intrinsic aerobic running capacity (LCR) and high intrinsic aerobic running capacity (HCR). After 19 generations of artificial selection, these rats differ greatly in their athletic ability. HCR rats can run much farther than LCR rats before reaching exhaustion. Although the researchers selected for running capacity, the rats differ in other ways. The table below shows some characteristics of the two lines. The figure below shows the time, in minutes, to heart failure in hypoxic conditions.

Trait	HCR	LCR
Heart Weight (g)	1.27 ± 0.0	1.46 ± 0.1
Body Weight (g)	416.8 ± 14.5	570.3 ± 23.7
HW/BW ratio ($\times 10^{-3}$)	3.1 ± 0.1	2.6 ± 0.2

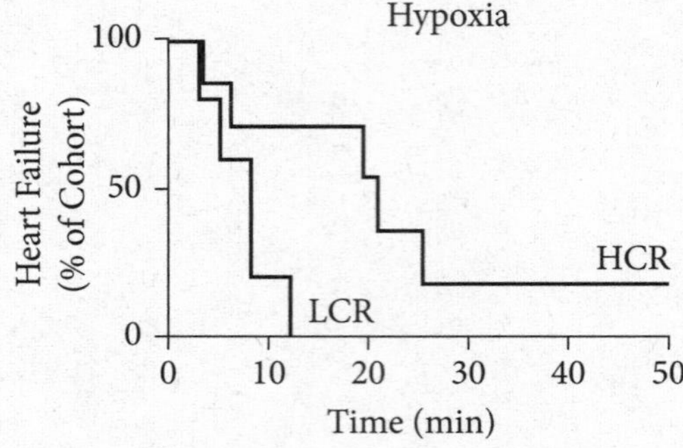

Table and figure adapted from Nathan J. Palpant et al., "Artificial Selection for Whole Animal Low Intrinsic Aerobic Capacity Co-Segregates with Hypoxia-Induced Cardiac Pump Failure," *PLoS ONE* 4, no. 7 (July 2009): e6117.

If the researchers selected for running capacity, why do the rats differ in these other ways?

(A) Artificial selection for a particular trait also affects related traits, such as cardiac function and muscle in these rats.

(B) The researchers deliberately selected multiple traits simultaneously when the lines were developed.

(C) All of these characteristics are associated with the same phenotype and therefore developed in association with limited running ability.

(D) The greater heart weight of LCR rats suggests that the researchers were selecting for overall reduced fitness.

9. Researchers are studying a population that has historically appeared to be in Hardy-Weinberg equilibrium based on the calculation of its allelic and genotypic frequencies. The population is unusual in that it is very large, does not experience immigration or emigration, does not show sexual selection, and is not exposed to mutagens. After one generation, the percentage of individuals with a particular dominant trait increases. After two generations, the percentage of individuals with the dominant trait slightly increases again. This pattern continues in the third generation.

   What is the most likely explanation of these findings?

   (A) The small changes are simply a result of random variation.

   (B) The population is experiencing a new selective pressure for the dominant trait.

   (C) The population is undergoing an increase in its rate of mutation.

   (D) The small changes are merely a result of genetic drift.

Answers to this quiz can be found at the end of this chapter.

TYPES OF EVOLUTIONARY CHANGE

16.1 Differentiate between types of selection affecting evolution

16.2 Describe the support for the common ancestry theory

16.3 Describe genetic drift and gene flow

16.4 Explain speciation and extinction

There is no biological concept more controversial and more misunderstood than biological evolution. The concept can be simply defined with the phrase "descent with modification." This means that over time, populations of organisms exhibit changes in characteristics that are passed on through inheritable (i.e., genetic) means. The important distinction between this simple definition and what is commonly accepted as biological evolution is mechanism. Mechanisms will be covered on the exam. You should also clarify the semantic distinction between evolution and biological evolution. Personalities evolve. Societies evolve. But only populations of organisms undergo biological evolution by means of a modification of inheritable characteristics. In this chapter, the term *evolution* will always refer to biological evolution.

✔ AP Expert Note

For more on evolution, see the discussions of life origin hypotheses and Darwin's theory of natural selection in chapter 4.

Evolutionary mechanisms

In the simplest sense of the term, evolution occurs whenever there is a change in gene frequencies within a population. Imagine a species of fish in which 50 percent of the population have green fins and 50 percent have blue fins. If, five years later, 51 percent of the population have green fins and 49 percent have blue fins, then that species has evolved. Evolution in this sense has two basic kinds of mechanisms: sources of variation (which increase genetic diversity) and selective pressures (which decrease genetic diversity).

Sources of variation have already been discussed in multiple places throughout this book. At the most foundational level, all heritable variations ultimately arise from mutations (discussed in chapter 16), direct changes to the genetic code, which impact the proteins produced by an organism's cells and, consequently, the phenotypes of that organism. Genetic variation can also arise from other sources, such as the shuffling of genes that occurs when sexual reproduction leads to a new organism with a novel combination of parental genes. For instance, immigration of new organisms into a population, followed by interbreeding, can lead to an influx of new genes in a population.

If sources of variation acted alone, then species would just become more and more diverse over time, but selective pressures serve to weed out some phenotypes (and their associated genotypes), acting as a force that decreases variation within a population. Selection effectively acts as a filter on the gene pool: organisms with phenotypes that allow them to survive and produce fertile offspring within their specific environmental niche pass through the filter successfully, while unfit organisms that are not well adapted to their environment are filtered out. This process of filtering is generally known as **natural selection**, a theory established by Darwin and discussed in chapter 4.

In addition to natural selection, Darwin also put forth a theory of **sexual selection** (in his 1871 book, *The Descent of Man, and Selection in Relation to Sex*). Sexual selection has specifically to do with selective pressures that impact the capacity for successful reproduction. Many male birds have bright mating plumage that makes them conspicuous in a forest and more prone to attack from predators. Male ungulates often have huge antlers that grow every year and make it more difficult to travel through dense woods. The bright feathers on the male bird and the large antlers on the male deer are characteristics that are **selective disadvantages** in the animals' natural environment but are **selective advantages** when it comes to courtship and mating. This is because females of those species have a strong preference for excessive adornment, refusing to mate with males who lack such augmentation. Because these female preferences are also often heritable, sexual selection can lead to rapid evolution of a population.

✔ **AP Expert Note**

The balance between survival and the ability to mate dictates how species evolve.

One other type of selection is *artificial selection*, a process that plays an active role in the domestication of plants and animals. Similar to sexual selection (in which mate preference shapes evolution), choice also plays a role in artificial selection but, in this case, it is the decisions of members of another species (usually human beings), which are able to exert control over which pairs of organisms mate. The wide variety of breeds of dogs, which share a common ancestor with wolves, is in large part due to extensive artificial selection.

These joint processes of variation and selection have spawned the tremendous amounts of diversity found in the natural world today. Indeed, biologists generally believe that life originated only once on this planet and that all living organisms ultimately share a single common ancestor. This theory is supported by a number of common biochemical processes that are shared by all living things, such as genetic information encoded in DNA or RNA, the transcription of RNA from DNA, and the translation of mRNA codons into specific amino acids. What's more, all extant (surviving) organisms use ATP as an energy carrier. Accordingly, ATP is known as the energy currency of life.

Two homeostatic processes that illustrate common ancestry are excretion and osmoregulation. Earthworms, arthropods, and vertebrates all use the same principles of filtration and active transport to excrete waste products from the body, and fish, protists, and bacteria use analogous osmoregulatory mechanisms to control water and solute concentrations in order to maintain internal water balance. Similar morphological features (for example, skeletal components of vertebrates) can also be seen across living organisms. Homologous structures, however, may or may not be used for the same function. Vestigial structures like the wings of a flightless bird have lost their original function. Phylogenetic trees, which represent evolutionary relationships, also provide evidence for common descent and are discussed in chapter 17.

High-Yield

Genetic Drift and Gene Flow

Biodiversity arises from **genetic drift**, a random sampling process that occurs in small populations. Two examples of genetic drift are the bottleneck effect and the founder effect. The bottleneck effect generally occurs when a catastrophe like a natural disaster leaves behind only a small group of individuals. Similarly, the founder effect occurs when a small group of individuals breaks off from the larger population and becomes isolated. In both cases, the allele frequencies in the small group differ from those of the original population. Over time, the changes in allele frequencies become prominent in the small population (but would have little effect in a large population). Genetic drift may lead to the formation of a new organism or species. However, since the new species is a result of random changes and not natural selection, it may not be well adapted to survive in its environment.

Unlike genetic drift, which takes place between two generations of one population and in only one species, **gene flow**, also known as gene migration, takes place between two populations of one species or between two species. Gene flow is the process of moving genes between populations via individuals entering or leaving the populations. Movement, for example from migration and pollination, may eliminate or introduce new alleles to the gene pool, causing significant changes in the offspring. On the other hand, geographical barriers such as oceans, mountain ranges, and deserts, restrict gene flow. Both genetic drift and gene flow can be analyzed using Hardy-Weinberg equilibrium, which is discussed later in this chapter.

Speciation and Extinction

High-Yield

Over a dozen different concepts of speciation have been promoted in scientific literature, but the most prevalent by far is the **biological species concept (BSC)**. This concept states that a species is defined by a naturally interbreeding population of organisms that produces viable, fertile offspring. In other words, two species are distinct if they can't breed with each other or don't naturally breed with each other due to certain barriers. There are two kinds of barriers to interbreeding: prezygotic and postzygotic.

Prezygotic barriers to interbreeding include **isolation** of species due to ecological, temporal, behavioral, or mechanical factors, or physiological incompatibility of gametes. **Postzygotic barriers** include ultimate inviability or sterility of **hybrid** organisms from the interbreeding of two species. Hybrid organisms may not die off in one generation, but ultimately the offspring of the mating of two species will die without producing offspring of their own.

Geographic isolation is not the only factor that can cause speciation, so additional terms have been defined to describe how species evolve due to isolating mechanisms. **Allopatric speciation** is when one population is separated into two distinct populations by some **geographic barrier** such as the movement of a tectonic plate or the elevation of a mountain range. After the original population is no longer able to share its alleles, it evolves into distinct populations that have a high probability of acquiring distinctive traits. In contrast, **sympatric speciation** occurs when individuals within a population acquire distinctively different traits while in the same geographic area. Sympatric speciation requires some other form of reproductive isolation, such as those previously mentioned. **Parapatric speciation** is less definitive. This occurs when two populations are able to interbreed along a border, but the exchange of alleles is negligible compared to the amount of genetic exchange occurring within each population. A narrow zone of hybridization exists at the meeting of the two populations, but the two populations never coalesce into one.

A variety of factors such as species interactions, environmental changes, and population size can lead to **extinction**—the termination of an organism or species. For two species whose niches overlap, typically the species that is better adapted will drive the other species to extinction. Human activities (for example, hunting, habitat destruction, introduction of disease, and changing the climate) have become a predominant cause of extinction.

Species that are unable to adapt to a changing environment may die out and become extinct. In particular, species with specialized diets or habitat requirements (like the giant panda that feeds mainly on bamboo) are more vulnerable to environmental changes. Generalist species like raccoons, on the other hand, are more able to survive because they feed on a wide variety of food and live in various habitats. With respect to population size, species with long generation times that produce few offspring (like rhinoceroses) are more subject to extinction than species with short generation times that produce many offspring (like rodents). The ability to increase in population quickly allows species to recover from low populations caused by disturbances and diseases.

The following are examples of species that became extinct in the wild but are being bred in captivity and then released back into the wild. Their low genetic diversity, however, keeps them at risk for extinction and makes them endangered species. California condors, prior to humans hunting and poisoning them to near extinction, were a genetically diverse population. Today, the genetic diversity is extremely low because all existing California condors are descended from just a handful of individuals. Likewise, black-footed ferrets, whose population once dropped to alarmingly low levels, have been bred from only a few individuals, making them especially vulnerable to threats. Tasmanian devils (alike on the brink of extinction due to low genetic diversity) are being bred on an island with disease-free animals to protect them from a contagious cancer. Due to habitat loss, prairie chickens live in small groups isolated from one another. The isolation of prairie chicken populations has in turn contributed to a loss in their genetic variance. On a related note, the Irish potato blight and southern corn rust (diseases caused by fungi) were the consequence of low genetic diversity.

Mass extinctions involve deadly events such as volcanic eruptions, asteroid collisions, or global warming/cooling that cause at least half of the species to die out in a relatively short amount of time. Following mass extinctions, new life-forms may emerge and evolve as new habitats open via **adaptive radiation**. Five major mass extinction events occurred on Earth in prehistoric times: Ordovician-Silurian Extinction (small marine organisms died out), Devonian Extinction (tropical marine species died out), Permian-Triassic Extinction (many vertebrates died out), Triassic-Jurassic Extinction (dinosaurs flourished after other species died out), and Cretaceous-Tertiary Extinction (dinosaurs died out). The extinction of the dinosaurs made room for other animals to diversify and evolve rapidly.

EVIDENCE FOR EVOLUTION

16.5 Explain how molecular biology and biochemistry support evolution

16.6 Explain how common structures and features support evolution

16.7 Describe how populations continue to evolve

Types of Evidence

Even though evolution is sometimes described as a "theory," it is more than a mere hypothesis. A wealth of evidence from a variety of fields indicates that evolution is a well-established scientific fact.

The Fossil Record

The geological layers in the Earth's crust stack on top of each other, with the oldest layers deeper and the youngest layers closer to the surface. Older geological layers hold more "primitive" fossils. Below is a table displaying different major biological events that have occurred through geologic time, the period in which these events occurred, and approximately how many millions of years ago they happened. You are not required to know the dates of these events, but be prepared to analyze given data and investigate certain claims.

The Fossil Record over Geologic Time

Period	Millions of Years Ago	Events
Precambrian	$> 3{,}500$	First prokaryotes
Precambrian	$> 1{,}000$	Earliest eukaryotes
Cambrian	540–490	Origin of all extant and some extinct animal phyla, including chordates
Ordovician	489–446	Continued evolution of ocean life
Silurian	445–415	First terrestrial organisms
Devonian	415–360	Diversification of bony fishes; first insects; first seed plants
Carboniferous	360–300	First gymnosperms
Permian	300–250	Diversification of reptiles
Triassic	250–200	First mammals and dinosaurs
Jurassic	200–145	Diversification of dinosaurs; first birds
Cretaceous	145–65	Origin and diversification of angiosperms; extinction of dinosaurs at end of period
Paleogene and Neogene	65–1.8	Diversification of all major living groups of birds and mammals, including hominids
Quaternary	1.8–present	Extinction of large land mammals; rise of humans

Transformations

Biogeography

Organisms are more like other organisms in their geographic vicinity. Organisms in adjacent dissimilar environments are more similar than organisms in similar environments on opposite sides of the Earth. This suggests that organisms in adjacent dissimilar environments are descended from recent common ancestors, rather than having evolved randomly and independently.

Comparative Anatomy

Organisms have very different structures that are composed of the same basic components. For example, the human arm has the same bones as the wing of a bat. These structures are called **homologous** structures because they are considered to have arisen from a common ancestor. **Analogous** structures are structures that may perform a similar function but have not arisen from the same ancestral condition.

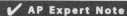

AP Expert Note

The wings of a bat and the wings of a butterfly are analogous structures.

Embryology or Ontogeny

Organisms that share a more recent common ancestor have similar modes of development. A classic example is that all vertebrate embryos have a stage of development in which they possess gills, whether they are aquatic or terrestrial. The presence of these more "primitive" characteristics in the embryos of "advanced" organisms suggests that these organisms share genetically controlled developmental physiologies that have been passed on from their common ancestors. The process through which an organism develops from an embryo to an adult is called **ontogeny**.

Taxonomy

Organisms are classified into smaller and smaller subgroups based on similar and dissimilar characteristics. This hierarchy is an implicit illustration of the tree of life, leading to common ancestry by linkage to a superseding group. For example, a plant in the family Euphorbiaceae is more closely related to other plants in Euphorbiaceae than it is to plants in the family Cactaceae. The housefly, *Musca domestica*, is more closely related to other flies in the genus *Musca* than it is to flies in the genus *Stomoxys*.

Phylogenetic trees and cladograms are illustrations that can be used to represent the relationships between similar and dissimilar organisms. They are constructed using morphological, molecular, or DNA evidence. For example, in the simplified cladogram that follows, you can see that Mollusca and Arthropoda are the most closely related phyla, as they shared the most recent common ancestor. However, the phyla Porifera, Cnidaria, Echinodermata, Arthropoda, and Mollusca all share a single common ancestor.

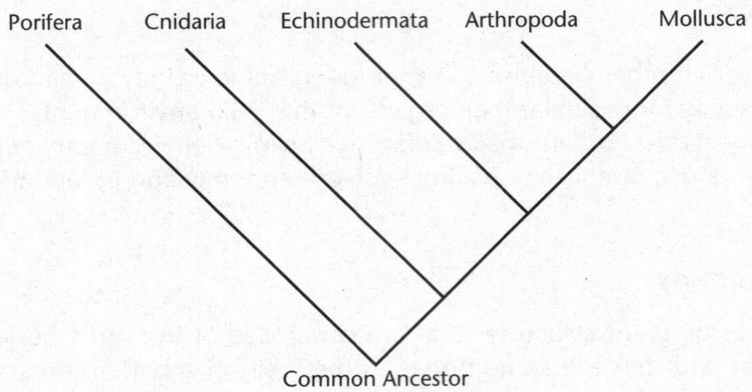

✔ AP Expert Note

For more on phylogenetic trees and cladograms, see chapter 17.

Molecular Biology

Siblings share more similar DNA with each other than they do with other members of the same species. Similar species have more similar DNA than do more distantly related species (in different genera, for example). Related genera share more similar DNA with one another than they do with genera in another family.

Rates of Evolution

Prior to Darwin, and even for scientists today, there is an underlying assumption that evolution takes a long time. The reasoning behind this theory is that because mutation is the ultimate source of variation and mutations that allow viable offspring are extremely rare, the probability of accumulating enough mutations to cause considerable change in organismal form requires a lot of time. Due to this assumption, prior to chemical dating techniques, geological layers were considered to be very old. Chemical dating has allowed modern scientists to assess the age of geological layers accurately.

Scientists have proposed two more hypotheses related to the rate of evolution. The first hypothesis is **punctuated equilibrium**. The hypothesis of punctuated equilibrium suggests that changes in organismal form did not take millions of years. Instead, very large changes in form happened relatively quickly (i.e., over thousands or tens of thousands of years) and were maintained thereafter over long periods of time. There is some evidence to suggest this hypothesis may be true, such as the Cambrian explosion in the fossil record, the discovery of cascading developmental genes, and the observation of large changes in phenotypic expression caused by single base-pair mutations. The second, the **molecular clock hypothesis**, is the notion that genetic mutations occur in a genome at a linear rate. Assuming the molecular clock hypothesis is true, one could extrapolate the age of divergence of two organisms by counting the number of genetic differences in their genomes. This latter hypothesis assumes that mutation rates are constant over time and between species; a likely incorrect pair of assumptions.

Populations and species continue to evolve today. Modern-day evolution can be seen in recent emergent diseases (those that appear for the first time or increase in incidence). Examples include severe acute respiratory syndrome (SARS), H1N1 influenza, and Ebola. Scientific evidence of chemical resistance also supports the claim that evolution continues to occur. Bedbugs today have evolved resistance to pesticides by developing thicker, waxier exoskeletons to protect them from toxins. Additionally, arising from a hybrid between the European house mouse and the Algerian mouse, which was introduced via human travel, the common house mouse can now survive rodenticides. Evolution in action can also be observed in artificial selection by humans, which has resulted in today's domestic animals, as well as crops with desired traits. Human populations who live in extremely high-altitude areas (for example, Tibet, the Andes, and Ethiopia) have evolved enhanced respiratory mechanisms that allow them to survive in low-oxygen environments.

AP BIOLOGY LAB 1: ARTIFICIAL SELECTION INVESTIGATION

16.8 Investigate the effects of artificial selection on evolution

In this inquiry investigation, you will explore real-time natural selection using Wisconsin Fast Plants (*Brassica*), just as Mendel did with pea plants. This is a long-term laboratory investigation in which you will select, plant, grow, and tend to a population of plants to observe how natural selection acts on phenotypic variations. This is a pretty simple investigation in terms of content and equipment. However, throughout the course of this seven-week investigation, you must keep detailed records in which you quantify variation, record images of that variation, and evaluate and explain your results. These skills will be helpful for you to cultivate for your responses on the free-response questions on the AP Biology exam.

Natural selection is the mechanism that describes the reproductive success or fitness of certain traits. Scientists and farmers use artificial selection to grow preferred agricultural crops. We also see natural selection occur inadvertently when diseases and pests grow resistance to the use of antibiotics or pesticides. In this lab, as a class and independently, you will select for a trait and investigate its reproductive success in two generations. Remember that directional selection tends to increase or decrease the trait in the next population. After growing your plants and recording your observations of the two different generations, you have many options for how to analyze your data. Deciding what to do with your data is the tricky part of this investigation, but these types of analyses will help you with the grid-in and free-response questions on your exam. Let's look at two ways we can analyze sets of data.

You could use different descriptive statistics such as mean, median, range, and standard deviation to describe your populations of study. Let's take a look at a couple of these descriptive statistics. The equation for solving for standard deviation *S* is:

$$S = \sqrt{\frac{\sum (x_i - \bar{x})^2}{n-1}}$$

where x_i is each data point, $\bar{x}$ is the mean of the data points, and n is equal to the size of the sample. The standard deviation calculates the difference between each of your data points and the mean value of your data, allowing you to see how much variability there is in the population.

For each set of data, you will construct a histogram. A histogram is a graphical representation of the distribution of the data. Using your histograms and other statistical analyses, such as a *t*-test or chi-square test, you can determine the validity of the differences in the populations. The formula for a chi-square test is:

$$\chi^2 = \sum \frac{(o-e)^2}{e}$$

where *o* is the number of individuals observed to have a specific phenotype and *e* is the number of individuals expected to have that particular phenotype.

All of this mathematical problem solving might seem intimidating, but it is essential for real-world biology practice and for the AP Biology exam. You can take a deep breath, though; all of these formulas will be provided to you on your exam. You will not be required to memorize these equations and formulas.

AP BIOLOGY LAB 2: MATHEMATICAL MODELING: HARDY-WEINBERG INVESTIGATION

High-Yield

16.9 Investigate allele distribution mathematical modeling

This lab is easy to perform, but it causes confusion for some students because it involves quantitative and analytical skills. In this investigation, you will develop a mathematical model to investigate the relationship between allele frequencies in populations of organisms. You will use a spreadsheet to build a model based on the Hardy-Weinberg relationship to determine how allele frequencies change from one generation to the next. This model will help you to see how selection, mutation, and migration can affect these inheritance patterns.

Now comes the hard part: putting information into equations. The way to determine the frequencies of the alleles in your breeding population is to assume the population is in **Hardy-Weinberg equilibrium.** If your population is in Hardy-Weinberg equilibrium, two things will be true. First, the addition of the frequencies of the alleles will equal one. The simple formula is:

$$p + q = 1$$

where the frequency of the dominant allele is indicated by *p* and the frequency of the recessive allele is indicated by *q*. For this experiment, $p = T$ and $q = t$. The frequencies of the phenotypes in a Hardy-Weinberg population follow the equation:

$$p^2 + 2pq + q^2 = 1$$

The genotypes in a Hardy-Weinberg population are indicated by each term on the left side of the previous equation. The frequency of the homozygous dominant phenotype in the population above is:

$$T \times T = T^2 = p^2$$

The frequency of the heterozygote phenotype is:

$$2 \times T \times t = 2Tt = 2pq$$

and the frequency of the homozygous recessive phenotype is:

$$t \times t = t^2 = q^2$$

Normally, success in mastering Hardy-Weinberg problems lies in the reading of the question. You usually have to figure out p and q in order to plug in the terms. If the question states "frequency of alleles in a population," then you should start with $p + q = 1$ and solve for p or q. If, on the other hand, the problem says "frequencies of organisms that express the trait (dominant or recessive)," then you start with $p^2 + 2pq + q^2 = 1$ and calculate your frequencies of p and q from this data. Once you have calculated p and q, you can plug those values into the equation where appropriate to figure out frequencies of homozygotes, heterozygotes, or carriers for a trait.

✔ **AP Expert Note**

Assumptions of Hardy-Weinberg Equilibrium

There are five assumptions that must be met for a population to be under Hardy-Weinberg equilibrium:

1. **The population is very large and not subject to small perturbations in the frequencies of alleles. There are no bottleneck effects.**
2. **The population is isolated from both immigration and emigration. There is no gene flow.**
3. **There is no mutation.**
4. **There is no selective breeding, and mating is random between individuals.**
5. **There is no genetic drift. All genotypes code for phenotypes that have an equal chance of viability and reproduction. There is no selection of phenotypes.**

Keep in mind that the theoretical Hardy-Weinberg population is the benchmark to which all naturally occurring populations are compared. There are probably few if any populations in complete equilibrium. Comparing the expected frequencies under the previously given assumptions to what actually occurs in a population gives insight into which of these assumptions is being violated. In this way, a scientist can determine which evolutionary or environmental forces prevent a population from maintaining equilibrium.

✔ **AP Expert Note**

There is a good likelihood that you will see some sort of Hardy-Weinberg question in either the grid-in or free-response section of the exam.

For this experiment, we can make assumptions about the theoretical alleles of our model population. Let's say that 5 percent of the population is homozygous recessive. This means that the frequency of the recessive allele can be determined by letting $q^2 = 0.05$, which means the frequency of the dominant allele can be determined with the equation:

$$p = 1 - q = 1 - 0.22 = 0.78$$

The frequencies of the homozygous dominant genotype and the heterozygous genotype, respectively, are:

$$p^2 = 0.78 \times 0.78 = 0.61$$
$$2pq = 2 \times 0.78 \times 0.22 = 0.34$$

You can check your work by adding all the genotypic frequencies and making sure that they total 1:

$$0.05 + 0.61 + 0.34 = 1$$

After developing your Hardy-Weinberg model and exploring how random events can affect allele frequencies over generations, you will identify and then test different factors that affect the evolution of allele frequencies. In addition to measuring allele frequencies, your model should track changes in population size, the number of generations, selection (fitness), mutation, migration, and genetic drift.

RAPID REVIEW

If you take away only 5 things from this chapter:

1. Evolution is shaped by a number of mechanisms, including selective pressures, sources of variation, and random effects, such as genetic drift.
2. Biological species concept: a species is a reproductively isolated population able to interbreed and produce fertile offspring.
3. In allopatric speciation, geographically separated populations develop into different species. Sympatric speciation occurs when populations in the same environment adapt to fill different niches. Parapatric speciation occurs with limited interbreeding between two groups.
4. Evidence for evolution comes from comparative anatomy (homologous and analogous structures), biogeography, embryology, the fossil record, biological classification, and molecular biology (relatives share DNA).
5. Hardy-Weinberg equilibrium occurs when genetic distribution remains constant in large, isolated, randomly mating populations with no mutation and no natural selection. These conditions rarely (if ever) occur together.

TEST WHAT YOU LEARNED

1. Although many metabolic reactions are different in distinct organisms, the pathways associated with energy release from nutrients are nearly identical in all organisms and depend on molecules such as ATP, NADH, and $FADH_2$. What is the most likely origin of these small molecules?
   (A) They probably arose very early in a common ancestor of all living organisms.
   (B) They probably came about through convergent evolution.
   (C) They probably exhibit small differences and cannot be considered identical.
   (D) They probably are derived from the breakdown of common chemical compounds.

2. The unique ecology of an island distant from the mainland has made it a tourist attraction, bringing visitors on boats. On the island, finches have large beaks that allow them to crack seeds. Finches on the mainland have a broad range of beak sizes, from very small to very large. Over time, ecologists notice the appearance of several new traits in the island finch population. One of these traits is a small beak, which is effective at obtaining food from sources that large-beaked finches overlook. Which of the following is the most reasonable hypothesis regarding the appearance of this new trait?
   (A) The visitors feed the birds, allowing the population of finches to expand and diversify.
   (B) The boats pollute the island, causing mutations in the finch population.
   (C) The boats also carry mainland finches, adding new genes to the gene pool.
   (D) Extensive tourism destroys the finches' habitat, exerting a selective pressure to adapt.

3. Researchers have been interested in understanding differences between Tibetan chickens, which live at very high altitudes, and lowland chickens, which do not. In one study, researchers examined the frequencies of several single nucleotide polymorphisms (SNPs) among these two populations. The data for one of these SNPs is shown in the table below. (Note that TT indicates an allele that features a genetic sequence with two thymines, while TC features a thymine and a cytosine.)

SNP2 rs14330062	TT	TC
Tibetan Chicken	152 (96.8%)	5 (3.2%)
Lowland Chicken	139 (100%)	0 (0%)

Adapted from Sichen Li et al., "A Non-Synonymous SNP with the Allele Frequency Correlated with the Altitude May Contribute to the Hypoxia Adaptation of Tibetan Chicken," *PLoS ONE* 12, no. 2 (February 2017): e0172211.

Based on this information, which of the following statements is accurate?

(A) The table shows the allelic frequencies of TT and TC, which can be represented as p^2 and q^2 in Hardy-Weinberg calculations.

(B) The table shows the genotypic frequencies of TT and TC, which can be represented as p^2 and q^2 in Hardy-Weinberg calculations.

(C) The table shows the allelic frequencies of TT and TC, which can be represented as p and q in Hardy-Weinberg calculations.

(D) The table shows the genotypic frequencies of TT and TC, which can be represented as p and q in Hardy-Weinberg calculations.

4. Many physiological activities in plants and animals oscillate according to a circadian rhythm. The circadian rhythm is driven by blue light receptors called cryptochromes. Cryptochromes consist of a protein attached to a pigment. The DNA sequences of the cryptochrome protein show extensive similarities among plants and animals. A search of genome databanks reveals that the cryptochromes share considerable sequence similarities with photolyases, which repair enzymes that repair UV radiation damage in DNA.

Which of the following conclusions is most justified based on these findings?

(A) The light receptor proteins are not homologous; they appeared independently in different lineages because of the selective pressure for all organisms to respond to blue light.

(B) The photolyases and cryptochromes are both receptors of blue light; therefore, their genes should share similar nucleotide sequences because of convergence.

(C) Photolyases appeared early during evolution in a common ancestor; photolyase genes mutated and were repurposed in descendants for circadian rhythm control and other blue light responses.

(D) Photolyases and cryptochromes have similar sequences; all light receptor proteins have the same structure.

5. Throughout human history, there have been considerable efforts to use artificial selection to develop breeds of animals that are useful in agriculture. For example, researchers have worked over the last 50 years to develop dairy cows that produce as much milk as possible. Which of the following best describes the methodology behind these efforts?

   (A) Researchers select for a single gene that enhances milk production, resulting in homozygous herds of dairy cattle.

   (B) Researchers only breed cows with greater than average milk production to increase production in future generations of cows.

   (C) Researchers cross cows with dominant alleles for high milk production, resulting in gradual improvements in milk production across each generation.

   (D) Researchers breed unrelated animals to increase genetic diversity and reduce the risk of inbreeding, which increases milk production.

6. A geological survey of seabed sediments revealed large and diverse populations of animal fossils dating from the Permian through the Triassic periods. Paleontologists compiled their data in the following diagram.

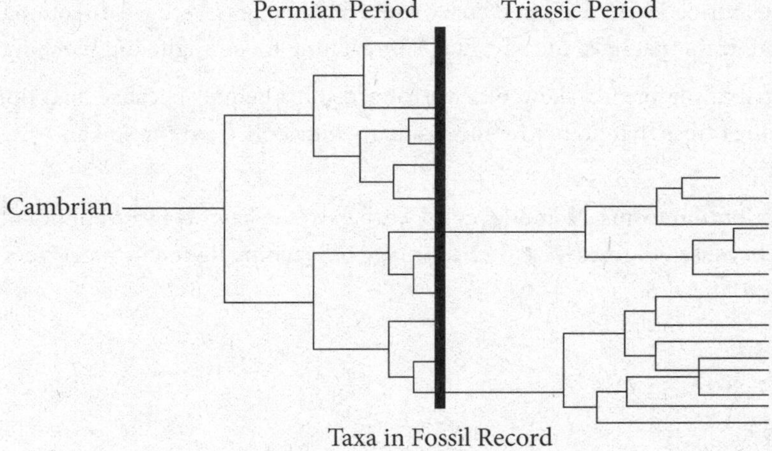

Adapted from P. Hull, "Life in the Aftermath of Mass Extinctions," *Current Biology* 25, no. 19 (October 2015): R941–R952.

The boundary between the Permian and the Triassic periods corresponds to the "Great Dying," when about 90–95% of marine life disappeared from the geological record. What is the most likely explanation for the fossil findings of the late Triassic period?

   (A) Brand-new animal life appeared after the mass extinction through spontaneous generation.

   (B) The surviving animals colonized newly available niches and underwent adaptive radiation.

   (C) Animal life never recovered after the Permian-Triassic extinction event.

   (D) A small percentage of each species survived and repopulated the environment.

7. Feathers and hair are associated with birds and mammals, respectively. The bodies of reptiles, on the other hand, are covered by scales. If birds and mammals are both descended from reptiles, scales would be an ancestral trait. Therefore, scales could have undergone modification during the evolution of birds and mammals. Which of the following best supports the hypothesis that scales represent an ancestral trait in birds and mammals?

   (A) Hair and feathers are modified skin structures and fulfill the same function as scales.

   (B) Hair and feathers are not found in present-day reptiles.

   (C) Scales present on bird legs are analogous to the scales on reptiles.

   (D) The same genes control the expression of hair, feathers, and scales.

8. Environmental scientists surveyed a population of spruce trees growing in a dense alpine forest. The population was in the path of high winds. They plotted the distribution of tree heights to investigate the selection pressures at work. Which of the following distributions of tree heights would be expected?

   (A) The distribution forms a bell curve because the selection stabilizes the population to a median height tall enough to reach for sunlight but not tall enough to be toppled by the winds.

   (B) The distribution forms a skewed curve toward tall heights because directional selection favors tall trees that have the advantage of reaching for sunlight and blocking gusts of wind.

   (C) The distribution forms a skewed curve toward short heights because directional selection favors short trees that do not require as many nutrients from the soil and are not uprooted by high winds.

   (D) The distribution forms a bimodal curve with extreme heights more abundant than average heights because tall trees have the advantage of reaching light and short trees are more resilient to the wind.

9. Avian brood parasites lay their eggs in the nests of other birds so that those other birds will incur the costs of raising their offspring. The figure below shows the results of a study of cuckoo parasitism on ashy-throated parrotbills. The x-axis shows a contrast score measuring the difference between the egg coloration of the host and parasite, with higher numbers representing greater color difference. The y-axis shows the percentage of parasite eggs rejected by the hosts.

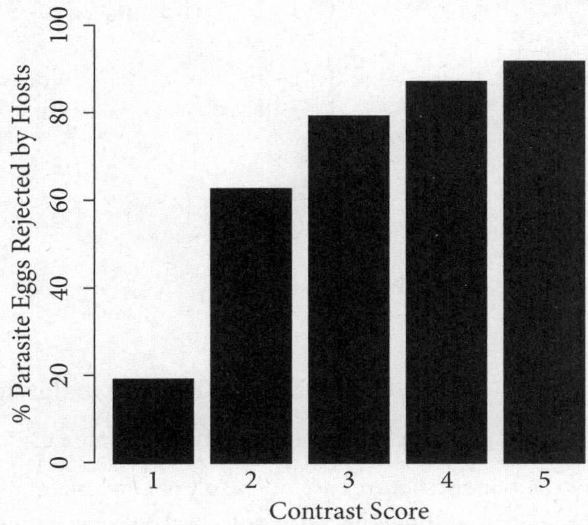

Adapted from Canchao Yang et al., "Coevolution in Action: Disruptive Selection on Egg Colour in an Avian Brood Parasite and Its Host," *PLoS ONE* 5, no. 5 (May 2010): e10816.

Which of the following conclusions is best supported by the data?

(A) There is evolution occurring in the population, but it is impossible to make specific predictions about the direction that evolution is taking.

(B) There is a stable situation in which closely-matching parasitic eggs are most likely to be raised by the host, without selection on coloration patterns.

(C) There is an increase in the amount of contrast with each generation, as evidenced by the increasing height of the bars on the right side of the graph.

(D) There is coevolution between the two species because only closely-matching parasitic eggs are likely to be raised by the host parent.

10. The figure below shows the underlying bone structure of three different animals.

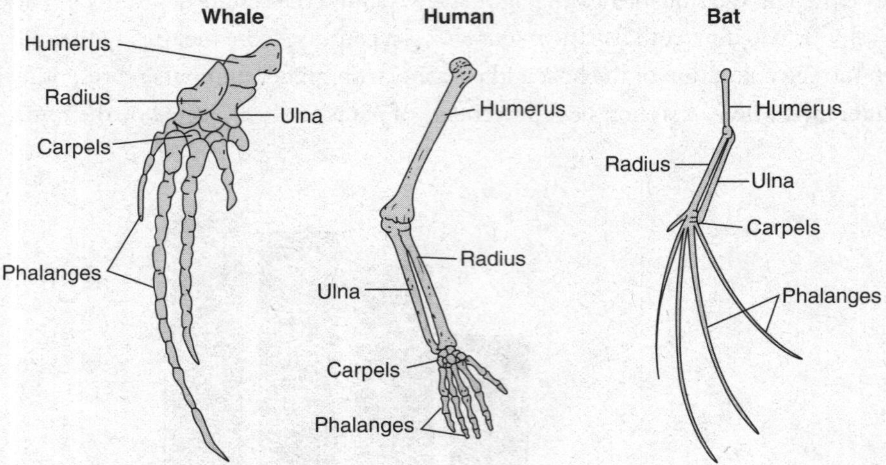

Which of the following conclusions can be reasonably drawn from this figure?

(A) All of these animals had a common ancestor that had the same underlying bone pattern.

(B) The similarities of the bone structures are the result of convergent evolution in which similar environmental forces caused different organisms to evolve similarly.

(C) The underlying bone structures are so different that we can conclude that each line evolved independently from different single-celled ancestors.

(D) The similarities in the bone structures suggest that all of these animals evolved from the whale.

11. *Archaefructus liaoningensis* is a fossil species of angiosperm (flowering plant) found in rock strata that is 125 to 130 million years old. Molecular data suggest that the divergence of angiosperms from earlier plants occurred 215 million years ago. Which of the following is a reasonable scientific approach to resolve this discrepancy?

(A) Manipulate the molecular data to make it match the fossil record.

(B) Search for fossil angiosperms in rocks older than 215 million years old.

(C) Search for fossil angiosperms in rock strata between 130 and 215 million years old.

(D) Search for fossil angiosperms in rock strata younger than 125 million years old.

The answer key to this quiz is located on the next page.

Answer Key

Test What You Already Know

1. D
2. B
3. B
4. C
5. C
6. A
7. D
8. A
9. B

Test What You Learned

1. A
2. C
3. C
4. C
5. B
6. B
7. D
8. A
9. D
10. A
11. C

REFLECTION

Test What You Already Know score: __________

Test What You Learned score: __________

Use this section to evaluate your progress. After working through the pre-quiz, check off the boxes in the "Pre" column to indicate which Learning Objectives you feel confident about. Then, after completing the chapter, including the post-quiz, do the same to the boxes in the "Post" column. Keep working on unchecked Objectives until you're confident about them all!

Pre	Post	
❑	❑	**16.1** Differentiate between types of selection affecting evolution
❑	❑	**16.2** Describe the support for the common ancestry theory
❑	❑	**16.3** Describe genetic drift and gene flow
❑	❑	**16.4** Explain speciation and extinction
❑	❑	**16.5** Explain how molecular biology and biochemistry support evolution
❑	❑	**16.6** Explain how common structures and features support evolution
❑	❑	**16.7** Describe how populations continue to evolve
❑	❑	**16.8** Investigate the effects of artificial selection on evolution
❑	❑	**16.9** Investigate allele distribution mathematical modeling

FOR MORE PRACTICE

Complete more practice online at kaptest.com. Haven't registered your book yet? Go to kaptest.com/booksonline to begin.

ANSWERS AND EXPLANATIONS

Test What You Already Know

1. D Learning Objective: 16.1

The circumstances described in the question stem provide an example of disruptive selection, in which there are selective pressures in favor of extreme phenotypes and against moderate phenotypes. This is likely to produce a body size distribution in which there are more small and large rabbits and fewer medium-sized rabbits. **(D)** is thus correct. (A) is incorrect because it suggests that only small body size is selected for, but large body size is also selected for. (B) is incorrect because it suggests that only large body size is selected for, but small body size is also selected for. (C) is incorrect because it presents an example of stabilizing selection, in which the mean is favored over the extremes, the opposite of the scenario described in the question stem.

2. B Learning Objective: 16.2

In order to be present in all three domains (Bacteria, Achaea, and Eukarya), glycolysis must be a primitive process. The presence of glycolysis in the common ancestor of the three domains would explain its prevalence among all organisms. Hence, **(B)** is correct. (A) is incorrect because the efficiency of glycolysis is not at issue in the question stem, but rather its prevalence in many organisms. (C) is incorrect because it is more likely that glycolysis appeared once during the early development of life than it is that it appeared independently many times during evolution. (D) is incorrect because glycolysis does not require the presence of mitochondria and because Bacteria and Archaea are domains of prokaryotes, which lack mitochondria.

3. B Learning Objective: 16.3

When the founders of an isolated population happen to have a particular trait in large abundance, subsequent generations of that population will likely continue to have that trait at a higher frequency due to genetic drift. Because more of the founding population had large beaks and because there were no selective pressures on beak size, it is reasonable to expect that more birds on the island will have large beaks. **(B)** is thus correct. (A) is incorrect because beak size does not experience selection in any particular direction since the question stem stated that food was available for many beak types. (C) is incorrect because the founding population had a different distribution of traits than the mainland population. (D) is incorrect because there is no selective advantage for large beak size as a result of the availability of food for any beak type; more large-beaked birds are expected due to a founder effect.

4. C Learning Objective: 16.4

According to the biological species concept, two populations are distinct species if they are unable to interbreed to produce fertile offspring. Because the resistant buffalo grass cannot interbreed with the nonresistant, the resistant grass has undergone speciation. Therefore, **(C)** is correct. (A) is incorrect because speciation does not require geographic isolation; distinct species can occupy the same geographical area. (B) is incorrect because distinct populations can look very similar while still being separate species. (D) is incorrect because the way that human beings use a population of organisms is irrelevant to whether that population is a distinct species.

5. C Learning Objective: 16.5

Although there is a lot of data in the table, the only relevant information concerns the comparison of yeast, a fungus, to other species. The table shows that yeast has fewer base substitutions in comparison to animal species (fruit flies, sea urchins, rats, and humans) than in comparison to plant species (soybeans and maize). Thus, the classification of Fungi as distinct from Plantae is appropriate and **(C)** is correct. (A) is incorrect because yeast are more closely related to humans than to maize, based on the table. (B) is incorrect because yeast are more closely related to animal species than to soybeans. (D) is incorrect because the revision was appropriate.

6. A Learning Objective: 16.6

The best way to support the hypothesis that snakes and lizards share a common ancestor would be to find more intermediate forms, such as snakes with hind limbs, in the fossil record. **(A)** is thus correct. (B) is incorrect because it provides an example of Lamarckian evolution, in which the use (or disuse) of a trait causes it to develop (or atrophy); this conception of evolution has been discredited. (C) is incorrect because it provides less solid evidence of common ancestry than does **(A)**; it is not much different than the evidence already provided in the question stem concerning embryonic hind limb buds in snakes. (D) is incorrect because this morphological similarity does not necessarily indicate a close relationship between the species; all reptiles are covered in scales.

7. D Learning Objective: 16.7

In the situation described in the question stem, parrotbills will experience a selective pressure in favor of being able to recognize the cuckoo eggs, so they can remove them from their nests and avoid the costs of parasitism. To counteract this, there will be a strong selective pressure on the cuckoos to have eggs as similar as possible to parrotbill eggs, to make it more difficult for the parrotbills to recognize the parasitic eggs. **(D)** is thus correct. (A) is incorrect because unique eggs would make it easier for parrotbills to recognize the intruders. (B) is incorrect because the size of the eggs is not at issue in the question stem; if anything, eggs that are the same size as the parrotbill eggs (thereby making them harder to recognize as parasitic) would be selected for. (C) is incorrect because a uniform, parrotbill-like appearance would be favored over greater variation in egg appearance.

8. A Learning Objective: 16.8

When researchers choose rats to breed based upon one characteristic, such as running ability, there is also selection on related traits and closely linked genes, making **(A)** correct. (B) is incorrect because the question stem states that the rats were specifically selected for running capacity and not multiple other traits. (C) is incorrect because some of the characteristics, such as body weight and susceptibility to heart failure, are distinct phenotypes. (D) is incorrect because the researchers were selecting for running ability, not reduced fitness.

9. B Learning Objective: 16.9

A consistent change in a specific direction across multiple generations suggests the existence of a selective pressure, which is making members of the species with the dominant trait fitter than those who lack the trait. **(B)** is thus correct. (A) is incorrect because random variation is equally likely to cause changes in any direction; three generations of changes in the same direction suggests something other than randomness. (C) is incorrect because it would be unlikely to see large numbers of mutations that happened to result in a higher frequency of the dominant trait. (D) is incorrect because only small populations are heavily affected by genetic drift, but the population in question is large.

Test What You Learned

1. A Learning Objective: 16.5

The most likely explanation for nearly identical pathways and identical compounds appearing in all living organisms is that these emerged early in the course of evolution in a common ancestor. **(A)** is thus correct. (B) is incorrect because convergent evolution would be unlikely to lead to the same result in all organisms. (C) is incorrect because the question stem stipulates that identical compounds (ATP, NADH, $FADH_2$) are used by all organisms. (D) is incorrect because the compounds listed are not the products of larger molecules being broken down.

2. C Learning Objective: 16.3

The question stem notes that the island features finches with large beaks, while the mainland includes finches with a wide array of beak sizes. It also notes that boats regularly carry travelers from the mainland. It is reasonable to conclude, therefore, that these boats could also bring finches with them, including small-beaked finches that can take advantage of food sources underutilized by the large-beaked population. This would be an example of gene flow. **(C)** is thus correct. (A) is incorrect because visitors feeding the finches would not necessarily cause the population to diversify. (B) is incorrect because a novel mutation is not necessary to introduce small beaks into the population; small beaks already exist on the mainland. (D) is incorrect because there is no connection suggested in the question between the destruction of the finch habitat and a selective pressure for smaller beaks.

3. C Learning Objective: 16.9

With such similar answer choices, this question can best be answered using the process of elimination. As noted in the question stem, TT and TC represent specific alleles, so the values in the table are allelic frequencies. This eliminates (B) and (D), which wrongly suggest that these are genotypic frequencies. In Hardy-Weinberg calculations, allelic frequencies are represented by the variables p and q, making (A) incorrect and **(C)** the correct answer.

4. C Learning Objective: 16.2

The function of photolyases, as described in the question stem, suggests that they would emerge relatively early in evolution, while cryptochromes would evolve later, most likely in a common ancestor of plants and animals.

The similarities between the two suggest that cryptochromes may have come about due to mutations in the genes for photolyases. **(C)** is thus correct. (A) is incorrect because the shared sequences between the two proteins suggest they are homologous, sharing a common lineage. (B) is incorrect because there is not extensive convergence between the functions of photolyases and cryptochromes. (D) is incorrect because not all light receptor proteins have identical structures.

5. B Learning Objective: 16.8

Artificial selection is typically used to select for polygenic traits (ones with multiple genetic influences) that experience a wide range of variation. The most straightforward way to do this is to breed only those animals that possess the desired traits (or that possess them to the greatest extent). Selectively breeding cows with above-average milk production would ensure that milk production increases in subsequent generations. **(B)** is thus correct. (A) is incorrect because it is unlikely that a characteristic as complex as milk production can be tied to a single gene. (C) is incorrect because milk production is unlikely to be governed only by a single gene with a simple Mendelian inheritance pattern. (D) is incorrect because increasing genetic diversity will not necessarily lead to higher milk production.

6. B Learning Objective: 16.4

Mass extinctions, such as the one described in the question stem, open a large number of niches for exploitation by the surviving species, often leading to adaptive radiation. **(B)** is thus correct. (A) is incorrect because organisms are not spontaneously generated; they evolve from ancestral organisms that survive and reproduce. (C) is incorrect because animal life eventually recovered; otherwise animals would not exist today. (D) is incorrect because some species went extinct, as indicated by the figure.

7. D Learning Objective: 16.6

If the same genes controlled the expression of scales, feathers, and hair, it would suggest that all three shared a common origin, the hypothesis presented in the question stem. **(D)** is thus correct. (A) is incorrect because distinct structures with analogous functions can be the result of convergent evolution and are not necessarily the product of divergence from a common ancestor. (B) is incorrect because it is a negative claim that provides no support for the hypothesis. (C) is incorrect because analogous structures often result from convergent evolution, so this does not necessarily suggest common ancestry.

8. A Learning Objective: 16.1

The question stem describes two selective pressures on the trees: high winds will select against any tree significantly taller than the others, which will be more susceptible to being blown down, while dense foliage selects against any tree significantly shorter than the others, which will be unable to get enough sunlight for photosynthesis. Because average heights are favored, the distribution should form a bell curve, as suggested in **(A)**. (B) is incorrect because tall heights make the trees susceptible to high winds that can uproot them. (C) is incorrect because short heights make it harder for the trees to get enough sunlight. (D) is incorrect because both tall and short trees have disadvantages, meaning that average values are favored over the extremes.

9. D Learning Objective: 16.7

The results of the study show that eggs with greater contrast are far more likely to be rejected than those with less contrast. This suggests that there is coevolution between the parrotbill and cuckoo populations, with a constant selective pressure for cuckoo eggs to resemble parrotbill eggs. **(D)** is thus correct. (A) is incorrect because there is an obvious selective pressure on the cuckoos to have eggs that resemble those of the parrotbills, so the effects are predictable. (B) is incorrect because there is selection on coloration: cuckoo eggs that resemble parrotbill eggs are favored over those that do not. (C) is incorrect because the taller bars indicate that contrasting eggs are more likely to be rejected; contrast should actually decrease with subsequent generations.

10. A Learning Objective: 16.2

(A) The bone structures shown are examples of homologous structures. The underlying pattern of bones suggests common ancestry. (B) is incorrect because convergent evolution produces structures that have a similar form but often different underlying structures. It is unlikely that separate evolutionary paths would all lead to bones, let alone the pattern of bones shown. Therefore, (C) is not correct. (D) is incorrect because there is no way to determine which of the species was the ancestral species from the figure shown. In fact, these species all shared a common ancestor that was none of the animals shown.

11. C **Learning Objective:** 16.6

(C) The molecular evidence suggests that there are undiscovered fossil angiosperms that are older than 130 million years old but younger than 215 million years old. Looking at rocks that span that range would be the best strategy for testing the hypothesis that angiosperms diverged earlier than 130 million years ago. (A) is incorrect and unethical. Scientists need to rely on data to drive conclusions rather than manipulate data to get the answer they want. (B) is incorrect because there is no reason to believe there would be any angiosperm fossils older than 215 million years old. Although scientists would likely find fossil angiosperms in rock strata younger than 125 million years old, this discovery would do nothing to resolve the difference between the fossil record and the molecular data. Therefore, (D) is not correct.

PART 6

Interactions of Life

CHAPTER 17

Biodiversity

LEARNING OBJECTIVES

In this chapter, you will review how to:

17.1 Explain how to read and test a phylogenetic tree

17.2 Explain how to read and test a cladogram

17.3 Investigate evolutionary changes with cladograms

TEST WHAT YOU ALREADY KNOW

1. The figure below shows a phylogenetic tree for birds.

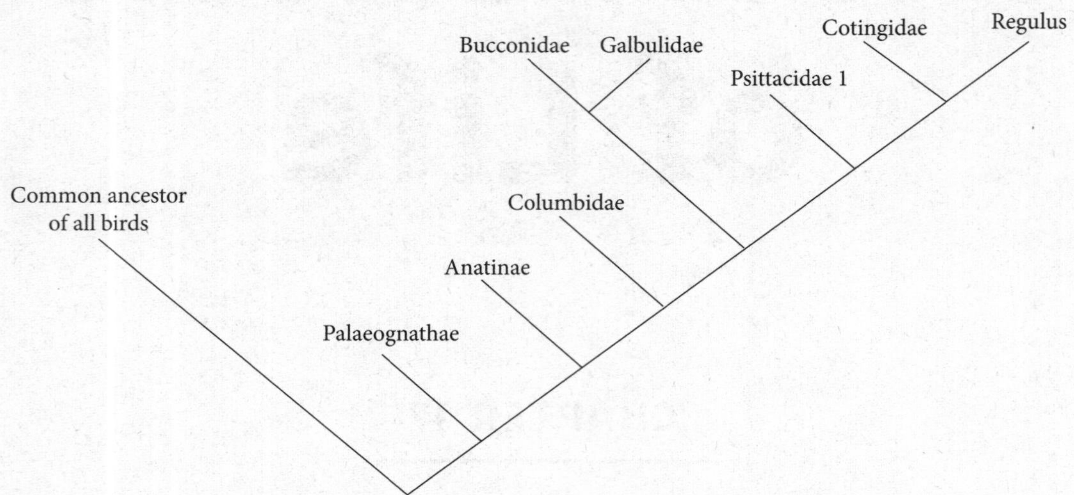

Adapted from Stephen A. Smith, Joseph W. Brown, and Cody E. Hinchliff, "Analyzing and Synthesizing Phylogenies Using Tree Alignment Graphs," *PLoS Comput Biol* 9, no. 9 (September 2013): e1003223.

Based on this tree, which of the following groups are most closely related?

(A) *Psittacidae* 1 and *Cotingidae*

(B) *Bucconidae* and *Galbulidae*

(C) *Palaeognathae* and *Regulus*

(D) *Anatinae* and *Columbidae*

2. The figure below shows a cladogram of fishes across time. The abbreviations at the top represent time periods and the bar at the bottom represents time.

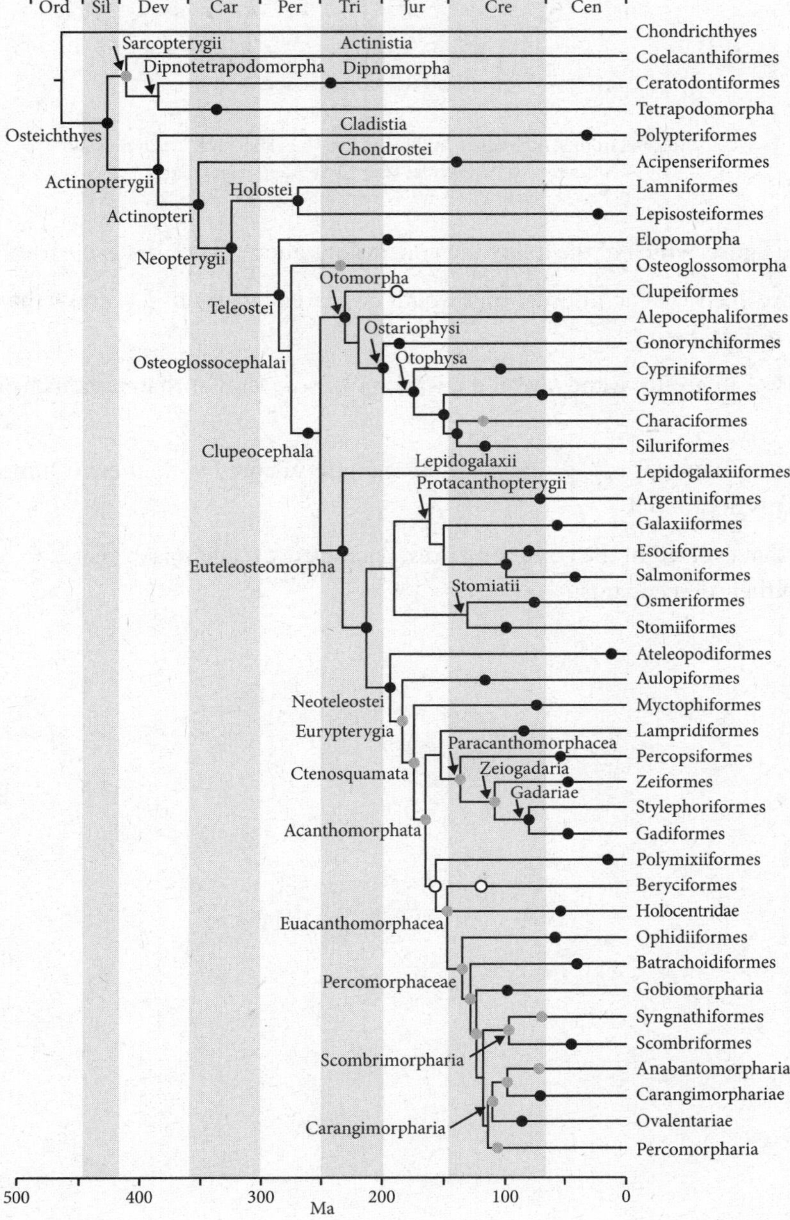

Adapted from Ricardo Betancur-R. et al., "The Tree of Life and a New Classification of Bony Fishes," *PLOS Currents* (April 2013).

Which of the following conclusions is most justified based on the cladogram?

(A) All of the *Sarcopterygians* are extinct.

(B) The most distantly related clades are the *Percomorpharia* and the *Ovalentariae.*

(C) The *Sarcopterygians* are a more diverse group than the *Actinopterygians.*

(D) An *Osteichthyean* was the common ancestor of the *Sarcopterygii* and the *Actinopterygii.*

Interactions

3. The figure below shows a simple cladogram with letters representing taxonomic groups.

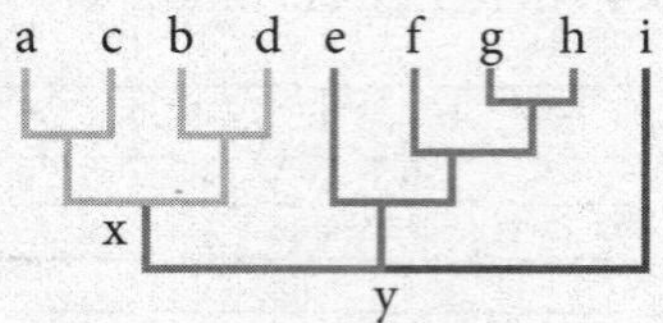

Adapted from S. A. Smith, J. W. Brown, and C. E. Hinchliff, "Analyzing and Synthesizing Phylogenies Using Tree Alignment Graphs," *PLoS Comput Biol* 9, no. 9 (September 2013): e1003223.

Based on the figure, which of the following conclusions about the clades is most valid?

(A) It is more likely that an unusual trait would be shared by groups *f*, *g*, and *h* than by groups *a*, *c*, and *e*.

(B) Organisms in groups *g* and *h* would always look more similar than organisms in groups *f* and *h*.

(C) Organisms in groups *a*, *c*, *b*, and *d* are more highly adapted to their environment than organisms in group *x*.

(D) A trait that evolved in the common ancestor for clades *a* and *d* is expected to be present in all of their descendants.

Answers to this quiz can be found at the end of this chapter.

PHYLOGENETIC TREES

High-Yield

17.1 Explain how to read and test a phylogenetic tree

Biologists maintain that all living organisms are descended from a single common ancestor, that all the great diversity of species existing today share a common lineage that can be traced back to the origin of life. Not only are all eukaryotic organisms descended from a single ancestor, but that first eukaryote is also a descendant of the first prokaryotes. Because of this fact, the emergence of biological diversity can be likened to the growth of a tree, in which the multitude of the tree's branches are ultimately derived from its trunk. This structural analogy explains why biologists sometimes use diagrams known as *phylogenetic trees* to model biodiversity.

✔ AP Expert Note

Until recently, the first level of classification of organisms was the kingdom. There were five recognized kingdoms (Monera, Protoctista, Fungi, Plantae, and Animalia). Now all of life is first divided into three domains: Archaea, Bacteria, and Eukarya.

A taxon is a grouping of related organisms, such as a **species**, **genus**, **phylum**, or **kingdom**. Phylogenetic trees show hypotheses of evolutionary relationships among various taxa, indicating common lines of descent from shared ancestors. The example on the following page shows evolutionary relationships among the phyla of kingdom Animalia, the animals.

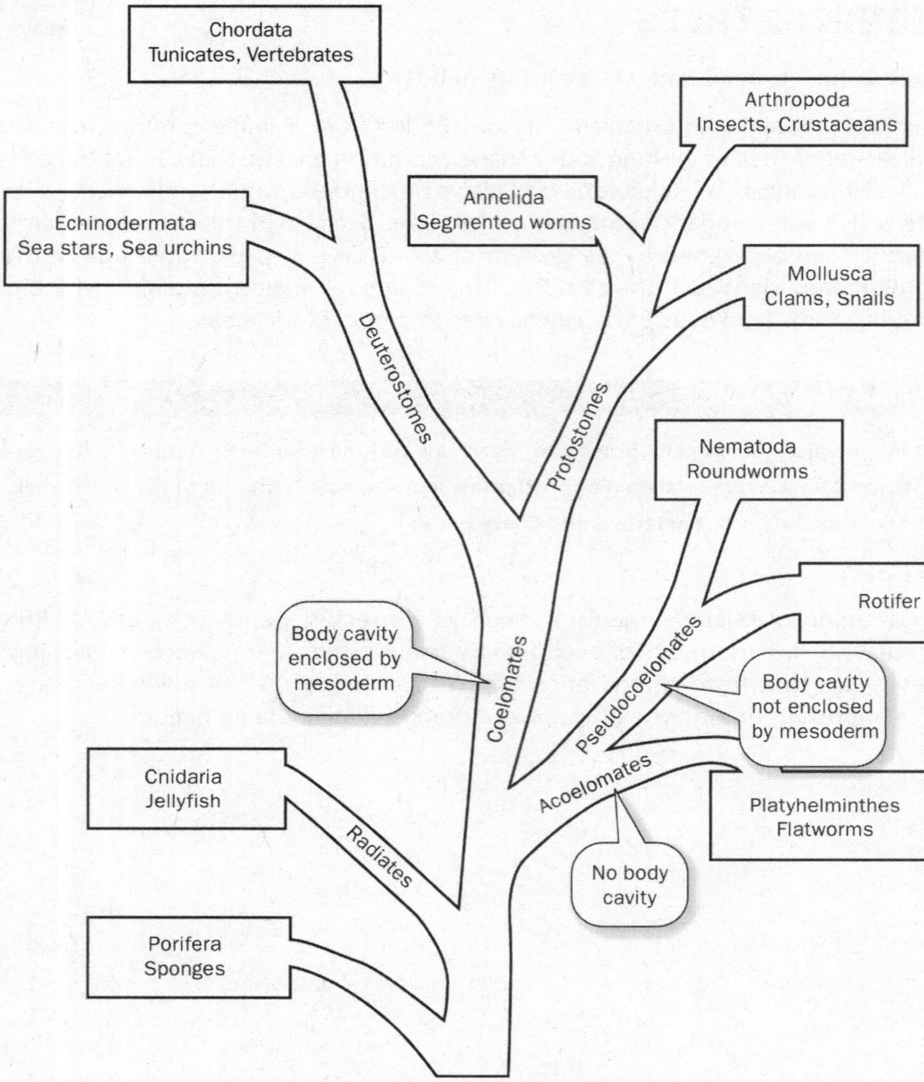

Chordata
Tunicates, Vertebrates

Arthropoda
Insects, Crustaceans

Annelida
Segmented worms

Echinodermata
Sea stars, Sea urchins

Mollusca
Clams, Snails

Deuterostomes

Protostomes

Nematoda
Roundworms

Rotifer

Body cavity
enclosed by
mesoderm

Coelomates

Pseudocoelomates

Body cavity
not enclosed
by mesoderm

Cnidaria
Jellyfish

Acoelomates

Platyhelminthes
Flatworms

Radiates

No body
cavity

Porifera
Sponges

Phylogenetic Tree of Major Animal Groups

✔ AP Expert Note

The names and characteristics of specific biological taxa, as well as the relationships between these taxa, are not tested on the AP Biology exam.

Some phylogenetic trees also give an indication of evolutionary timescale, by keeping the distances between branchings in the tree proportional to the amount of time between species divergences. On a phylogenetic tree, as you move from the root (ancestor) to the tips of the branches (descendants), you move forward in time. Each branch represents speciation. The following example shows the lineage of primates.

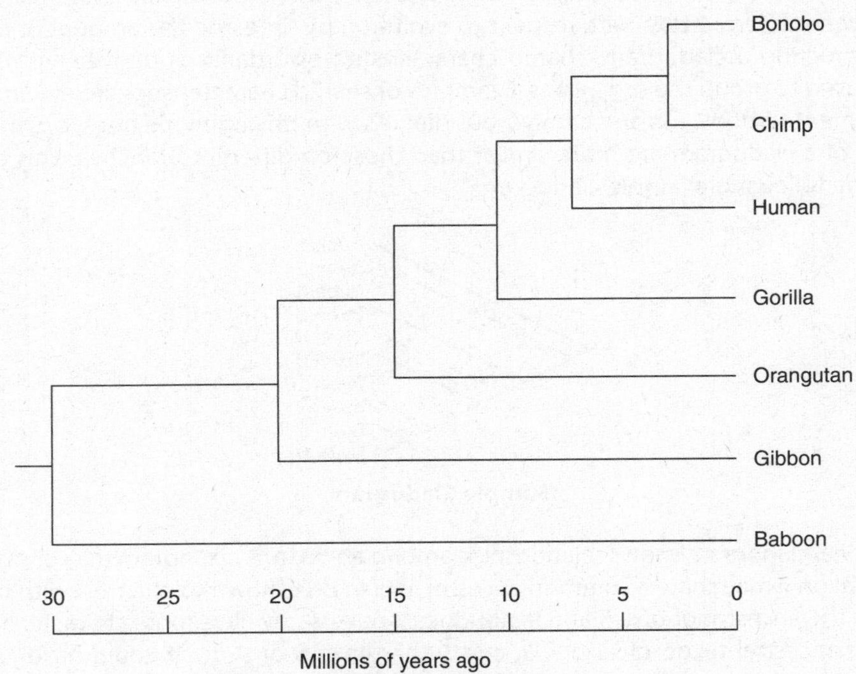

Phylogenetic Tree of Primates Over Time

CLADOGRAMS

17.2 Explain how to read and test a cladogram

Also used to represent hypotheses on evolutionary relationships in a group of organisms, a *cladogram* is similar in appearance to a phylogenetic tree. Cladograms demonstrate similarities between types of organisms with respect to a common ancestor, but they do not show relationships of descent between different taxa with respect to evolutionary time and the amount of change with time. In constructing a cladogram, shared characteristics of organisms derived from a common ancestor are used to group the organisms. Examples of shared characteristics are eukaryotic, warm-blooded, segmented body, fur, and carnivorous diet. As with phylogenetic trees, organisms on the same branch of a cladogram are more similar than those on different branches. This can be illustrated with the following example cladogram.

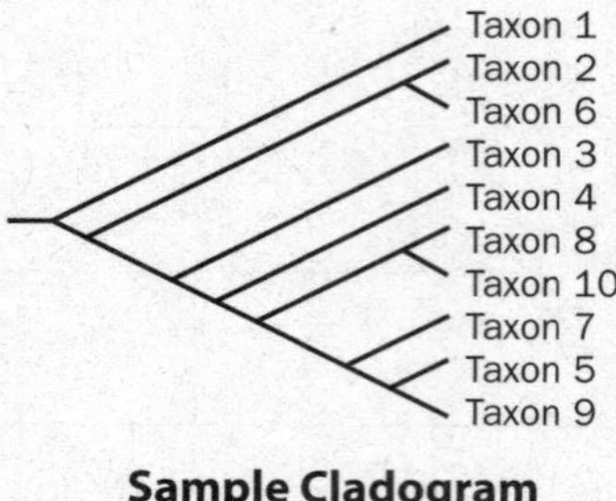

Sample Cladogram

This hypothetical cladogram shows relationships among ten taxa. A cladogram may or may not show a time scale. Taxon 7 may share a common ancestor, which isn't shown on the tree, with taxa 5 and 9. Taxon 7 is not the ancestor of taxa 5 and 9, nor does it necessarily have more characteristics in common with the ancestral taxon of taxa 7, 5, and 9 than taxa 5 or 9 do. It could have just as many *anagenetic* changes (evolutionary events without speciation) along its branch as *cladogenetic* changes (evolutionary events that lead to speciation). Taxon 4 is more closely related to taxa 8, 10, 7, 5, and 9 than it is to taxon 3. Even though taxon 3 is close to taxon 4 on the tree, it is not on the same branch that includes taxa 4, 8, 10, 7, 5, and 9. Likewise, taxon 2 is more closely related to taxon 6 than it is to taxon 1. Taxa 2 and 6 belong to the same branch that does not include taxon 1.

The following is an example of a cladogram showing the relationships among primates.

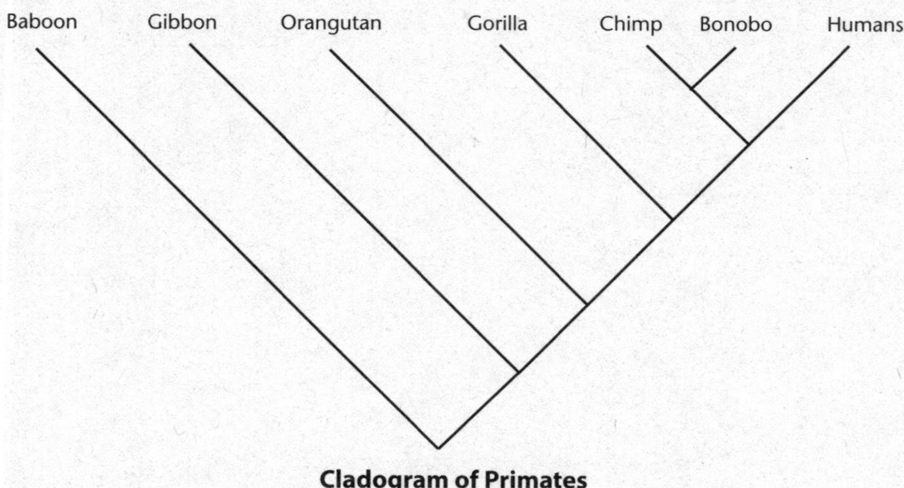

Cladogram of Primates

AP BIOLOGY LAB 3: COMPARING DNA SEQUENCES TO UNDERSTAND EVOLUTIONARY RELATIONSHIPS WITH BLAST INVESTIGATION

17.3 Investigate evolutionary changes with cladograms

Since 1990, scientists have been working to develop a comprehensive library of genes from several species, including humans, mice, fruit flies, and *E. coli.* BLAST (Basic Local Alignment Search Tool) is a powerful bioinformatics program that helps scientists to compare genes from different organisms cataloged in this library. Information derived from BLAST can, therefore, show scientists the evolutionary relationships between different organisms.

In this investigation, you will use BLAST to compare several genes from different organisms and then construct a cladogram to represent the evolutionary relationships among these different species. While cladograms can be constructed using many different factors, including the presence of different morphological traits (wings, gills, etc.), in this investigation you will use DNA evidence to establish the similarity among different organisms.

Locating and sequencing genes in different organisms not only helps us to better understand evolutionary relationships among organisms, but it can also provide important insights into genetic diseases. Species with smaller genomes, such as the fruit fly and mouse, are easier for scientists to study than humans. When scientists locate a disease-causing gene in a fruit fly or mouse, they can then use BLAST to see if there is a similar sequence in the human genome.

Using BLAST requires several specific steps and an abundance of information, none of which you need to memorize for the AP Biology exam. Rather, you should be able to apply similarities and differences in morphology and genetics to determine the evolutionary relationships between different species. Drawing and analyzing cladograms (as well as phylogenetic trees) is a skill you should practice and be able to perform successfully.

Let's use some data from an example in this investigation to construct a cladogram based on genetic data. In humans, the GAPDH gene produces a protein that catalyzes a step in glycolysis. The following table shows the percentage similarity between the GAPDH gene and protein in humans compared with four different species.

Species	**Gene Similarity Percentage**	**Protein Similarity Percentage**
Chimpanzee (*Pan troglodytes*)	99.6%	100%
Dog (*Canis lupus familiaris*)	91.3%	95.2%
Fruit fly (*Drosophila melanogaster*)	72.4%	76.7%
Roundworm (*Caenorhabditis elegans*)	68.2%	74.3%

Just as with the cladogram shown earlier in this chapter, we can use this data to construct a cladogram that tells us how these species are related to each other based on GADPH. There are five species being compared here, so our cladogram will have five branches.

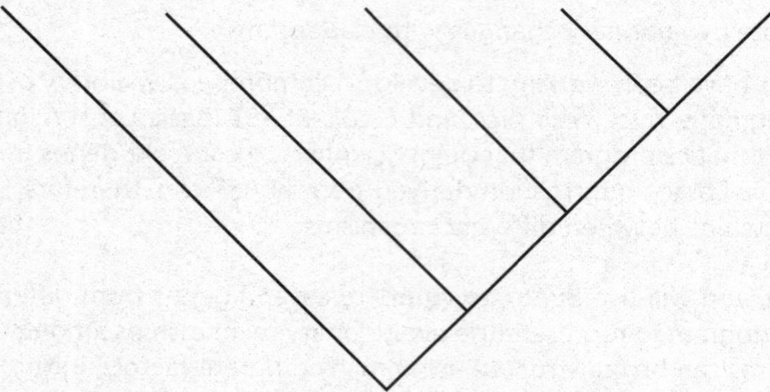

Reading the table, we see that humans and chimpanzees have the most similarities in their GADPH genes and proteins. Therefore, humans and chimpanzees would be placed closest together on our cladogram as shown.

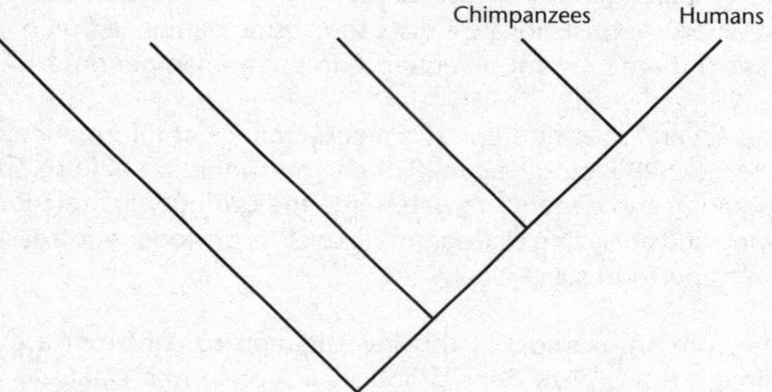

We can make deductions for the remaining three species based on their similarity to humans. Roundworms' GADPH genes have the least in common with humans, so roundworms should be placed farthest from humans on the cladogram. Dogs are closer to humans than fruit flies but farther away than chimpanzees.

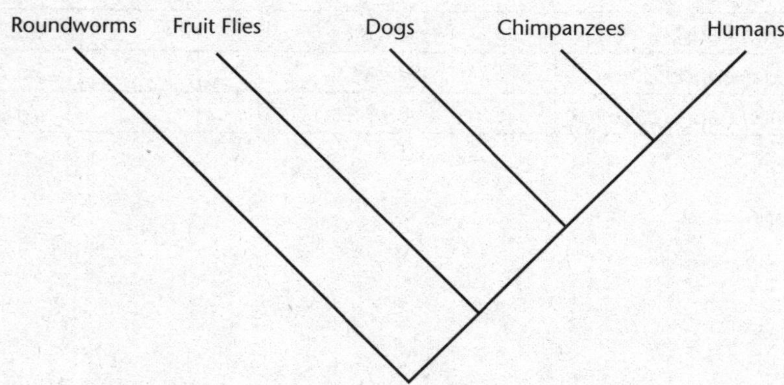

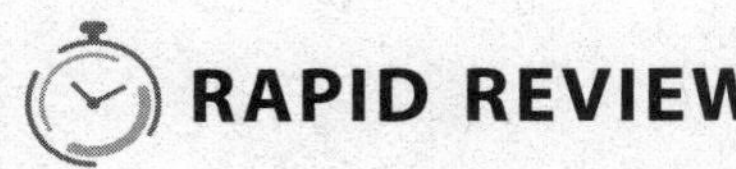

RAPID REVIEW

If you take away only 3 things from this chapter:

1. A phylogenetic tree is a diagram indicating evolutionary relationships between taxa of organisms. It depicts an ancestral species with other species branching off from it. More closely related taxa are nearer to one another in the diagram. Phylogenetic trees are intended to show ancestral connections between species and some include information about the timescale of species divergences.
2. A cladogram is typically a simplified version of a phylogenetic tree, which shows degrees of relatedness without necessarily offering information about the ancestral relationships between species or the timescale of divergences.
3. Phylogenetic trees and cladograms can show anagenetic evolutionary changes (which do not lead to speciation), as well as cladogenetic evolutionary changes (which do).

TEST WHAT YOU LEARNED

1. The figure below shows a recent reptile cladogram developed using molecular, morphological, and paleontological data. The *Gekkota*, *Anguimorpha*, *Scincidae*, *Teiidae*, and *Iguania* are lizard groups. The *Mosasauria* is a group of extinct marine reptiles. The *Serpentes* are the snakes.

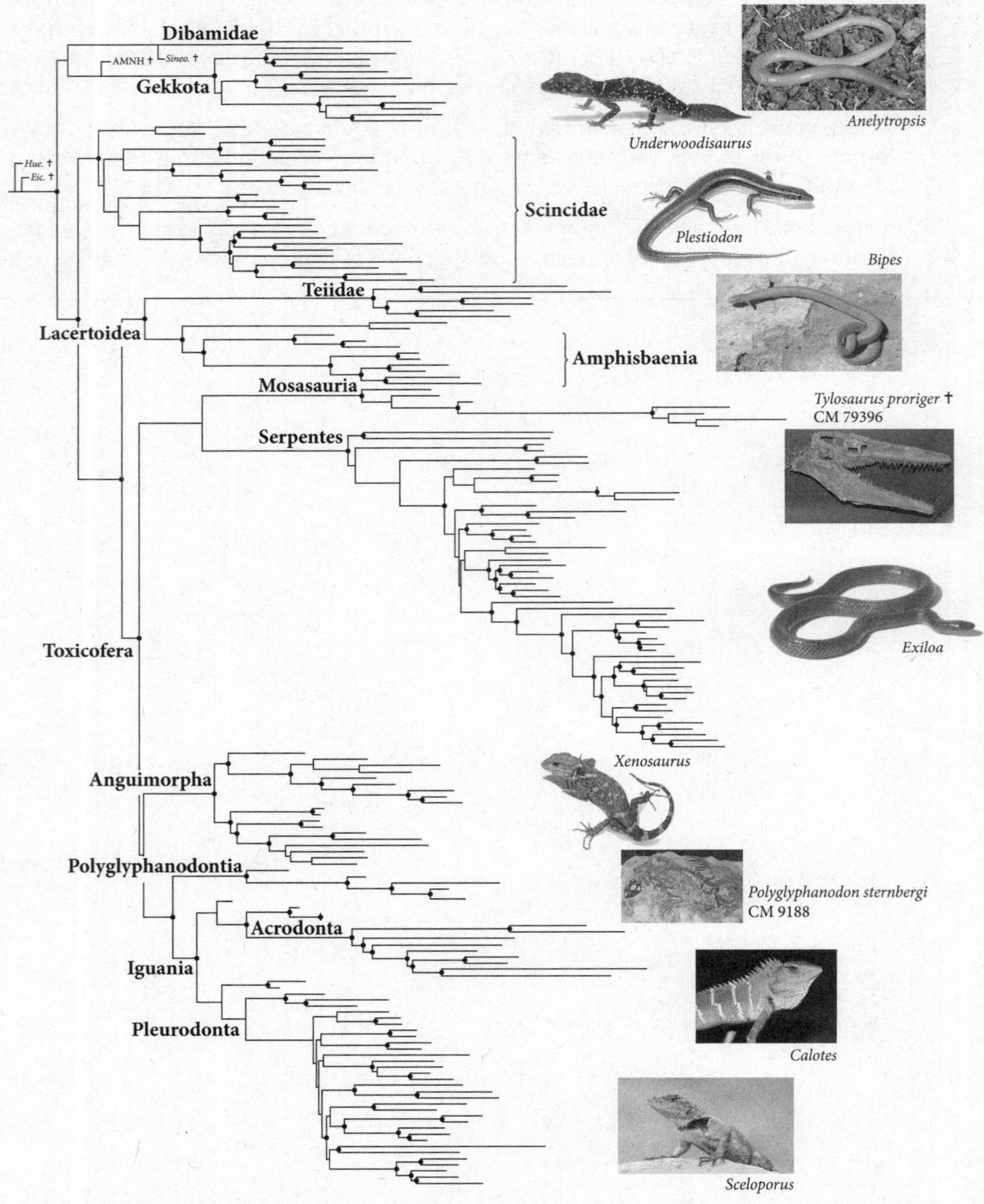

Adapted from Tod W. Reeder et al., "Integrated Analyses Resolve Conflicts over Squamate Reptile Phylogeny and Reveal Unexpected Placements for Fossil Taxa," *PLoS ONE* 10, no. 3 (March 2015): e0118199.

Which of the following statements is best supported by the cladogram?

(A) The snakes are more closely related to the *Iguania* than to the *Gekkota*.

(B) The *Scincidae* is polyphyletic, containing descendants from multiple different ancestors.

(C) All of the lizards are more closely related to each other than to the snakes and mosasaurs.

(D) Extinct groups branch toward the left whereas modern groups are clustered toward the right of the cladogram.

2. The figure below shows a phylogeny for amphipods, a type of crustacean. There is information on two traits, temperature tolerance and tube building. The hour values represent survival times at increased temperature, with larger circles representing greater temperature tolerance.

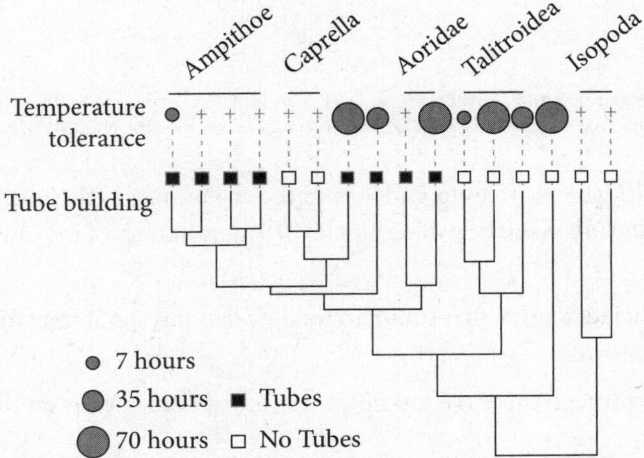

Adapted from Rebecca J. Best and John J. Stachowicz, "Phylogeny as a Proxy for Ecology in Seagrass Amphipods: Which Traits Are Most Conserved?" *PLoS ONE* 8, no. 3 (March 2013): e57550.

Based on this data, which of the following conclusions is most likely to be accurate?

(A) Tube building was a trait possessed by the common ancestor of all of these groups but was secondarily lost in many groups, including *Isopoda*.

(B) Tube building evolved in a common ancestor of the *Ampithoe*, *Caprella*, and *Aoridae*, but was secondarily lost in the *Caprella*.

(C) Temperature tolerance was a trait possessed by the common ancestor of all of these groups but was secondarily lost in some groups, including *Isopoda*.

(D) Temperature tolerance evolved in a common ancestor of the *Caprella*.

Interactions

3.

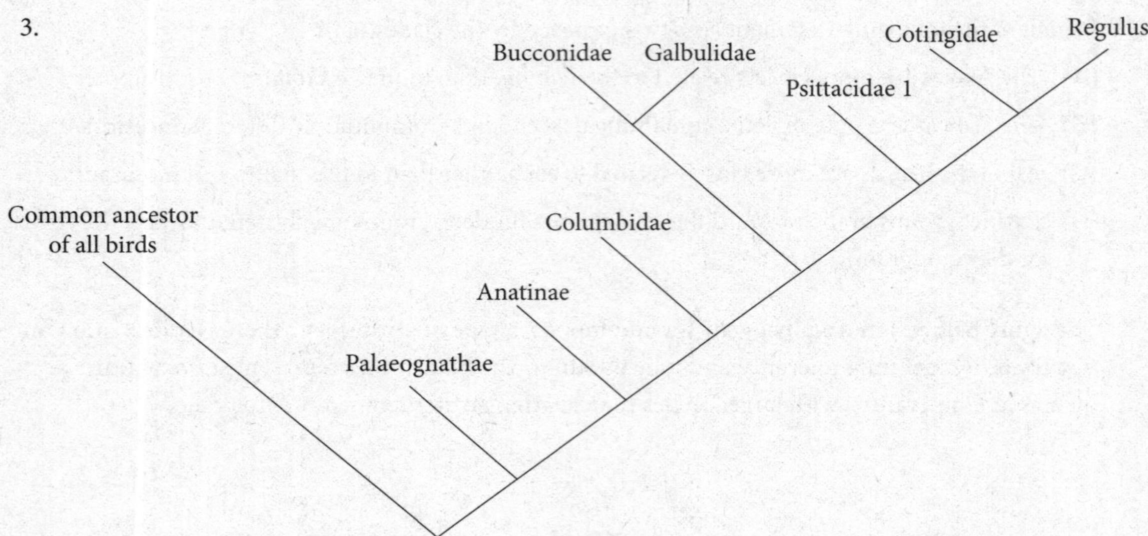

Adapted from S. A. Smith, J. W. Brown, and C. E. Hinchliff, "Analyzing and Synthesizing Phylogenies Using Tree Alignment Graphs," *PLoS Comput Biol* 9, no. 9 (September 2013): e1003223.

Researchers use phylogenetic trees to understand the evolutionary relationships between different species. Which of the following best describes the leftmost branch of the phylogenetic tree shown above?

(A) This branch includes a title that summarizes the relationships among the other branches of the phylogenetic tree.

(B) This branch represents all of the unknown and unresolved species on the bird phylogenetic tree.

(C) This branch indicates the most ancestral group on the tree, which evolved from non-bird groups such as reptiles.

(D) This branch represents extinct groups of birds that cannot be included in the main phylogenetic tree.

Questions 4–5

The question "Which came first, the chicken or the egg?" is presented by many people as an unsolvable conundrum. The phylogenetic tree shown below is supported by molecular analysis of many different proteins.

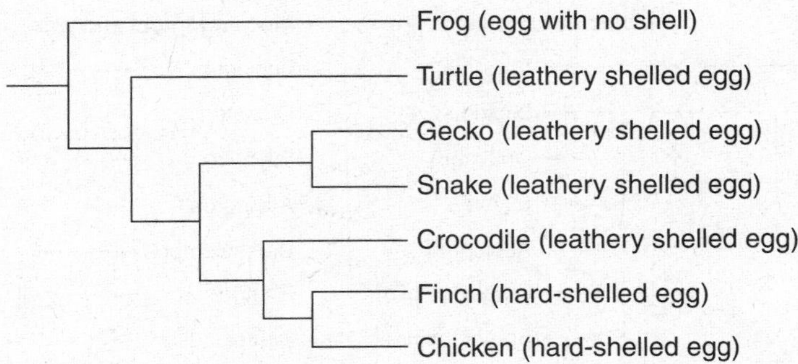

4. Which of the following provides the best scientific answer to the question "Which came first, the chicken or the egg"?
   (A) Since all of the animals shown on the tree make eggs, the egg must have evolved after the animals diverged from a common ancestor.
   (B) The chicken and the egg evolved at the same time because you cannot have one without the other.
   (C) Since all of the existing species on the tree make eggs, egg production must be an ancestral trait and must have existed in the common ancestor. Therefore, the egg came before the chicken.
   (D) It is impossible to draw a conclusion based on the information presented.

5. The crocodile, finch, and chicken share a more recent common ancestor than do any of the other species shown on the tree. Which of the following is the most likely description of the type of egg their most recent common ancestor would have made?
   (A) The most recent common ancestor of the crocodile, finch, and chicken likely made an egg without a shell because frogs are ancestors to all of them and made that type of egg.
   (B) The ancestor of the crocodile, finch, and chicken likely made a hard-shelled egg because two of the three make hard-shelled eggs.
   (C) The ancestor of the crocodile, finch, and chicken likely made an egg with a leathery covering because species more distantly related to all of them also produced leathery shelled eggs.
   (D) It is impossible to draw that kind of conclusion from the data presented.

6. The diagram below shows two phylogenetic trees that represent the hypothetical evolutionary histories of fireflies. The tree on the left was based on molecular analysis of DNA sequences. The tree on the right was based on morphological characteristics. Lines between the two trees connect individual species.

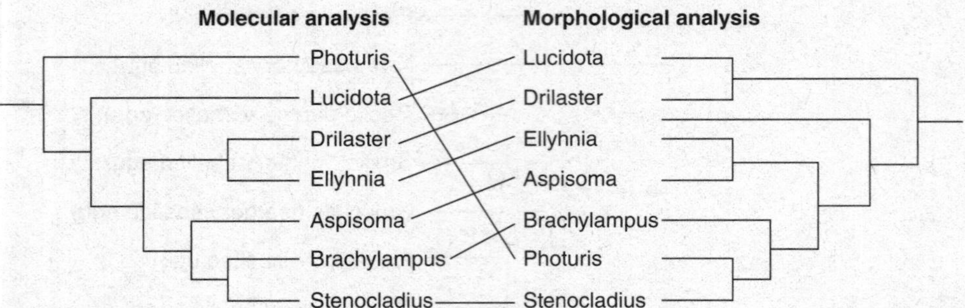

Which of the following is a reasonable conclusion from the phylogenetic trees shown?

(A) No significant differences exist between the phylogenetic relatedness predicted by the molecular analysis of DNA sequences and the morphological analysis.

(B) Molecular analysis of DNA sequences incorrectly predicted the phylogenetic relatedness of firefly species.

(C) Morphological analysis incorrectly predicted the phylogenetic relatedness of firefly species.

(D) Significant differences in the hypothesized phylogenies exist. Additional morphological and molecular studies should be done to resolve the positions of different species.

Interactions

7. The figure below shows phylogenetic relationships of several groups of bees and their social behavior.

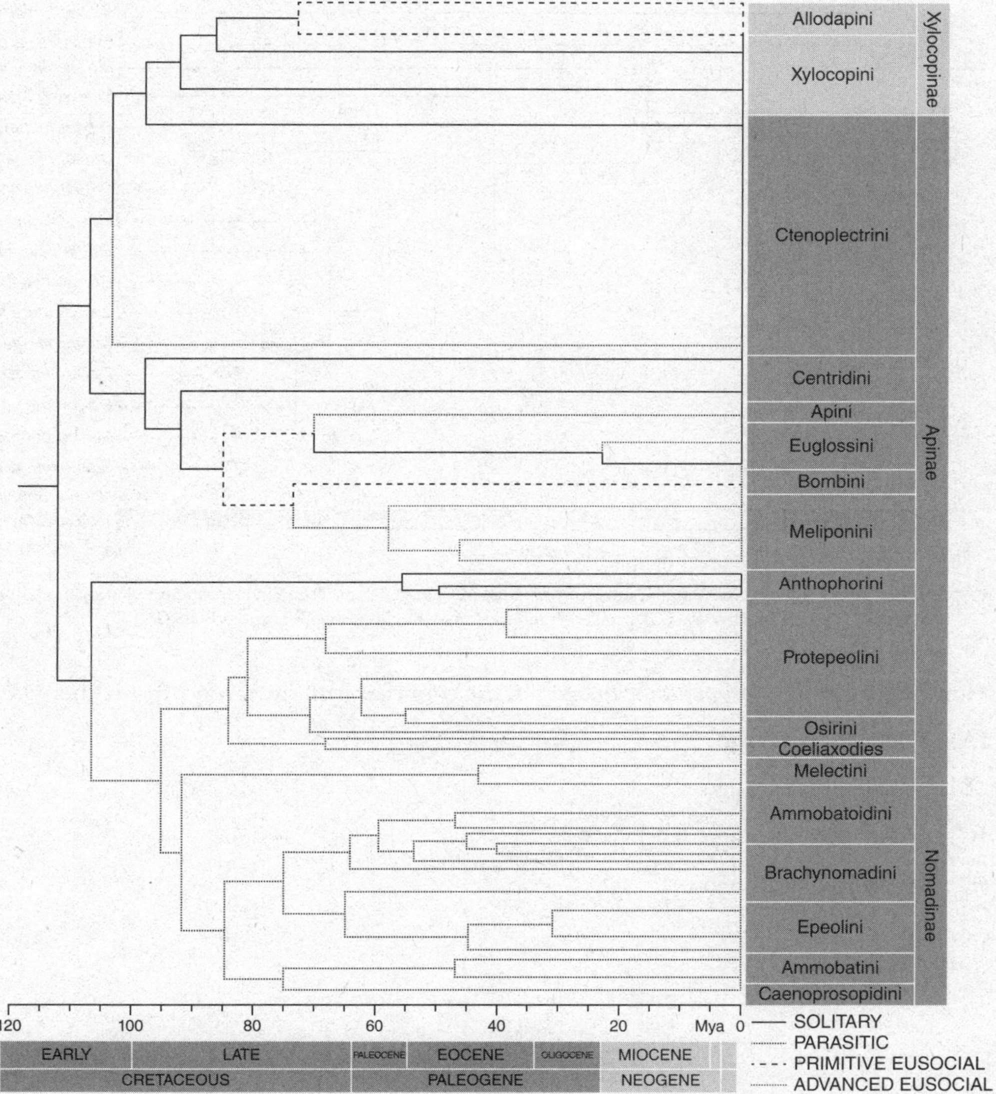

Adapted from Cardinal S, Danforth BN. "The Antiquity and Evolutionary History of Social Behavior in Bees." *PLoS ONE* 6(6)(2011): e21086.

Which of the following conclusions can most reasonably be drawn from the information in the cladogram?

(A) Parasitism arose only once and is present in all descendants of the parasitic common ancestor.

(B) Advanced eusocial bees evolved from primitively eusocial bees.

(C) Parasitic bees evolved from advanced eusocial bees.

(D) Most solitary bees evolved from a eusocial common ancestor.

8. Consider the phylogenetic tree of flies, shown below. The numbers on the phylogeny represent times at which divergence occurred.

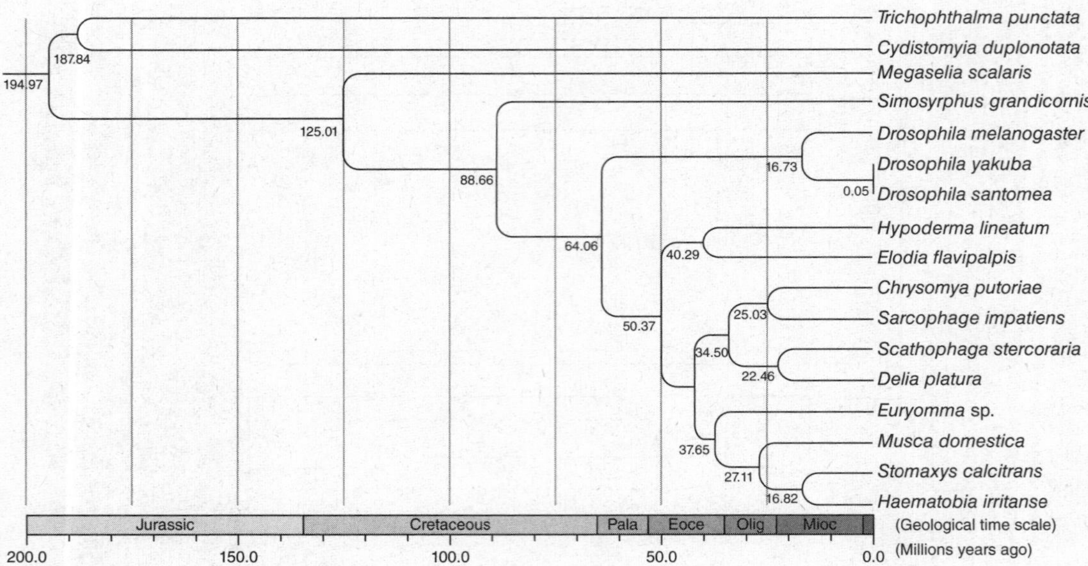

Adapted from Ding S, Li X, Wang N, Cameron SL, Mao M, Wang Y, et al. "The Phylogeny and Evolutionary Timescale of Muscoidea (Diptera: Brachycera: Calyptratae) Inferred from Mitochondrial Genomes." *PLoS ONE* 10(7)(2015): e0134170.

How long ago did *Drosophila melanogaster*, the common fruit fly, diverge from other *Drosophila*?

(A) 5,000,000 years ago

(B) 16,780,000 years ago

(C) 64,060,000 years ago

(D) 88,660,000 years ago

Interactions

9. Consider the figure below, which shows two possible cladograms of the early gnathosomes (jawed vertebrates).

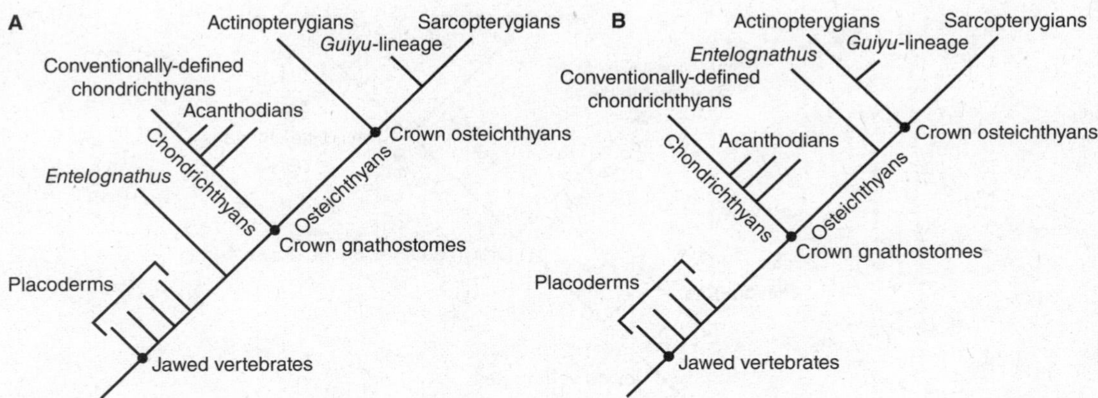

Adapted from Qiao T, King B, Long JA, Ahlberg PE, Zhu M. "Early Gnathostome Phylogeny Revisited: Multiple Method Consensus." *PLoS ONE* 11(9)(2016): e0163157.

Which of the following best describes a major difference in the evolutionary relationships suggested by the cladograms?

(A) In cladogram A only, the crown gnathostomes and crown osteichthyans both diverged into more than one clade.

(B) In cladogram A only, the crown gnathostomes were a common ancestor for the conventionally-defined chondrichthyans.

(C) In cladogram B only, *Entelognathus* forms its own clade.

(D) In cladogram B only, the *Guiyu* lineage shares a closer ancestor with the actinopterygians than with the sarcopterygians.

10. A researcher identifies a characteristic found in crown osteichthyans but not in the placoderms. Examine the cladogram of early Gnathostomes shown below.

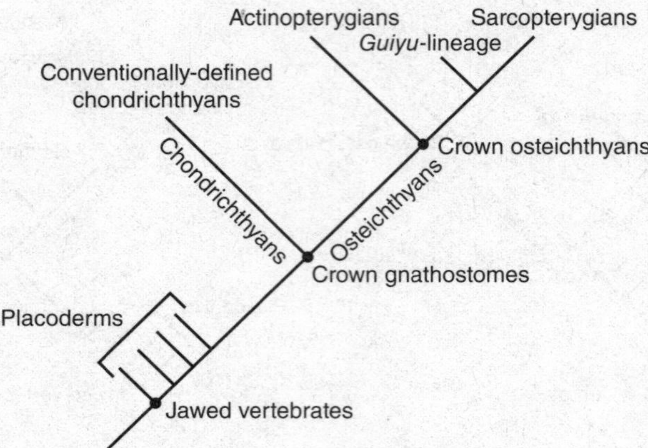

Adapted from Qiao T, King B, Long JA, Ahlberg PE, Zhu M "Early Gnathostome Phylogeny Revisited: Multiple Method Consensus." *PLoS ONE* 11(9)(2016): e0163157.

Assuming the simplest possible explanation, how many additional clades named on the cladogram would likely share a unique characteristic found in crown osteichthyans?

(A) 0 (B) 1 (C) 2 (D) 3

11. While developing a cladogram from genetic data, a researcher discovers that one organism, organism X, has different physical characteristics from those clustered nearby on the cladogram. All of the nearby organisms originate from the same node as organism X. Which of the following is most likely true regarding organism X?

(A) Organism X must have diverged long before the nearby organisms.

(B) Organism X does not share a common ancestor with the nearby organisms.

(C) Organism X may have diverged in appearance, but its recent common ancestry with nearby organisms suggests they are still closely-related.

(D) It is likely that organism X has been placed incorrectly on the cladogram because those placed close together should have similar characteristics.

12. The figure below shows a theoretical phylogenetic tree for the early gnathosomes (jawed vertebrates).

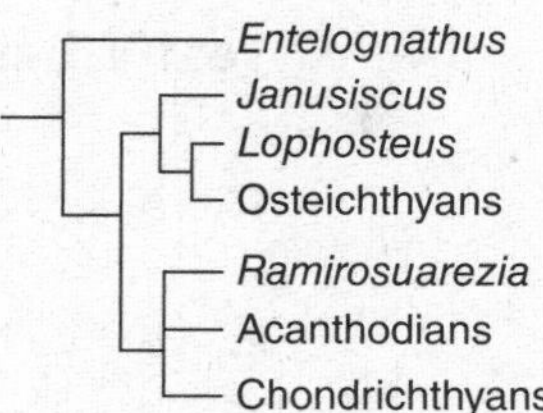

Adapted from Qiao T, King B, Long JA, Ahlberg PE, Zhu M.
"Early Gnathostome Phylogeny Revisited: Multiple Method Consensus."
PLoS ONE 11(9)(2016): e0163157.

Which of the following conclusions is best supported by the tree?

(A) *Lophosteus* and the Osteichthyans are more closely related than *Janusiscus* and the Osteichthyans.

(B) The Acanthodians and the Chondrichthyans are more closely related than *Lophosteus* and the Osteichthyans.

(C) *Entelognathus* and the Chondrichthyans are more closely related than the Acanthodians and the Chondrichthyans.

(D) *Janusiscus* and the Osteichthyans are more closely related than *Ramirosuarezia* and the Chondrichthyans.

Answer Key

Test What You Already Know

1. B
2. D
3. A

Test What You Learned

1. A
2. B
3. C
4. C
5. C
6. D
7. B
8. B
9. D
10. D
11. C
12. A

Test What You Already Know score: ___________

Test What You Learned score: ___________

Use this section to evaluate your progress. After working through the pre-quiz, check off the boxes in the "Pre" column to indicate which Learning Objectives you feel confident about. Then, after completing the chapter, including the post-quiz, do the same to the boxes in the "Post" column. Keep working on unchecked Objectives until you're confident about them all!

Pre	Post	
❑	❑	**17.1** Explain how to read and test a phylogenetic tree
❑	❑	**17.2** Explain how to read and test a cladogram
❑	❑	**17.3** Investigate evolutionary changes with cladograms

FOR MORE PRACTICE

Complete more practice online at kaptest.com. Haven't registered your book yet? Go to kaptest.com/booksonline to begin.

ANSWERS AND EXPLANATIONS

Test What You Already Know

1. B Learning Objective: 17.1

In the phylogenetic tree, the two most closely related groups will have the closest shared branch, representing the most recent common ancestor. Among the options listed in the answer choices, the groups with the closest common ancestor are the *Bucconidae* and the *Galbulidae*, **(B)**. All the other choices feature groups with less recent common ancestors.

2. D Learning Objective: 17.2

In the cladogram, the branch that diverges to form the *Sarcopterygii* and the *Actinopterygii* is labeled *Osteichthyes*, so it is true that an *Osteichthyean* was the common ancestor of those two clades. **(D)** is thus correct. (A) is incorrect because the *Sarcopterygii* branch extends to the end of the cladogram, suggesting that members of this clade still exist. (B) is incorrect because the *Percomorpharia* and the *Ovalentariae* are relatively close on the cladogram, with a recent common ancestor. (C) is incorrect because the *Actinopterygii* has many more branches in the cladogram (and thus greater diversity) than the *Sarcopterygii*.

3. A Learning Objective: 17.3

Unusual traits are more likely to develop once in a common ancestor than to develop independently on multiple occasions. Thus, it is more likely that a common ancestor led to a shared trait in groups *f*, *g*, and *h* than that the same unusual trait would evolve independently for group *e* and for the common ancestor of *a* and *c*. **(A)** is thus correct. (B) is incorrect because a cladogram shows phylogenetic relationships but does not necessarily reflect morphological similarities. (C) is incorrect because cladograms do not represent "progress"; the organisms in group *x* were likely just as well adapted to their environment as any of the later organisms. (D) is incorrect because not all traits are preserved in the descendants of particular groups.

Test What You Learned

1. A Learning Objective: 17.2

According to the cladogram, *Serpentes* and *Iguania* share a relatively recent common ancestor (the point labeled *Toxicofera*), while the common ancestor between *Gekkota* and *Serpentes* is only found at the first branching of the tree. **(A)** is thus correct. (B) is incorrect because *Scincidae* is monophyletic, deriving not from multiple ancestors but from a single common ancestor. (C) is incorrect because some lizard clades (such as the *Iguania*) are more closely related to snakes and mosasaurs than they are to other lizards, such as the *Gekkota*. (D) is incorrect because this cladogram is not organized according to timescale, but rather according to common ancestry; extinct species are not localized to a specific part of the diagram.

2. B Learning Objective: 17.3

The groups that branch earliest lack tube building, suggesting that tube building evolved in a common ancestor of the *Ampithoe*, *Caprella*, and *Aroidae* but was secondarily lost in the *Caprella*. **(B)** is thus correct. (A) is incorrect because tube building is only found in groups that emerge from a later common ancestor. (C) is incorrect because more than half of the groups in the diagram lack temperature tolerance, so it seems improbable that it was possessed by the progenitor of all of these groups. (D) is incorrect because the *Caprella* lack temperature tolerance.

3. C Learning Objective: 17.1

Phylogenetic trees always branch out from a common ancestor, so the leftmost branch represents the common ancestor of all birds. Because it is the common ancestor, it represents the point at which birds first evolved from other vertebrate ancestors (most likely reptiles, based on fossil and molecular evidence to date). **(C)** is thus correct. (A) is incorrect because the branch does not indicate relationships; the structure of the tree itself is what indicates the relationships between taxa. (B) is incorrect because unresolved and unknown species are not featured on the tree. (D) is incorrect because this tree provides no information about whether particular species are extinct.

4. C Learning Objective: 17.1

(C) If all the organisms in a phylogenetic tree share a trait, the simplest explanation is that the common ancestor also shared that trait. (A) has this reasoning backward. (B) relies on the misunderstanding that the only type of egg is a chicken egg. If (B) had read "the chicken and the chicken egg evolved at the same time," it would have been correct. (D) is wrong because a conclusion can be drawn.

5. C Learning Objective: 17.3

(C) This explanation requires the fewest changes to occur in the evolutionary history of all of the species shown and is therefore the best choice. (A) and (B) each require more changes to have occurred during the evolutionary history of all of the species shown in order to result in the pattern shown. Therefore, these choices are incorrect. (D) is incorrect because a conclusion can be drawn from the data presented.

6. D Learning Objective: 17.1

(D) Since the two phylogenetic trees don't match, we know that there are differences. Additional studies might reveal changes that could resolve the differences so that the trees would match. The fact that the trees don't correspond to each other makes (A) incorrect. Since the actual phylogenetic relatedness of the firefly species is not known, it is impossible to know whether the molecular analysis or the morphological analysis is closer to reality. Therefore, (B) and (C) are incorrect.

7. B Learning Objective: 17.3

Evaluate each answer choice. According to the figure, parasitism did not arise only once because there are multiple clades with parasitic bees. In addition to those under Nomadinae, there are those under Apinae such as Ctenoplectrini, Euglossini, Protepeolini, and Melectini, so (A) is incorrect. The eusocial bees shown all had ancestors that were primitively eusocial. Thus, **(B)** is the correct answer. (C) is incorrect because the majority of the parasitic clades shown did not arise from a eusocial common ancestor. (D) is also incorrect because the majority of the solitary clades shown did not arise from a eusocial common ancestor.

8. B Learning Objective: 17.1

According to the tree, *Drosophila melanogaster* diverged from other *Drosophila* (*D. yakuba* and *D. santomea*) at the node labeled 16.78 million years ago, so **(B)** is correct.

9. D Learning Objective: 17.2

The differences between cladograms A and B are the locations of *Entelognathus* and *Guiyu*-lineage. *Entelognathus* diverges before the crown gnathostomes in cladogram A and after the crown gnathostomes in cladogram B. *Guiyu*-lineage diverges from the sarcopterygian lineage in cladogram A and from the actinopterygian lineage in cladogram B. Thus, the correct answer is **(D)**. (A) is incorrect because the crown gnathostomes and crown osteichthyans diverged into more than one clade in both cladograms, as indicated by the lines branching from nodes (the dark, filled-in circles). (B) is incorrect because, in both cladograms, the crown gnathostomes were a common ancestor for the conventionally-defined chondrichthyans (found at the end of the line extending from the crown gnathostome node). (C) is incorrect because *Entelognathus* forms its own clade in both cladograms, but diverges at different points.

10. D Learning Objective: 17.3

(D) A characteristic present in the crown osteichthyans would be present in all of their descendants, so it would be present in three additional clades (actinopterygians, *Guiyu*-lineage, and sarcopterygians). Even though characteristics are often secondarily lost in descendants as species evolve, the question stem says to assume the simplest explanation, in which case it would be likely that all three clades share the unique characteristic.

11. C Learning Objective: 17.2

Organisms are placed on a cladogram based on evolutionary relationships rather than by appearance. Hence, all of the organisms that share a recent common ancestor are relatively closely related, evolutionarily, even if some appear quite different. The correct answer is **(C)**. (A) is incorrect because cladograms sometimes include information about timescales. but not always, and this is not the most likely explanation. (B) is incorrect because all of the organisms originate at a single node in every cladogram, which represents their common ancestor. (D) is incorrect because, even though organisms that are close together on a cladogram generally have similar characteristics since they are often closely related, organisms are placed based on evolutionary relationships.

12. A Learning Objective: 17.1

Evaluate each answer choice. According to the phylogenetic tree, *Lophosteus* and the Osteichthyans are adjacent on the tree, with *Janusiscus* branching from both. This indicates that *Lophosteus* and the Osteichthyans are more closely related than *Janusiscus* and the Osteichthyans, so **(A)** is the correct answer. The tree shows that the Acanthodians and Chondrichthyans are more closely related to each other than to other groups on the tree, as are *Lophosteus* and the Osteichthyans. However, there is no scale given that indicates whether one of these pairs is more closely related than the other, so (B) is incorrect. (C) is incorrect because *Entelognathus* and the Chondrichthyans are the two most distant groups on the tree since they are separated by multiple branch points. (D) is also incorrect because *Ramirosuarezia* and the Chondrichthyans branch from a single point, so they are more closely related than *Janusiscus* and the Osteichthyans.

CHAPTER 18

Behavior

LEARNING OBJECTIVES

In this chapter, you will review how to:

18.1 Differentiate between innate and learned behaviors

18.2 Contrast how competition and cooperation affect survival

18.3 Explain environmental and invasive species effects

18.4 Explain how communication improves survival

18.5 Investigate how behavior affects survival

TEST WHAT YOU ALREADY KNOW

1. A plant biologist is investigating the response of a Venus flytrap to stimulation using a probing needle. When she lightly touches an open insect-catching leaf, she observes no response. She increases the pressure of the stimulus, but the leaf still does not move. When she touches the leaf quickly in several spots, the leaf slams shut on her probing needle.

   Which of the following conclusions is best supported by her results?

   (A) Chemical compounds accumulate in the tissue causing the leaf to snap shut.

   (B) The stimulus must reach a threshold of pressure before the leaf can respond by shutting.

   (C) The leaf shuts when a rapid stimulus in several locations mimics the presence of an animal.

   (D) The plant becomes sensitized to her presence and responds to her irritating behavior.

2. A vineyard represents a complex ecosystem containing not just the plants, but also many microorganisms associated with the plants and soil. These microorganisms can be both beneficial and detrimental to the vines. For example, beneficial mycorrhizae in the soil are symbiotic fungi that grow in or on plant roots. Conversely, molds in the air will settle on fruit and rot the grapes. Viticulturists can treat vineyards with fungicides to destroy the spores of damaging mold. However, the yield from plots treated with fungicides is usually low, and viticulturists must heavily fertilize to increase the harvest.

   Which is the best hypothesis to test in determining why fertilizers are needed after fungicide use?

   (A) The fungicides deplete the soil of nutrients, which must then be replaced by fertilizers.

   (B) The fungicides damage the roots, so they extract nutrients from the soil less efficiently.

   (C) The fungicides destroy the mycorrhizae, which aid nutrient uptake by the roots.

   (D) The fungicides kill the mold on the fruit, slowing nutrient uptake by the plant.

3. Himalayan blackberries, a small and sour fruit, were introduced in the Pacific Northwest as a potential cash crop that requires low maintenance. The plants spread easily through rhizomes and seeds in the wild. Their brambles cover the ground with thick, thorny tangles that are impenetrable to wildlife and prevent native plants from germinating and growing. Which of the following statements most accurately describes the probable long-term impact of the blackberries?

   (A) The blackberries represent a steady source of revenue because the berries are harvested as food.

   (B) The blackberries decrease biodiversity because they outcompete native plants and are impenetrable.

   (C) The blackberries increase biodiversity because the brambles provide food and shelter for many species.

   (D) The blackberries do not readily spread and cannot be considered an invasive species.

4. In Belding's ground squirrel populations, females rear most of the young, while males mate with several females and wander from territory to territory. They use alarm calls to alert other members of their species to approaching land predators. A researcher investigated the frequency of alarm calls in a population of Belding's ground squirrels and plotted the results as a function of age and sex. The diagram compares the expected frequency of calls (which assumes the calls were random) to the observed frequency.

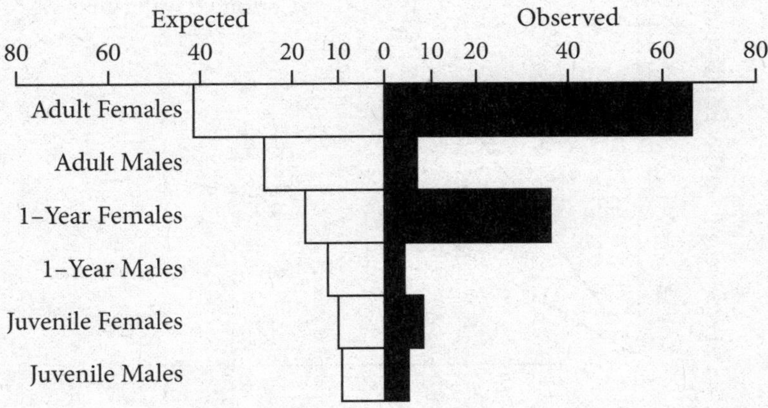

Adapted from Paul W. Sherman, "Nepotism and the Evolution of Alarm Calls," *Science* 197, no. 4310 (September 1977): 1246.

According to the data, which of the following best explains the alarm calls of Belding's ground squirrels?

(A) Alarm calls are altruistic behaviors adapted to protect next of kin.

(B) Alarm calls are intended to alert all members of the group to take cover.

(C) Alarm calls distract the predator's attention and redirect it to other squirrels nearby.

(D) Alarm calls warn the predator that the caller is going to defend itself.

5. In order to study fruit fly behavior, students built a chamber made of two empty plastic bottles taped together. At one end, they added a cotton tip coated with a chemical stimulus. At the opposite end, they introduced the fruit flies. They record the distribution of fruit flies as a function of distance to the cotton tip. Their results are summarized in the following graph.

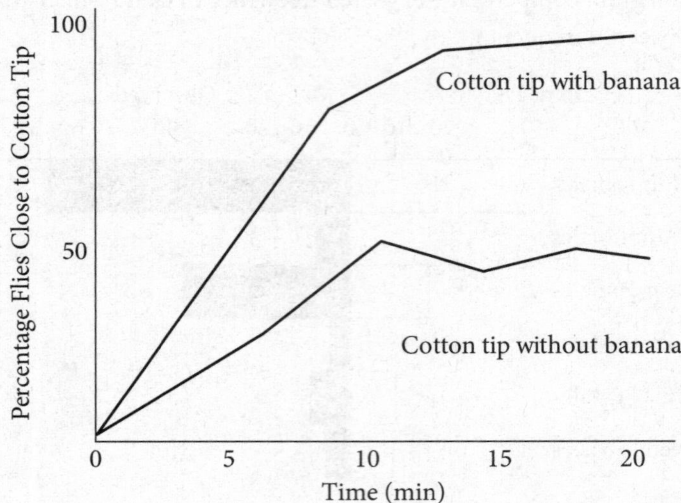

Adapted from Lar L. Vang, Alexei V. Medvedev, and Julius Adler, "Simple Ways to Measure Behavioral Responses of *Drosophila* to Stimuli and Use of These Methods to Characterize a Novel Mutant," *PLoS ONE* 7, no. 5 (May 2012): e37495.

What is the best interpretation of the results of the experiment?

(A) Fruit flies distribute randomly in the bottles because flies have a poor sense of smell.

(B) Flies gather away from the cotton tip because the smell acts as a negative chemotactic stimulus.

(C) Although there are more flies in the chamber with the banana stimulus, a chi-square analysis would show that the movement is random.

(D) Fruit flies gather in the chamber with the banana cotton tip because the smell acts as a positive chemotactic stimulus.

Answers to this quiz can be found at the end of this chapter.

INNATE VERSUS LEARNED BEHAVIOR

18.1 Differentiate between innate and learned behaviors

Ethology is the study of behavior—the way organisms act. Behaviors in organisms are triggered by internal and external stimuli, such as hunger and danger. Natural selection favors behaviors that promote reproduction and survival. Two types of behavior are innate and learned.

Innate behaviors are responses encoded in the genes of organisms, meaning organisms are genetically programmed how to respond to particular stimuli. Birds instinctively know how to build nests, and nest building occurs when both internal stimuli (hormonal signals) and external stimuli (proper nest building materials) are present. Innate behaviors that increase the organism's ability to survive and reproduce will persist in a population and be passed from one generation to the next. A change in innate behavior occurs only if a genetic mutation arises, and the altered behavior would develop over many generations. Other examples of innate behavioral responses include taxes (singular: **taxis**) (movements by the entire body) and reflexes (movements by individual muscles). For instance, chemotaxis involves the response to chemical gradients, and a reflex to the sensation of pain is avoidance.

Learned behaviors, the opposite of innate behaviors, are acquired or lost through interacting with the world or through teaching. Young geese and other birds learn via imprinting, in which they follow an object (usually a parent) that they have become attached to during a receptive period after birth or hatching. Compared to innate behavior, learned behavior is largely independent of inheritance and dependent on the environmental context, so if the context disappears, the behavior will also disappear. As a result, learned behavior spreads through a population, passes within a single generation, and develops/degrades more rapidly than innate behavior. By obtaining knowledge and/or skills, organisms can modify learned behaviors to survive in unpredictable environments. For example, a mouse can learn how to run through a maze to reach a piece of cheese.

Innate and learned behaviors are not mutually exclusive and may influence behavior together. In honeybees, genes associated with foraging behavior determine whether bees are foragers or workers and whether they forage for pollen or for nectar. However, foraging performance is also correlated with learning. The ability to adjust foraging behavior in response to the constantly changing environment allows honeybees to minimize energetic cost and maximize food acquisition.

COMPETITION AND COOPERATION

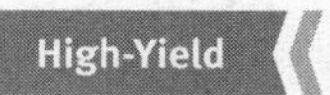

18.2 Contrast how competition and cooperation affect survival

18.3 Explain environmental and invasive species effects

Innate and learned behaviors may be cooperative and/or competitive. **Cooperation** is the process of acting together for common benefits, whereas **competition** is the process of striving for limited resources. Both cooperation and competition affect survival and occur at various levels of organization: cells, tissues, organs, organ systems, organisms, populations, communities, and ecosystems.

On the cellular level, cooperation and competition increase the efficiency of using matter and energy. Chemical reactions catalyzed by enzymes are affected by the cooperation between enzymes and their substrates, as well as by the competition between inhibitors and substrates. Examples of cooperation between organs and organ systems include the coordination of roots and shoots of plants and of the digestive and excretory systems of animals. In plants, to replace water lost via transpiration from shoots, roots absorb water. In animals, organs in the digestive system work together to process food, and the digestive system works in parallel with the excretory system to eliminate undigested waste.

Contributing to the survival of individuals and the population, cooperation and competition occur between individuals of the same species (intraspecific) and/or between individuals of different species (interspecific). Cooperative behavior within a population can provide protection from predators, acquisition of prey and resources, recognition of offspring, and transmission of learned responses. Cooperation can be seen in schools of fish, in which living together decreases the chance of a predator attack, and among wolves in a pack, in which each wolf has a specific role in the pack's hunting strategy.

Competition arises when resources such as food, territories, and mates that two or more individuals depend upon become limited. Direct competition occurs when individuals compete for the same resource (for example, two males competing to mate with a single female), and indirect competition occurs when organisms use the same resource (for example, a waterhole) but may not interact with each other. The use of resources by one individual decreases the availability for use by others, so the less competitive individual will be forced to go elsewhere, resulting in competitive exclusion. When sunlight is limited, plants unable to outgrow surrounding plants may use energy to produce many seeds in order to spread their offspring to areas with more sunlight. Species may coexist if intraspecific competition is stronger than interspecific competition because then each species limits their own growth and population size.

In addition to cooperation and competition, ecological relationships that affect population dynamics include predation and symbiosis (**mutualism, commensalism**, and **parasitism**). **Predation** involves one organism feeding on another and typically occurs interspecies. For instance, some protozoans prey on bacteria and some plants can trap and digest insects. **Symbiosis** describes the relationship between two different organisms. In mutualism, both organisms in the relationship benefit. Examples include bacteria in digestive tracts of animals and pollination. Bacteria in cows help cows digest cellulose and in turn cows provide bacteria with nutrients and a hospitable place to live. During pollination, plants provide insects with food and insects help plants spread their pollen from one plant to another. In commensalism, one organism benefits and the other is unharmed, whereas in parasitism one benefits while the other is harmed. Remoras attached to sharks obtain protection and leftover food, while the shark is unaffected. Fleas and ticks, in contrast, are external parasites that feed on the blood of their hosts and cause itching and transmit disease.

Environmental Changes and Invasive Species

Changes in the environment, such as environmental catastrophes, geological events, and the sudden influx/depletion of abiotic resources, also affect the behavioral responses of organisms. An environmental catastrophe or disaster due to human activity (like deforestation or overhunting) puts extraordinary pressures on organisms and may result in the loss of species or the selection for altered behavioral traits. Geological events such as volcanic eruptions and earthquakes may also result in the extinction of species. Geologically separating a population (for instance, by the formation of a mountain range) can lead to changes in diversity. Changing the abiotic environment can alter the nutrient availability of a community, which determines whether a species can invade and/or persist in a community. A flood may bring an influx of nutrients, while a drought will diminish the availability of resources and may thus kill the species.

Changes in an ecosystem, even the slightest, can elicit an adaptive response and lead to changes in the behavior of native populations. When a nonnative species is introduced into an ecosystem, it can disrupt the balance of the ecosystem and have a negative impact. Nonnative species that cause harm to the environment, the health of native species, or the local economy are termed **invasive species**. An invasive species can be any kind of living organism—a bird, fish, insect, fungus, bacterium, plant, seed, or egg.

Invasive species can be a direct or indirect threat to the native population in the ecosystem. Direct threats include preying on native species, out-competing for food and resources, out-competing for habitats or breeding sites, and causing or carrying toxins or disease that native species are not adapted to deal with. In response to invasive predatory species and competition, native organisms may change their anti-predator, foraging, and feeding behaviors. For instance, crickets suppress their calling behavior to avoid acoustically-hunting parasites. Invasive species may dominate in a new ecosystem because they grow and reproduce rapidly and/or lack natural enemies or pests. As a result, the invasive species will displace the native species, as illustrated by both kudzu and Dutch elm disease. Kudzu is an invasive plant that has replaced the diverse ecosystem in the southeastern United States with a monoculture. Dutch elm disease, caused by the fungus *Ophiostoma ulmi* and transmitted to trees by elm bark beetles, has killed native populations of elms that are not resistant to the disease. Indirect threats include changing the food web by destroying or replacing native food sources, decreasing the abundance and diversity of native species, and changing the environment. For example, invasive plants can modify nutrient availability (making an environment less inhabitable) by affecting soil pH or causing erosion (because they may not hold soil as well as native species).

COMMUNICATION

18.4 Explain how communication improves survival

In response to stimuli, signals are transmitted and received to exchange information. Individuals can then act on the information and communicate it to others (individuals of the same or different species). Communication among organisms produces changes in behavior that are vital to reproductive success, natural selection, and evolution.

Signals to improve survival may stem from environmental cues like temperature and oxygen levels. Hibernation is regulated by temperature, level of food supply, and/or photoperiod (length of day). When it gets cold or when the food supply gets low, animals get ready to hibernate to save energy. Likewise, animals estivate, or slow their activity in response to high temperatures and arid conditions, to save energy.

Organisms communicate using visual, audible, tactile, electrical, and chemical (pheromone) signals. These signals may be used to locate resources, show dominance, defend territory, coordinate group behavior, and care for young. The following examples illustrate communication in organisms. Male lizards use visual signals to show their territorial dominance by standing up high off the ground and swallowing air to increase their size. Frogs and toads produce auditory signals to attract mates, and the bright colors of toxic species warn predators not to eat them. On the other hand, the bright coloration of flowers and fruits signal animals to pollinate their flowers or disperse their fruits. Honeybees use tactile signals to communicate the location of nectar sources via a waggle dance, and minnows release an alarm pheromone when their skin is damaged to warn other minnows of a predatory fish.

AP BIOLOGY LAB 12: FRUIT FLY BEHAVIOR INVESTIGATION

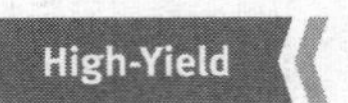

18.5 Investigate how behavior affects survival

Taxis, mentioned earlier in this chapter, will be explored in this investigation of the relationship between a model organism, *Drosophila* (fruit fly), and its response to different environmental conditions. You can examine the behavior of fruit flies by placing them in an apparatus called a choice chamber, which presents two (or more) choices to the fruit flies. Experiments performed in the choice chamber will investigate chemotaxis (movement in response to a chemical stimulus), geotaxis (movement in response to gravity), and/or phototaxis (movement in response to light). Fruit flies that move toward a stimulus exhibit positive taxis (attract), whereas fruit flies that move away from a stimulus exhibit negative taxis (repel). Undirected, random movement in response to an external stimulus is **kinesis**.

You will design a controlled experiment to explore environmental factors that trigger fruit fly responses. A simple choice chamber may be constructed by joining together two clear plastic bottles with the bottoms cut off. After exposing fruit flies in the chamber to a stimulus (one end) or stimuli (both ends), give them time to respond and then determine the number of flies at each end of the chamber.

From observations, generate your own hypotheses to explore other factors (sex, age, different colors of light, ripeness of fruit, mutations) that affect fruit fly behavior. To verify the results, you need to conduct several trials and change the positions of the stimuli in the chamber. Quantify and express your results graphically. Then complete a chi-square analysis and construct a preference table to identify and compare the preferences of fruit flies. Your goal is to identify a pattern in the behavior of fruit flies and which of the responses (geotaxis, chemotaxis, or phototaxis) is the strongest. Potential challenges may include difficulty in counting fruit flies accurately and fruit flies escaping the choice chamber.

RAPID REVIEW

If you take away only 5 things from this chapter:

1. Behaviors in organisms are responses that are triggered by stimuli (internal and external), and the behaviors that promote reproduction and survival are favored by natural selection.
2. Innate behaviors are inherited and instinctive, whereas learned behaviors are acquired through interactions with the environment. The behaviors are not mutually exclusive.
3. Cooperation involves organisms working together for mutual benefits, while competition involves organisms contending with each other for limited resources.
4. Both environmental changes (such as deforestation and earthquakes) and invasive species (nonnative species introduced into an ecosystem) disrupt the balance of an ecosystem and affect the behavior of native populations.
5. Communication among organisms involves the transmission of signals to produce changes in behavior that are vital to reproductive success, natural selection, and evolution.

TEST WHAT YOU LEARNED

1. The gut microbiome is a crowded ecosystem in which large amounts of bacteria compete for resources from the host. Microbiologists have found that infection by a parasite that is resistant to the host's nonspecific immune system actually elicits a strong nonspecific immune response from the host.

   Which of the following best describes how a resistant parasite would most likely benefit from stimulating its host's immune system in this way?

   (A) Stimulation of the host immune system depletes the energy of the host and makes it more vulnerable to infection.

   (B) Stimulation of the immune system reduces competition by decreasing the population of other parasites susceptible to the immune system.

   (C) Stimulation of the immune system is automatic and does not present an advantage to the parasite.

   (D) Stimulation of the host's nonspecific defenses at the site of infection results in a flow of nutrients which supports parasite growth.

2. Students built a chamber to study fruit fly behavior. The chamber consisted of two empty plastic bottles taped together. At one end, they added a cotton tip coated with a chemical stimulus. They introduced the fruit flies at the opposite end. They record the distribution of fruit flies as a function of distance to the cotton tip.

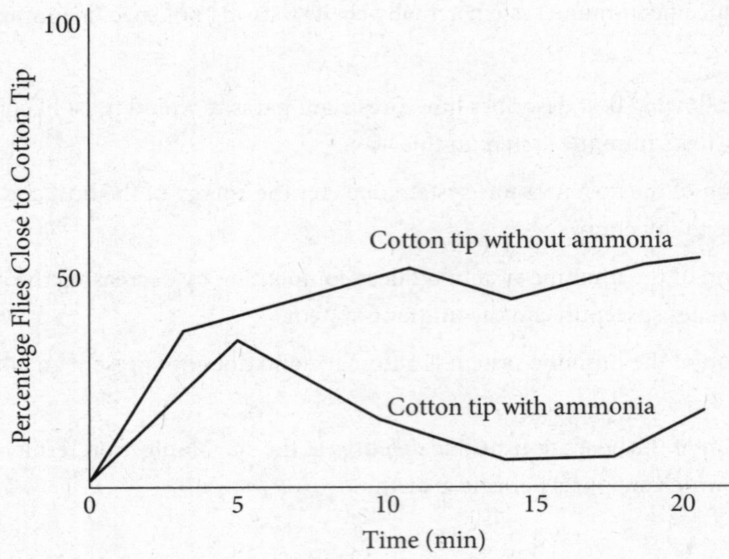

Adapted from Lar L. Vang, Alexei V. Medvedev, and Julius Adler, "Simple Ways to Measure Behavioral Responses of *Drosophila* to Stimuli and Use of These Methods to Characterize a Novel Mutant," *PLoS ONE* 7, no. 5 (May 2012): e37495.

According to the graph, what can be concluded about the reaction of fruit flies to ammonia, a strong base?

(A) Ammonia does not elicit a strong response from fruit flies.

(B) Ammonia paralyzes fruit flies as it diffuses through the chamber.

(C) Ammonia appears to attract and stimulate the fruit flies.

(D) Ammonia is a volatile compound that repels fruit flies.

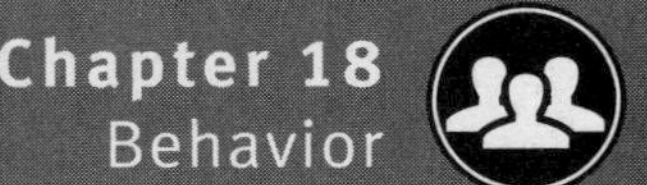

3. Plants are exposed to many threats from the environment, including predators and diseases caused by infectious agents. Warding off injury presents a distinct survival advantage. While investigating the response of marigolds to predation by golden beetles, researchers detected a noxious chemical compound that signals to the beetle that the plant is not a good source of food. The researchers hypothesized that plants synthesize this compound in response to communication with neighboring plants.

   Which of the following compounds is the most likely means by which the marigolds communicate?

   (A) Jasmonate, a volatile organic compound that is produced when the plant is stressed
   (B) Oligosaccharides, large sugars that act as signaling molecules in response to infection
   (C) Abscisic acid, a water soluble plant hormone that is produced as a response to stress
   (D) Salicylic acid, a water soluble acid that mediates defense against pathogens

4. One of the ecosystem changes that followed the disappearance of wolves from Yellowstone National Park was the substantial erosion of riverbanks along streams, which followed a decrease in the number of trees and other plants. After wolves were reintroduced in the park, the effects of erosion were reversed. Which of the following best explains how the disappearance of wolves would most likely affect the erosion of riverbanks?

   (A) Wolves destroy vegetation along the riverbanks while hunting. The loss of vegetation can cause erosion.
   (B) Wolves hunt elk, which graze on vegetation that protects riverbanks from erosion. Without wolves, elk populations increase, reducing vegetation.
   (C) Wolves competitively exclude other predators. These predators eat plants and thus contribute to the degradation of the riverbanks.
   (D) Wolves eat beavers, which cut down trees to build dams. The dams slow down the flow of rivers, which decreases the amount of erosion.

5. An octopus in an aquarium would occasionally splash water on several summer volunteers. One of the volunteers, named Michael, claims that the octopus does not like him because he always gets drenched when he walks by. The other volunteers insist that the drenching is purely random. Which of the following experimental treatments would help to test the hypothesis that the octopus learned to recognize faces?

   (A) Have each of the volunteers except Michael walk past the tank once to see if any get drenched.
   (B) Arm Michael with a water gun to drench the octopus back if the octopus drenches him.
   (C) Cover the faces of each of the volunteers with masks before they walk past the tank and compare those results to walks without masks.
   (D) Randomize the time at which different volunteers walk by the octopus tank and determine whether the octopus is more likely to splash at specific times.

6. Adelgids are insects that feed on the sap of trees. A particular species of adelgid was accidentally imported into North America from Japan. It was able to feed on native hemlock and spruce trees, which are abundant in the forests of eastern North America. Which of the following likely had the greatest impact on native forests?

   (A) Since the adelgid did not have any native predators or pathogens in the new environment, it likely spread uncontrollably.

   (B) North American birds ate this new resource and kept the adelgid population small.

   (C) A virus that infects native adelgids evolved to attack the foreign species.

   (D) The adelgid species interbred with native adelgids, eventually demonstrating a founder effect.

7. Researchers have found that many animals, even ones that do not appear very social, can distinguish between familiar and unfamiliar animals of their species. Which of the following provides the LEAST likely explanation for why some asocial species can recognize other individuals of their species?

   (A) This reduces the risk of injury from repeated fights.

   (B) This allows them to cooperate with the individuals they recognize.

   (C) This reduces the risk of wasting energy by repeatedly fighting the same individuals.

   (D) This allows them to spend more time on other activities, such as foraging.

8. Researchers studied whether the Sahmalaza Sportive Lemur, a solitary species, responded to the alarm calls of other species to protect itself from possible predators. The table below shows the percentage of time that the lemurs spent in rest, vigilance, or self-grooming before and after exposure to different types of sounds played through speakers.

Sound	Rest % before	Rest % after	P	Vigilance % before	Vigilance % after	P	Self-grooming % before	Self-grooming % after	P
Crested coua alarm	68	22.5	<0.05	27.5	74.5	<0.05	4.5	3	>0.1
Crested coua song	61	51	>0.1	31	40	>0.1	8	9	>0.1
Madagascar magpie-robin alarm	68	24	<0.05	28.5	75	<0.05	3.5	1	>0.1
Madagascar magpie-robin song	72	61	>0.1	23	35	>0.1	5	4	>0.1
Blue-eyed black lemur aerial alarm	67.5	23	<0.05	24	77	<0.05	8.5	0	>0.1
Blue-eyed black lemur terrestrial alarm	72	62	>0.1	24	36	>0.1	4	3	>0.1
Blue-eyed black lemur agitation call	73.5	55	>0.1	19	41.5	>0.1	7.5	2.5	>0.1
Blue-eyed black lemur contact call	78	77	>0.1	19	20	>0.1	2.5	4	>0.1

Adapted from Seiler M, Schwitzer C, Gamba M, Holderied MW (2013) Interspecific Semantic Alarm Call Recognition in the Solitary Sahamalaza Sportive Lemur, *Lepilemur sahamalazensis*. PLoS ONE 8(6): e67397.

Why did the researchers play these varied sounds rather than just specific alarm calls?

(A) Using varied sounds shows that the lemurs respond to all vocalizations by other species by increasing vigilance, decreasing grooming, and resting less.

(B) Using varied sounds shows that lemurs do not respond to either alarm calls or other vocalizations by other species.

(C) Using varied sounds shows that lemurs specifically respond to alarm calls with increased vigilance, distinguishing those calls from other songs and calls.

(D) Using varied sounds shows that lemurs show significantly more self-grooming behavior when songs are played than when alarm calls are played.

9. Researchers have extensively studied the way birds learn birdsong. In one study, researchers raised spotted antbirds in soundproof aviaries ("No Tutor") or with an individual of a different species with a distinct song ("Heterospecific Tutor"). The graphs below show the changes in pitch of each bird's song.

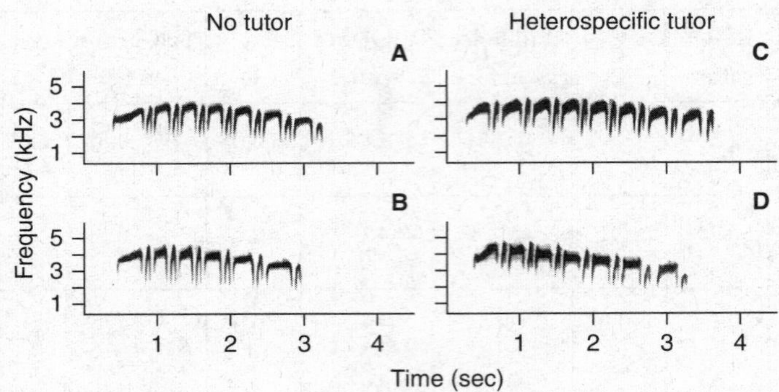

Adapted from Touchton JM, Seddon N, Tobias JA (2014) Captive Rearing Experiments Confirm Song Development without Learning in a Tracheophone Suboscine Bird. PLoS ONE 9(4): e95746.

Which of the following conclusions can the researchers most reasonably draw based on these findings?

(A) Because the songs appear different in the "No Tutor" and "Heterospecific Tutor" birds, spotted antbirds must learn to produce the typical spotted antbird song.

(B) Because the songs appear similar in the "No Tutor" and "Heterospecific Tutor" birds, spotted antbirds must learn to produce the typical spotted antbird song.

(C) Because the songs appear different in the "No Tutor" and "Heterospecific Tutor" birds, song characteristics are innate in spotted antbirds and do not require learning.

(D) Because the songs appear similar in the "No Tutor" and "Heterospecific Tutor" birds, song characteristics are innate in spotted antbirds and do not require learning.

The answer key to this quiz is located on the next page.

Answer Key

Test What You Already Know

1. C
2. C
3. B
4. A
5. D

Test What You Learned

1. B
2. D
3. A
4. B
5. C
6. A
7. B
8. C
9. D

REFLECTION

Test What You Already Know score: __________

Test What You Learned score: __________

Use this section to evaluate your progress. After working through the pre-quiz, check off the boxes in the "Pre" column to indicate which Learning Objectives you feel confident about. Then, after completing the chapter, including the post-quiz, do the same to the boxes in the "Post" column. Keep working on unchecked Objectives until you're confident about them all!

Pre	Post	
☐	☐	**18.1** Differentiate between innate and learned behaviors
☐	☐	**18.2** Contrast how competition and cooperation affect survival
☐	☐	**18.3** Explain environmental and invasive species effects
☐	☐	**18.4** Explain how communication improves survival
☐	☐	**18.5** Investigate how behavior affects survival

FOR MORE PRACTICE

Complete more practice online at kaptest.com. Haven't registered your book yet? Go to kaptest.com/booksonline to begin.

ANSWERS AND EXPLANATIONS

Test What You Already Know

1. C Learning Objective: 18.1

Venus flytraps have adapted to detecting small animal movements, with the innate behavior of snapping shut when an animal is detected, so prey might be consumed. The plant will thus only respond when the researcher simulates animal movements by rapidly stimulating the plant in several locations. **(C)** is thus correct. (A) is incorrect because snapping shut must be a rapid response in order to capture prey effectively; the accumulation of compounds would take too much time, giving prey an opportunity to escape. (B) is incorrect because increasing the pressure had no effect. (D) is incorrect because the Venus flytrap is exhibiting an innate behavior; lacking a proper nervous system, it is too simple of an organism to learn and become sensitized to specific responses.

2. C Learning Objective: 18.2

As noted in the question stem, mycorrhizae are a class of fungi that are beneficial to the vines. Using a fungicide will harm the damaging mold but also the helpful mycorrhizae. The fertilizers are necessary to counteract the loss of the mycorrhizae, which help plants to acquire nutrients. **(C)** is thus correct. (A) is incorrect because the nutrients in the soil (nitrogen, phosphorus, and minerals) are simple compounds and ions that would not be affected by a chemical treatment. (B) is incorrect because fungicides only damage fungus, not plants. (D) is incorrect because mold does not assist in nutrient uptake; on the contrary, it is a parasite that consumes nutrients in the vines' grapes.

3. B Learning Objective: 18.3

By making the land impenetrable to wildlife and preventing native plants from flourishing, the blackberries function as an invasive species. Their most probable effect would be to decrease biodiversity, as invasive species tend to do. **(B)** is thus correct. (A) is incorrect because the negative characteristics of the plants, such as the difficulty that thorny brambles pose for harvesting, make them less valuable as a food source. (C) is incorrect because the blackberries are likely to have the opposite effect, as described in the question stem. (D) is incorrect because the plants function as an invasive species.

4. A Learning Objective: 18.4

As noted in the question stem, females are far more likely to rear offspring than males, and the data in the figure shows that females are responsible for most of the alarm calls. Because making an alarm call not only alerts other squirrels to the presence of a predator, but also reveals the location of the squirrel making the call (thereby making the caller a potential target for the predator), it is an example of an altruistic behavior. Since females are making most of these calls, the function is most likely to protect the young that females tend to raise. **(A)** is thus correct. (B) is incorrect because the frequency of calling behavior is affected by age and sex, so it is unlikely to be a behavior intended to benefit the entire group; more probably, it is a behavior that females use to alert the young they are rearing. (C) is incorrect because the alarm calls would only divert the predator's attention to the squirrel making the call and not to nearby squirrels (who would likely run away and hide in response to the call). (D) is incorrect because the calls are adapted to alert other squirrels, not to warn the predator. Squirrels are small animals; rather than defending themselves, they can better avoid being eaten by running and hiding.

5. D Learning Objective: 18.5

According to the graph, 100% of the fruit flies eventually gathered close to the cotton tip with banana (while far fewer gathered close to the tip without it). This suggests that the banana smell serves as a positive chemotactic stimulus. **(D)** is thus correct. (A) and (C) are incorrect because the movement is clearly not random, irrespective of whether a chi-square analysis is performed; all of the flies gathered near the banana-coated tip, but only half near the non-coated tip. (B) is incorrect because the results show the opposite, that the fruit flies were attracted to the smell.

Test What You Learned

1. B Learning Objective: 18.2

If a parasite is resistant to a host's nonspecific immune system, then stimulating an immune response can aid the parasite by eliminating competing parasites. **(B)** is thus correct. (A) is incorrect because stimulating the immune system will not make the host significantly more vulnerable to infection, even though it does require energy. (C) is incorrect because this stimulation does, in fact, present an indirect advantage to the parasite by allowing it to outcompete parasites that are vulnerable to the host's nonspecific immune system. (D) is incorrect because stimulation of the nonspecific immune system usually leads to an inflammatory response, which will not increase the amount of nutrients at the site of inflammation.

2. D Learning Objective: 18.5

According to the data in the graph, fruit flies stay farther away from an ammonia-soaked cotton tip than from a non-soaked tip. This suggests that ammonia is a volatile compound that repels the flies, **(D)**. (A) is incorrect because the data shows that the flies do respond to ammonia, moving away from it. (B) is incorrect because the data shows that the flies are still capable of moving; the percentage close to the ammonia-soaked tip fluctuates with time. (C) is incorrect because more flies prefer to approach the non-soaked tip, indicating that the ammonia actually repels them.

3. A Learning Objective: 18.4

A volatile compound is one that evaporates easily, allowing it to become airborne and spread out over significant distances. Thus, jasmonate is the most likely candidate for a communication molecule, making **(A)** correct. (B) is incorrect because large sugars are unlikely to become airborne, so there is no way for them to allow for communication between plants. (C) and (D) are incorrect because these water soluble compounds are unlikely to become airborne, so they would not make for effective communication between plants.

4. B Learning Objective: 18.3

Wolves are predators, so if they disappear from an ecosystem, it should lead to an increase in the populations of animals that they preyed upon. The cause of the erosion most likely has to do with effects caused by one of these prey species, as they increased in number. **(B)** is correct because it presents a plausible scenario, in which an increase in elk populations leads to a decrease in vegetation, which in turn would contribute to riverbank erosion. (A) is incorrect because it gives an explanation of why the presence of wolves could cause erosion, but the question stem specifies that it was the disappearance of wolves that caused this. (C) is incorrect because predators hunt other animals; they do not generally feed on vegetation. (D) is incorrect because erosion occurred after the wolves disappeared; if the dams help combat erosion, then erosion should not be a problem after a beaver predator is removed and more dams can be built.

5. C Learning Objective: 18.1

In order to detect whether the octopus recognizes faces, the volunteers could conduct an experiment in which none of their faces are visible (by using masks to cover them) and observe whether this has any impact on drenching frequency. If the results showed that Michael is drenched less often when wearing a mask, it suggests that the octopus can in fact recognize faces. **(C)** is thus correct. (A) is incorrect because it excludes the volunteer whose face is most likely recognizable by the octopus. (B) is incorrect because it only serves to provide feedback to the octopus, but does not reveal whether it might recognize faces. (D) is incorrect because it tests the hypothesis that the octopus splashes at specific times, not the hypothesis that it recognizes faces.

6. A Learning Objective: 18.3

(A) An invasive species spreading uncontrollably would have a large impact on the native forest. (B) and (C) would lead to a smaller impact on the forest. Interbreeding with native species, while unlikely, would only lead to more diversity in the adelgid population. There is no clear reason for this to have a large impact on the native forest, so (D) is incorrect.

7. B Learning Objective: 18.5

Evaluate each answer choice. By recognizing individuals that they have previously fought, animals can weigh the risks of fighting, thereby reducing their risk of injury. Thus, (A) is a likely explanation. Individuals that were regularly observed cooperating with other individuals would no longer be considered asocial, so **(B)** is an unlikely explanation and the correct answer. (C) and (D) are also likely explanations, and therefore incorrect answers, because recognizing individuals rather than fighting them repeatedly would save the animals energy and time, so they can spend more time doing other important activities, such as foraging.

8. C Learning Objective: 18.4

By including other vocalizations as controls, the researchers showed that lemurs specifically recognized alarm calls and responded to them with increased vigilance and reduced resting behavior, as shown in the table. A P-value less than 0.05 (meaning a less than 5% probability that a difference is due to chance) is generally recognized as significant. The table shows multiple P-values less than 0.05. In these cases, researchers conclude that the lemurs behaved differently after the sound was played compared with before the sound was played. Thus, **(C)** is the correct answer, and (A) and (B) are incorrect. (D) is incorrect because none of the P-values for self-grooming were less than 0.05 according to the table, suggesting that this behavior did not change when vocalizations were played.

9. D Learning Objective: 18.1

The songs appear to be very similar in the "No Tutor" and "Heterospecific Tutor" birds, so eliminate (A) and (C), which state that the songs appear different. The similarity in song, regardless of the experimental conditions tested, suggests that the song is innate rather than learned from exposure to conspecific tutors. Thus, the correct answer is **(D)**. (B) is incorrect because, if the songs were learned, the songs would almost certainly appear different.

CHAPTER 19

Ecology

LEARNING OBJECTIVES

In this chapter, you will review how to:

19.1 Contrast exponential and logistic growth models

19.2 Explain how biotic and abiotic factors affect population

19.3 Explain how environmental factors affect organisms

19.4 Contrast how matter is recycled and energy flows

19.5 Explain how diversity affects stability

19.6 Describe the human impact on biodiversity and climate

19.7 Investigate energy flow through a system

TEST WHAT YOU ALREADY KNOW

1. When species are introduced into a new area, they can sometimes find the environment suitable and establish new populations. If a new species is accidentally released in an area in which there are appropriate climate conditions, abundant food, and no predators, which of the following best describes the most likely growth of its population?

   (A) The population will follow an exponential growth curve as there are no limitations on food and other resources in this new location.

   (B) The population growth curve will rapidly level off and then decline because this species does not belong in this area.

   (C) The population will remain at a steady level since the species has become established but cannot become abundant in a new location that differs from its native habitat.

   (D) The population growth curve may initially appear to be exponential, but will then level off over time to form a logistic growth curve as resources become more limited.

2. A study examined the response of bird communities to Hurricane Iris. Hurricanes can affect communities in a variety of ways, particularly by knocking down trees and causing structural damage. The figure below shows the mean number of birds captured for every 100 hours of effort using a net. Values are given for (1) captures over 58 days before Hurricane Iris ("pre-Iris"), (2) captures over about a month beginning 11 days after Iris ("Post-I"), and (3) captures over 69 days one year after Iris ("Post-II").

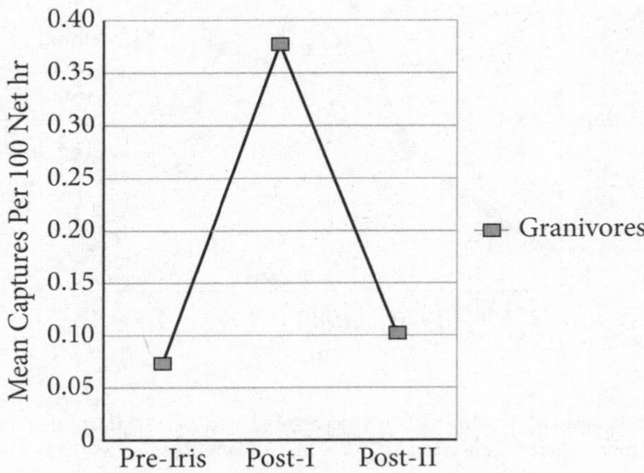

Adapted from Andrew B. Johnson and Kevin Winker, "Short-Term Hurricane Impacts on a Neotropical Community of Marked Birds and Implications for Early-Stage Community Resilience," *PLoS ONE* 5, no. 11 (November 2010): e15109.

Which of the following hypotheses is NOT consistent with these findings?

(A) Birds living in the area experienced high mortality due to the hurricane.

(B) Reduced canopy cover after the hurricane made it easier to capture birds.

(C) Birds had to change their foraging habits due to changes in the forest.

(D) Because the habitat was damaged, birds needed larger home ranges and territories.

3. The figure below shows data from a study in a Serengeti ecosystem. Rinderpest is a virus that causes disease in wildebeest, herbivores that are the dominant grazing animal in this ecosystem. A vaccination program was used to reduce the prevalence of rinderpest in the wildebeest population. The prevalence of the virus is shown with open squares in the figure below. The wildebeest population is represented by closed circles.

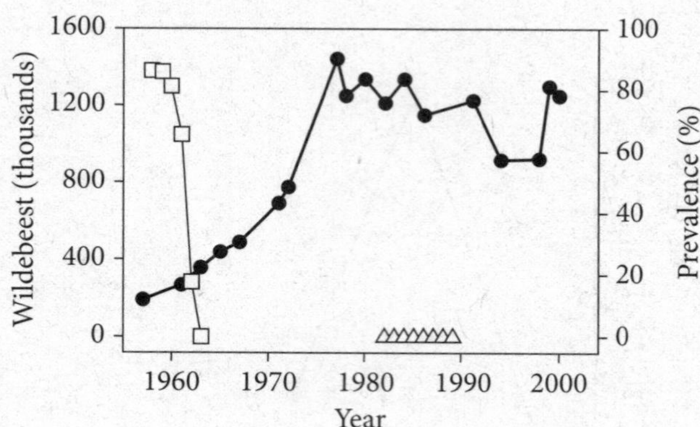

Adapted from Ricardo M. Holdo et al., "A Disease-Mediated Trophic Cascade in the Serengeti and its Implications for Ecosystem C," *PLoS Biol* 7, no. 9 (September 2009): e1000210.

Based on this data, how will the vaccination program most likely affect the Serengeti ecosystem?

(A) The wildebeest population will increase due to the reduction in rinderpest prevalence, which will lead to an increase in grass cover and tree cover.

(B) The wildebeest population will decrease due to the reduction in rinderpest prevalence, which will cause the rinderpest to develop immunity to the vaccination protocol.

(C) The wildebeest population will increase due to the reduction in rinderpest prevalence, which will result in less grass available to generate and spread fires.

(D) The wildebeest population will decrease due to the reduction in rinderpest prevalence, which will result in the emergence of new parasites to lower the population size further.

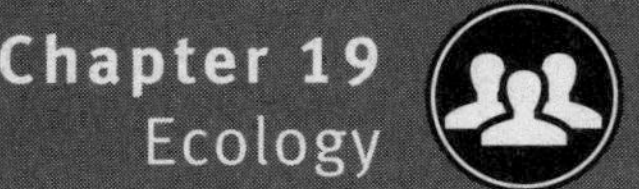

4. Which of the following best describes the effects, in terms of energy and matter flow and cycling, when plants are fertilized with nitrogen-rich fertilizer?

    (A) Plants take up energy from the fertilizer, which can be recycled. Nitrogen is consumed during plant metabolism.

    (B) Plants obtain energy from the Sun, which flows to primary consumers. Nitrogen in fertilizer is used by plants to make organic molecules that can be recycled through the nitrogen cycle.

    (C) Plants take up energy from fertilizer, which is eventually lost as heat. Nitrogen in fertilizer is used to make organic molecules that can be recycled through the nitrogen cycle.

    (D) Plants obtain energy from the Sun and lose it again as heat. Nitrogen fertilizer slows this process, allowing the plants to conserve more energy.

5. In 1995, there were fewer than 30 Florida panthers remaining in their native ecosystem. Researchers introduced eight panthers from Texas, members of a different subspecies, to southern Florida that year. Why would it be beneficial to introduce panthers from another population to interbreed with the Florida panthers?

    (A) The existing small population was very inbred, increasing the risk that they could go extinct due to a lack of genetic diversity in the population.

    (B) Increasing the size of the population is very important for survival, and subspecies are sufficiently similar that the genetic differences did not matter.

    (C) Texas populations were likely better adapted to the southern Florida habitat, and introducing their genes would increase the survival of the population.

    (D) The population was so small that it was impossible to save, making it a good location to experiment with interbreeding panther subspecies.

6. Wetlands are of considerable environmental importance and researchers actively study wetland restoration methods. Figure A shows the ratio of vertebrate density, vertebrate richness, and macroinvertebrate density in restored wetlands versus natural wetlands. Figure B shows plant density and richness in restored wetlands. The dotted line represents the values for reference wetlands. Negative values represent numbers below those of the reference wetlands.

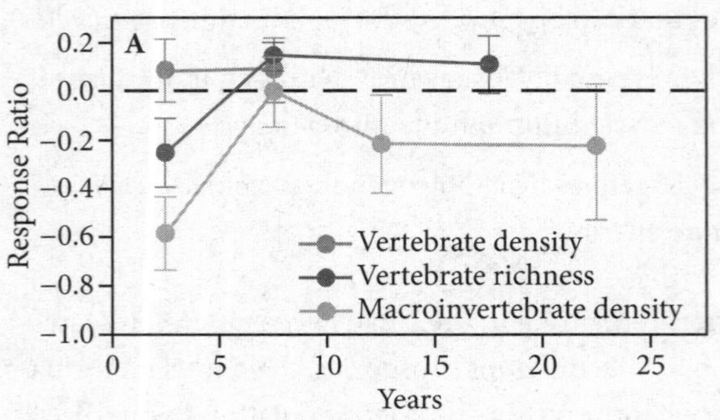

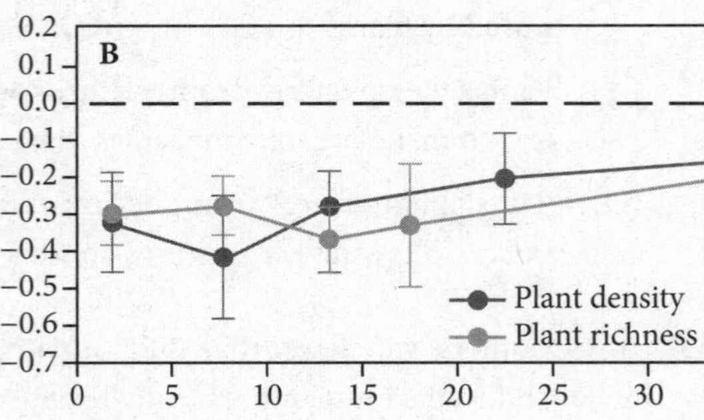

Adapted from David Moreno-Mateos et al., "Structural and Functional Loss in Restored Wetland Ecosystems," *PLoS Biol* 10, no. 1 (January 2012): e1001247.

Which of the following conclusions is best supported by the data provided?

(A) Although plants have the lowest mean response ratios initially, they stabilize at a similar level to macroinvertebrates after extended periods of time.

(B) Although plants all reach relatively stable levels equivalent to those found in reference wetlands, animals in the studied locations do not.

(C) Macroinvertebrate density in manipulated wetlands never reaches the levels found in reference wetlands.

(D) Although vertebrates may be able to colonize wetlands relatively easily, macroinvertebrates and plants do not successfully stabilize at the levels found in natural wetlands.

7. Scientists are interested in understanding how climate change can affect forest biomass. The figures below show the amount of above-ground biomass at three forest locations, with dates on the *x*-axis, growth rate on the *y*-axis, and a dotted line representing no net change in biomass.

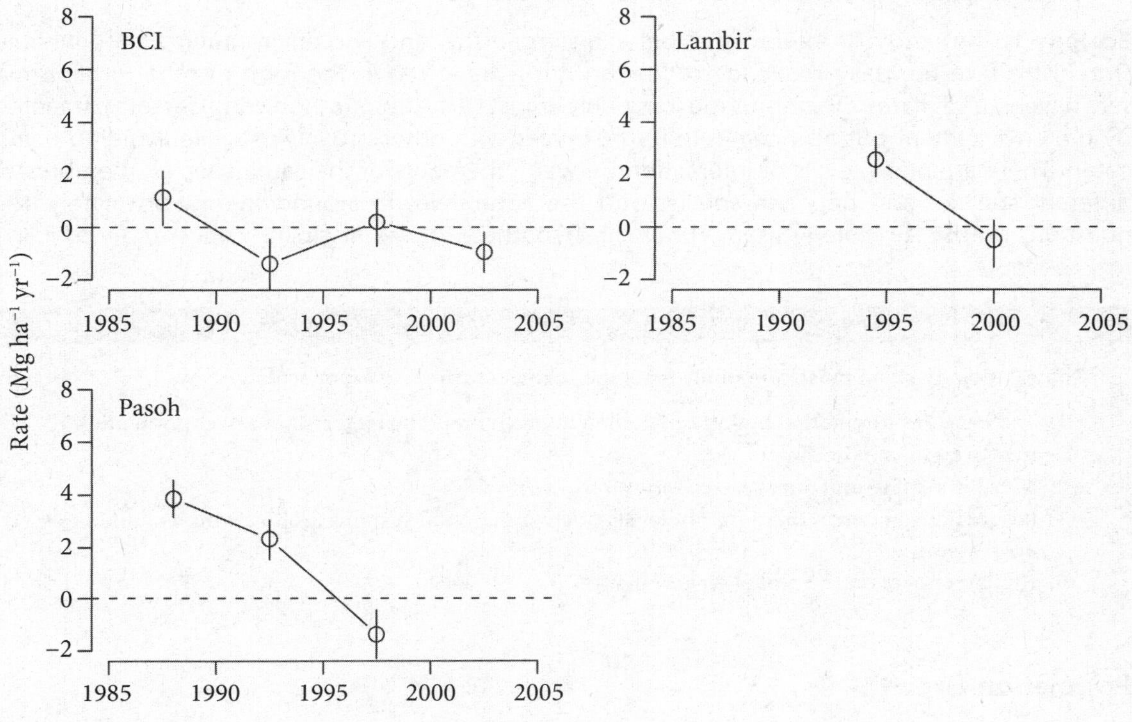

Adapted from Jérôme Chave et al., "Assessing Evidence for a Pervasive Alteration in Tropical Tree Communities," *PLoS Biol* 6, no. 3 (March 2008): e45.

Which of the following conclusions can most reasonably be drawn from the results?

(A) There is an increase in biomass growth rate in these locations. This data may be specific for the study locations and may not be generalizable.

(B) There is no change or a slight decrease in biomass growth rate in these locations. This data may be specific for the study locations and may not be generalizable.

(C) There was a net increase in biomass growth rate in some plots but not in others. More research is needed to determine whether there is a consistent pattern.

(D) There is not a change in biomass growth rate in these locations. More research is needed to determine whether there is a change in underground biomass growth rate.

Answers to this quiz can be found at the end of this chapter.

POPULATION DYNAMICS

19.1 Contrast exponential and logistic growth models

19.2 Explain how biotic and abiotic factors affect population

Ecology is the study of interactions between organisms and the environments they inhabit. Organisms face the daily challenge of finding nutrients (often in the form of other organisms), safe havens, and mates. Organisms must not only adjust to the abiotic (nonliving) factors presented by the environment but also compete for resources with other organisms while avoiding being eaten. There are many levels of interaction between members of the same species, members of different species, and between species and the natural world around them. Fortunately, the questions on the AP Biology exam about this hypothesis-dense discipline concentrate on just a few concepts.

✔ AP Expert Note

The AP Biology exam most commonly tests the following concepts from ecology:

1. **The basics of population biology, including abiotic and biotic factors that affect population size, growth, and decline**
2. **Nutrient cycling and energy exchange**
3. **Interactions between the different levels of environmental systems (populations, communities, ecosystems, etc.)**
4. **Human impact on the global environment**

Population Growth

What do you get if you put one bacterium in a lot of space and give it an unlimited amount of resources and time to reproduce? You get a whole lot of bacteria, that's what! As a matter of fact, you get an exponential rise in the number of bacteria according to the equation $N_t = N_0e^{rt}$, where

N_t = the number of bacteria at time t,

N_0 = the number of bacteria at the beginning,

r = the rate of population growth (a difference of the reproduction and death rates),

t = the number of chronological steps of reproduction (seconds, minutes, years, etc.), and

e = Euler's number (approximately 2.71828).

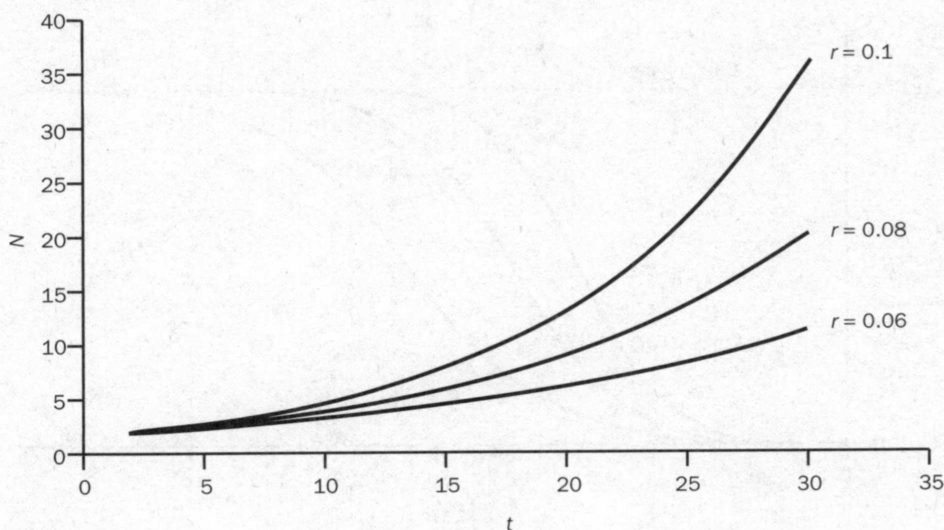

Exponential Population Increase

The three lines on the above graph show the exponential increase in population size with no limits according to the equation $N_t = N_0e^{rt}$. Each line represents a population increasing at the rate indicated, having started with two individuals.

This graph shows how rapidly populations can grow when the increase is exponential. Remember that *t* is a measure of the population's rate of producing another generation, which is very fast for bacteria. Some bacteria can reproduce every 10 to 30 minutes, so 30 time steps would only take between 5 and 15 hours. The upshot is that bacteria can grow to numbers in the millions in only a matter of hours!

In reality, populations do not grow exponentially without limits. Most reach a size at which they plateau, called the **carrying capacity**. This type of limited increase is known as *logistic growth*. The rate of change in size of natural populations can be estimated by the equation $\frac{\Delta N}{\Delta t} = rN(\frac{K - N}{K})$, where

$\frac{\Delta N}{\Delta t}$ = the change in the population size over the given time,

N = the size of the population at the beginning of time *t*,

r = the rate of population growth (again, a difference of the reproduction and death rates), and

K = the carrying capacity.

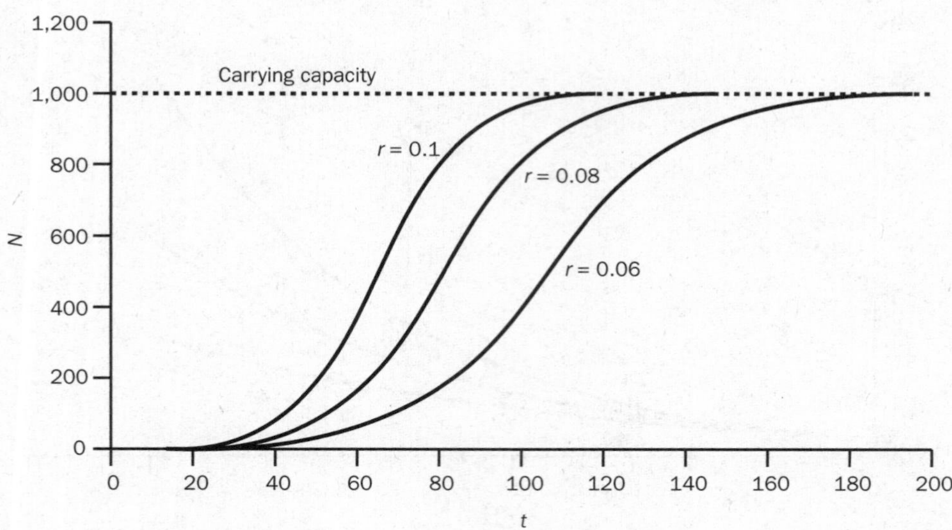

Population Increase to Carrying Capacity

These three lines show the increases in population size with the same rates of population growth as in the previous graph, but with a limited carrying capacity of 1,000 individuals. This graph shows that carrying capacity is the ultimate limit to population size, while the rate of population growth affects how quickly the population increases. A change in carrying capacity will change where the plateau occurs.

> **✔ AP Expert Note**
>
> **There is a good possibility that you will see a graph illustrating exponential growth and carrying capacity on the AP Biology exam. Be sure that you are able to interpret major trends such as exponential growth, logistic growth, *K* and *r* selection, and factors that can alter these trends.**

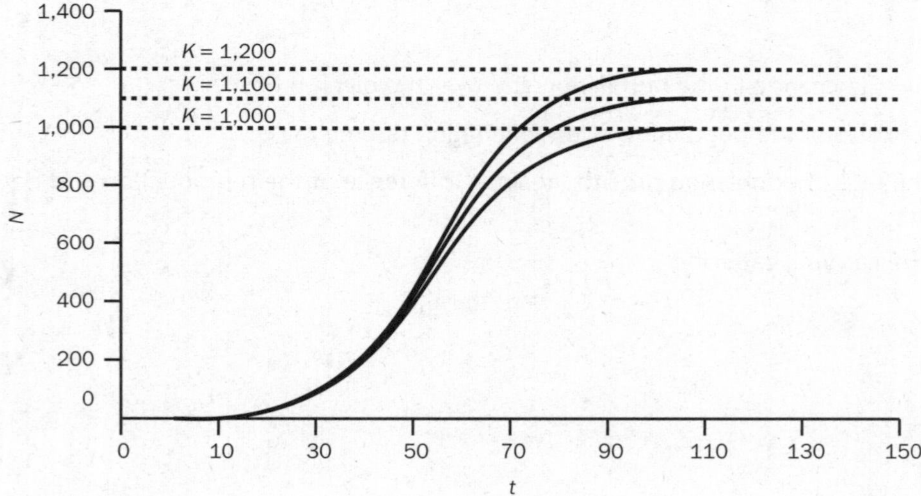

Populations with Different Carrying Capacities

The three lines in the previous graph show the increases in population size with a constant rate of population growth (r = 0.10) for three populations whose carrying capacities are not the same. Carrying capacities are indicated on the graph. When the carrying capacity increases for a population with a constant growth rate, the shape of the **growth curve** hardly changes. It is the ultimate population size that is really affected.

✔ **AP Expert Note**

Population Growth Factors

Factors that affect population growth fall into two categories, density-independent and density-dependent.

- **Density-independent factors affect populations in the same way regardless of how many individuals are in the population at that given time. These factors tend to be more catastrophic, abiotic factors, such as flood, drought, hurricane, or fire.**
- **The impact of density-dependent factors on populations increases as the number of individuals in the population, and therefore the density in a given area, increases. These factors include variables such as competition among or between species, predation, or emigration.**

Different **biotic** and **abiotic** factors in the environment have contributed to adaptive strategies that allow organisms to maximize their reproduction. Species that monopolize rapidly changing environments, such as disturbed habitats, produce many offspring quickly. These species are called ***r*-selected**, or *r*-strategists, because their strategy is to increase their *r*-value. Species that are *r*-strategists generally have short life spans, begin breeding early in life, and produce large numbers of offspring.

Species in more stable environments tend to produce fewer offspring and invest more resources into the success of each offspring. These species are called ***K*-selected**, or *K*-strategists. *K*-strategists have longer life spans, begin breeding later in life, have longer generation times, and produce fewer offspring. They take better care of their young than *r*-strategists; they have also evolved to be better at exploiting limited parts of their environments. There is a continuum between *r*-selected and *K*-selected species, as few communities are made up of strictly *r*-strategists or *K*-strategists.

Every species has a set of conditions that are optimal for its reproductive strategies; in any ecosystem, the most abundant species will be the one for which the environment most closely approximates this set of optimal conditions. If these conditions remain constant, all other things being equal, the distribution of organisms in an ecosystem should remain roughly the same.

✔ **AP Expert Note**

Without important changes to the environment or available resources, there is little reason to expect a population to boom or crash. However, these kinds of significant changes are inevitable over time, and they determine which species can thrive or dominate in an ecosystem. This change in the species makeup of an ecosystem is called ecological succession.

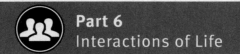

ENVIRONMENTAL DYNAMICS

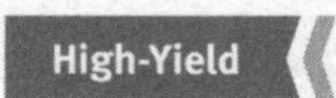

19.3 Explain how environmental factors affect organisms

19.4 Contrast how matter is recycled and energy flows

Environmental dynamics deals with the flow of energy and other nutrients through the environment. **Energy flow** is traditionally shown in the form of a **food web**. Arrows are drawn between different species of organisms, with the direction of the arrow indicating the direction of energy flow.

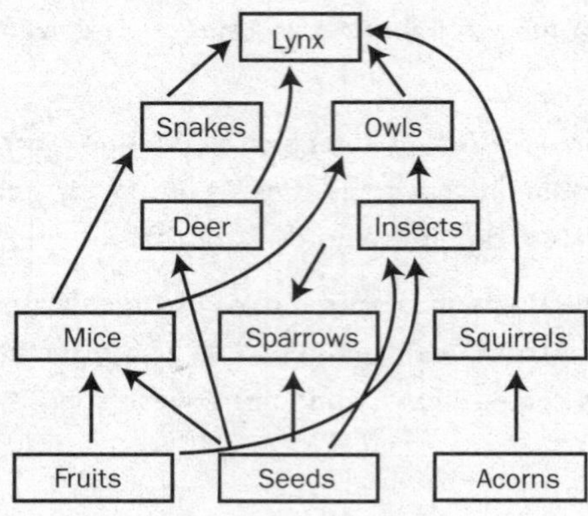

Food Web

In the illustration of a terrestrial food web, the arrows indicate the direction of energy flow between the organisms. The fruits, seeds, and acorns only have arrows pointing away from them, so they are primary producers. The lynx is the top predator (tertiary **consumer**) and only has arrows pointing toward it. There are no decomposers shown in this food web.

> **✔ AP Expert Note**
>
> **Energy is lost as it moves through the various food webs. Only about 10 percent of the energy received is passed from one trophic level to the next.**

Food webs can quickly become quite complicated because there are so many different organisms interacting in any given ecosystem. Food webs are rarely entirely comprehensive for the same reason. As one kind of organism feeds another, nutrition moves through a food web from one trophic level to the next. Organisms included in a food web may be **primary producers** (**autotrophs**), which produce energy from sunlight, **primary consumers** and **secondary consumers**, and **decomposers**. Because they make their own food, primary producers only have arrows pointing away from them in a food web. Organisms that get energy by consuming other living things (**heterotrophs**) have arrows pointing toward them, indicating the flow of energy from another trophic level. If decomposers are part of the food web, there will also be arrows leading from the various levels of heterotrophs to the decomposers. Primary consumers eat plants, while secondary consumers, tertiary consumers, and quaternary consumers feed on the levels below them, such as the primary consumers. **Herbivores** eat only plants (thus, they are always primary consumers).

However, some secondary and tertiary consumers are omnivores, which eat both animals and plants. Omnivores in a food web are indicated by arrows pointing toward them from both primary producers and primary consumers.

✓ AP Expert Note

It is important to understand productivity as it relates to energy flow in food webs—many AP Biology exam questions will deal with differences between gross and net primary productivity and factors that influence productivity within a food chain or food web.

Other simplified models illustrating how nutrients are cycled include those for water, nitrogen, carbon, and phosphorous. These systems are known as biogeochemical cycles.

In the *water cycle*, the Sun's energy drives the movement of water through the environment. The Sun causes surface water to evaporate and plants to transpire, releasing water vapor into the atmosphere. In the atmosphere, the water vapor cools and condenses into clouds and precipitation. Precipitation returns water to the Earth's surface, where it runs off into lakes and rivers and percolates into groundwater.

There is more nitrogen in the air than oxygen: nitrogen gas makes up over two-thirds of Earth's air. However, most living things cannot get the nitrogen they need from nitrogen gas. Certain groups of bacteria in the soil, called *nitrogen-fixing bacteria*, convert nitrogen gas into nitrates and nitrites; plants then take up the nitrites and nitrates from the soil and from fertilizers. Animals, including humans, get nitrogen by eating plants and other living or dead organic matter. When animals and plants die, groups of bacteria and fungi break down these organic remains and release nitrogen back into the soil. Animal waste contains nitrogen as well. *Denitrifying bacteria* in the soil then change the nitrates and nitrites back into nitrogen gas, which returns to the air. Together, all of these steps make up the *nitrogen cycle*.

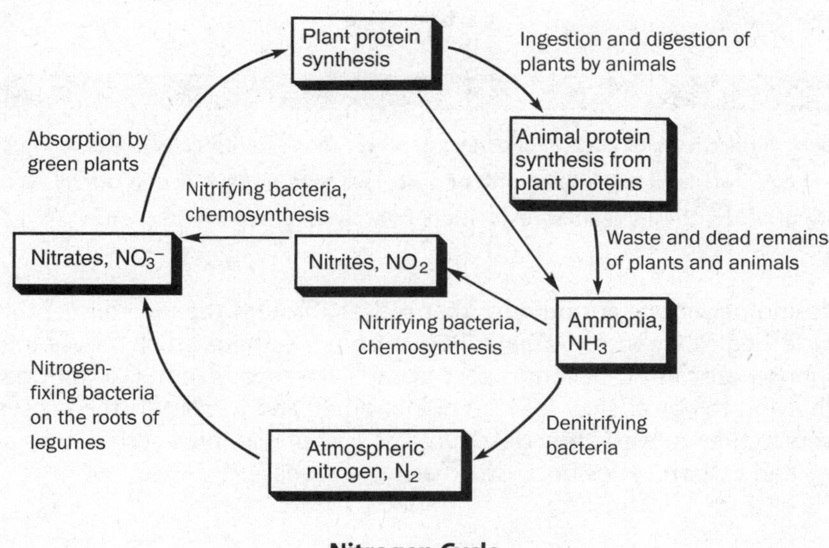

Nitrogen Cycle

Photosynthesis and respiration are the complementary reactions of the **carbon cycle**. In photosynthesis, carbon dioxide and water combine to form sugars and oxygen; the Sun's energy is stored in the bonds among carbon atoms in the sugars. Respiration uses carbohydrates and oxygen to produce carbon dioxide, water, and energy; it releases the energy that is stored by photosynthesis. As animals eat the plants and one another and then breathe, carbon moves through the food web; the carbon that plants and animals contain is returned to the soil when they decompose. The carbon in the soil enters the cycle again, perhaps as part of a microorganism. A great deal of carbon is stored in the oceans and in minerals. Combustion is another way that carbon enters the atmosphere, in the form of carbon dioxide. The widespread and heavy use of fossil fuels has created concerns about how all of this combustion is affecting the environment.

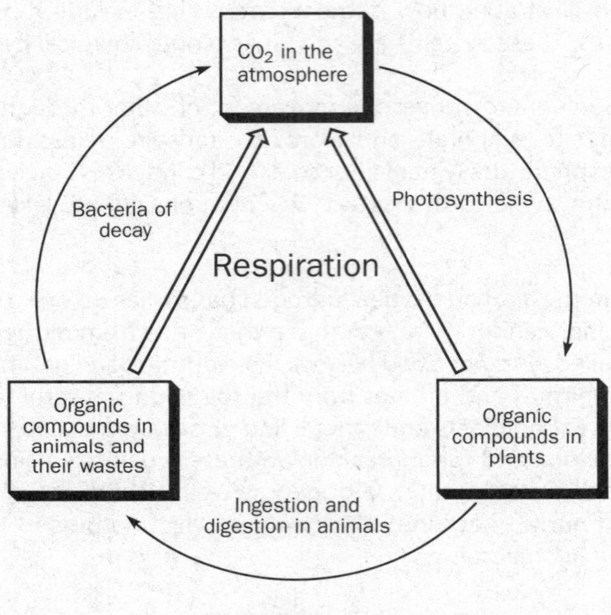

Carbon Cycle

> **✔ AP Expert Note**
>
> **The biogeochemical cycles of carbon and nitrogen are good examples of areas in which you can relate several concepts in biology. Examples of processes that are involved in the recycling of these elements are photosynthesis, cellular respiration, hydrolysis, and condensation.**

Phosphorous is another important nutrient that moves through the environment in a cycle. The phosphorus cycle begins when phosphate (PO_4^{3-}) from weathered rock moves into soils. Plants then take up phosphate, and it becomes part of the living ecosystem. Like nitrogen, phosphate moves through living things as they feed on one another, and it reenters the ecosystem through the decomposers' action on waste and dead remains. The phosphorus cycle stands apart from the water, nitrogen, and carbon cycles because it has no gas phase.

ECOSYSTEMS

19.5 Explain how diversity affects stability

In ecology, the levels of organization from lowest complexity to highest are organism, population, community, ecosystem, biome, and biosphere. **Populations** are all the organisms that occur in a specific **habitat**, and **communities** are all the populations that live in that habitat. **Ecosystems** encompass communities and their abiotic environment, including their interactions through which nutrients and energy flow. The role and position a species has in an ecosystem and its interactions with abiotic and biotic factors of the environment is called an ecological **niche**. Ecosystems within a specific geographic region form **biomes**, and together all ecosystems form a complex web that make up the biosphere.

An ecosystem can be a tropical rain forest, an entire mountain range, or a coral reef, but can also be an ephemeral pool of water in the middle of the desert. Due to limiting factors, such as food, nutrients, land, water, and sunlight, ecosystems are unable to support an unlimited number of organisms. The health and balance of ecosystems are sustained by the biodiversity of the ecosystem. Biodiversity includes the genetic diversity of the species, as well as the various ways species interact with each other and their environment. All species in an ecosystem depend on one another. As seen in chapter 16, the level of variation in a population affects population dynamics (size and age). Similarly, the diversity within an ecosystem may influence the stability of the ecosystem. Providing robustness to the ecosystem, diversity increases the ability of ecosystems to tolerate and respond to changes in the environment, as well as prevents widespread diseases. Keystone species, producers, and abiotic and biotic factors all contribute to maintaining the diversity of an ecosystem. Human activities, however, have strained the diversity of ecosystems and could eventually lead to another mass extinction.

Keystone species are species that help to increase the diversity of an ecosystem. Many species are dependent upon keystone species. Thus, the disappearance of keystone species from an ecosystem would drastically affect the balance of the ecosystem and may lead to the disappearance of other species. For example, sea otters, a keystone species, feed on sea urchins, which feed on kelp. The regulation of sea urchin populations, in turn, maintains sufficient kelp for food and habitats of other species like fish, which use kelp to hide from predators. The removal of sea otters would drastically affect the marine ecosystem since, without a predator, sea urchins would consume all the kelp.

Other predators, like jaguars, which have a diet of many different species, keep species in balance by controlling distribution and population of prey species. Without predators, prey would increase and lead to competition with each other for resources and possibly the decline of a species. Beavers alter the environment by building dams and creating wetlands upon which the prevalence and activities of many species depend. Bees and hummingbirds are also keystone species because they contribute to the survival of several plant species through pollination. Plants provide shelter for insects, which are food for other species.

Plants not only provide habitats for a variety of species, they also provide food. Plants, or producers, convert energy from the environment into organic compounds, which begins the energy flow through organisms living in ecosystems. Energy is required for a stable ecosystem and, without producers, ecosystems would have no basis of energy flow.

Abiotic factors including climate (temperature, humidity, day length, rainfall) and physical conditions (pH, water, ions) affect the amount of environmental stress on an ecosystem and, in turn, the stability of the ecosystem. Abiotic factors that increase environmental stress decrease ecosystem stability, whereas abiotic factors that decrease environmental stress increase ecosystem stability.

GLOBAL ISSUES

19.6 Describe the human impact on biodiversity and climate

Other than the major extinction events recorded every 100 million years or so in Earth's geological history, the only variables that seem to have a global impact are related to human activity. You should be familiar with two of these variables.

The Earth is surrounded by a layer of ozone (O_3). The layer lies in the stratosphere, 10–17 kilometers above the Earth's surface, and protects the surface from harmful UV rays emitted by the Sun. Over the past few decades, scientists have determined that the ozone layer is being depleted by compounds containing chlorine, fluorine, bromine, carbon, and hydrogen (halocarbons) that seem to be produced by human activity. One particular group of chemicals, CFCs (chlorofluorocarbons), has received a considerable amount of the blame for depletion of the ozone layer.

✔ AP Expert Note

The ozone layer may be depleted by as much as 60 percent over the Antarctic in the spring, causing the notorious hole in the ozone layer. Scientists worry that without changes in human activity, the ozone layer will be depleted beyond recovery, preventing certain forms of life from surviving on Earth.

The other major global impact of human activity is the release of greenhouse gases (water vapor, carbon dioxide, methane, and nitrous oxide) into the atmosphere from the inefficient burning of fossil fuels (there are sources from manmade aerosols as well). As these gases become denser in the Earth's atmosphere, they prevent heat from escaping the Earth's surface, causing global warming. Scientists believe that the Earth's atmospheric temperature is steadily rising and that this will cause major shifts in seawater levels, local climates, and world weather patterns.

Habitat destruction occurs as people clear natural areas for natural resources, housing, and recreational areas. This has resulted in the extinction of many species of plants and animals and has an adverse impact on our worldwide water resources, global temperatures, and food availability.

AP BIOLOGY LAB 10: ENERGY DYNAMICS INVESTIGATION

High-Yield

19.7 Investigate energy flow through a system

Using a model ecosystem, this investigation explores how matter and energy flow, the roles of producers and consumers, and the complex interactions between organisms. First, estimate the net primary productivity of Wisconsin Fast Plants growing under lights, and then determine the flow of energy from the plants to cabbage white butterflies as the larvae consume cabbage-family plants.

Remember that the source of almost all energy on Earth is the Sun. Free energy from sunlight is captured by producers that convert the energy to oxygen and carbohydrates through photosynthesis. The net amount of energy captured and stored by the producers in a system is the system's net productivity. Net productivity is calculated by assuming the change in biomass of a plant is due to uptake and use of energy.

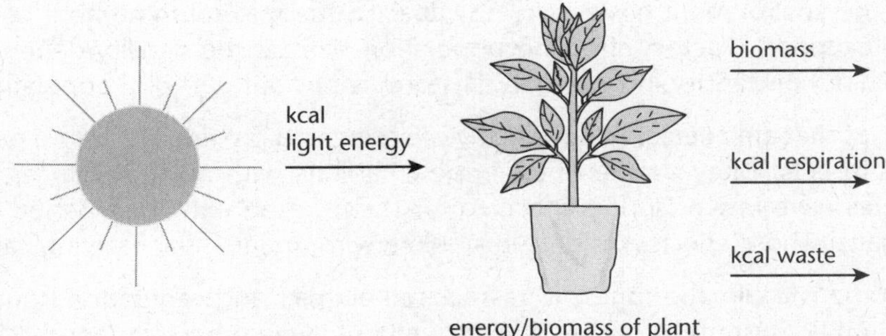

To determine the net primary productivity of your plants, convert the difference in biomass over the growing time to energy according to the following equation:

$$\text{Energy} = \text{Biomass in grams} \times 4.35\ \text{kcal/gram}$$

Let's look at a sample set of data for 10 plants grown over seven days. In seven days, the 10 plants gain 4.2 grams of dry mass. Using the equation, you can determine that 4.2 grams of dry mass is equivalent to 18.27 kcal of energy captured from the Sun. To determine the net primary productivity per plant per day, you must divide that total amount of energy by 10 plants and by seven days. In this example, the net primary productivity is 0.26 kcal per plant per day.

Age of Plants	Dry Mass Gained	Energy (g biomass × 4.35 kcal/g)	Net Primary Productivity per Plant per Day
7 days	4.2 grams	18.27 kcal	0.26 kcal/day

The efficiency of energy transfer from producer to primary consumers varies with the type of organism and with the characteristics of the ecosystem. In the second part of the investigation, determine the biomass change in butterfly larvae that eat your plants to evaluate the energy use of primary consumers and ultimately the energy efficiency of the relationships.

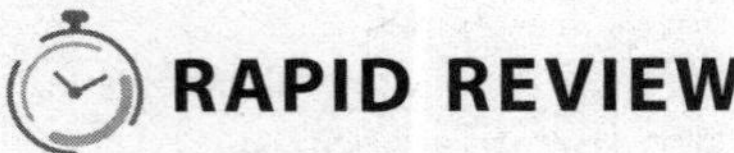

RAPID REVIEW

If you take away only 6 things from this chapter:

1. Populations with infinite resources increase exponentially in a J-shaped curve. Usually, the environment has a carrying capacity, the maximum number of individuals it will support. Factors affecting population size can be density-dependent (overcrowding) or density-independent (a storm wipes out part of a population).
2. Species that are *r*-selected have many offspring and provide almost no parental care (fish, insects). They succeed best in new habitats with many resources. *K*-selected species have few offspring and care for them for an extended period (elephants, humans). Those species do best in stable environments near carrying capacity.
3. A biome is a climatic zone with associated animals and vegetation (tundra, desert, etc.). An ecosystem comprises a community of living organisms and its habitat. Different populations of organisms make up the community. Each organism is adapted to a specific niche or role.
4. Food webs trace energy flow in a community. Different trophic levels consist of primary producers (plants), primary consumers (herbivores), secondary consumers (carnivores), and decomposers (bacteria).
5. Materials such as water, nitrogen, carbon, and phosphorus travel through the environment and are recycled in biogeochemical cycles involving different life forms.
6. Human activity has affected the environment significantly (e.g., ozone depletion, habitat destruction, and global warming).

TEST WHAT YOU LEARNED

1. Oligotrophic lakes are nutrient poor and have very clear water. However, large quantities of fertilizer runoff can cause more biological growth in these lakes. Continuing input of fertilizer can transform oligotrophic lakes into eutrophic lakes that are rich in organic material. How does the addition of fertilizer ingredients, such as ammonia, lead to changes in the lakes?

   (A) Ammonia in fertilizer can be broken down to release energy through cellular respiration, making it important for organisms to be able to carry out metabolism.

   (B) Ammonia can be broken down to release heat, making it important in thermoregulation within the lake.

   (C) Ammonia plays a role in the nitrogen cycle, making nitrogen available for organisms, such as plants and animals.

   (D) Ammonia is a major source of energy for producers and a component of the nitrogen cycle, so its presence is sufficient to promote the growth of photoautotrophs.

2. A study investigated how various disturbances affect coral reef recovery. Data from the study is shown in the graphs below. The asterisks represent disturbances that may affect the amount of coral.

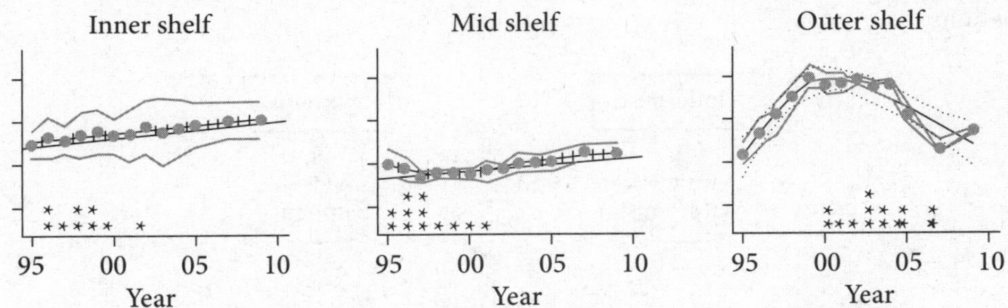

Percent coral cover over the years shown. The average linear trend is depicted with a straight line, whereas the dashed lines represent subregions. Each asterisk shows the timing of a disturbance.

Adapted from Kate Osborne et al., "Disturbance and the Dynamics of Coral Cover on the Great Barrier Reef (1995–2009)," *PLoS ONE* 6, no. 3 (March 2010): e17516.

Which of the following is most justified based on this data?

   (A) The coral populations on the inner shelf and mid shelf increased slightly when the disturbances ceased. The outer shelf population had a period approximating exponential growth followed by a period of disturbance and decrease in growth.

   (B) The coral populations on the inner shelf and mid shelf were relatively stable during the study due to lack of disturbance. The outer shelf population was affected by more disturbances and exhibited a classic logistic growth curve.

   (C) All three figures show classic exponential growth curves.

   (D) All three figures show classic logistic growth curves.

3. A researcher decided to create a trophic pyramid by modeling an ecosystem with plants, caterpillars that eat the plants, and birds that eat the caterpillars. Each part of the pyramid represents the biomass of that level, with primary producers on the bottom. The plants were first grown in low light, and the light intensity was raised over time to the maximum intensity that doesn't harm the plants. How would the trophic pyramid most likely change over the course of this experiment?
   (A) The pyramid would become inverted due to the rapid movement of energy and materials from one level to another.
   (B) The level representing the primary producers would increase, but the upper levels would stay the same size as before because of the limits on energy transfer.
   (C) The pyramid would stay the same because the amount of energy that could be transferred from one level to another would be unchanged.
   (D) There would be an increase in the size of each piece of the pyramid due to gradually increasing biomass at each level.

4. The figure below shows a model generated from a study of interactions in a Serengeti ecosystem. This figure describes hypothesized relationships between a parasite that causes disease in wildebeest, rinderpest, and other important components of the ecosystem. Thick arrows represent dominant effects, and "grass" is in a dotted circle because that variable was not directly measured in the study.

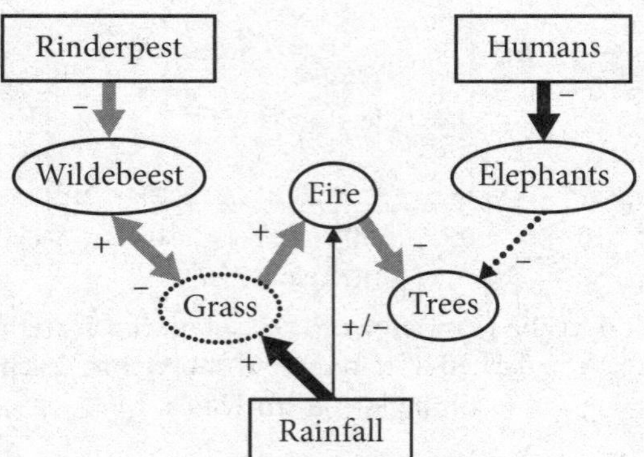

Adapted from Ricardo M. Holdo et al., "A Disease-Mediated Trophic Cascade in the Serengeti and its Implications for Ecosystem C," *PLoS Biol* 7, no. 9 (September 2009): e1000210.

Which of the following can most reasonably be predicted based on this figure?
   (A) Increasing the amount of rinderpest would decrease the amount of grass.
   (B) Increased fire would lead to higher tree and elephant populations.
   (C) Increasing the amount of rinderpest could indirectly lower tree numbers.
   (D) Increased human populations have a net positive effect on the elephant population.

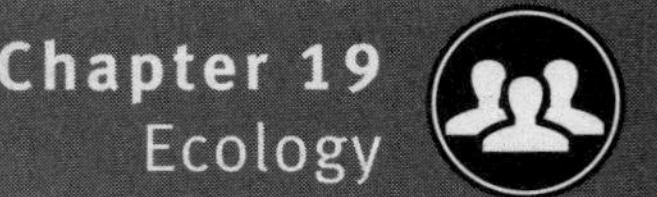

5. A researcher is interested in studying how biotic and abiotic factors influence the population of a particular species. Which of the following experimental designs would most effectively test whether one abiotic and one biotic factor influenced a population?

   (A) The researcher conducts three experiments. In one, she manipulates prey density. In a second, she manipulates population density of the study species. In the third, she manipulates both prey and study species population density to examine the effects of interactions between prey and study species density.

   (B) The researcher conducts three experiments. In one, she manipulates light levels. In a second, she manipulates prey availability. In a third, she manipulates light levels and prey availability to examine the effects of interactions between light levels and prey availability.

   (C) The researcher conducts two experiments. In one, she manipulates the number of cover objects available for the animals to use for shelter. In another, she manipulates the amount of light available in the study areas.

   (D) The researcher conducts a single experiment in which she compares animals with high light levels and high prey density to animals with low light levels and low prey density.

6. As it has become easier for humans to travel around the world, there have been some unintended consequences. Which of the following is NOT an unintended consequence of increased human movement and trade around the world?

   (A) Species have been introduced into new areas with food shipments, allowing them to compete with native species and disrupt ecosystems.

   (B) Pathogens can rapidly reach new locations via airplanes, potentially harming human and animal populations.

   (C) Ship ballast water can contain living organisms, and these can be transported to new destinations between ballast water discharges.

   (D) Bacteria can develop resistance after being exposed to antibiotics, making the illnesses they cause harder to treat.

7. Figure A below shows the relative fitness of a prey species when one or two predators are present. Figure B shows how the prey phenotype is predicted to change depending on which predators are present.

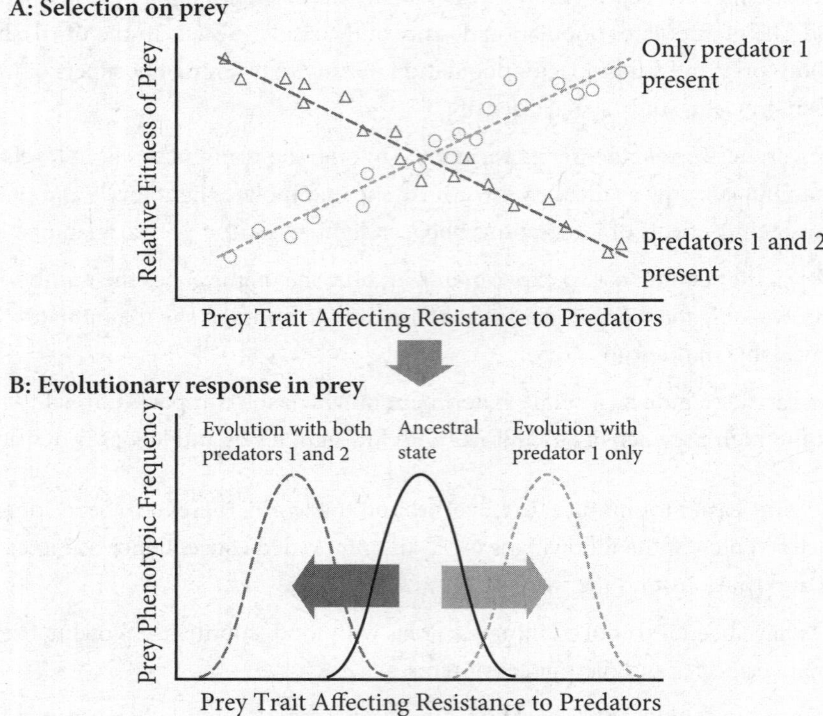

Adapted from Martin M. Turcotte, Michael S. C. Corrin, and Marc T. J. Johnson, "Adaptive Evolution in Ecological Communities," *PLoS Biol* 10, no. 5 (May 2012): e1001332.

Which of the following conclusions can most reasonably be drawn regarding this system?

(A) The change in phenotypic frequency shown can only occur if the trait is ubiquitous in the population.

(B) The change in phenotypic frequency shown can only occur if there is existing genetic variation for the trait in the population.

(C) The trait under selection must be inherited in a Mendelian manner in order for changes in phenotypic frequency to occur.

(D) The plot lines of figure A must have a different shape in order for the changes in phenotypic frequency to occur.

Questions 8–9

The graph below shows lynx and snowshoe hare population levels as estimated by pelt harvest records. The snowshoe hare is an herbivore and is the predominant food source for the lynx.

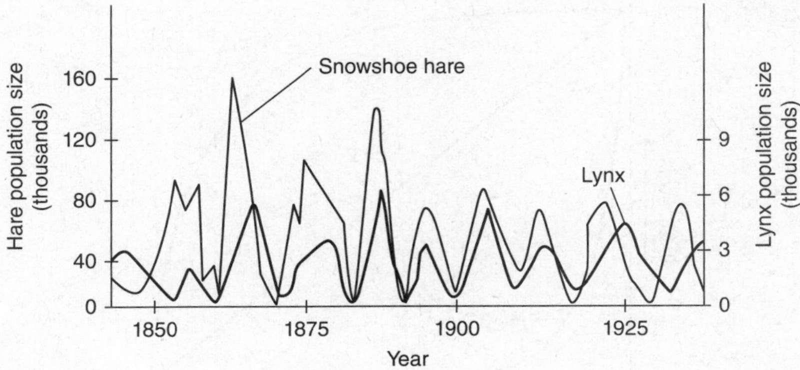

8. One hypothesis to explain the cycling is that the predator population size drives the population size of the prey. When the predator population increases beyond a certain point, the lynx eat more snowshoe hares than the snowshoe hares can replace through reproduction. The snowshoe hare population crashes as a result of predation by the overabundant lynx. The lower availability of snowshoe hares to serve as food for the lynx causes many lynx to die from starvation. Which of the following would be a reasonable experiment to test the hypothesis that the lynx population levels would fall due to a lack of food?

   (A) Maintain the lynx and snowshoe hare populations in an enclosure to see if the same cycles occur.

   (B) Supply the lynx with alternative food sources when the snowshoe hare population falls.

   (C) Remove the lynx, and see if the snowshoe hare population grows beyond previously established maximum levels.

   (D) Provide the snowshoe hares with alternative food sources when the lynx population falls.

9. One hypothesis suggests that as the snowshoe hare population increases, plants produce more antiherbivore toxins. This leads to poorer nutrition and increased mortality in the snowshoe hares. This drop in the snowshoe hare population causes a decrease in the lynx population. Which of the following experiments would test this hypothesis?

   (A) Remove the lynx from the environment.

   (B) Provide the snowshoe hares with food sources that do not contain these toxins.

   (C) Sample the plants for toxins in the years after the snowshoe hare population crashes.

   (D) Trap snowshoe hares, and remove them from this environment to keep their population levels low.

Questions 10-12

The graph below shows three different types of survivorship curves. Different species of organisms have different patterns of survival over their lifetimes.

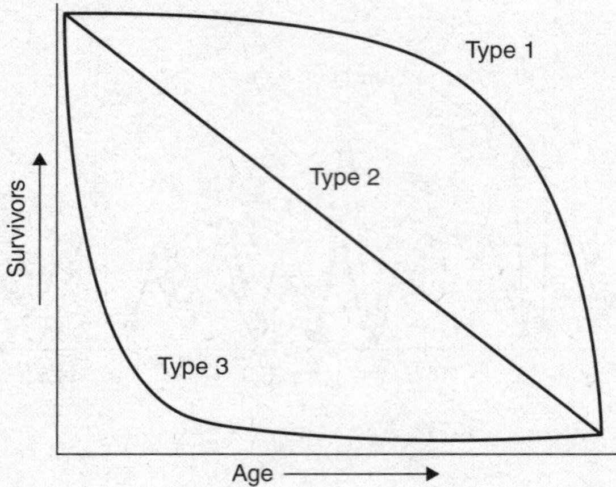

10. Approximately 1 out of every 1,000 green sea turtle eggs that are laid will reach adulthood. Adults can live for 100 years. According to the graph, which type of survivorship curve do green sea turtles best match?

    (A) Type 1

    (B) Type 2

    (C) Type 3

    (D) Green sea turtles do not follow any of the survivorship curves shown on this graph.

11. Giant sequoia (*Sequoiadendron giganteum*) trees produce millions of very small seeds. Very few of those seeds reach adulthood. According to the graph, which type of survivorship curve do giant sequoia trees best match?

    (A) Type 1

    (B) Type 2

    (C) Type 3

    (D) Giant sequoia trees do not follow any of the survivorship curves shown on this graph.

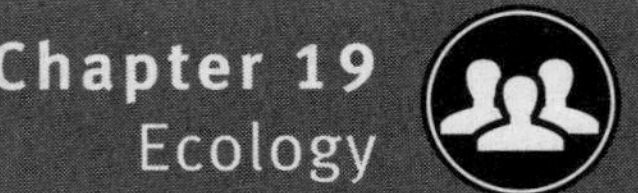

12. Elephants have relatively few young, have a gestation period of approximately 2 years, and provide extensive care for their offspring. As a result, a large percentage of young elephants live to adulthood. The death rate becomes high in old elephants. According to the graph, which type of survivorship curve do elephants best match?

    (A) Type 1

    (B) Type 2

    (C) Type 3

    (D) Elephants do not follow any of the survivorship curves shown on this graph.

Answer Key

Test What You Already Know

1. D
2. A
3. C
4. B
5. A
6. D
7. B

Test What You Learned

1. C
2. A
3. D
4. C
5. B
6. D
7. B
8. B
9. B
10. C
11. C
12. A

REFLECTION

Test What You Already Know score: __________

Test What You Learned score: __________

Use this section to evaluate your progress. After working through the pre-quiz, check off the boxes in the "Pre" column to indicate which Learning Objectives you feel confident about. Then, after completing the chapter, including the post-quiz, do the same to the boxes in the "Post" column. Keep working on unchecked Objectives until you're confident about them all!

Pre	Post	
☐	☐	**19.1** Contrast exponential and logistic growth models
☐	☐	**19.2** Explain how biotic and abiotic factors affect population
☐	☐	**19.3** Explain how environmental factors affect organisms
☐	☐	**19.4** Contrast how matter is recycled and energy flows
☐	☐	**19.5** Explain how diversity affects stability
☐	☐	**19.6** Describe the human impact on biodiversity and climate
☐	☐	**19.7** Investigate energy flow through a system

FOR MORE PRACTICE

Complete more practice online at kaptest.com. Haven't registered your book yet? Go to kaptest.com/booksonline to begin.

ANSWERS AND EXPLANATIONS

Test What You Already Know

1. D Learning Objective: 19.1

The population may grow very rapidly at first as it accesses new resources but, eventually, a larger population will make resources more limited. This will form a logistic growth curve, as suggested in correct **(D)**. (A) is incorrect because exponential growth cannot continue forever; given a large enough population, resources will start to become limited. (B) and (C) are incorrect because species are capable of thriving in new environments if conditions are suitable (and the circumstances in the question stem suggest that they are).

2. A Learning Objective: 19.2

Because many more birds were captured in the time immediately following Hurricane Iris, it is not reasonable to conclude that the birds experienced a high rate of mortality. **(A)** is thus correct. (B) is incorrect because this could explain why more birds were captured immediately after the hurricane. (C) is incorrect because a change in foraging habits could lead to more birds being captured. (D) is incorrect because a tendency to spread out further could also lead to more of the birds being captured.

3. C Learning Objective: 19.3

The question stem explains that rinderpest makes wildebeest sick, so a reduction in rinderpest should lead to an increase in the wildebeest population. Because the wildebeest are grazers, an increase in their population should lead to a decrease in the amount of grass, which in turn will make it harder for fires to spread. **(C)** is thus correct. (A) is incorrect because more wildebeests will result in less grass, not more grass. (B) and (D) are incorrect because the wildebeest population should increase when rinderpest prevalence decreases.

4. B Learning Objective: 19.4

Plants obtain energy from the Sun, so the role of fertilizer is not to provide energy but to provide other nutrients that the plant needs, such as nitrogen, which is used to build proteins and other organic molecules. **(B)** is thus correct. (A) is incorrect because plants do not derive energy directly from the fertilizer and because the nitrogen is not consumed, but is recycled through the nitrogen cycle. (C) is incorrect because plants do not take up energy directly from the fertilizer. (D) is incorrect because nitrogen does not slow down plant metabolism.

5. A Learning Objective: 19.5

Small populations tend to experience a lot of inbreeding, which increases their susceptibility to deleterious recessive alleles. By introducing the Texas panthers, the researchers were able to increase the genetic diversity in the population, increasing the chances that the population will survive. **(A)** is thus correct. (B) is incorrect because the genetic differences from the Texas subspecies actually helped to promote the survival of the population, so the genetic differences did matter. (C) is incorrect because there is no reason to suspect that the Texas panthers would be better adapted to the environment than the Florida panthers that had already evolved there. (D) is incorrect because the population was not beyond saving.

6. D Learning Objective: 19.6

According to the data provided, vertebrates are able to reach levels above the reference values, while macroinvertebrates and plants always remain at or below the reference values. This suggests that vertebrates can colonize wetlands easily, but that it is not so easy for plants and macroinvertebrates. **(D)** is thus correct. (A) is incorrect because the macroinvertebrates actually have lower initial mean response ratios than plants. (B) is incorrect because the animal populations represented in Figure A do eventually reach stable levels. (C) is incorrect because macroinvertebrate density did reach the reference levels, even though it did drop off later.

7. B Learning Objective: 19.7

The plots showed relatively stable or decreasing net biomass growth, but three plots is probably not sufficient to make generalizations. **(B)** is thus correct. (A) and (C) are incorrect because none of the locations showed an increase in biomass growth. (D) is incorrect because there was some change in the biomass growth rate in the three locations.

Test What You Learned

1. C Learning Objective: 19.4

Ammonia can be taken up by plants and transformed into other forms of nitrogen, which can then be taken up by other organisms. Because ammonia plays a role in the nitrogen cycle, **(C)** is correct. (A) is incorrect because ammonia is not broken down in cellular respiration. (B) is incorrect because organisms do not break down ammonia to release heat. (D) is incorrect because ammonia is not a major source of energy for producers.

2. A Learning Objective: 19.1

According to the graphs, the inner shelf and mid shelf coral populations were relatively stable with slight increases during the period without disturbances. However, the outer shelf population had a period of rapid growth followed by a drop after disturbances. **(A)** is thus correct. (B) is incorrect because the outer shelf does not exhibit a logistic (S-shaped) growth curve since the population drops off rather than plateauing. (B) also says that the inner and mid shelf lacked disturbance, which is false based on the data. (C) and (D) are incorrect because none of the figures showed classic exponential or logistic growth curves.

3. D Learning Objective: 19.7

Increasing the light intensity will cause the plants to engage in additional photosynthesis, increasing their biomass. This, in turn, will cause increases in the biomass of the caterpillars and birds. **(D)** is thus correct. (A) is incorrect because trophic pyramids are only very rarely inverted. (B) is incorrect because the other two levels would also increase in biomass. (C) is incorrect because all the levels would increase in biomass.

4. C Learning Objective: 19.3

Increased rinderpest prevalence would lower wildebeest populations, increasing the amount of grass (because wildebeests feed on the grass) and providing more material to start and spread fires, thereby decreasing the number of trees. **(C)** is thus correct. (A) is incorrect because more rinderpest means fewer wildebeests, which in turn leads to more grass. (B) is incorrect because fire reduces the tree population. (D) is incorrect because the figure suggests that humans have only an adverse effect on the elephant population.

5. B Learning Objective: 19.2

Light levels are an abiotic factor, while prey availability is a biotic factor. **(B)** thus presents an experimental design that investigates exactly one abiotic and one biotic factor. (A) is incorrect because it only investigates biotic factors. (C) is incorrect because it only investigates abiotic factors. (D) is incorrect because it manipulates the biotic and abiotic factors simultaneously, making it impossible to determine the effects of each factor individually.

6. D Learning Objective: 19.6

The spread of antibiotic resistance is typically a direct result of excessive use of antibiotics in medicine, rather than a result of human travel and trade. Antibiotics kill nonresistant bacteria, leaving behind only those bacteria able to resist them. The surviving bacteria then reproduce, passing on antibiotic resistance genes to all offspring. **(D)** is thus correct. The remaining answers all present unintended consequences of human transportation.

7. B Learning Objective: 19.5

In order for selection to favor some individuals in the prey species over others, there must already be genetic variation with respect to the trait being selected for. **(B)** is thus correct. (A) is incorrect because selection can occur even when a trait is relatively rare; it need not be ubiquitous. (C) is incorrect because non-Mendelian traits are also subject to selection. (D) is incorrect because the plot lines depicted in Figure A are sufficient to produce the phenotypic changes in Figure B.

8. B Learning Objective: 19.3

(B) If the drop in the lynx population levels was due to a lack of food, providing the lynx with alternative food sources when the snowshoe hare population falls should remove that effect and prevent the cycling. (A) is incorrect because a lack of cycling might simply be caused by the fact that the animals are in enclosure. (C) is incorrect because this experiment would not test the hypothesis stated in the question, which is that the lynx population levels fall due to a lack of food. (D) would not test the effect of a lack of food on the lynx population.

9. B Learning Objective: 19.2

(B) If the drop in the snowshoe hare population was due to toxins in the snowshoe hares' food, providing an alternative food source should prevent the crash. Removing the lynx from the environment would not test the effects of these toxins, so (A) is incorrect. Sampling the plants for toxins after the snowshoe hare population crashes would give no information about why the population crashed, so (C) is incorrect. (D) is incorrect because simply removing the hares from this environment would give no information about the toxicity of the plants or the effect of plant toxicity on the snowshoe hare population levels.

10. C Learning Objective: 19.5

(C) Lots of green sea turtles die young, but the adults live a long time. Therefore, the curve should plummet initially and then level off with increasing age. Thus, green sea turtle survivorship is a Type 3 curve, according to this graph. In a Type 1 survivorship curve, young individuals have a relatively low death rate compared to old individuals. The curve stays relatively level and then plummets with old age. This is the opposite of what happens with green sea turtles, so (A) is incorrect. In a Type 2 survivorship curve, individuals are equally likely to die at any age. However, green sea turtles have a much greater mortality rate when they are young than when they are old. Therefore, (B) is incorrect. (D) is incorrect because a Type 3 survivorship curve accurately describes the situation for green sea turtles.

11. C Learning Objective: 19.5

(C) Although giant sequoias are majestic trees, the amount of seeds and seedling mortality clearly makes this a species with a Type 3 survivorship curve. (A) is incorrect because a Type 1 curve is the opposite of what happens with giant sequoias. For Type 1 to be correct, the seeds would have to have a large chance of survival and the older trees would have to have a much greater chance of dying. This is not the case for giant sequoia trees. (B) is incorrect because a Type 2 curve applies to organisms that have a constant death rate, regardless of the age of the organism. (D) is incorrect because giant sequoia trees do show a Type 3 survivorship curve.

12. A Learning Objective: 19.5

(A) Elephants bear few young, most of which survive. Therefore, their curve begins high and stays high, only dropping off with old age. (B) is incorrect because elephants do not have a constant death rate regardless of age. (C) is the opposite of what is true for elephants. Young animals die infrequently. The death rate increases as elephants get older. (D) is incorrect because elephants do follow a Type 1 survivorship curve.

PART 7

Practice

CHAPTER 20

Free Response

LEARNING OBJECTIVES

In this chapter, you will learn how to:

20.1 Use the reading period to outline your responses

20.2 Craft effective responses for long free-response questions

20.3 Craft effective responses for short free-response questions

20.4 Use a scoring rubric to self-score your responses

FREE RESPONSE STRATEGY

20.1 Use the reading period to outline your responses

20.2 Craft effective responses for long free-response questions

20.3 Craft effective responses for short free-response questions

Approaching the 10-Minute Reading Period

There is a 10-minute reading period sandwiched between the multiple-choice and grid-in section and the free-response portion of the AP Biology exam. Note that this is a "reading" period, not a "nap" period. Ten minutes isn't much time, but it does give you an opportunity to read the essay questions. You can write notes in the question booklet, but these will not be seen by graders.

This gives you just over one minute per question to plan your response. Take at least 30 seconds to read and reread each question. Make sure you understand what is being asked. Next, you should jot down any thoughts you have about the answer on a piece of scratch paper or in the test booklet. Write down keywords you want to mention. At some point or other, you have probably brainstormed ideas when writing an essay for your English class. This is exactly what you want to do here as well: Brainstorm ideas about the best way to answer each free-response question.

With the remaining time, make a quick outline of how you would answer each question. You don't need to write complete sentences; just jot down notes that you can understand. If drawings or diagrams are requested, make a brief, crude version of what they will look like.

The following is an example of a free-response question and some notes that could be taken. The actual answer should not be in this outline form, but in a coherent essay, with more detailed descriptions of each concept where appropriate.

Transcription is the process of generating RNA from a DNA blueprint. It occurs in both prokaryotes and eukaryotes, but the mechanism and results are different.

(a) Transcription in both prokaryotes and eukaryotes involves enzymes and other molecules and sequences. For each (prokaryote and eukaryote), **describe** the function of TWO such components needed for transcription.

(b) Messenger RNA produced by transcription in prokaryotes differs from that in eukaryotes produced by transcription and post-transcriptional processing. **Describe** TWO post-transcriptional modifications in eukaryotes that lead to these differences.

(c) **Explain** how differences in the location of transcription between prokaryotes and eukaryotes affect the translation of a transcript.

Things to Cover:

(a) Prok — same RNA pol for all RNA (sigma factor), −10 and −35 region of promoter

Euk — RNA pol I, II, and III (rRNA, mRNA, tRNA), transcription factors

(b) Euk — Introns removed to form single gene, cap/poly A tail added

Prok — many genes on one mRNA (operon/polycistronic), no introns, cap, or tail

(c) Prok — cytoplasm, translation begins before transcription done → Christmas-tree structure

Euk — nucleus (rRNA in nucleolus), no translation until exits

By the time you've written something like this, your one-to-two minutes on that question should be up. Move on to the next question and repeat the process. When the time comes to start writing your answers, you'll have a good set of notes on which to base your answers to each question.

Approaching Free-Response Questions

For Section II, you'll have 80 minutes (after the reading period) to answer six questions. That's an average of 13.3 minutes per question, but you'll want to spend about 22 minutes each on the two long questions and about 9 minutes each on the four short ones. Take the time to make your answers as precise and detailed as possible while managing the allotted time.

Important Distinctions

Each free-response question will, of course, be about a distinct topic. However, this is not the only way in which these questions differ from one another. Each question will also need a certain kind of answer, depending on the type of question it is. Part of answering each question correctly is understanding what general type of answer is required. There are several important signal words that indicate the rough shape of the answer you should provide. Here are five of the most common:

- Describe
- Discuss
- Explain
- Compare
- Contrast

Each of these words indicates that a specific sort of response is required; none of them mean the same thing. Questions that ask you to *describe*, *discuss*, or *explain* are testing your comprehension of a topic. A description is a detailed verbal picture of something; a description question is generally asking for "just the facts." This is not the place for opinions or speculation. Instead, you want to create a precise picture of something's features and qualities. A description question might, for example, ask you to describe the results you would expect from an experiment. A good answer here will provide a rich, detailed account of the results you anticipate.

A question that asks you to discuss a topic is asking you for something broader than a mere description. A discussion is more like a conversation about ideas, and—depending on the topic—this may be an appropriate place to talk about the tension between competing theories and views. For example, a discussion question might ask you to discuss which of several theories offers the best explanation for a set of results. A good answer here would go into detail about why one theory does a better job of explaining the results, and it would talk about why the other theories cannot address the results as thoroughly.

A question that asks you to explain a topic is asking you to take something complicated or unclear and present it in simpler terms. For example, an explanation question might ask you to explain why an experiment is likely to produce a certain set of results, or how one might measure a certain sort of experimental result. A simple description of an experimental setup would not be an adequate answer to the latter question. Instead, you would need to describe that setup *and* talk about why it would be an effective method of measuring the result.

✔ **AP Expert Note**

Compare vs. Contrast

Questions that ask you to *compare* or *contrast* are asking you to analyze a topic in relation to something else. A question about comparison needs an answer that is focused on similarities between the two things. A question that focuses on contrast needs an answer emphasizing differences and distinctions.

Three Points to Remember about Free-Response Questions

1. *Most questions are stuffed with smaller questions.* You usually won't get one broad question like, "Are penguins really happy?" Instead, you'll get an initial setup followed by questions labeled (a), (b), (c), and so on. Expect to spend a paragraph writing about each lettered question.

2. *Writing smart things earns you points.* For each subquestion on a free-response question, points are given for saying the right thing. The more points you score, the better off you are on that question. The AP Biology people have a rubric, which acts as a blueprint for what a good answer should look like—we feature similar rubrics in the explanations to our free-response questions, so you have an idea of what these might look like. Every subsection of a question has one to five key ideas attached to it. If you write about one of those ideas, you earn yourself a point. There's a limit to how many points you can earn on a single subquestion, but it boils down to this: Writing smart things about each question will earn you points toward that question.

✔ **AP Expert Note**

Don't forget—you *only* receive points for relevant correct information; you receive *no points* for incorrect information or for restating the question, which also eats up valuable time!

3. *Mimic the Data questions*. Data questions often describe an experiment and provide a graph or table to present the information in visual form. On at least one free-response question, you will be asked about an experiment in some form or another. To score points on this question, you must describe the experiment well and perhaps present the information in visual form.

Beyond these points, there's a bit of a risk in the free-response section because there are only eight questions. If you get a question on a subject you're weak in, things might look grim. Still, take heart. Quite often, you'll earn some points on every question because there will be some subquestions or segments that you are familiar with.

Remember, the goal is not perfection. If you can ace four of the questions and slog your way to partial credit on the other four, you will put yourself in a position to get a good score on the entire test. That's the Big Picture—don't lose sight of it just because you don't know the answer to any particular subquestion!

✔ **AP Expert Note**

Maximize Your FRQ Score

1. **Only answer the number of subsections the long free-response questions call for. For example, if the question has four sections (a, b, c, and d) and says to choose three parts, then choose *only* three parts.**
2. **There are almost always easy points that you can earn. State the obvious and provide a brief but accurate explanation for it.**
3. **In many instances, you can earn points by defining relevant terms. (Example: Writing *osmosis* would not get you a point, but mentioning "movement of water down a gradient across a semi-permeable membrane" would likely get the point).**
4. **While grammar and spelling are not assessed on the free-response portion, correct spellings of words and legible sentences will increase your chances of earning points.**
5. **You do not have to answer free-response questions in the order in which they appear on the exam. It's a good strategy to answer the questions you are most comfortable with first and then answer the more difficult ones.**
6. **The length of your response does not determine your score—a one-paragraph written response containing accurate, succinct, yet detailed information can score the maximum amount of points, while other essays spanning three to four paragraphs of vague, inaccurate materials may not earn any.**
7. **Be careful that you do not over-explain a concept. While the initial explanation gets you points, contradictions can cause points to be taken away.**
8. **Keep personal opinions out of free responses. Base your response on factual researched knowledge.**
9. **Relax and do your best. You know more than you think!**

SAMPLE QUESTIONS

20.4 Use a scoring rubric to self-score your responses

Attempt the following questions on scratch paper. At the end of this chapter, you will find rubrics and sample responses you can use to self-score your work.

1. Students performed an experiment to examine how hypertonic, hypotonic, and isotonic conditions affect osmosis. Eggs soaked in vinegar overnight were placed in solutions containing mixtures of distilled water and corn syrup. Initial and final mass measurements were recorded, and the results are shown in Table 1.

Table 1. Changes in the mass of eggs immersed in varying percent corn syrup solutions

Percent Corn Syrup (%)	0	10	20	30	40	50	60	70	80	90	100
Initial Mass (g)	71.1	67.4	71.7	72.3	69.4	70.3	72.9	71.6	77.0	77.5	76.1
Final Mass (g)	75.9	67.7	65.6	59.4	51.3	47.8	46.4	44.1	46.2	45.8	44.3
Change in Mass (g)	4.8	0.3	−6.1	−12.9	−18.1	−22.5	−26.5	−27.5	−30.8	−31.7	−31.8
Percent Change in Mass (%)	6.8	0.4	−8.5	−17.8	−26.1	−32.0	−36.4	−38.4	−40.0	−40.9	−41.8

(a) **Describe** whether the 0% and 100% solutions are hypertonic or hypotonic relative to the eggs. **Provide** reasoning for your response.

(b) The reaction of egg shells (calcium carbonate) with vinegar (4% acetic acid) produces a water-soluble compound (calcium acetate) and carbon dioxide gas. **Explain** why, prior to the experiment, students placed eggs in vinegar overnight.

(c) According to the data, **determine** the approximate percentage of corn syrup at which the solution would be isotonic to the egg. **Justify** your response.

(d) Suppose an egg from the 100% corn syrup solution was then immersed into pure distilled water. **Predict** what would happen to the egg, and **justify** your response.

2. To investigate the relatedness of four different species, researchers determined the nucleotide sequences for four homologous genes. The table below shows a 20-nucleotide section of the gene for each species.

Table 1. Nucleotide sequences

	1	2	3	4	5	6	7	8	9	10	11	12	13	14	15	16	17	18	19	20
Species A	C	T	C	A	T	G	A	A	A	A	T	T	C	**A**	T	A	G	A	T	**T**
Species B	C	T	C	**G**	T	G	A	A	A	A	T	T	C	T	T	A	G	A	T	A
Species C	C	T	C	A	T	G	A	**C**	A	A	T	T	C	T	T	A	G	**T**	T	A
Species D	C	T	C	A	T	G	A	**C**	A	A	T	**G**	C	T	T	A	G	A	T	A

(a) On the template provided, use the information in Table 1 to **construct** a table scoring the number of nucleotide differences in the sequences between each pair of species.

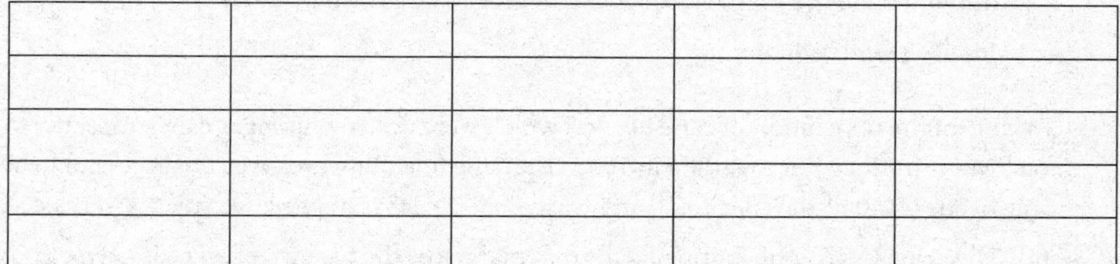

(b) Using the data in the table you constructed in part (a), **create** a phylogenetic tree on the template provided to reflect the evolutionary relationships of the four species.

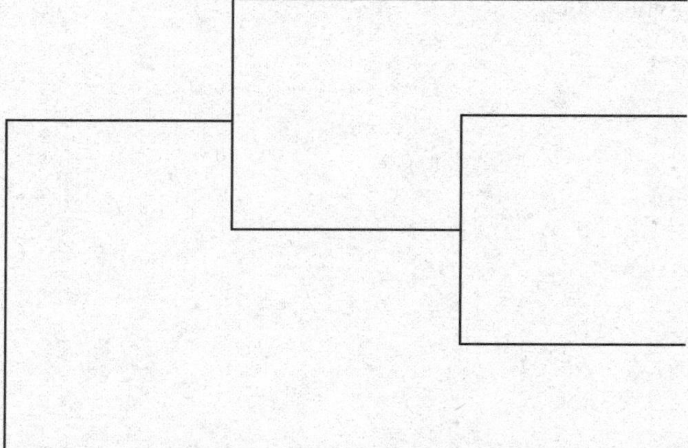

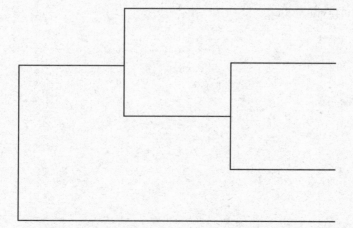

(c) **Provide** reasoning for the placement on the tree of the two species that are most closely related and for the species that is most distantly related to the others.

(d) Using the data in the phylogenetic tree you created in part (b) and the information in Table 1, **identify** the nucleotide position that corresponds with the branch point leading to Species C. **Justify** your answer.

3. Scientists collected beetles from three grassy locations within a 25-mile radius. The table below shows the number of green and brown beetles found at each location.

location	beetle color	
	green	brown
1	77	48
2	48	26
3	67	34

(a) **Describe** how the Hardy-Weinberg equation can be used to show the evolution of a population.

(b) **Identify** the null and alternative hypotheses for the evolution of the beetles.

(c) For several years, the 25-mile radius area experienced severe drought. **Predict** the effect the drought will have on the frequencies of green and brown beetles.

(d) **Justify** your prediction.

4. To control the population of cane beetles, which were destroying sugar cane crops, the toxic cane toad was introduced in Australia in 1935. The cane toad, however, was unsuccessful in controlling cane beetles. Today, the cane toad, an invasive species, threatens native species in the ecosystem.

(a) **Describe** ONE contributing factor that maintains the balance of an ecosystem.

(b) **Explain** why the introduction of the cane toad failed to control the cane beetle population.

(c) **Predict** the change in the cane toad population in Australia after its introduction in 1935.

(d) **Justify** your prediction.

5. Scientists discovered that during aerobic exercise, muscle fibers obtain ATP (required for muscle contraction) via four sources. The figure shows the four energy pathways.

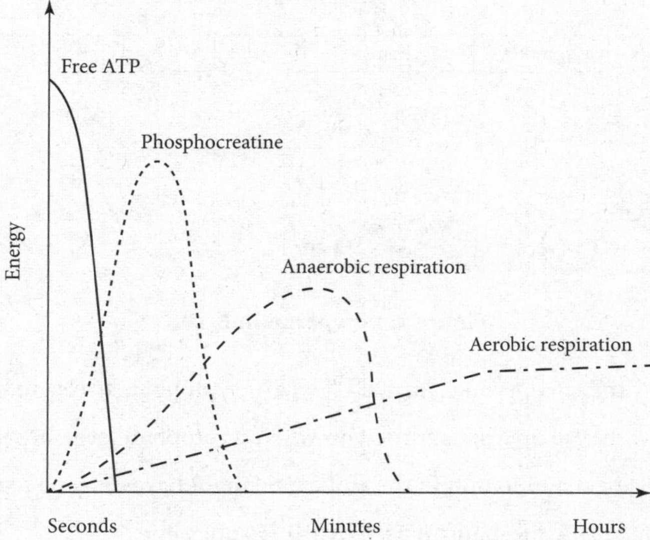

Figure 1. Energy production in muscle fibers during aerobic exercise

(a) **Describe** why cellular respiration is required for long periods of exercise.

(b) Based on the figure, **explain** the four sources of ATP during muscle metabolism with respect to time.

(c) Using the template, **graph** the predicted shape of the aerobic respiration energy curve as oxygen becomes limited after a few hours of exercise.

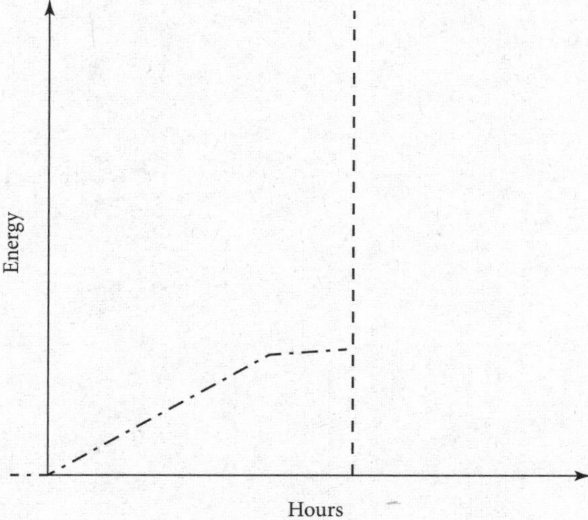

(d) The reaction of phosphocreatine + ADP to ATP + creatine is reversible, and during periods of rest the store of phosphocreatine is regenerated. **Explain** how cellular respiration returns muscles to normal after exercise.

6. The *trp* operon encodes biosynthetic enzymes to produce the amino acid tryptophan in *E. coli*. The expression of the structural genes (*trpE* to *trpA*) is regulated by a repressor protein (encoded by *trpR*) that binds the operator. When tryptophan is present, the operon is not transcribed.

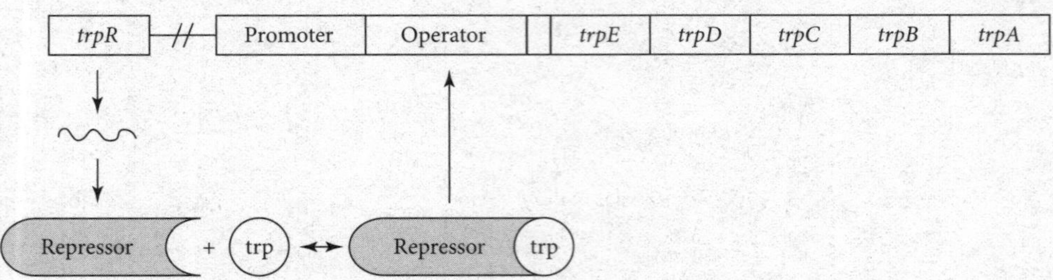

Figure 1. *trp* operon in *E. coli*

(a) **Describe** how the *trp* operon is turned off when tryptophan levels are high.

(b) **Describe** how the *trp* operon is turned on when tryptophan levels are low.

(c) **Predict** ONE effect a mutation in the *trpR* gene might have on gene regulation.

(d) **Explain** the feedback mechanism used by the *trp* operon.

ANSWERS AND EXPLANATIONS

1. **Scoring Guidelines for Free-Response Question 1**

   (a) **Describe** whether the 0% and 100% solutions are hypertonic or hypotonic relative to the eggs. **Provide** reasoning for your response. **(4 points)**

Description (1 point each, 2 points maximum)	Reasoning (1 point each, 2 points maximum)
• The 0% corn syrup (100% distilled water) solution is hypotonic relative to the egg • The 100% corn syrup (0% distilled water) solution is hypertonic relative to the egg	• Egg gained mass in 0% corn syrup • Egg lost mass in 100% corn syrup

Here is a possible response that would receive full credit:

The egg gained mass in the 0% corn syrup solution (100% distilled water), indicating water moved into the egg because the water concentration was greater outside the egg than inside. On the other hand, in the 100% corn syrup solution (0% distilled water), the egg lost mass, indicating water moved out of the egg because the water concentration was greater inside the egg than outside. Therefore, the 0% corn syrup solution is hypotonic, and the 100% corn syrup solution is hypertonic relative to the egg.

   (b) The reaction of egg shells (calcium carbonate) with vinegar (4% acetic acid) produces a water-soluble compound (calcium acetate) and carbon dioxide gas. **Explain** why, prior to the experiment, students placed eggs in vinegar overnight. **(2 points)**

Explanation (2 points)
• The reaction of calcium carbonate with vinegar dissolved the eggshells (exposing the semi-permeable membrane of the eggs) so osmosis through the egg could be examined

Here is a possible response that would receive full credit:

Vinegar dissolved the eggshells and exposed the semi-permeable membrane of the eggs. The semi-permeable eggs could then be used to examine how hypertonic, hypotonic, and isotonic conditions affect osmosis.

(c) According to the data, **determine** the approximate percentage of corn syrup at which the solution would be isotonic to the egg. **Justify** your response. **(2 points)**

Determination (1 point)	Justification (1 point)
• The solution will be isotonic between 10% and 20% corn syrup	• The solution is isotonic when there is no change in the mass of the egg (0 g falls between 0.3 g and −6.1 g, which corresponds to 10% and 20% corn syrup)

Here is a possible response that would receive full credit:

A solution is isotonic when movement of water in and out of a cell is equal, meaning in this case that the mass of the egg would not change. According to the data, a zero change in mass would occur between the concentrations that caused changes of 0.3 g and −6.1 g, which correspond to the 10% and 20% corn syrup solutions, respectively. Thus, the percentage of corn syrup for the isotonic solution would fall somewhere between 10% and 20%.

(d) Suppose an egg from the 100% corn syrup solution was then immersed into pure distilled water. **Predict** what would happen to the egg, and **justify** your response. **(2 points)**

Prediction (1 point)	Justification (1 point)
• The egg from the 100% corn syrup solution would increase in mass when placed in pure distilled water	• The concentration of water is greater outside the egg than inside, so water will move into the egg

Here is a possible response that would receive full credit:

When the egg from the 100% corn syrup solution is immersed into pure distilled water, the concentration of water is greater outside the egg than inside. As a result, water will move into the egg and cause it to swell.

2. **Scoring Guidelines for Free-Response Question 2**

(a) On the template provided, use the information in Table 1 to **construct** a table scoring the number of nucleotide differences in the sequences between each pair of species. **(2 points)**

Table components (1 point each, 2 points maximum)
• Columns and rows correctly labeled
• Data correctly entered in table

Here is a sample table that would earn full credit:

Species	A	B	C	D
A	0	3	4	4
B		0	3	3
C			0	2
D				0

(b) Using the data in the table you constructed in part (a), **create** a phylogenetic tree on the template provided to reflect the evolutionary relationships of the four species. **(3 points)**

Phylogenetic tree components (1 point for each species, 4 points maximum)
• Species A is the most distantly related species
• Species B is equally related to the other species
• Species C and D are the most closely related species (Note: C and D can be interchanged on the tree due rotation about the branch point)

Here is a sample tree that would earn full credit:

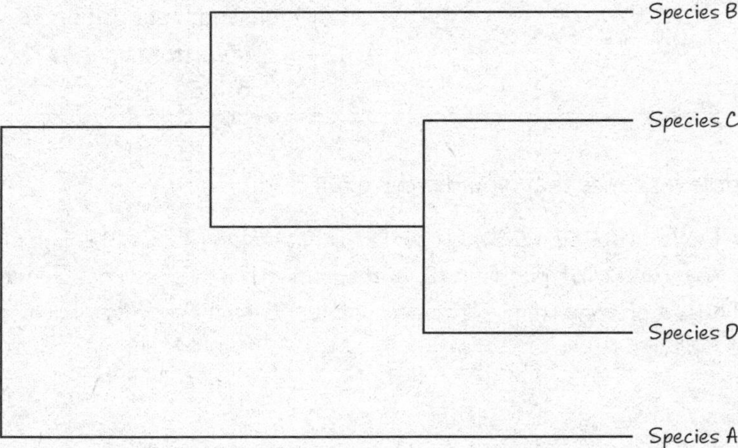

(c) **Provide** reasoning for the placement on the tree of the two species that are most closely related and for the species that is most distantly related to the others. **(3 points)**

Reasoning (1 point each, 2 points maximum)
• The number of differences in nucleotides between Species C and D is the smallest (2), so Species C and D are the most closely related species (Note: C and D can be interchanged on the tree due to rotation about the branch point)
• The number of differences in nucleotides between Species A and C and between Species A and D are the largest (4), so Species A is most distantly related to Species C and D

Here is a possible response that would receive full credit:

Since the number of differences in nucleotides between Species C and D is 2, which is the smallest difference on the table, Species C and D are the most closely related. Therefore, the middle two branches on the phylogenetic tree should be labeled Species C and D. Species A is the most distantly related species because the number of differences in nucleotides between Species A and C, as well as between Species A and D, is the largest on the table. Thus, the bottom branch should be labeled A. Species B is equally different from Species A, C, and D, so it belongs on the top branch of the tree.

(d) Using the data in the phylogenetic tree you created in part (b) and the information in Table 1, **identify** the nucleotide position that corresponds with the branch point leading to Species C. **Justify** your answer. **(2 points)**

Identification (1 point)	Justification (1 point)
• A change at nucleotide position 18 in the branch leads to Species C	• Species C and D differ at positions 18 and 12; Species C differs from Species A, B, and D at position 18, while Species D differs from Species A, B, and C at position 12

Here is a possible response that would receive full credit:

According to Table 1, Species C and D differ at positions 18 and 12. Species C differs from Species A, B, and D at position 18, and Species D differs from Species A, B, and C at position 12. Thus, a change at nucleotide position 18 corresponds with the branch point that leads to Species C.

3. **Scoring Guidelines for Free-Response Question 3**

(a) **Describe** how the Hardy-Weinberg equation can be used to show the evolution of a population. **(1 point)**

Description (1 point)
• The Hardy-Weinberg equation for a population in equilibrium is $p + q = 1$; changes in the allele frequencies between generations indicate evolution

Here is a possible response that would receive full credit:

The Hardy-Weinberg equation is $p + q = 1$, where p and q represent the allele frequencies. A population in Hardy-Weinberg equilibrium is not evolving and allele frequencies will stay the same from generation to generation. The five conditions required to maintain the Hardy-Weinberg equilibrium are no mutation, random mating, no gene flow, large population, and no natural selection. Thus, changes in allele frequency between generations would indicate evolution of the population.

(b) **Identify** the null and alternative hypotheses for the evolution of the beetles. **(1 point)**

Identification (1 point)
• The null hypothesis is that the population is in Hardy-Weinberg equilibrium; the alternative hypothesis is that the population is not in Hardy-Weinberg equilibrium

Here is a possible response that would receive full credit:

The null hypothesis is that the population is in Hardy-Weinberg equilibrium; the alternative hypothesis is that the population is not in Hardy-Weinberg equilibrium. A statistically significant deviation from the equilibrium allele frequencies (p-value < 0.5) indicates evolution. Thus, the null hypothesis is rejected and the alternative hypothesis is accepted; the population is not in Hardy-Weinberg equilibrium. If the deviation is not significant, the null hypothesis is not rejected and the population is in Hardy-Weinberg equilibrium.

(c) For several years, the 25-mile radius area experienced severe drought. **Predict** the effect the drought will have on the frequencies of green and brown beetles. **(1 point)**

Prediction (1 point)
• The frequencies of the brown and green beetles will change; the frequency for brown beetles will be higher

Here is a possible response that would receive full credit:

The change in the environment most likely will cause a change in the frequencies of brown and green beetles. The drought will probably cause an increase in the number of brown beetles and a decrease in the number of green beetles.

(d) Justify your prediction. **(1 point)**

Justification (1 point)
• The drought dried up the grass, exposing the green beetles to predators and in turn lowering their reproduction rate

Here is a possible response that would receive full credit:

According to the data, prior to the drought, there were more green beetles than brown beetles. The reason may be because brown beetles against green grass are more easily spotted by predators. The drought may cause the grass to dry up, which could result in the green beetles to be more easily spotted than the brown beetles by predators. Thus, the rate of reproduction for the brown beetles would be higher than that of the green beetles. This would shift the frequencies of green and brown beetles such that the frequency of the brown beetles would be higher than that of the green ones.

4. **Scoring Guidelines for Free-Response Question 4**

(a) **Describe** ONE contributing factor that maintains the balance of an ecosystem. **(1 point)**

Description (1 point)
• The balance of an ecosystem is maintained via keystone species, producers, or other abiotic and biotic factors

Here is a possible response that would receive full credit:

One contributing factor that maintains the balance of an ecosystem is keystone species, which help to preserve the diversity in an ecosystem. In turn, the ecosystem is able to tolerate and respond to changes in the environment and thus maintain balance.

(b) **Explain** why the introduction of the cane toad failed to control the cane beetle population. **(1 point)**

Explanation (1 point)
• Cane toads do not strictly feed on cane beetles but feed on a variety of foods, and if given the choice, cane toads prefer feeding on other insects than cane beetles

Here is a possible response that would receive full credit:

The introduction of the cane toad failed to control the cane beetle population because the cane toad did not feed solely on cane beetles. The cane toads found other food sources they preferred.

(c) **Predict** the change in the cane toad population in Australia after its introduction in 1935. **(1 point)**

Prediction (1 point maximum)
• The population of cane toads skyrocketed to threatening levels

Here is a possible response that would receive full credit:

After the introduction of cane toads in Australia in 1935, the population of cane toads most likely skyrocketed to threatening levels.

(d) **Justify** your prediction. **(1 point)**

Justification (1 point maximum)
• Cane toads have no natural predators to limit the toads' population growth, or • Cane toads are able to feed on a variety of foods and outcompete native species, or • Cane toads produce toxins that kill native animals, or • Cane toads have evolved longer legs, which allows them to migrate greater distances and spread

Here is a possible response that would receive full credit:

Cane toads, being non-native to Australia, have no natural predators in Australia. Thus, without predators limiting their growth, the toad population continuously increases.

5. **Scoring Guidelines for Free-Response Question 5**

(a) **Describe** why cellular respiration is required for long periods of exercise. **(1 point)**

Description (1 point)
• Cells consume more energy during long periods of exercise, and aerobic respiration produces more ATP than anaerobic respiration to meet that energy need

Here is a possible response that would receive full credit:

During a long period of exercise, cells consume more energy. To meet that energy need, cells undergo aerobic respiration, which produces a higher ATP yield than anaerobic respiration, approximately 36 molecules of ATP per molecule of glucose, compared to only 2 ATP per glucose for anaerobic.

(b) Based on the figure, **explain** the four sources of ATP during muscle metabolism with respect to time. **(1 point)**

Explanation (1 point)
• The four sources of ATP are free ATP, phosphocreatine, anaerobic respiration, and aerobic respiration; free ATP is used first, then ATP from phosphocreatine, and then ATP from anaerobic and aerobic respiration.

Here is a possible response that would receive full credit:

Muscles obtain ATP, which is required for muscle metabolism, from free ATP, phosphocreatine, anaerobic respiration, and aerobic respiration. First, a small amount of free ATP is available for muscles to use immediately. After exhausting the free ATP, muscles obtain ATP from phosphocreatine, which can rapidly provide ATP for a few seconds. Then, anaerobic respiration (glycolysis) can provide ATP for about a minute, and aerobic respiration can generate ATP needed for long periods of exercise.

(c) Using the template, **graph** the predicted shape of the aerobic respiration energy curve as oxygen becomes limited after a few hours of exercise. **(1 point)**

Graph characteristics (1 point)
• The aerobic respiration energy curve will decrease as oxygen becomes limited

Here is a sample graph that would earn full credit:

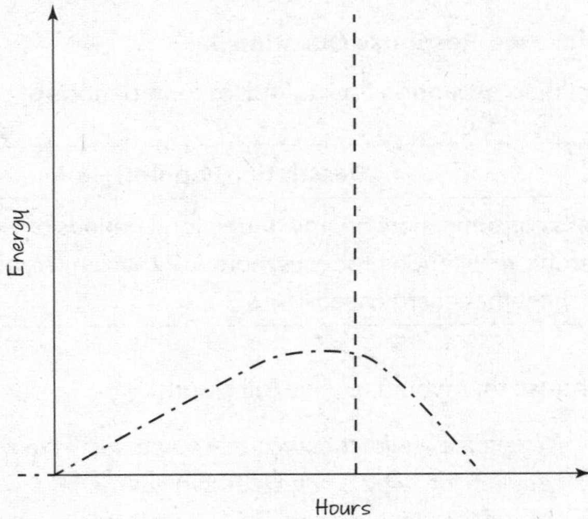

Here is a possible response that would receive full credit:

When oxygen levels decrease, pyruvate in the Krebs cycle becomes insufficient for cellular respiration. For energy, muscle fibers will switch from aerobic respiration to anaerobic respiration, which produces lactic acid. Buildup of lactic acid triggers a cramp during exercise.

(d) The reaction of phosphocreatine + ADP to ATP + creatine is reversible, and during periods of rest the store of phosphocreatine is regenerated. **Explain** how cellular respiration returns muscles to normal after exercise. **(1 point)**

Explanation (1 point)
• During periods of rest, cellular respiration uses excess ATP to replenish stores of ATP and phosphocreatine in muscle fibers

Here is a possible response that would receive full credit:

Since during periods of rest after exercise cells do not require as much ATP, excess ATP produced via cellular respiration is used to replenish stores of ATP and phosphocreatine in muscle fibers.

6. **Scoring Guidelines for Free-Response Question 6**

(a) **Describe** how the *trp* operon is turned off when tryptophan levels are high. **(1 point)**

Description (1 point)
• At high tryptophan levels, tryptophan activates the repressor protein, which binds to the operator and blocks RNA polymerase from binding to the promoter, inhibiting gene expression

Here is a possible response that would receive full credit:

When tryptophan levels are high, *E. coli* does not need to synthesize tryptophan. Tryptophan activates the repressor, which binds to the operator and blocks RNA polymerase from binding to the promoter and thereby blocks gene expression.

(b) **Describe** how the *trp* operon is turned on when tryptophan levels are low. **(1 point)**

Description (1 point)
• At low tryptophan levels, the repressor protein is inactive, so RNA polymerase is able to bind to the promoter, promoting gene expression

Here is a possible response that would receive full credit:

When tryptophan levels are low, *E. coli* needs to synthesize tryptophan. Thus, the repressor is inactive and RNA polymerase binds to the promoter, which promotes gene expression.

(c) **Predict** ONE effect a mutation in the *trpR* gene might have on gene regulation. **(1 point)**

Prediction (1 point)
• A mutation in the *trpR* gene could lead to a nonfunctional repressor protein (*trp* operon always on), or • A mutation in the *trpR* gene could lead to an always active repressor protein (*trp* operon always off), or • A mutation may have no effect

Here is a possible response that would receive full credit:

A mutation in the *trpR* gene could lead to a nonfunctional repressor protein, so the *trp* operon would always be expressed and tryptophan would build up in the cell.

(d) **Explain** the feedback mechanism used by the *trp* operon. **(1 point)**

Explanation (1 point)
• The *trp* operon is a repressible operon that uses negative feedback.

Here is a possible response that would receive full credit:

In negative feedback, the output of the system inhibits the system; whereas in positive feedback, the output of the system amplifies the system. Since when tryptophan is present, the operon is not transcribed, the *trp* operon is a repressible operon that uses negative feedback.

Practice Exam 1

Section I

1 Ⓐ Ⓑ Ⓒ Ⓓ	12 Ⓐ Ⓑ Ⓒ Ⓓ	23 Ⓐ Ⓑ Ⓒ Ⓓ	34 Ⓐ Ⓑ Ⓒ Ⓓ	45 Ⓐ Ⓑ Ⓒ Ⓓ	56 Ⓐ Ⓑ Ⓒ Ⓓ
2 Ⓐ Ⓑ Ⓒ Ⓓ	13 Ⓐ Ⓑ Ⓒ Ⓓ	24 Ⓐ Ⓑ Ⓒ Ⓓ	35 Ⓐ Ⓑ Ⓒ Ⓓ	46 Ⓐ Ⓑ Ⓒ Ⓓ	57 Ⓐ Ⓑ Ⓒ Ⓓ
3 Ⓐ Ⓑ Ⓒ Ⓓ	14 Ⓐ Ⓑ Ⓒ Ⓓ	25 Ⓐ Ⓑ Ⓒ Ⓓ	36 Ⓐ Ⓑ Ⓒ Ⓓ	47 Ⓐ Ⓑ Ⓒ Ⓓ	58 Ⓐ Ⓑ Ⓒ Ⓓ
4 Ⓐ Ⓑ Ⓒ Ⓓ	15 Ⓐ Ⓑ Ⓒ Ⓓ	26 Ⓐ Ⓑ Ⓒ Ⓓ	37 Ⓐ Ⓑ Ⓒ Ⓓ	48 Ⓐ Ⓑ Ⓒ Ⓓ	59 Ⓐ Ⓑ Ⓒ Ⓓ
5 Ⓐ Ⓑ Ⓒ Ⓓ	16 Ⓐ Ⓑ Ⓒ Ⓓ	27 Ⓐ Ⓑ Ⓒ Ⓓ	38 Ⓐ Ⓑ Ⓒ Ⓓ	49 Ⓐ Ⓑ Ⓒ Ⓓ	60 Ⓐ Ⓑ Ⓒ Ⓓ
6 Ⓐ Ⓑ Ⓒ Ⓓ	17 Ⓐ Ⓑ Ⓒ Ⓓ	28 Ⓐ Ⓑ Ⓒ Ⓓ	39 Ⓐ Ⓑ Ⓒ Ⓓ	50 Ⓐ Ⓑ Ⓒ Ⓓ	
7 Ⓐ Ⓑ Ⓒ Ⓓ	18 Ⓐ Ⓑ Ⓒ Ⓓ	29 Ⓐ Ⓑ Ⓒ Ⓓ	40 Ⓐ Ⓑ Ⓒ Ⓓ	51 Ⓐ Ⓑ Ⓒ Ⓓ	
8 Ⓐ Ⓑ Ⓒ Ⓓ	19 Ⓐ Ⓑ Ⓒ Ⓓ	30 Ⓐ Ⓑ Ⓒ Ⓓ	41 Ⓐ Ⓑ Ⓒ Ⓓ	52 Ⓐ Ⓑ Ⓒ Ⓓ	
9 Ⓐ Ⓑ Ⓒ Ⓓ	20 Ⓐ Ⓑ Ⓒ Ⓓ	31 Ⓐ Ⓑ Ⓒ Ⓓ	42 Ⓐ Ⓑ Ⓒ Ⓓ	53 Ⓐ Ⓑ Ⓒ Ⓓ	
10 Ⓐ Ⓑ Ⓒ Ⓓ	21 Ⓐ Ⓑ Ⓒ Ⓓ	32 Ⓐ Ⓑ Ⓒ Ⓓ	43 Ⓐ Ⓑ Ⓒ Ⓓ	54 Ⓐ Ⓑ Ⓒ Ⓓ	
11 Ⓐ Ⓑ Ⓒ Ⓓ	22 Ⓐ Ⓑ Ⓒ Ⓓ	33 Ⓐ Ⓑ Ⓒ Ⓓ	44 Ⓐ Ⓑ Ⓒ Ⓓ	55 Ⓐ Ⓑ Ⓒ Ⓓ	

SECTION I
90 Minutes—60 Questions

Directions: Section I of this exam contains 60 multiple-choice questions to be answered in 90 minutes. Each question is followed by four suggested answers. Using the information provided and your own knowledge of biological systems, select the best answer choice and fill in the corresponding letter on your answer grid or a sheet of scratch paper.

1. Natural selection ensures that the teeth of particular animal species are specialized to fit the diet of that species. Members of the species that lack the appropriate dentition will be selected against, because it will be more difficult for them to satisfy their nutritional needs. If an animal's jaw contains teeth with broad, rigid surfaces, it is reasonable to conclude that the animal is most likely which of the following?

   (A) An herbivore

   (B) A producer

   (C) A top consumer

   (D) A carnivore

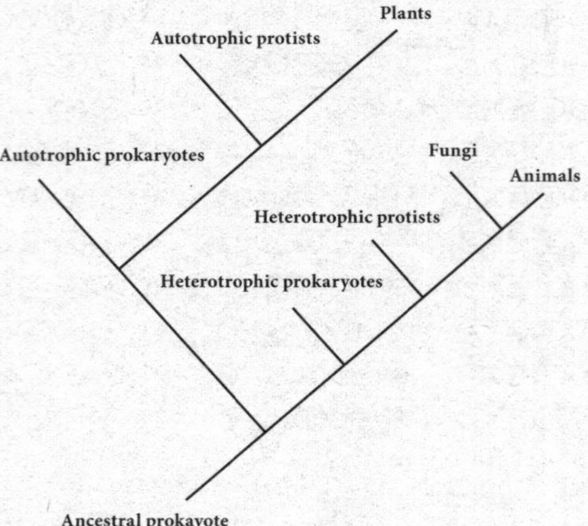

2. Based on the cladogram, which of the following is the most reasonable conclusion?

   (A) Animals are descended from fungi

   (B) Fungi are more similar to animals than to plants

   (C) All protists are closely related

   (D) The first organisms were eukaryotes

3. In humans, short-term energy supplies are stored as glycogen and long-term energy supplies are stored as fat. Which of the following statements best explains this phenomenon?

   (A) Lipids are more readily soluble in water than carbohydrates.

   (B) Lipids are more easily broken down by mitochondria than carbohydrates.

   (C) Lipids are more oxidized and produce less energy per gram.

   (D) Glycogen is more readily converted to glucose than fat.

GO ON TO THE NEXT PAGE.

Site	Plants		Amphibians		Reptiles		Mammals		Total	
	Species	Individuals	Species	Individuals	Species	Individuals	Species	Individuals	Species	Individuals
Bluewater Swamp	15	113	2	8	3	8	5	7	25	136
Papago Buttes	5	27	0	0	2	4	2	3	9	34
Beaver's Bend	8	121	2	2	0	0	3	18	13	141
Sherwood Forest	4	159	1	1	0	0	6	24	11	184
Tortilla Flats	4	63	0	0	3	24	1	5	8	92

4. Based on the table above, which site has the greatest species diversity?
   (A) Bluewater Swamp
   (B) Papago Buttes
   (C) Beaver's Bend
   (D) Sherwood Forest

5. Plants form close associations with mycorrhizae, fungi that colonize plant roots. The plant benefits because the fungus makes soil phosphorus available to the plant. The fungus benefits because the plant provides it with sugars. What is this an example of?
   (A) Commensalism
   (B) Competition
   (C) Parasitism
   (D) Mutualism

6. A man who has a sex-linked recessive disorder carries the gene for the condition on his X chromosome. If he marries a woman who does not have the gene on either of her X chromosomes, what are the chances that their first son will have the disease?
   (A) 0%
   (B) 25%
   (C) 50%
   (D) 100%

7. Transpiration is a process that transports water from roots to leaves in trees and other plants. Transpiration against the force of gravity is possible in trees 100 meters tall for which of the following reasons?
   (A) Water is actively transported from roots to leaves
   (B) Evaporation from stomata pulls water up through the tree
   (C) High pressure in the soil pushes the water up
   (D) Gravity creates pressure in the xylem, squeezing water out of the stomata

8. During complete aerobic cellular respiration, each molecule of glucose broken down in mitochondria can yield 36 molecules of ATP. Which of the following conditions would most likely lead to a decrease in the amount of ATP produced in a given system?
   (A) An increase in the amount of glucose added to the system
   (B) A decrease in the amount of light the system is exposed to
   (C) A decrease in the amount of oxygen available in the system
   (D) A decrease in the amount of carbon dioxide available in the system

GO ON TO THE NEXT PAGE.

9. Archaea are unique prokaryotes that are thought to be more closely related to eukaryotes than they are to bacteria. Which of the following characteristics best supports this idea?
   (A) Many archaea are adapted to extreme environments, such as deep-sea thermal vents.
   (B) The cell walls of archaea lack peptidoglycans.
   (C) Archaea have introns in some genes.
   (D) Archaea lack a membrane-bound nucleus and organelles.

Questions 10–12

Recent geological changes in the landscape of a coastal region in southern California allowed multiple species of animals to gain entry to a section of the shoreline region that they previously did not have access to. One of these species is the *Himantopus mexicanus*, more commonly known as the black-necked stilt.

The black-necked stilt consumes a diet consisting of a range of aquatic invertebrates (such as crustaceans and other arthropods), mollusks, small fish, and tadpoles. They have also been known to consume plant seeds, though they do so rarely.

As part of an investigation to observe the effects of the access to the new territory on the black-necked stilt community, researchers tracked the number of individuals in the population over the span of eight years. The results from their data collection are summarized in the graph below.

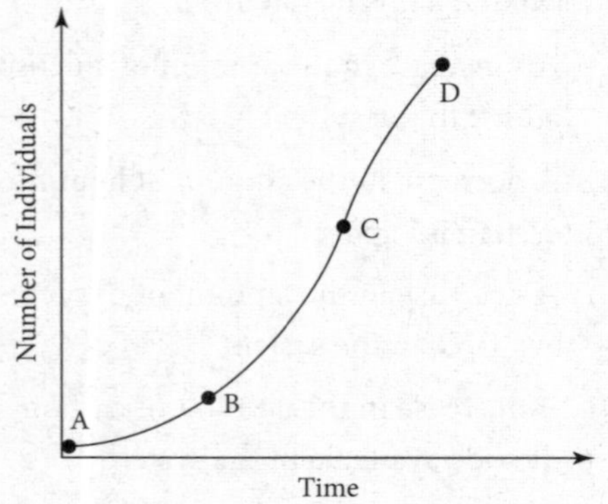

10. The graph indicates that the black-necked stilt population is most likely to exhibit which of the following?
    (A) Stabilizing and plateauing within the next year as limits from density-dependent and density-independent factors are imposed
    (B) Growing at a rapid rate due to the lack of density-dependent limiting factors and environmental constraints
    (C) Progressing toward extinction because the data shows their reproduction rate is not reflective of their environmental constraints
    (D) Growing in excess of its carrying capacity, since there are no notable fluctuations in population size

11. In addition to tracking the population growth rate, researchers also mapped out the food pyramid (shown below) during the first year of immigration into the new region:

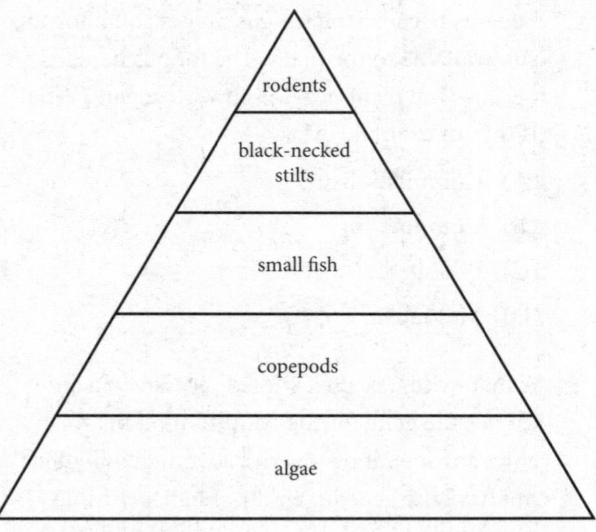

GO ON TO THE NEXT PAGE.

If researchers later realize they need to amend the food pyramid to include large birds of prey, which of the following would most likely be true of the species in the habitat?

(A) The rodent population is larger than the black-necked stilt population.

(B) The large birds of prey have the highest concentration of fat-soluble toxins.

(C) The total biomass of the copepods is greater than that of the algae.

(D) The black-necked stilts can provide more food for the birds of prey than the small fish can.

12. A few years after the initial migration of the black-necked stilt population into the newly accessible coastal region, another species of coastal bird called *Charadrius nivosus*, also known as the snowy plover, began to enter and nest in the new territory as well. After monitoring both populations for five years, researchers were surprised to find that the population size and growth rate of both populations appear unchanged. What is the most likely explanation for this occurrence?

(A) The snowy plover and black-necked stilt are closely related species occupying different niches.

(B) The snowy plover and black-necked stilt are related species occupying different habitats.

(C) The snowy plover and black-necked stilt are unrelated species occupying different niches.

(D) The snowy plover and black-necked stilt are unrelated species occupying the same niche.

13. In a given population, 7,192 people have attached earlobes and 4,881 people have unattached earlobes. The allele for unattached earlobes is recessive. Assuming that the population is in Hardy-Weinberg equilibrium, what fraction of the population is heterozygous?

(A) 0.23

(B) 0.36

(C) 0.46

(D) 0.64

14. In the cell cycle, G_1 and G_2 are names for the phases in which cell growth occurs, M refers to mitosis, and S is the name for the phase in which DNA replication occurs. Assuming that the cycle proceeds continuously without interruption (returning to the first phase after completing the last), which of the following sequences depicts a correct ordering of the events in the cell cycle?

(A) $G_1 \rightarrow$ cytokinesis $\rightarrow G_2 \rightarrow M \rightarrow S$

(B) $G_1 \rightarrow G_2 \rightarrow S \rightarrow M \rightarrow$ cytokinesis

(C) $S \rightarrow G_2 \rightarrow$ cytokinesis $\rightarrow M \rightarrow G_1$

(D) $S \rightarrow G_2 \rightarrow M \rightarrow$ cytokinesis $\rightarrow G_1$

GO ON TO THE NEXT PAGE.

15. For *E. coli* to utilize lactose as a carbon and energy source, the protein β-galactosidase must be translated. In the presence of both lactose and glucose, *E. coli* will preferentially utilize glucose, conserving the resources necessary to produce β-galactosidase. However, when glucose is absent, lactose will functionally induce the expression of β-galactosidase. What does this most strongly suggest?
    (A) Lactose represses expression of the *lac* operon.
    (B) Lactose is sufficient to cause an increase in β-galactosidase.
    (C) Glucose decreases expression of the *lac* operon in the presence of lactose.
    (D) The presence of glucose is necessary for β-galactosidase production.

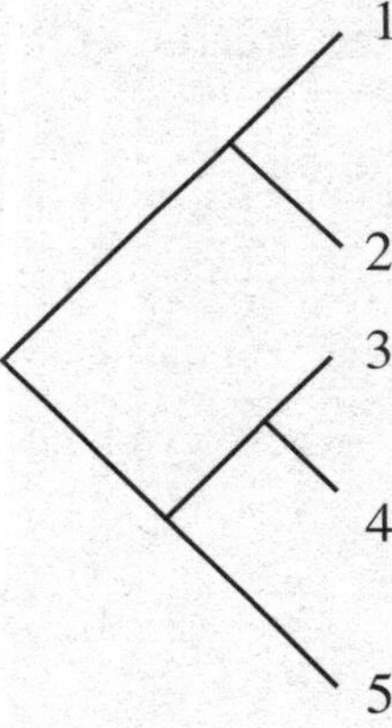

16. The phylogenetic tree above depicts five newly-discovered bacterial species. The branch point separating species 1 and 2 from 3, 4, and 5 is the presence of a gene for enzyme X, required for the metabolism of glucose. The branch point separating species 1 from species 2, and species 3 and 4 from species 5 is the presence of protein Y, a pump that removes a certain antibiotic from cells.

    Which of the following findings, if true, would require the most extensive redrawing of the tree?
    (A) Molecular studies indicate that protein Y evolved 250,000 years before enzyme X.
    (B) Molecular studies indicate that enzyme X evolved 1,000,000 years before protein Y.
    (C) Only species 3 expresses protein Z, a pump that regulates entry of sodium ions into the cell.
    (D) A newly discovered species 6 expresses protein Y but not enzyme X.

17. Nondisjunction, the failure of homologous chromosomes or sister chromatids to separate, results in aneuploidy (an irregular number of chromosomes in a cell). The figure shows the result of one form of nondisjunction.

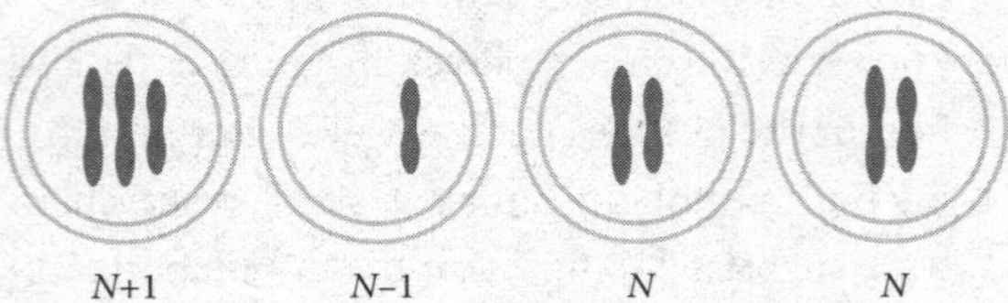

    Based on the figure, which of the following conclusions would be most justified?
    (A) Two homologous chromosomes failed to separate in interphase.
    (B) Sister chromatids failed to separate in mitosis during anaphase.
    (C) Two homologous chromosomes moved to the same pole in meiosis I.
    (D) Sister chromatids moved to the same pole in meiosis II.

GO ON TO THE NEXT PAGE.

18. The ocean food web below shows how carbon flows between organisms and the surrounding environment. The arrows at the far left indicate how much carbon is lost via respiration. Arrows pointing upward indicate how much carbon is consumed by species at higher trophic levels, while the curved arrows pointing down indicate how much carbon is lost to decomposition by bacteria. Note that some of the sources of carbon loss for bacteria and small fish are not depicted.

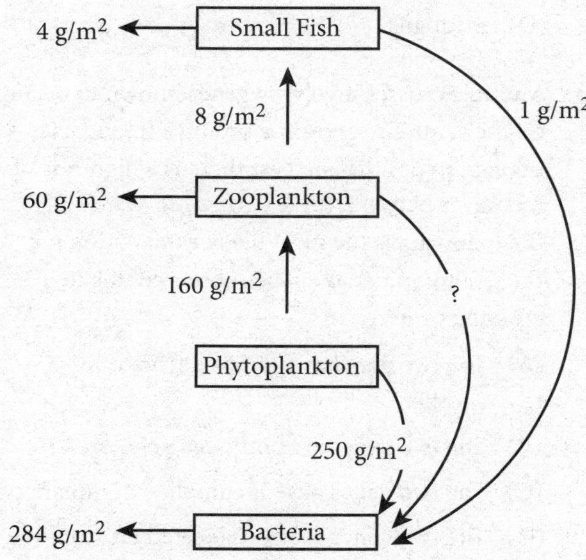

How much carbon from zooplankton is used by decomposers?

(A) 22 g/m^2

(B) 33 g/m^2

(C) 51 g/m^2

(D) 92 g/m^2

19. In a certain terrestrial ecosystem, over a given period of time, producers had 1.1 MJ (megajoules) of energy available in the form of sunlight, whereas primary consumers had approximately 12 kJ (kilojoules) and secondary consumers, 1.1 kJ. Which of the following is the most reasonable conclusion from this data?

(A) The majority of the energy in a trophic level is transferred to the next higher level.

(B) The transfer of energy from producers to primary consumers is more efficient than the transfer between consumers.

(C) Energy within an ecosystem is constantly cycled through the various levels.

(D) Approximately 90% of the energy available to primary consumers is not transferred to the next highest level.

Practice

GO ON TO THE NEXT PAGE.

Questions 20–23

The following diagram shows energy transformations within a cell. Each form of energy is represented by the symbols E I, E II, E III, and E IV. Two cellular organelles are represented by the letters A and B.

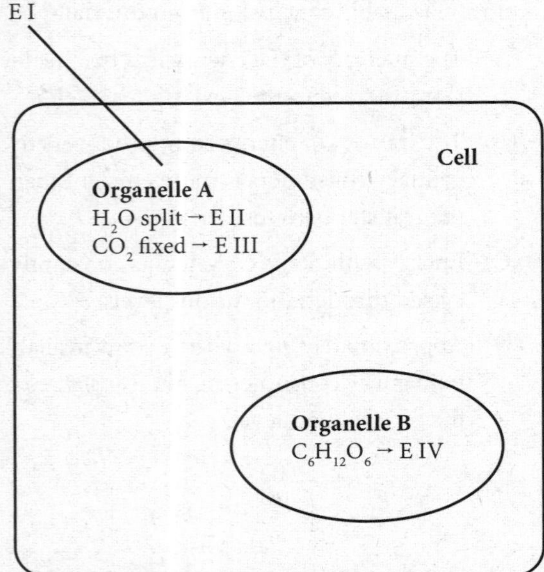

20. What form of energy is represented by E II?
    (A) Radiant energy in the form of photons
    (B) Chemical energy being stored in the bonds of glucose
    (C) Chemical energy in the form of ATP
    (D) Chemical energy released by glycolysis

21. If the transformation depicted in organelle B requires oxygen, what form of energy is represented by E IV?
    (A) Radiant energy in the form of photons
    (B) Chemical energy being stored in the bonds of glucose
    (C) Chemical energy in the form of ATP
    (D) Chemical energy released by glycolysis

22. What cellular organelles are represented as A and B, respectively?
    (A) The nucleus and a ribosome
    (B) A mitochondrion and a chloroplast
    (C) A mitochondrion and a ribosome
    (D) A chloroplast and a mitochondrion

23. Which of the following organisms could the cell shown belong to?
    (A) A photosynthetic bacterium
    (B) A photosynthetic protist
    (C) A heterotroph
    (D) A fungus

24. Multiple crosses involving genes known to occur on the same chromosome produce frequencies of phenotypes that suggest there is a high rate of crossover between these two genes. Which of the following is the most likely explanation for the phenotypic frequencies observed due to crossing over?
    (A) The two genes are far apart from one another.
    (B) The two genes are both recessive.
    (C) The two genes have incomplete dominance.
    (D) The two genes are both located far from the centromere.

25. Which of the following is the likely source of energy for the synthesis of the small organic molecules that presumably predated the first forms of life on Earth?
    (A) Fermentation by bacteria
    (B) Photosynthesis by microscopic algae
    (C) Lightning from frequent storms
    (D) Shifts in ocean currents

GO ON TO THE NEXT PAGE.

26. A scientist places free strands of DNA, which contain a gene that codes for a protein enabling the metabolism of glucose, in a medium containing bacteria that can only survive on the sugar lactose. The scientist heat shocks the bacteria in $CaCl_2$ and lets them recover before plating them in several petri dishes with only glucose as a nutrient source. After several days, there are no signs of bacterial growth in the glucose medium.

    Which of the following is NOT a plausible explanation for the results of the experiment?

    (A) All of the bacteria died from the heat shock treatment.

    (B) The gene for glucose metabolism was not successfully incorporated into any of the bacterial genomes.

    (C) The petri dishes were contaminated with a powerful antibiotic from a previous experiment.

    (D) The DNA strands came from sheep, so the glucose metabolism genes could not be translated in the bacteria.

Questions 27–29

The graph below depicts the distribution of a trait in an animal population.

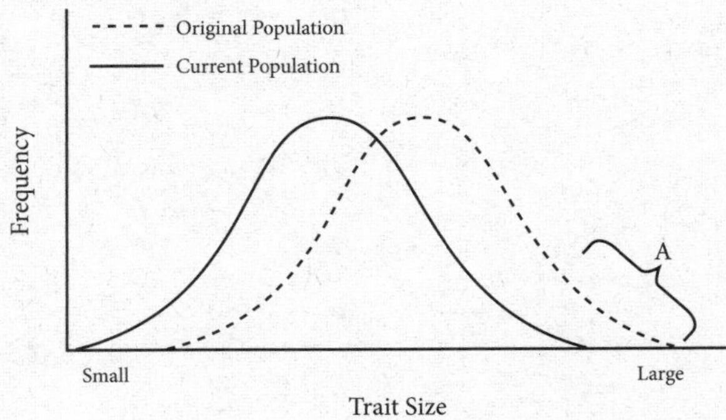

27. What does the region indicated by the A in the figure depict?

    (A) The variability of sizes in the population

    (B) The mean size of the population

    (C) The phenotypes that were successful in recent environmental conditions

    (D) The phenotypes that fared poorly in recent environmental conditions

28. What type of selection does this figure depict?

    (A) Disruptive

    (B) Directional

    (C) Stabilizing

    (D) Artificial

29. Suppose that in subsequent generations, individuals with the smallest and largest trait sizes had lower reproductive success than individuals with moderate trait sizes. What would the new population curve most probably look like?

    (A) The curve would resemble the original population curve in shape and location.

    (B) The curve would be shifted to the left of the current one.

    (C) The curve would have two peaks, one at smaller sizes and another at larger sizes.

    (D) The curve would be narrower and higher than the current one but mean size would remain the same.

GO ON TO THE NEXT PAGE.

30. In *Vibrio fischeri*, quorum sensing regulates bioluminescence, the ability to produce light. *V. fischeri* produce and secrete autoinducers, which at high levels and at high cell density causes a regulatory response and the expression of the luxCDABE genes. Which of the following illustrations correctly depicts quorum sensing?

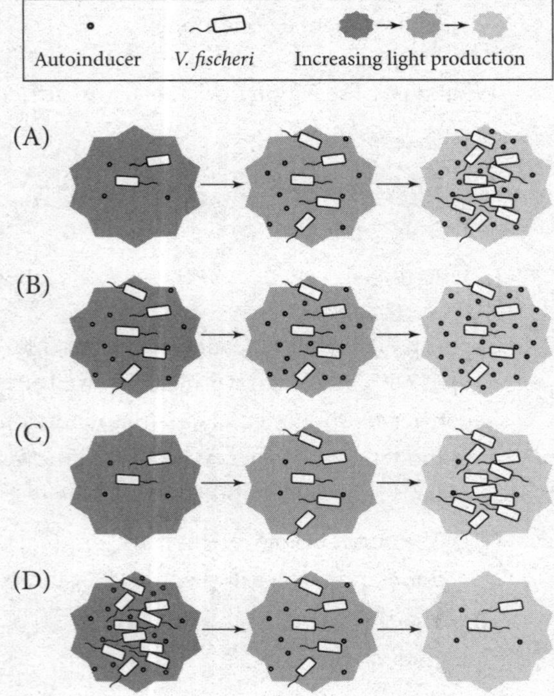

31. Animals in the phylum *Echinodermata*, such as sea stars and sand dollars, are thought to be more closely related to the phylum *Chordata* (which includes humans and other vertebrates) than to other animal phyla. Which of the following observations provides the best justification for this conclusion?

    (A) *Echinodermata* and *Chordata* are the most abundant animal phyla.

    (B) During development, the anus forms prior to the mouth in both phyla.

    (C) All species in both phyla have a common ancestor.

    (D) Neither phyla includes obligate anaerobes.

32. Most pharmacological agents, including common drugs like acetaminophen, can exhibit toxic effects at sufficiently high doses, while at lower doses they may be harmless or even therapeutically useful. Which of the following best characterizes why most drugs exert toxic effects at sufficiently high dosages?

    (A) Artificial substances are inherently toxic, so high enough doses of any such substance will result in negative effects.

    (B) At sufficiently high doses the metabolic processing of most drugs changes, thereby creating toxic metabolites that are not present at lower doses.

    (C) Most drugs interfere with normal biological functioning, and higher doses can cause an excessive degree of interference that manifests as toxicity.

    (D) Even pharmaceutically pure drugs contain trace amounts of impurities, and at high doses these impurities are present in sufficiently great quantities to cause toxic side effects.

GO ON TO THE NEXT PAGE.

33. Both skin cells and nerve cells arise from embryonic cells with the same genetic information, though these two cell types are radically different in both form and function. How is it that two cell types within an organism can arise from the same progenitor cell, yet perform vastly different specialized functions?

(A) Because the cells contain the same DNA, they also contain the same mRNA transcripts. However, these transcripts are expressed by ribosomes in different ways, leading to differences between cells.

(B) The cells contain all the same DNA, but different transcription factors and other regulatory proteins promote the synthesis of different sets of mRNA in each cell type.

(C) During development, genetic information is lost as cells develop into their mature forms, thereby silencing genes that will not be useful to the cell's intended function.

(D) The proteins created within the cell are ultimately the same, but they perform different functions based on the signals that the cell receives from neighboring cells.

34. The illustration below depicts the regulation of erythropoiesis (formation of red blood cells).

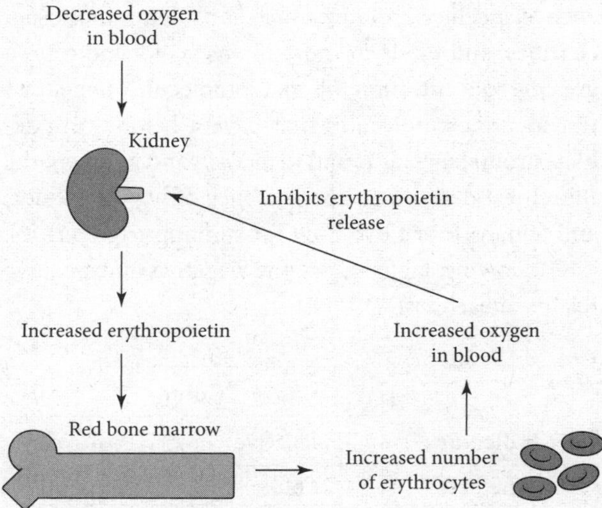

Suppose kidney cells are NOT able to detect low oxygen levels in the blood due to pathological conditions. Which of the following would be the most likely result?

(A) An increase in erythropoietin and an increase in the number of erythrocytes

(B) Tissue hypoxia due to cells not receiving enough oxygen

(C) The return to homeostasis by relieving the initial stimulus

(D) Inhibition of erythropoietin release

Practice

GO ON TO THE NEXT PAGE.

Questions 35–38

Indicators are chemicals that change color in the presence of specific substances such as proteins, monosaccharides, and lipids. A more intense color indicates a greater concentration of macromolecule. Benedict's test identifies molecules that have a ketone or aldehyde group (such as found in aldoses and ketoses), the Biuret test determines the presence of peptide bonds, and Sudan III is a dye used for staining triglycerides. The following table shows the negative and positive results of each test.

Indicator	Color	
	Negative	**Positive**
Benedict's	blue	orange
Biuret	blue	purple
Sudan III	dark red	orange

Researchers used Benedict's solution, Biuret reagent, and Sudan III to identify unknown solutions. The following data were obtained.

	Benedict's	Biuret	Sudan III
Solution 1	orange	purple	orange
Solution 2	blue	blue	dark red
Solution 3	orange	blue	dark red
Solution 4	blue	purple	dark red
Solution 5	blue	blue	orange
Solution 6	yellow-orange	purple	dark red
Solution 7	blue	pink	dark red

35. Which indicator(s) test(s) for the presence of lipids?
    (A) Benedict's
    (B) Biuret
    (C) Sudan III
    (D) Benedict's and Sudan III

36. Which of the following conclusions is best supported by the data?
    (A) Solution 3 contains glucose because Benedict's test is positive.
    (B) Solution 3 contains albumin because Benedict's test is positive.
    (C) Solution 4 contains glucose because the Biuret test is positive.
    (D) Solution 5 contains albumin because the Biuret test is positive.

37. Which of the following statements best explains why solution 7 turned pink when the Biuret reagent was added?
    (A) The concentration of peptide bonds is low, indicating the solution contains short-chain peptides.
    (B) The concentration of peptide bonds is high, indicating the solution contains long-chain peptides.
    (C) The concentration of peptide bonds is low, indicating the solution contains monosaccharides.
    (D) The concentration of peptide bonds is high, indicating the solution contains polysaccharides.

38. Whole milk is composed of water, lactose, fat, protein, and minerals. Based on the data, which of the solutions could be whole milk?
    (A) Solution 1
    (B) Solution 2
    (C) Solution 6
    (D) Solution 7

GO ON TO THE NEXT PAGE.

39. Researchers use vectors to introduce new genetic material into target cells. Which of the following best explains the advantage of plasmids over viruses as DNA delivery vectors?

(A) Plasmids deliver only a limited amount of genetic material but can provide cells with antibiotic resistance.

(B) Plasmids have the capacity to deliver larger genes and have a lower probability of toxic effects on non-target cells.

(C) Plasmids are more efficient in transferring DNA to host cells and generally do not elicit an immune response.

(D) Plasmids are able to target specific types of cells and can be manufactured on a large scale.

40. A cell at equilibrium has no free energy for metabolic reactions that keep it alive. The diagram below shows a metabolic pathway in a cell in which one reaction feeds the next.

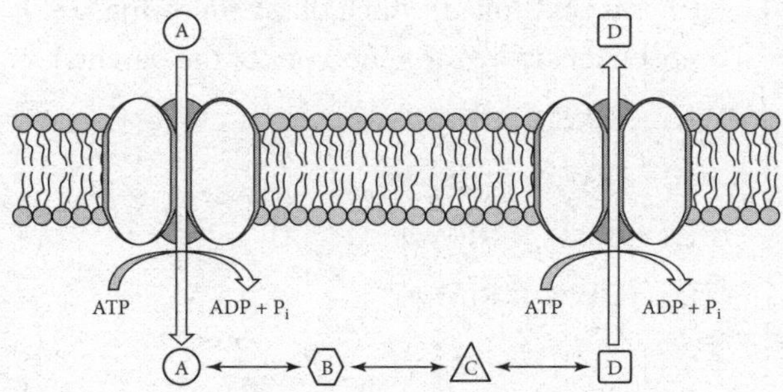

Which of the following best explains how the cell stays out of equilibrium?

(A) The cell consumes ATP to increase the concentration of molecule A in the cell.

(B) The cell produces ATP to increase the concentration of molecule B in the cell.

(C) The cell increases the concentration of molecule C in the cell to generate ATP.

(D) The cell uses energy to increase the concentration of molecule D in the cell.

41. A certain drug acts by binding to enzymes at positions other than the active site, causing enzymatic activity to decrease. What is this is an example of?

(A) Noncompetitive inhibition

(B) Competitive inhibition

(C) Allosteric inhibition

(D) Non-allosteric inhibition

42. The sequence of nucleotides in DNA codes for production of proteins, and that code is carried to the ribosome as mRNA. Which of the following protein structures represents the longest strand of nucleotide bases?

(A) The gene for hemoglobin, a protein, which consists of 20 amino acids

(B) The normal chloride channel protein gene associated with cystic fibrosis, which consists of 63 nucleotide bases

(C) The mutated sickle cell form of the hemoglobin gene, which has a single substituted base

(D) The mutated cystic fibrosis mRNA strand, which has a base insertion

43. A genetic map is a diagram of the positions of genes on a particular chromosome. What assumption underlies the methods used to construct a genetic map?

(A) Recombination frequencies are directly proportional to the distance between genes on a chromosome.

(B) Recombination frequencies are inversely proportional to the distance between genes on a chromosome.

(C) Linked genes never cross over.

(D) Recessive genes are less common than dominant genes.

Practice

GO ON TO THE NEXT PAGE.

Questions 44–45

Free energy (G) is defined as $\Delta G = \Delta H - T\Delta S$ where H is enthalpy, T is temperature in Kelvin, and S is entropy. The ΔG during a reaction indicates whether or not the reaction occurs spontaneously.

Living organisms use free energy to grow, maintain organization, and reproduce. Examples of reactions that occur in cells include breaking down glucose, converting pyruvate to lactate, and/or synthesizing ATP:

glucose + $O_2 \rightarrow CO_2 + H_2O$	$\Delta G = -2880$ kJ/mol of glucose
glucose $\rightarrow$ lactate + H^+	$\Delta G = -195.8$ kJ/mol
pyruvate + NADH + $H^+ \rightarrow$ lactate + NAD^+	$\Delta G = -25.1$ kJ/mol
ADP + $P_i \rightarrow$ ATP + H_2O	$\Delta G = 57$ kJ/mol

44. Which of the following statements best describes how cells use energy from the breakdown of glucose via cellular respiration to increase organization in the system?
    (A) Cells can couple the endergonic reaction with a process that increases entropy.
    (B) Cells can couple the endergonic reaction with a process that decreases entropy.
    (C) Cells can couple the exergonic reaction with a process that increases entropy.
    (D) Cells can couple the exergonic reaction with a process that decreases entropy.

45. Living organisms do not violate the second law of thermodynamics. Which of the following pairs of cellular processes can a cell couple to maintain order?
    (A) Breakdown of glucose via cellular respiration and formation of lactate from glucose
    (B) Formation of lactate from glucose and formation of lactate from pyruvate
    (C) Formation of lactate from pyruvate and breakdown of glucose via cellular respiration
    (D) Breakdown of glucose via cellular respiration and ATP synthesis

46. In *Drosophila*, the alleles for red eye color (R) and straight wings (W) are dominant, and the alleles for white eye color (r) and curly wings (w) are recessive. A cross between two flies produces progeny comprised of 607 red-eyed flies with straight wings and 202 red-eyed flies with curly wings. Assuming neither of these two genes is sex-linked, which of the following are most likely to be the genotypes of the parents?
    (A) RRWW × RRWW
    (B) RRWW × RRWw
    (C) RrWw × RRWw
    (D) RrWw × RrWw

GO ON TO THE NEXT PAGE.

47. According to Darwin's theory of natural selection, which of the following illustrates the evolution of bacteria in the presence of antibiotics?

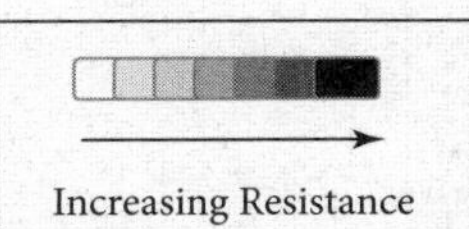

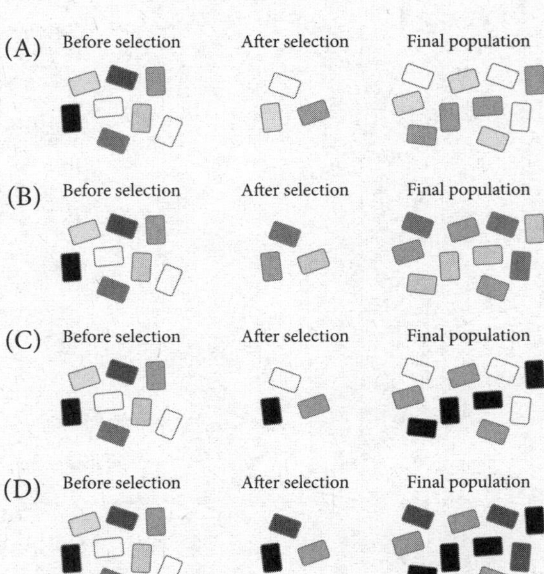

48. Transpiration is a process by which water moves through plants from soil to atmosphere. Water potential defines the flow of water in a system (the tendency of water to move from one area to another). The table lists water potential for a plant system.

Component	Water Potential (MPa)
Soil	−0.3
Roots	−0.3 – −0.6
Stem	−0.6 – −0.8
Leaves	−0.8 – −7.0
Atmosphere	−10.0 – −100.0

Which of the following statements is most consistent with the information in the table?

(A) When water evaporates from a leaf into the atmosphere, it moves from higher water potential to lower water potential.

(B) As water molecules evaporate from a leaf, they increase the water potential in the stem and cause the uptake of water.

(C) Water moves up a gradient of water potential from soil to roots.

(D) Lower water potential in the soil than in the roots causes water to enter the roots.

Practice

GO ON TO THE NEXT PAGE.

49. The graph shows the growth of *Paramecium aurelia* and *Paramecium caudatum* when they are grown separately.

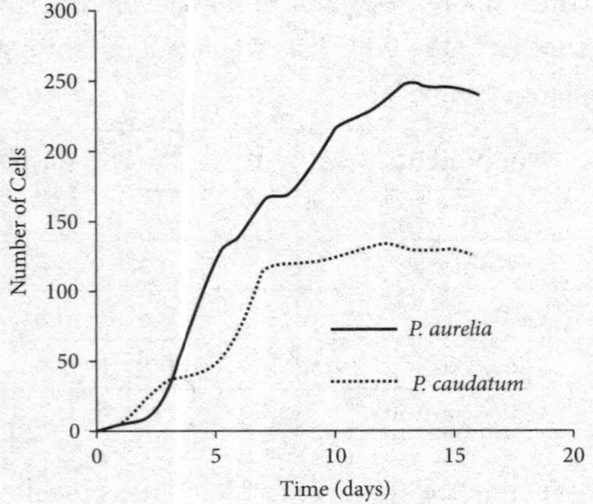

Which of the following best illustrates how competition affects the growth of *P. aurelia* and *P. caudatum* when they are grown together in the same niche?

(A)

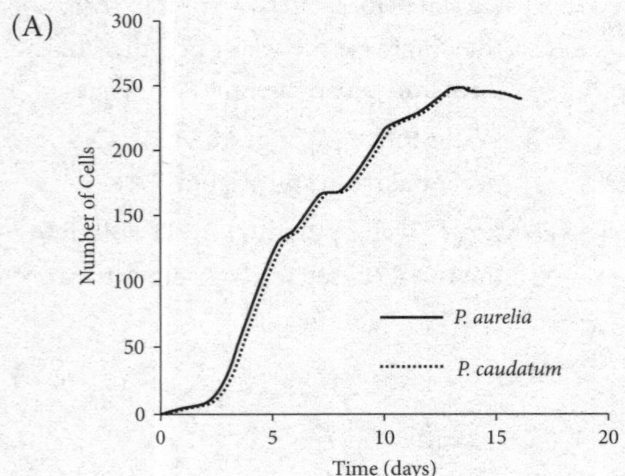

(B)

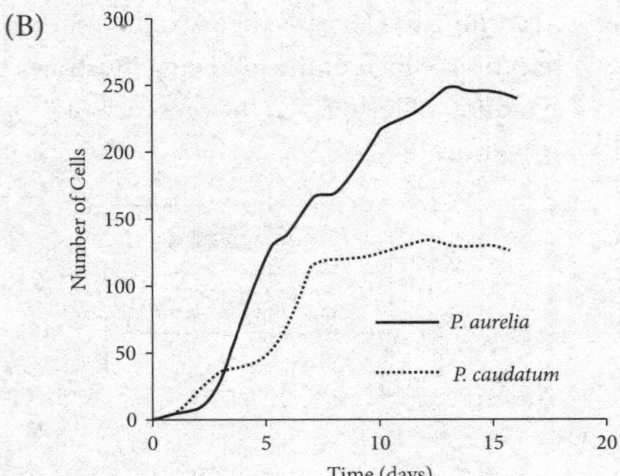

(C)

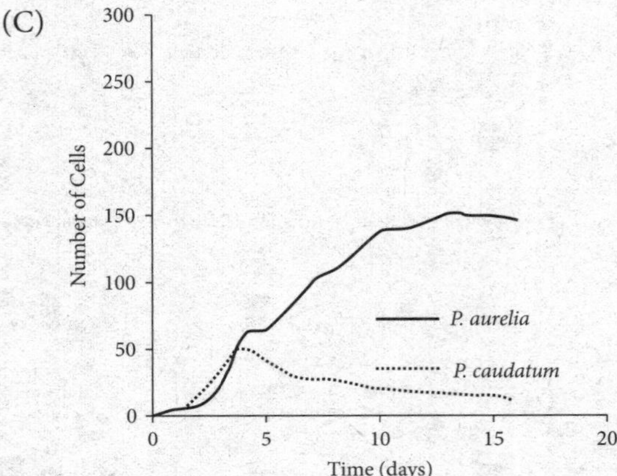

(D)

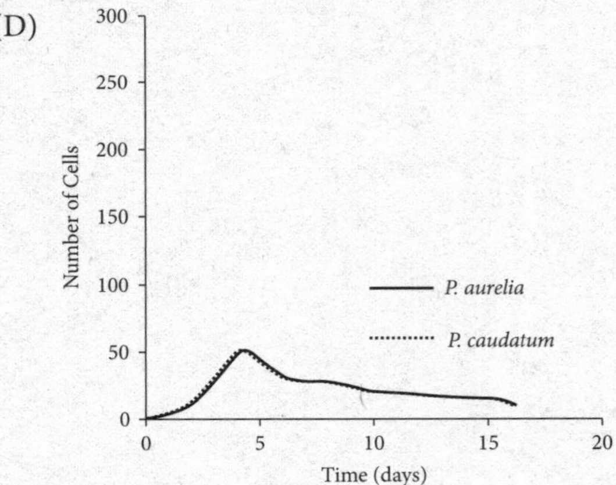

GO ON TO THE NEXT PAGE.

Questions 50–53

In a drug efficacy experiment, samples of *E. coli*, *P. chrysogenum*, *P. bursaria*, and *S. cerevisiae* were exposed to the drug, allowing researchers to observe its effects on bacteria, fungi, protozoa, and yeast, respectively. Samples were plated on their respective growth plates. An additional set of samples that were not treated with the drug was plated as well. The results of the experiment are summarized below. The shaded area represents extensive growth.

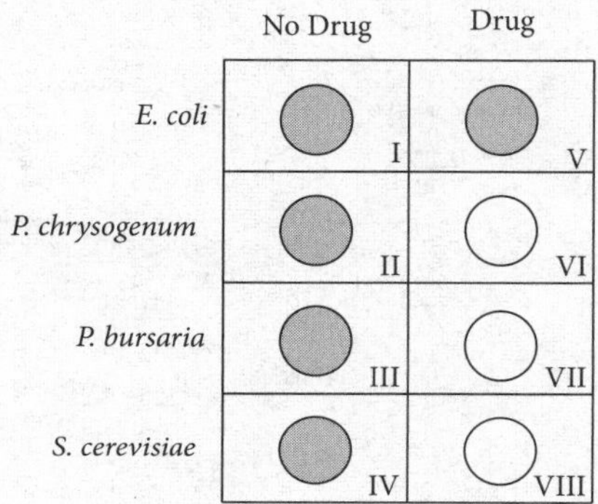

50. Various drugs function by targeting a specific structure within a cell. Because different types of cells contain different components, this allows drugs to act with increased selectivity. Which of the following cellular structures could be targeted specifically to affect eukaryotes while not affecting prokaryotes?
    (A) Cell membrane
    (B) Chromosome
    (C) Endoplasmic reticulum
    (D) Ribosome

51. Based on the results of the experiment, the least likely cellular target of the drug is the organism's
    (A) mitochondria
    (B) Golgi apparatus
    (C) lysosomes
    (D) cell wall

52. The researchers perform a second set of experiments, focusing on multicellular organisms. They collect data on the effects of the drug on different types of cells in each organism. They find that the drug has the greatest effect on cells with a high concentration of rough ER and Golgi bodies, a moderate amount of mitochondria and smooth ER, few lysosomes, a cell wall, and no cilia. The likely primary function of these cells is which of the following?
    (A) Locomotion
    (B) Protein synthesis
    (C) Storage
    (D) Transport

53. Which of the following is an advantage that eukaryotic organisms have over prokaryotic organisms?
    (A) Eukaryotes have specialized organelles that increase the efficiency of their specific processes.
    (B) Eukaryotes possess flagella that allow them to propel themselves toward specific stimuli.
    (C) Eukaryotes do not possess a cell wall and so can better control what enters and exits the cell.
    (D) Eukaryotes do not possess a nuclear envelope around their DNA, decreasing the likelihood of harmful mutations.

Practice

GO ON TO THE NEXT PAGE.

54. The data table below displays the population growth for a population of bacteria that reproduces every 12 hours. Calculate the mean rate of population growth (individuals per hour) between 36 and 48 hours. Give your answer to the nearest whole number.

Hours	Number of Individuals
0	20
12	60
24	180
36	540
48	1,620
60	4,860
72	14,580
84	43,740
96	44,299
108	44,800

(A) 90
(B) 144
(C) 1,080
(D) 1,432

Questions 55–56

Starling's law states that transport across capillary walls is dependent on the balance between hydrostatic pressure and osmotic pressure. Filtration occurs when the hydrostatic pressure is greater than the osmotic pressure, and reabsorption occurs when the osmotic pressure is greater than the hydrostatic pressure.

55. Which of the following most accurately depicts capillary fluid exchange?

(A) 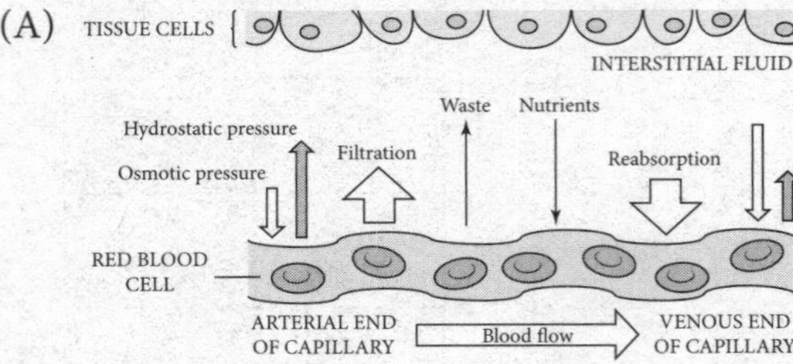

(B) 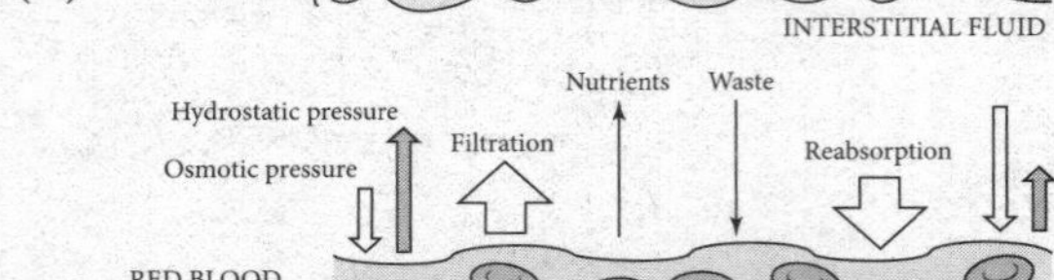
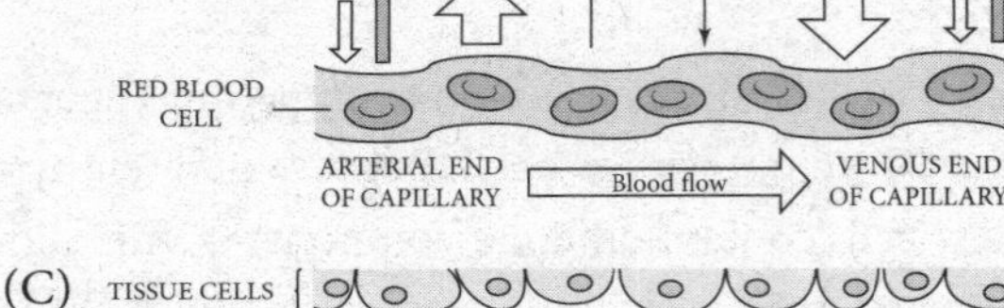

(C) 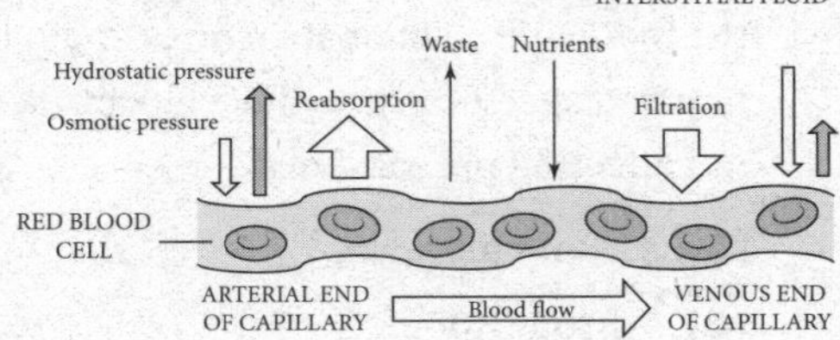

(D) 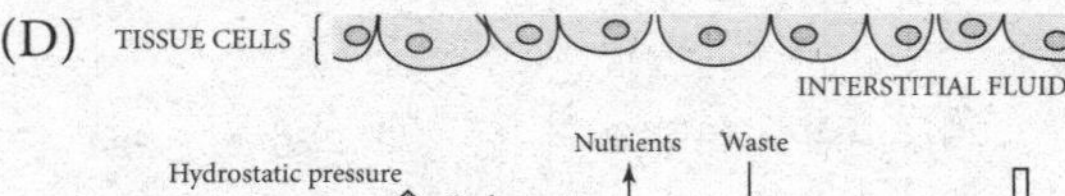
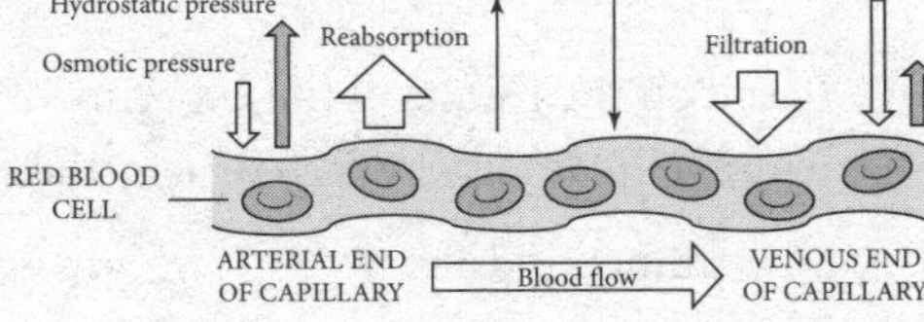

GO ON TO THE NEXT PAGE.

56. Increasing blood pressure is the body's response to tissues not receiving sufficient oxygen and nutrients. Methylsulfonylmethane (MSM) increases the permeability of cell membranes. Which of the following best explains how MSM affects cell pressure and fluid exchange?

(A) MSM increases blood pressure to deliver sufficient oxygen and nutrients to tissues.

(B) MSM decreases blood pressure to balance cell pressure and promotes transport of oxygen and nutrients to tissues.

(C) MSM decreases blood pressure, causing tissues to become overloaded with carbon dioxide and waste.

(D) MSM restores cell pressure balance by hardening cell membranes to restrict transport of oxygen and nutrients.

57. A scientist performs an experiment under laboratory conditions to study the activity of a specific enzyme. Data about substrate and product concentration over time is collected and analyzed. Which of the following statements correctly explains why the indicated experimental variable must be carefully maintained to ensure that accurate data is obtained?

(A) The cofactor concentration must be maintained at low levels because an overabundance will cause them to competitively inhibit the substrate.

(B) The temperature must be held as high as possible because enzymes have their greatest catalytic activity at high temperatures.

(C) The pH must be held constant because changes in pH can cause the enzymes to undergo a structural change, rendering them unusable.

(D) The enzyme concentration must be maintained at high levels because increasing enzyme concentration will linearly increase the rate of reaction.

GO ON TO THE NEXT PAGE.

Questions 58–60

The following diagram presents a simplified version of the nitrogen cycle.

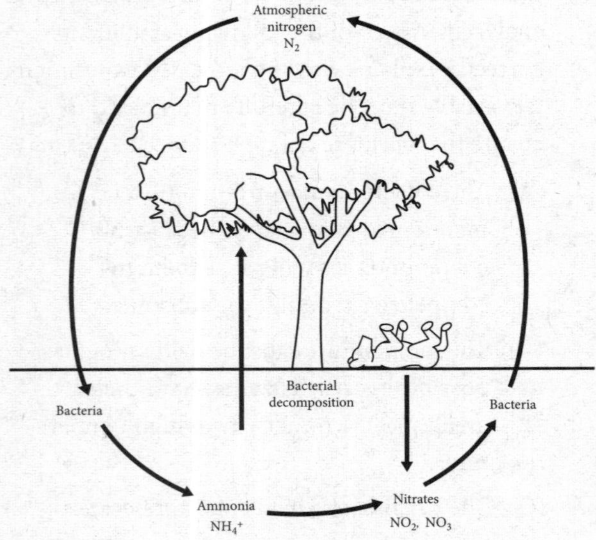

58. Unlike some species of bacteria and plants, animals are more limited in the forms of nitrogen that they can utilize in their cells. Which of the following is a form of nitrogen that animals are capable of utilizing?

    (A) N_2

    (B) NH_4^+

    (C) Amino acids

    (D) NO_3^-

59. Nitrifying bacteria are a group of organisms that are capable of converting ammonia (NH_3) or ammonium (NH_4^+) to nitrate (NO_3^-) through a series of enzyme-catalyzed reactions. This conversion is most analogous to

    (A) the aerobic metabolism of glucose to carbon dioxide

    (B) the reduction of carbon dioxide in photosynthesis

    (C) the synthesis of polypeptides in translation

    (D) the hydrolysis of ATP in muscle cells

60. Nitrogen is the mineral nutrient that contributes the most to plant growth and crop yields. Which of the following best explains why plants can suffer from nitrogen deficiencies, despite the fact that the atmosphere is nearly 80% nitrogen?

    (A) Free nitrogen in the atmosphere cannot be used by plants.

    (B) Plants can only utilize nitrogen in its gaseous state.

    (C) The nitrogen available to plants is dependent upon the breakdown of rock.

    (D) 80% of the atmosphere is only a fraction of the total nitrogen demand of plants.

END OF SECTION I

IF YOU FINISH BEFORE TIME IS CALLED, YOU MAY CHECK YOUR WORK ON SECTION I ONLY.

DO NOT GO ON TO SECTION II UNTIL YOU ARE TOLD TO DO SO.

SECTION II
80 Minutes—6 Questions

Directions: Begin by taking a 10-minute reading period, during which time you may sketch graphs, make notes, and plan your answers. You then have 80 minutes to complete your responses to the 6 free-response questions. Questions 1 and 2 are long free-response questions that should require about 22 minutes each to answer. Questions 3 through 6 are short free-response questions that should require about 9 minutes each to answer. Read each question carefully and write your response on scratch paper. Answers must be written out in paragraphs. Outline form is not acceptable. It is important that you read each question completely before you begin to write.

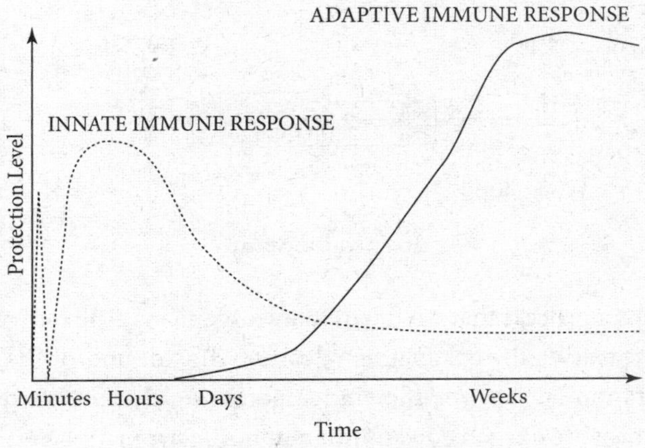

Figure 1. Immune System Response

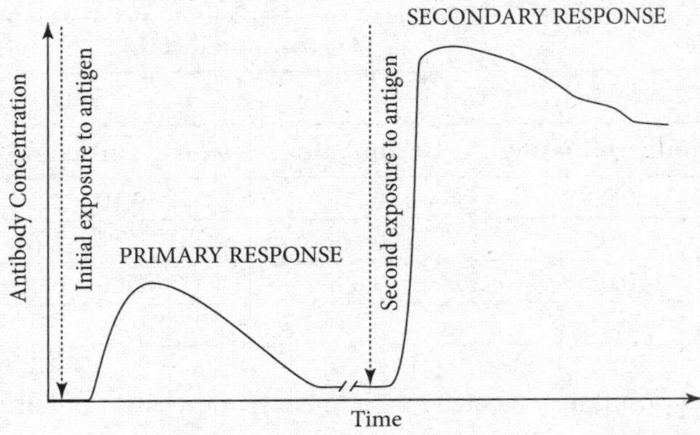

Figure 2. Immunological Memory

1. The mammalian immune system consists of innate immunity and adaptive immunity (Figure 1). Innate immunity involves nonspecific defense mechanisms such as physical and chemical barriers that respond to antigen (bacteria or virus) exposure. Adaptive immunity involves antigen-specific immune response and occurs if the innate immune system is unsuccessful in destroying the

Practice

GO ON TO THE NEXT PAGE.

antigen. The adaptive immune system has the unique ability to remember an encounter with an antigen. Upon a subsequent encounter with the same antigen, immunological memory allows the immune system to mount a response that is faster and greater (Figure 2). Artificially acquired immunity can be induced by vaccination.

(a) **Describe** how the innate and the adaptive immune systems respond to a laceration, such as an injury from falling on a barbed-wire fence.

(b) **Identify** two reasons why the flu vaccination has a fairly low success rate.

(c) **Describe** how vaccination is based on immunological memory.

(d) **Explain** the purpose of a booster dose (the re-exposure to an antigen after initial immunization).

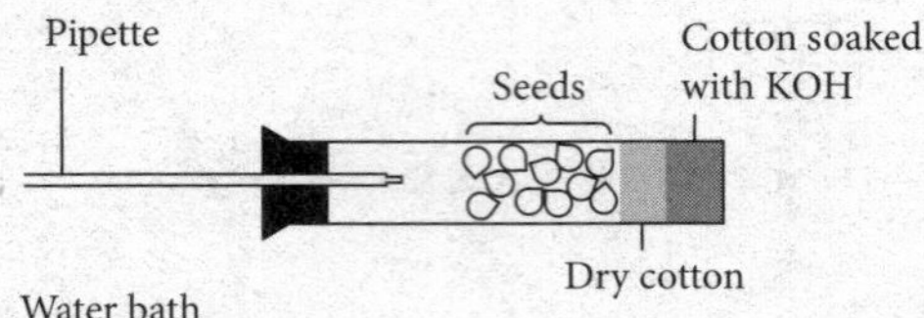

Respirator Setup

2. Cellular respiration is a process that occurs in many organisms. The rate of cellular respiration can be determined by measuring the production of carbon dioxide and/or the consumption of oxygen. To investigate respiration rate at two different temperatures, students used a respirator (see diagram above) to measure oxygen consumption of non-germinating and germinating corn seeds, recording the following results:

Oxygen Consumption of Corn Seeds

Time (min)	O_2 consumed (mL)			
	10°C		25°C	
	Non-germinating	Germinating	Non-germinating	Germinating
5	0.02	0.20	0.10	0.40
10	0.05	0.40	0.15	0.80
15	0.08	0.60	0.20	1.20
20	0.11	0.80	0.25	1.60

(a) **Construct** an appropriately labeled graph to analyze the effect of temperature over time on oxygen consumption of germinating corn seeds.

(b) Based on the data, **explain** the differences in oxygen consumption between non-germinating and germinating corn seeds. **Describe** the most likely effect of temperature on the rate of cellular respiration. **Predict** the likely oxygen consumption for germinating corn at 20°C after 20 min.

GO ON TO THE NEXT PAGE.

(c) **Propose** an appropriate control treatment for the experiment, and **describe** how the control treatment would increase the validity of the results.

(d) The tetrazolium test is an alternative test for seed viability, in which seeds are incubated in tetrazolium, rinsed, and then evaluated for color. In the presence of hydrogen ions, tetrazolium is reduced from a colorless compound to a red compound. **Predict** the color of non-germinating and germinating corn seeds. **Justify** your response.

3. A scientist conducts an experiment with penicillin-sensitive bacteria in which he adds a plasmid containing a gene that confers penicillin resistance. Following a protocol that elicits normal growth and uptake of the plasmid DNA, the scientist then adds bacteria to four new plates, as shown here.

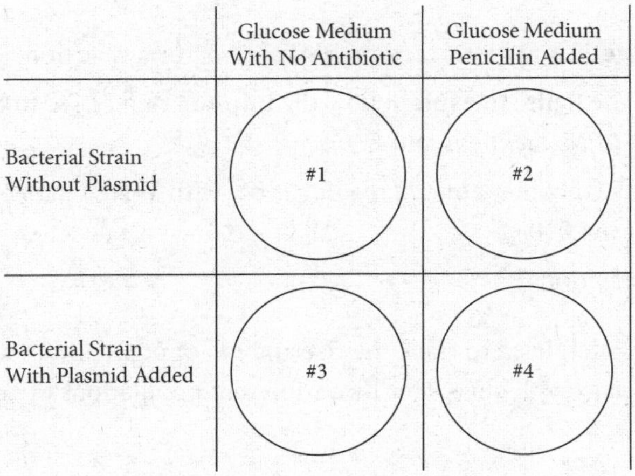

	Glucose Medium With No Antibiotic	Glucose Medium Penicillin Added
Bacterial Strain Without Plasmid	#1	#2
Bacterial Strain With Plasmid Added	#3	#4

(a) **Describe** plasmids and their involvement in passing hereditary information from generation to generation.

(b) **Identify** the protocol the scientist likely used to encourage the uptake of the plasmid DNA.

(c) **Predict** the growth patterns the scientist should expect to see on the plates.

(d) **Justify** your prediction.

4. During late summer and early fall, bats increase their body fat from approximately 10% to 30% in preparation for hibernation. White-nose syndrome (WNS), which causes over-winter mortality, first appeared in the United States in the winter of 2007. Caused by infection with the fungus *Pseudogymnoascus destructans (Pd.)*, WNS disrupts torpor patterns and causes more frequent arousal episodes (periods of high metabolic rate) such that bats use energy twice as fast as healthy bats. The figure shows the relative population of two bat species, *Eptesicus fuscus* and *Myotis lucifugus*, hibernating from 2001 to 2010.

GO ON TO THE NEXT PAGE.

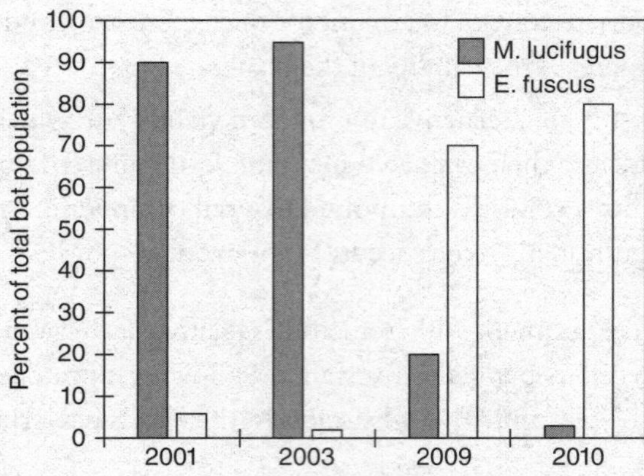

This data is adapted from *PLoS ONE*, Vol. 9, December 2014.

(a) **Describe** the abiotic and biotic factors that cause natural selection.

(b) Use the data in the figure to **explain** how the introduction of *Pd.* to the United States affected the populations of *M. lucifugus* and *E. fuscus*.

(c) **Predict** the body fat content of a *M. lucifugus* bat with WNS relative to that of a healthy *M. lucifugus* bat in 2010.

(d) **Justify** your prediction.

5. Growth curves are models used to track the rise and fall of populations. The graph shown here depicts the population growth curves for three different populations of an organism.

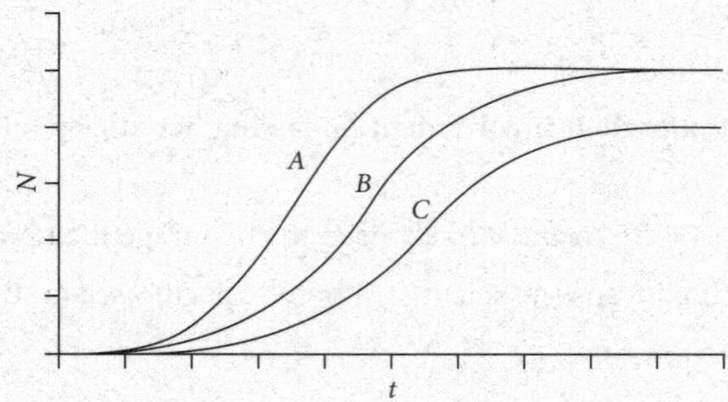

(a) **Describe** the components of the logistic growth model given by $\frac{N}{t} = rN\frac{(K-N)}{K}$.

(b) Using the information presented in the graph, **determine** the carrying capacities of the three populations relative to one another and give support for your answer.

(c) Based on the graph, **determine** the relative growth rates of the three populations.

(d) **Explain** factors that affect population growth.

Practice

GO ON TO THE NEXT PAGE.

6. Students examined the effects of pH on the reactivity of three enzymes (amylase, arginase, and pepsin). The results are shown in the figure.

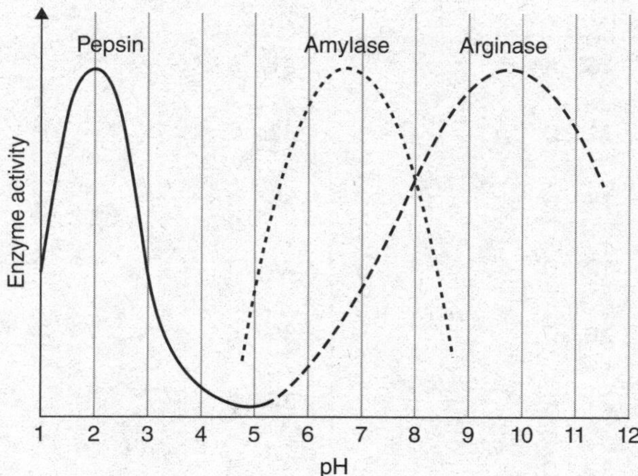

Figure 1. Enzyme activity as a function of pH

(a) Using the data, **determine** the optimum pH of pepsin.

(b) **Explain** why the optimum pH is not the same for each enzyme.

(c) Based on the data, **predict** the optimum pH of chymotrypsin, which has a working pH range from 5.2 to 9.8.

(d) **Explain** how pepsin aids in the digestion of proteins in the stomach.

Practice

STOP—END OF EXAM

ANSWER KEY

Section I

1. A	16. A	31. B	46. C
2. B	17. D	32. C	47. D
3. D	18. D	33. B	48. A
4. A	19. D	34. B	49. C
5. D	20. C	35. C	50. C
6. A	21. C	36. A	51. D
7. B	22. D	37. A	52. B
8. C	23. B	38. A	53. A
9. C	24. A	39. B	54. A
10. B	25. C	40. A	55. B
11. B	26. D	41. C	56. B
12. C	27. D	42. D	57. C
13. C	28. B	43. A	58. C
14. D	29. D	44. D	59. A
15. C	30. A	45. D	60. A

Section II

1. See Answers and Explanations
2. See Answers and Explanations
3. See Answers and Explanations
4. See Answers and Explanations
5. See Answers and Explanations
6. See Answers and Explanations

SCORING

Section I Number Correct: ______________

Section II Points Earned: ______________

Enter your results to your Practice Exam 1 assignment to see your 1–5 score and view detailed answers and explanations by logging in at kaptest.com.

Haven't registered your book yet? Go to kaptest.com/booksonline to begin.

PRACTICE EXAM 1 ANSWERS AND EXPLANATIONS

Section I

1. A

Herbivores tend to have broad teeth, which are good for grinding up plant material; they are primary consumers that feed on producers. **(A)** is thus correct. (B) is incorrect because producers consist of plants and simpler photosynthetic organisms that lack teeth entirely. (C) and (D) are incorrect because top consumers and carnivores tend to have pointed teeth, used for tearing meat.

2. B

Cladograms demonstrate how similar different organisms are to each other, but they do not necessarily indicate ancestral relationships. **(B)** is correct because the fungi and animal branches are closer together than the fungi and plant branches. (A) is incorrect because relationships of descent and ancestry cannot be inferred from a cladogram. At best, it would be reasonable to conclude that fungi and animals shared a common ancestor, but not that animals directly come from fungi. (C) is incorrect because autotrophic protists and heterotrophic protists occupy two separate branches with considerable distance between them. (D) is incorrect not only because it is factually inaccurate, but also because this kind of information cannot be inferred from a cladogram.

3. D

Glycogen is a complex carbohydrate consisting of glucose monomers. Because of its structure, it can be easily converted to or from glucose, which allows it to serve as a quick, short-term energy source. **(D)** is thus correct. (A) is incorrect because carbohydrates are actually more water-soluble than lipids, which tend to be more nonpolar. (B) is incorrect because mitochondria tend to break down carbohydrates more easily than fats. (C) is incorrect because carbohydrates are usually more oxidized (contain more oxygen) than lipids, which tend to contain long hydrocarbon chains.

4. A

Species diversity is a measure of the number of different kinds of organisms present at a site. With 25 different species, including plants, amphibians, reptiles, and mammals, Bluewater Swamp has the greatest diversity in the table, making **(A)** correct. (B) is incorrect because Papago Buttes does not have very many species or many individuals. (C) is incorrect because, even though Beaver's Bend has a lot of different species, it has fewer than Bluewater Swamp and most of its animals are mammals. (D) is incorrect because, although Sherwood Forest has the most individual plants and animals, it contains relatively few species.

5. D

The question stem describes a form of symbiosis in which two distinct species each benefit from the interaction. This type of relationship is known as mutualism, making **(D)** correct. (A) is incorrect because commensalism is a form of symbiosis in which one organism benefits and the other is neither benefited nor harmed. (B) is incorrect because competition occurs when two organisms both use the same limited resources, such as food or sunlight. (C) is incorrect because parasitism is another form of symbiosis, one in which one organism benefits while the other is harmed.

6. A

According to the question stem, the disorder is sex-linked and X-linked, passed on the X chromosome. The father can pass the gene on to his daughters, but not to his sons. Because boys have one X and one Y chromosome and mothers lack Y chromosomes, a boy always gets his Y chromosome from his father. Sons would therefore never get the X-linked gene. **(A)** is correct. In this instance, no daughters would have the disorder either, because they would all get one healthy X chromosome from their mother and the disorder is recessive.

7. B

Transpiration is able to counter the force of gravity by depending on some of the distinctive properties of water. Because water is cohesive and adhesive, evaporation from stomata in the leaves literally pulls water up the tree, much like sucking up liquid through a straw. **(B)** is thus correct. Transpiration is a passive process, one that does not require energy expenditure by the plant, ruling out (A). Although positive pressure in the roots can cause some water loss, it is not the typical case and does not occur in very tall trees, so (C) is incorrect. (D) is incorrect because gravity could only exert pressure downward.

8. C

Complete cellular respiration is an aerobic (oxygen-requiring) process. In the absence of oxygen, fermentation occurs and the amount of ATP yielded decreases. **(C)** is thus correct. Adding glucose to the system, (A), would not decrease the amount of ATP produced, nor would it increase the amount produced per glucose molecule. (B) and (D) are incorrect because light and carbon dioxide are required for photosynthesis, not cellular respiration.

9. C

All of the answers state something true about archaea, but only **(C)** supports the idea that archaea are closer to eukaryotes: the presence of introns is a distinctive characteristic of eukaryotes that is not shared by bacteria. (A) is incorrect because few organisms, either prokaryotic or eukaryotic, are adapted to extreme environments, so this characteristic reveals little about what archaea are more closely related to. (B) is incorrect because many eukaryotes lack cell walls entirely and because not all bacteria contain peptidoglycans in their cell walls. (D) is incorrect because eukaryotes do contain nuclei and other membrane-bound organelles, while bacteria do not.

10. B

Population size is highly dependent on two opposing forces. The first is the biotic potential, which is defined as the maximum rate at which a population can increase when the birth rate is at its maximum and the death rate is at its minimum. The second is environmental resistance, which includes the limits set by the living and nonliving environment that decrease birth rates. These limits can be either density-dependent, such as availability of food, predation, and disease, or density-independent, such as weather and climate. When a population is first being established and these limits are not yet a problem, ideal conditions encourage exponential growth, as seen in the figure above and described by **(B)**. (C) is incorrect because the data is simply suggesting that environmental constraints have not yet begun to affect the population growth rate. However, when constraints do become a limiting factor, this rapid growth rate will be unable to continue indefinitely. As a result, all populations that exhibit exponential growth eventually stabilize or crash, resulting in a plateau and a more sustainable logistic growth model. The number of individuals at this point is known as the carrying capacity, or the maximum population size that can be sustained by an ecosystem for an extended period of time. However, though it is known that this population will eventually reach its carrying capacity, it is not possible to know when that will occur, eliminating (A), and there is no data to suggest that it has already been reached and exceeded, as stated in (D).

11. B

Organisms are often grouped into trophic levels of feeding relationships to allow the transformation of energy in a certain environment to be studied. Because energy transfer between trophic levels is usually less than 20 percent efficient, organisms at higher levels of the pyramid must consume a greater quantity of the organisms under them in the pyramid to fulfill their energy demands. As a result, most food pyramids narrow sharply, to reflect corresponding biomass, from the primary producers (in this case, algae) to the top-level carnivores (in this case, rodents or large birds of prey). (A), (C), and (D) all violate this narrowing principle, so they are incorrect. Bioaccumulation is the process by which certain toxins build up in animal tissue as they are ingested. Toxicants that bioaccumulate are easily transferred to other organisms as they move up the food chain because when one organism eats another, the consumer also ingests all the toxicants stored in its food source. Since, as stated previously, each organism must eat a greater amount of the organism under it on the food chain to meet its energy needs, biomagnification occurs, magnifying the toxicant concentration. Thus, if the large birds of prey are at the top of this food pyramid, as is likely for such organisms, then they would contain the highest concentrations of fat-soluble toxins, as in **(B)**.

12. C

The competitive exclusion principle, also known as Gause's law, states that if two species are competing for the same resource, the species with even the slightest advantage will dominate in the long term, resulting in either the extinction of the competitor or a shift toward a different ecological niche. In this case, however, both species were able to coexist with each other immediately, and no change in population or growth rate was observed. This means that the two species are not in competition for resources and, therefore, must occupy different niches, eliminating (D). (B) is incorrect because it is stated in the question stem that both the black-necked stilt and the snowy plover are occupying the newly accessible coastal space and, therefore, the same habitat. Given that the two birds do not share the same genus, it is unlikely that the two species are closely related, as suggested in (A). Rather, the species are more likely to be unrelated, as in **(C)**.

13. C

The frequencies of alleles in a Hardy-Weinberg population can be determined using the following equations:

$$p^2 + 2pq + q^2 = 1$$

$$p + q = 1$$

Use *A* and *a* to represent the alleles for attached and unattached earlobes, respectively. According to the information provided, the frequency of the unattached phenotype (*aa* genotype) is 4,881/(4,881 + 7,192) = 0.404. Therefore, $q^2 = 0.404$ and $q = 0.64$. Because $p + q = 1$, $p = 1 - 0.64 = 0.36$.

The frequency of the heterozygous *Aa* genotype is equal to $2pq$, or $2(0.64)(0.36) = 0.46$. Therefore, **(C)** is correct.

14. D

The cell cycle begins with the Gap 1 (G_1) phase, in which the cell grows, until it passes a checkpoint and proceeds to the Synthesis (S) phase, in which new DNA is synthesized. Subsequently, the cell proceeds to the Gap 2 (G_2) phase, passing another checkpoint before beginning the Mitotic (M) phase. The M phase is followed by cytokinesis, in which the parent cell is divided into two daughter cells, each of which will then begin G_1 anew. Only **(D)** reflects this order, making it correct.

15. C

The *lac* operon is a genetic region in *E. coli* that includes the gene that encodes for β-galactosidase. According to the information provided, the *lac* operon will only be expressed when glucose is absent and lactose is present. Consequently, it follows that glucose must decrease the expression

Practice

of the *lac* operon, even when lactose is present. **(C)** is thus correct. (A) is incorrect, because lactose induces the expression of the *lac* operon when glucose is not present. (B) is incorrect because the presence of lactose is insufficient; glucose must also be absent for the cell to produce β-galactosidase. (D) states the opposite of what is true: the *absence* of glucose is necessary for β-galactosidase production.

16. A

A phylogenetic tree shows the evolution of organisms. Here, the horizontal axis of the diagram represents time. Thus, to argue that the diagram shown here is correct, it must be assumed that enzyme X evolved *before* protein Y. If the reverse were true, as (A) suggests, the entire tree would need to be redrawn, because the branch points would be in the wrong order. **(A)** is correct. (B) is incorrect because it reflects the tree as drawn, with X emerging before Y. (C) is incorrect because it is consistent with the tree as drawn; the emergence of protein Z could simply be what caused 3 and 4 to occupy separate branches. (D) is incorrect because, even though it would require adding a new branch to the existing diagram, it would not require the fundamental rearrangement necessary with (A).

17. D

Nondisjunction can occur when (1) sister chromatids fail to separate in mitosis during anaphase, (2) homologous chromosomes fail to separate in meiosis I, and (3) sister chromatids fail to separate during meiosis II. The formation of one (N+1) gamete, one (N−1) gamete, and two normal (N) gametes occurs when sister chromatids fail to separate in meiosis II during anaphase II. One daughter cell divides normally and produces two gametes with the normal number of chromosomes (N). In the other daughter cell, the sister chromatids fail to separate, resulting in two cells with an irregular number of chromosomes: one (N+1) and one (N−1). Thus, the correct answer is **(D)**. (A) is incorrect because nondisjunction does not occur in interphase. (B) is incorrect because nondisjunction in mitosis results in diploid (2N) cells, not gametes, which are haploid (N). (C) is incorrect because nondisjunction in meiosis I will result in two (N+1) gametes and two (N−1) gametes.

18. D

Matter and energy are conserved in ecosystems. When consuming phytoplankton, zooplankton take in 160 g/m^2. The amount of carbon released into the atmosphere through cellular respiration is 60 g/m^2, and 8 g/m^2 are transferred to the small fish that consume zooplankton. Therefore, the amount of carbon that is transferred back to decomposers is $160 - 60 - 8 = 92$ g/m^2, so **(D)** is correct.

19. D

According to the information provided, producers pass on about 1% of their available energy to primary consumers, who in turn pass on about 10% of their available energy to secondary consumers. That means that about 90% of the energy available for primary consumers is not transferred, **(D)**. (A) is incorrect because only a small minority of the energy is transferred between levels. (B) is incorrect because about 99% of energy is lost between producers and primary consumers, but only about 90% between primary and secondary consumers. (C) is incorrect because energy does not cycle through the various levels; rather, it flows from the lowest levels to the highest, with most energy being lost with each transfer.

20. C

The reaction occurring in organelle A is photosynthesis, as indicated by the splitting of water and fixation of CO_2. The first reaction in this organelle represents the light reactions, in which the Sun's energy is converted to chemical energy in the form of ATP, labeled E II in the diagram. **(C)** is thus correct. (A) corresponds to E I and (B) corresponds to E III. (D) is not represented in the figure because glycolysis occurs in the cytoplasm, not within an organelle.

21. C

The reaction occurring in organelle B is cellular respiration, as indicated by the release of energy stored in the bonds of glucose. Because the question specifies that oxygen is required, the process depicted must show aerobic respiration rather than glycolysis. The form of energy released in aerobic respiration becomes chemical energy in the form of ATP, labeled E IV in the diagram. **(C)** is thus correct. (A) corresponds to E I in the diagram. (B) is incorrect because the energy is not being stored (as it is in E III), but rather the energy in the bonds of glucose is being *released*. (D) is incorrect because glycolysis does not require oxygen and because glycolysis occurs in the cytoplasm, not in an organelle.

22. D

The organelle labeled A is transforming energy via the process of photosynthesis, a process that occurs in chloroplasts. The organelle labeled B is transforming energy via cellular respiration, which takes place in mitochondria. **(D)** is correct.

23. B

Although both bacteria and protists are capable of photosynthesis, only photosynthetic protists possess both chloroplasts (organelle A) and mitochondria (organelle B). **(B)** is thus correct. (A) is incorrect because bacteria lack membrane-bound organelles. (C) and (D) are incorrect because heterotrophs and fungi (which are a type of heterotroph) lack chloroplasts.

24. A

High rates of crossover occur between genes on the same chromosome only when they are located far apart from each other. **(A)** is correct. (B) and (C) are incorrect because the dominance or recessivity of the genes has no impact on their recombination frequency. (D) is incorrect because it does not rule out the possibility that the two genes are both far from the centromere *on the same side*, which would mean the two genes are relatively close to one another.

25. C

Early Earth is believed to have been a site of frequent thunderstorms and, indeed, one of the first experiments that synthesized organic molecules from inorganic precursors relied upon electric current as a catalyst. Thus, it is reasonable to believe that lightning was a source of energy for the synthesis of the first organic molecules, as in **(C)**. (A) and (B) both involve organisms, so they could not explain the rise of organic molecules that predated life. (D) is incorrect because ocean currents do not factor into any of the currently accepted hypotheses about the origin of life on Earth.

26. D

DNA and the genetic code that translates nucleotide sequences to amino acids is universal across organisms. Thus, it makes no difference whether a gene comes from bacteria, from sheep, or even from artificial synthesis using biotechnology—irrespective of origin, a cell can translate a gene into a protein. **(D)** is thus correct. All of the other choices present plausible reasons for the lack of bacterial growth.

27. D

The section of the graph indicated by A includes individuals with the largest trait size in the original population. In the current population, none of the existing individuals possess a trait size so large, because the curve has shifted to the left. Thus, it is reasonable to infer that region A denotes individuals who were selected against by recent conditions in the environment, who died out because they fared poorly. **(D)** is thus correct.

28. B

The graph shows the movement of the mean trait value in the direction of one of the extremes (in this case, toward smaller size). This is known as directional selection, **(B)**. (A) is incorrect because disruptive (or diversifying) selection favors phenotypes at both extremes and disfavors intermediate values. (C) is incorrect because stabilizing selection favors intermediate phenotypes, causing there to be relatively fewer phenotypes at the extremes. (D) is incorrect because there is no evidence that this shift was the result of artificial selection (i.e., breeding).

29. D

The situation from the question stem describes stabilizing selection, which favors intermediate phenotypes over extremes. The resulting curve would be higher and narrower with the same mean as the current curve, **(D)**. (A) and (B) are incorrect because they describe directional selection (in opposite directions). (C) is incorrect because it describes disruptive or diversifying selection.

30. A

According to the information provided, bioluminescence is expressed at high cell density when a high level of autoinducers promotes the transcription of the luxCDABE genes. Thus, the correct illustration will show greater light production when there are more *V. fischeri* and more autoinducers, such as can be seen in **(A)**. (B) is incorrect because the cell density is shown to remain constant, instead of increasing with increasing light production. (C) is incorrect because the level of autoinducers is shown to remain constant, instead of increasing with increasing light production. (D) is incorrect because it shows cell density and autoinducer levels both decreasing with increasing light production.

31. B

To support a close relationship between *Echinodermata* and *Chordata*, it is necessary to find a shared characteristic that would differentiate the two phyla from others. Of the nine major animal phyla, only *Echinodermata* and *Chordata* are deuterostomes, in which the anus forms prior to the mouth in the gut cavity. **(B)** is thus correct. (A) is not true but, even if it were, a relative abundance of species would not support a close relationship. (C) and (D) are characteristics common among all *Animalia* and would not explain why *Echinodermata* are thought to be more closely related to *Chordata* than to other animals.

32. C

Most drugs exert therapeutic effects by interfering with the normal functioning of a biological system, and the toxicity of a drug often depends on the extent of interference and not whether such interference is present or absent. **(C)** is thus correct. (A) is incorrect because many artificial substances are inherently inert and non-toxic (e.g., polyethylene), and many drugs with toxic side-effects can be obtained from natural sources (e.g., aspirin). (A) also fails to explain why high doses of a drug would be toxic while lower doses would be potentially therapeutic. (B) is incorrect because, even though changes in metabolic pathways may occur with very large doses of some drugs, this is not generally true of most drugs, nor does it account for the fact that, in many cases, it is the drug itself and not its metabolites that exert toxic effects. As such, (B) would account only for a very small minority of cases of toxicity. (D) is incorrect because impurities are typically present at such low concentrations that they never become physiologically relevant.

33. B

Differential gene expression is controlled at the level of mRNA synthesis, which is controlled by transcription factors and other mechanisms, such as DNA methylation and chromatin condensation. Thus, while all of a multicellular organism's cells will contain the same DNA, these differences in expression result in different types of cells. **(B)** is correct. (A) is incorrect because the mRNA that is transcribed varies widely between cell types; gene regulation happens mostly at the level of mRNA synthesis, not protein synthesis. (C) is incorrect because, under normal conditions, very little genetic information, if any, is lost during development. Nearly all cell types will retain the full genome of the organism by the time that they have developed into their mature forms. (D) is incorrect because, even though some proteins can perform different functions in different contexts, they generally perform the same function across different cell types.

34. B

Erythropoiesis would not be stimulated if kidney cells were unable to detect low oxygen levels in the blood. Thus, oxygen levels in the blood would continue to decrease. In turn, cells would not receive enough oxygen, resulting in tissue hypoxia (a condition in which available oxygen reaching the tissue is inadequate). **(B)** is correct. (A) is incorrect because erythropoietin and the number of erythrocytes would decrease as a result of kidney cells being unable to detect low oxygen levels in the blood. (C) is incorrect because the initial stimulus (decreased oxygen in the blood) that disrupted homeostasis is not relieved. (D) is incorrect because erythropoietin production is inhibited by the detection of increased oxygen levels in the blood, which would not be possible in the scenario described.

35. C

A subgroup of lipids is triglycerides, or fats. Since Sudan III stains triglycerides, it tests for the presence of lipids. The correct answer is **(C)**. (A), (B), and (D) are incorrect because Benedict's tests for the presence of monosaccharides (aldose and ketose), and Biuret tests for the presence of proteins (chains of amino acids linked by peptide bonds).

36. A

Glucose is an aldose (monosaccharide), so a solution containing glucose will test positive (orange) for the Benedict's solution. Albumin is a protein (contains peptide bonds), so a solution containing albumin will test positive (purple) for the Biuret reagent. Solution 3 is positive for Benedict's test, so it contains glucose. The correct answer is **(A)**. (B) is incorrect because a solution containing albumin would test positive for Biuret, not for Benedict's. (C) is incorrect because a solution containing glucose would test positive for Benedict's, not for Biuret. (D) is incorrect because solution 5 did not test positive for Biuret.

37. A

Since a more intense color indicates a higher concentration in solution, a pink color (less intense) in the Biuret test indicates the concentration of peptide bonds is low. Thus, solution 7 contains short-chain peptides. The correct answer is **(A)**. (B) is incorrect because a high concentration of peptide bonds would result in a more intense purple color. (C) and (D) are incorrect because peptide bonds are not found in monosaccharides and polysaccharides.

38. A

Since whole milk contains fat, protein, and lactose (an aldose), it will test positive for all three tests. Of the solutions listed in the answer choices, only Solution 1 tested positive for all three tests: Benedict's (orange),

Biuret (purple), and Sudan III (orange). The correct answer is **(A)**. (B), (C), and (D) are incorrect because they did not test positive for all three tests.

39. B

Plasmids are less efficient than viruses at delivering genes into host cells but are less likely to elicit an immune system response. Plasmids have the capacity to carry larger genes that may be too big to fit into viruses and can be amplified quickly (millions of copies can be produced within bacteria in hours). The correct answer is **(B)**. (A) is incorrect because, although plasmids may provide cells with antibiotic resistance, they can deliver large amounts of genetic material. (C) is incorrect because, although plasmids generally do not elicit an immune response, they are less efficient in transferring DNA to host cells. (D) is incorrect because, although plasmids can be mass produced, they are unable to target specific types of cells.

40. A

Cells stay out of equilibrium by using energy to manipulate concentrations of reactants and products. To keep the reactions running forward, cells use energy to import reactant molecules (keeping them at high concentration) and/or use energy to export product molecules (keeping them at low concentration). In the figure, the reactions are linked by shared intermediates. Thus, increasing the concentration of molecule A or decreasing the concentration of molecule D in the cell will drive the pathway forward. The correct answer is **(A)**. (B) is incorrect because an increase in concentration of molecule B in the cell would result from the cell consuming ATP to increase the concentration of molecule A in the cell. (C) is incorrect because an increase in the concentration of molecule C results from consuming ATP. (D) is incorrect because, to drive the reactions forward, the cell would use energy to decrease the concentration of molecule D in the cell (increasing the concentration out of the cell).

41. C

The question stem states that the drug binds to a location that is not the active site and subsequently deactivates the enzyme, which is an example of allosteric inhibition, **(C)**. Upon binding, the drug causes the enzyme to change shape, which stops it from being able to bind the substrate. In noncompetitive inhibition, the enzyme possesses two active sites that bind two substrates that do not resemble each other. If one substrate binds to its active site, it blocks the other substrate from binding to the other active site. Here, the question stem tells us the drug is not binding to an active site at all. Thus, this is not an example of noncompetitive inhibition, (A). Because it does not bind at the active site itself, the drug is not in direct competition with the substrate. Thus, this is not an example of competitive inhibition, eliminating (B). (D) is incorrect because the inhibition is allosteric, as noted above.

42. D

Both the nonmutated hemoglobin and cystic fibrosis gene sequences consist of 20 amino acids plus the stop codon. Because each amino acid has a three-nucleotide codon, both are composed of 63 bases, so (A) and (B) would be the same length. (C) is incorrect because the mutation associated with sickle cell is a substitution; one nucleotide replaces another, so the length of the molecule is the same. The cystic fibrosis mutation involves an insertion—the addition of a nucleotide—making it the longest molecule, with 64 nucleotide bases. **(D)** is correct.

43. A

Linkage maps are based on recombination frequencies, which increase the farther apart genes are on a chromosome. In other words, the more often a particular pair of alleles are passed on together (the lower the recombination frequency), the closer the genetic loci of those alleles are (the smaller the distance). Because recombination frequency and distance are directly correlated, **(A)** is correct. (B) suggests the opposite. (C) is too extreme. (D) is simply irrelevant.

44. D

A $\Delta G < 0$ indicates the reactants have more free energy than the products; the breakdown of glucose is exergonic (energy is released). Since ΔG is large and negative (−2880 kJ/mol), the Gibbs free energy equation suggests the reaction increases entropy or decreases order ($\Delta S > 0$). Cells maintain order by coupling cellular processes that increase entropy (and $\Delta G < 0$) with those that decrease entropy (and $\Delta G > 0$). The correct answer is **(D)**. (A) and (B) are incorrect because the reaction is exergonic, not endergonic. (C) is incorrect because the opposite is true: the reaction is coupled with a process that increases order (decreases entropy).

45. D

Energetically favorable exergonic reactions ($\Delta G < 0$) can be used to maintain or increase order in a living system by being coupled with endergonic reactions ($\Delta G > 0$). The breakdown of glucose via cellular respiration ($\Delta G = -2880$ kJ/mol), the formation of lactate from glucose ($\Delta G = -195.8$ kJ/mol), and the formation of lactate from pyruvate ($\Delta G = -25.1$ kJ/mol) are all exergonic reactions. ATP synthesis ($\Delta G = 57$ kJ/mol) is an endergonic reaction. The breakdown of glucose via cellular respiration and ATP synthesis is the only pair that couples an exergonic reaction with the endergonic reaction, so the correct answer is **(D)**. (A), (B), and (C) are incorrect because they couple two exergonic reactions.

46. C

The cross produces only red-eyed flies, which means that at least one of the parents must have had a homozygous dominant genotype for eye color (RR). This rules out (D), a dihybrid cross that would produce some flies with white eyes. With wings, the ratio of straight to curly is about 3 to 1, which suggests that both parents must have been heterozygous for wings (Ww). This eliminates (A) and (B), meaning that **(C)** must be correct. This can be confirmed by a Punnet square of (C), as shown below, which produces all red-eyed flies and a 3 to 1 ratio of straight to curly.

RrWw × RRWw	RW	rW	Rw	rw
RW	RRWW red, straight	RrWW red, straight	RRWw red, straight	RrWw red, straight
Rw	RRWw red, straight	RrWw red, straight	RRww red, curly	Rrww red, curly

47. D

According to Darwin's theory of natural selection, individuals with more favorable phenotypes are more likely to survive and pass those traits on to subsequent generations. In the presence of antibiotics, bacteria that exhibit greater antibiotic resistance (indicated by darker shading) will be selected for, because those that lack resistance will be killed by the antibiotics. The offspring of those surviving bacteria will inherit the resistance trait, giving the final population greater resistance. The correct answer is **(D)**. (A), (B), and (C) are incorrect because in the presence of antibiotics, the bacteria with less resistance will be killed by the antibiotics.

48. A

Water flows down a gradient of water potential from high (less negative) to low (more negative). During transpiration, water flows from the soil (−0.3 MPa) to the roots, then to the stem, then the leaves, and finally evaporates into the atmosphere (−100 MPa). Therefore, the correct answer is **(A)**. (B) is incorrect because as water molecules evaporate into the air at the surface of leaves, they reduce the water potential (create a more negative pressure) and cause the uptake of water. (C) is incorrect because water moves down a gradient of water potential, not up. (D) is incorrect because higher water potential in the soil than in the roots causes water to enter the roots.

49. C

When resources are limited within a habitat, species may compete for the nutrients. According to Gause's law, two species that compete for the same resource cannot coexist in the same community. When *P. aurelia* and *P. caudatum* are grown individually, they both thrive. However, when they are grown together, *P. aurelia* more efficiently uses the resources and drives *P. caudatum* to death. The correct answer is **(C)**. (A) and (D) are incorrect because *P. aurelia* will outcompete *P. caudatum*. (B) is incorrect because *P. aurelia* and *P. caudatum* will not grow at the same rate when grown together as they did when grown separately.

50. C

All prokaryotes contain ribosomes, cytoplasm, and a circular chromosome, all of which are enclosed in a single plasma membrane. Although the eukaryotic versions of these structures differ, eukaryotes also contain ribosomes, cytoplasm, chromosomes, and a cell membrane, which eliminates (A), (B), and (D). **(C)** is correct because only eukaryotes contain membrane-bound organelles, such as the endoplasmic reticulum.

51. D

The results show that only eukaryotic organisms appear to be affected by the drug, so the cellular target is most likely an organelle that prokaryotes do not possess. Both prokaryotes and eukaryotes contain ribosomes, DNA, cytoplasm, and a cell membrane. Most prokaryotes and some eukaryotes, such as plants, fungi, and protists, have evolved to possess a cell wall as well. Thus, **(D)** is the least likely target because it is shared by both eukaryotes and prokaryotes. Mitochondria, the Golgi apparatus, and lysosomes are all examples of membrane-bound organelles that eukaryotes possess and are thus possible targets for this drug.

52. B

Because multicellular organisms consist of more than one cell, their cells usually take on specialized functions. Smooth ER play a large role in the production of lipids and hormones, rough ER are important for protein synthesis, Golgi bodies are important for protein packaging and dissemination, lysosomes contain enzymes that break down different kinds of biomolecules, cell walls provide structural support and rigidity, and cilia are important for cellular motility. Therefore, it is possible to form hypotheses about the function of a specific cell based on its concentration of these different organelles. A cell that has an abundant amount of rough ER and Golgi bodies is most likely to be involved in protein production and packaging, which makes **(B)** correct. The lack of cilia and moderate amount of mitochondria, which can provide energy for movement, make it less likely to be involved in locomotion, (A). The lack of abundant space in the cell, due to the high concentration of rough ER and Golgi bodies, make it unlikely that this cell is involved in storage or transport, eliminating (C) and (D).

53. A

A major structural difference between eukaryotes and prokaryotes is that eukaryotes possess organelles, which are structures that have evolved to perform specialized functions. This means that different cells can have differing concentrations of particular organelles, corresponding to their function. This increases efficiency by ensuring that resources are not wasted and that every cell has the tools it needs to perform its function. Therefore, **(A)** is correct. (B) is incorrect because both eukaryotes and prokaryotes may possess flagella, so it is not an advantage for either group. (C) is incorrect because some eukaryotes do, in fact, possess cell walls. Furthermore, the cell wall mainly provides mechanical support to a cell; it is the cell membrane that plays the main role in controlling what enters and leaves the cell. (D) is incorrect because eukaryotes do have a nuclear envelope around their DNA, while prokaryotes do not.

54. A

From 36 to 48 hours, the population of bacteria grew from 540 to 1,620 individuals. The mean growth rate is the change in population over the change in time:

$$\frac{1{,}620 - 540 \text{ bacteria}}{48 - 36 \text{ hours}} = \frac{1{,}080 \text{ bacteria}}{12 \text{ hours}}$$
$$= 90 \text{ bacteria/hour}$$

The correct answer is 90 bacteria per hour. Therefore, **(A)** is correct.

55. B

In capillary fluid exchange, filtration pushes fluid out of the capillary when hydrostatic pressure is greater than osmotic pressure, and reabsorption pulls fluid into the capillary when osmotic pressure is greater than hydrostatic pressure. Blood pumped through capillaries supplies oxygen and nutrients to tissue cells and takes up carbon dioxide and waste in the interstitial fluid. The correct answer is **(B)**. (A) is incorrect because the directions of the arrows for waste and nutrients are reversed. (C) and (D) are incorrect because the directions of the arrows for filtration and reabsorption are reversed.

56. B

By increasing the permeability of cell membranes, MSM allows the diffusion of oxygen and nutrients. As tissues receive sufficient oxygen and nutrients, blood pressure will decrease, which balances the cell pressure. The correct answer is **(B)**. (A) is incorrect because MSM decreases, not increases, blood pressure. (C) is incorrect because MSM increases permeability, which would enable the diffusion of carbon dioxide and waste from the tissue into the capillary. (D) is incorrect because MSM does not harden the cell membrane, but makes it more flexible by increasing its permeability.

57. C

The activity of an enzyme is affected by its environment. Factors like temperature, pH, and concentration play a large role in determining the rate of a reaction and whether the substrate will even be able to bind to the enzyme. Different enzymes have different optimum pH values, where the active site is best able to match the shape of the substrate, resulting in the fastest rate of reaction. Therefore, as **(C)** correctly states, any change in pH above or below this optimum value would cause the enzyme to change shape, thereby decreasing the reaction rate. (A) is incorrect because an overabundance of cofactor has no known detrimental effects; cofactors and substrates bind to the enzyme at different sites, so they would not be competing against each other. Increasing the temperature increases the rate of the reaction to a certain extent. However, raising the temperature too high will cause bond breakage, causing the enzyme to denature and the active site to change shape, thereby lowering the rate of reaction. Thus, (B) is incorrect. Similarly, increasing enzyme concentration increases the reaction rate until the point when the enzyme is no longer the limiting factor. Therefore, the rate does not increase linearly but rather increases rapidly before plateauing, making (D) incorrect.

58. C

Animals can only utilize organic forms of nitrogen. The only organic form of nitrogen among the answer choices is found in **(C)**.

59. A

The conversion of ammonia or ammonium to nitrate is an oxidation reaction. The only other reaction among the options provided that is also oxidation is the metabolism of glucose to carbon dioxide (CO_2). **(A)** is thus correct.

60. A

From the figure, it can be seen that free nitrogen (N_2) does not flow directly from the atmosphere into plants. This makes **(A)** a viable explanation and the correct answer. Indeed, plants must obtain nitrogen through absorption of ammonium and nitrate, which rules out (B). Unlike other minerals, ammonium and nitrate are not derived from the breakdown of rock, so (C) is incorrect as well. (D) is untrue because there is more than enough nitrogen in the atmosphere to facilitate all plant needs; it just happens to be in a form (N_2) that plants cannot utilize directly.

ANSWERS AND EXPLANATIONS

Section II

1. **Scoring Guidelines for Free-Response Question 1**

   (a) **Describe** how the innate and the adaptive immune systems respond to a laceration, such as an injury from falling on a barbed-wire fence. **(4 points)**

Description (1 point each, 4 points maximum)
• The innate immune system provides immediate/initial defense against an antigen.
• The body responds with an inflammatory response and releases histamine.
• White blood cells capture and engulf antigens via phagocytosis.
• When the adaptive immune system is activated later, lymphocytes produce antibodies to target the antigen specifically.

Here is a possible response that would receive full credit:

The innate immune system provides immediate defense against a foreign invader. The first line of defense is the skin, which provides a protective barrier around the body. When a laceration is made in the skin, the protective barrier is broken. The body responds with an inflammatory response, releasing the chemical histamine. This release increases the permeability of blood vessels and signals blood vessels to dilate, which in turn causes a rush of blood to the area. White blood cells called neutrophils surround and engulf any foreign organisms via phagocytosis.

The innate immune system does not provide long-lasting immunity, however. If after days, the innate immune system is unable to fight off the invasion, it will activate the adaptive immune system. The adaptive immune response is carried out by white blood cells called lymphocytes. B lymphocytes and T cells are natural killer cells, which attack any foreign object that enters the body. B lymphocytes produce antibodies that specifically target the antigens and prevent them from sticking to the host cell or releasing chemicals that can damage the host cell. The information about the invading organisms is passed on to T cells, which activate macrophages to digest pieces of the antigen. The entire immune response causes clotting at the site of the laceration, as white blood cells and lymphocytes die and blood coagulates on the surface of the skin, preventing organisms from further entering the bloodstream.

(b) **Identify** two reasons why the flu vaccination has a fairly low success rate. **(2 points)**

Reasoning (1 point each, 2 points maximum)
• The flu virus can rapidly mutate into different strains.
• It is difficult to predict how the flu virus will mutate.

Here is a possible response that would receive full credit:

Flu vaccines can protect against certain strains of the flu virus that are known. The problem with the vaccine is that the flu virus can rapidly mutate into different strains, some of which may be resistant to the vaccine. The vaccine is only effective two or more weeks after a person receives it. That amount of time gives a person's body the chance to develop immunity to the strains of flu virus found in the vaccine. There is no way for the manufacturer of a vaccine to predict how the flu virus will mutate, so by the time the vaccine is made, it may be ineffective against a variety of new strains of flu or may only partially work against them. An "arms race" is thereby constantly occurring between host and pathogen. The host is constantly trying to fight off pathogens through the development of active immunity, and the pathogen (virus) is constantly trying to evolve so that it is effectively able to invade and attack the host organism.

(c) **Describe** how vaccination is based on immunological memory. **(3 points)**

Description (3 points)
• Immunological memory leads to an enhanced response to subsequent encounters with an antigen.
• Vaccines contain inactive virus strains or parts.
• Vaccination evokes a mild, primary immune response to enable a greater secondary response.

Here is a possible response that would receive full credit:

The immunological memory of an immune system can remember antigens and mount a greater and faster immunological response upon re-encounter of the same antigens. For instance, if the body has been invaded by the virus before, it will "remember" how to fight off the virus because the specific protein has been encoded in the immune system's memory cells. The body then has specific immunity toward the virus that has already invaded.

Vaccination works by stimulating the adaptive immune system to produce antibodies against invading organisms. This vaccine contains inactivated virus strains or parts, which activate the production of cells and antibodies that remove the foreign invaders from the body. Now the body knows how to fight off that particular invading organism and has immunization against that virus. If the body is exposed to an active strain of the same virus, the immune system elicits a rapid and large secondary response against the antigen. Thus, the body is protected from the viral infection.

(d) **Explain** the purpose of a booster dose (the re-exposure to an antigen after initial immunization). **(1 point)**

Explanation (1 point)
• A booster dose increases immunity against an antigen back to protective levels.

Here is a possible response that would receive full credit:

If memory against the antigen has declined through time, a booster dose re-exposes the immune system to the antigen in order to help it "remember" the antigen. The booster dose increases the immunity against the antigen by stimulating antibody production back to protective levels.

2. **Scoring Guidelines for Free-Response Question 2**

(a) **Construct** an appropriately labeled graph to analyze the effect of temperature over time on oxygen consumption of germinating corn seeds. **(3 points)**

Graph components (1 point each, 3 points maximum)
• Axes correctly labeled and scaled • Data separated into two temperature curves with oxygen consumption as the dependent variable • Curves correctly labeled and points plotted accurately

Here is a sample graph that would earn full credit:

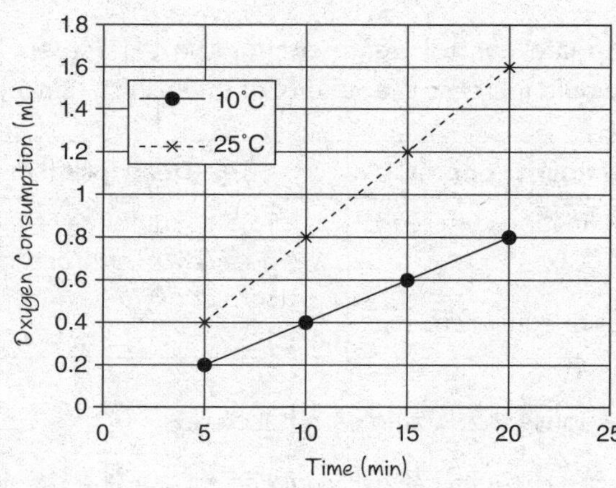

(b) Based on the data, **explain** the differences in oxygen consumption between non-germinating and germinating corn seeds. **Describe** the most likely effect of temperature on the rate of cellular respiration. **Predict** the likely oxygen consumption for germinating corn at 20°C after 20 min. **(3 points)**

Explanation (1 point)	Description (1 point)	Prediction (1 point)
• Oxygen consumption is higher in germinating corn seeds because germinating seeds have higher energy needs and thus higher rates of cellular respiration than non-germinating corn seeds.	• As temperature increases, cellular respiration increases (oxygen consumption increases). OR • As temperature decreases, cellular respiration decreases (oxygen consumption decreases).	• Between 0.8 and 1.6 mL (or a selected value between 0.8 and 1.6 mL)

Here is one possible response that would earn full credit:

According to the data, at each temperature, oxygen consumption for the germinating corn seeds is higher than that for the non-germinating corn seeds. Germinating corn seeds require more energy than non-germinating corn seeds, so the rate of cellular respiration would be greater. Thus, oxygen consumption, which indicates the rate of cellular respiration, is greater for germinating corn seeds. Compared to seeds at 10°C, seeds at 25°C have higher oxygen consumption (higher rate of cellular respiration). As temperature increases, cellular respiration increases. Oxygen consumption for germinating corn seeds at 20°C for 20 min would be approximately 1.4 mL.

(c) **Propose** an appropriate control treatment for the experiment, and **describe** how the control treatment would increase the validity of the results. **(2 points)**

Control Treatment (1 point)	Description (1 point)
• Boiled germinating corn seeds OR • Glass beads or equivalent	• To account for any changes in volume due to KOH and/or temperature and to correct oxygen consumption readings

Here is one possible response that would earn full credit:

A control treatment for the experiment would be to use the respirator to measure oxygen consumption of glass beads in place of seeds over time. This control would show the amount of carbon dioxide in the container that is decreasing because of KOH, which absorbs carbon dioxide. Subtracting these values from those measured from the non-germinating and germinating seeds would give a more accurate value for oxygen consumption.

(d) The tetrazolium test is an alternative test for seed viability, in which seeds are incubated in tetrazolium, rinsed, and then evaluated for color. In the presence of hydrogen ions, tetrazolium is reduced from a colorless compound to a red compound. **Predict** the color of non-germinating and germinating corn seeds. **Justify** your response. **(2 points)**

Prediction (1 point)	Justification (1 point)
• Germinating corn seeds will be redder than non-germinating corn seeds.	• Germinating seeds have higher rates of cellular respiration and thus a higher concentration of hydrogen ions, which reduce tetrazolium to a red compound that stains living cells.

Here is one possible response that would earn full credit:

Germinating seeds will be red because they have a higher rate of cellular respiration than non-germinating seeds, which will be white. At a higher rate of cellular respiration, the germinating seeds produce more hydrogen ions, which will reduce tetrazolium to a red compound, and thus turn the seeds red. Non-germinating seeds will not reduce tetrazolium, and thus remain white.

3. **Scoring Guidelines for Free-Response Question 3**

(a) **Describe** plasmids and their involvement in passing hereditary information from generation to generation. **(1 point)**

Description (1 point)
• Plasmids are small extrachromosomal, circular DNA molecules that may enhance survival.

Here is a possible response that would receive full credit:

Plasmids are circular bits of DNA that are distinct from the chromosomal DNA and replicate independently. Plasmids may provide bacteria with genetic advantages like antibiotic resistance that enhance survival, growth, and reproduction.

Practice

(b) **Identify** the protocol the scientist likely used to encourage the uptake of the plasmid DNA. **(1 point)**

Identification (1 point each, 1 point maximum)
• Used heat shock • Added chemicals such as $CaCl_2$ • Performed sonication

Here is a possible response that would receive full credit:

Scientists promoted transformation in bacteria using heat shock to increase the permeability of the plasma membrane, which enabled the uptake of the plasmid DNA.

(c) **Predict** the growth patterns the scientist should expect to see on the plates. **(1 point)**

Prediction (1 point)
• No growth on plate 2 • Growth on plates 1, 3, and 4

Here is a possible response that would receive full credit:

There will be no growth on plate 2 and growth on plates 1, 3, and 4 with the most growth on plates 1 and 3.

(d) **Justify** your prediction. **(1 point)**

Justification (1 point)
• No growth on plate 2: penicillin will kill the bacteria • Growth on plate 4: transformed bacteria are penicillin resistant • Most growth on plates 1 and 3: bacteria grow in the absence of penicillin

Here is a possible response that would receive full credit:

On plate 2, which has penicillin and bacteria that have not acquired resistance from the plasmid, there will be no growth because the penicillin will kill the bacteria. On plate 4, there will likely be bacterial growth because some of the bacteria will be transformed with the plasmid and will acquire penicillin resistance. There should be the most growth on plates 1 and 3, where penicillin resistance is not necessary for growth.

4. **Scoring Guidelines for Free-Response Question 4**

(a) **Describe** the abiotic and biotic factors that cause natural selection. **(1 point)**

Description (1 point)
• Abiotic factors include sunlight, water, nutrients, temperature, and natural disasters. • Biotic factors include predators, prey, competitors, mating, and disease.

Here is a possible response that would receive full credit:

Ecosystems are made up of abiotic and biotic factors, which serve as selective pressures. Abiotic factors include, sunlight, water, nutrients, temperature, and natural disasters. Biotic factors include predator, prey, competitors, mating, and disease. These factors select individuals with an increased chance of surviving.

(b) Use the data in the figure to **explain** how the introduction of *Pd.* to the United States affected the populations of *M. lucifugus* and *E. fuscus*. **(1 point)**

Explanation (1 point)
• *M. lucifugus* population went down. • *E. fuscus* population went up.

Here is a possible response that would receive full credit:

Pd. first appeared in the United States in the winter of 2007. According to the figure, from 2003 to 2009, the population of M. lucifugus decreased while the population of E. fuscus population increased. This suggests that E. fuscus is resistant to the fungus that causes white-nose syndrome (WNS).

(c) **Predict** the body fat content of a *M. lucifugus* bat with WNS relative to that of a healthy *M. lucifugus* bat in 2010. **(1 point)**

Prediction (1 point)
• Body fat of a healthy *M. lucifugus* bat is twice that of a *M. lucifugus* bat with WNS.

Here is a possible response that would receive full credit:

In 2010, the body fat content of a M. lucifugus bat with WNS is most likely half of that of a healthy M. lucifugus bat.

(d) **Justify** your prediction. **(1 point)**

Justification (1 point)
• Bats with WNS use energy twice as fast as healthy bats.

Here is a possible response that would receive full credit:

Since bats with WNS use energy twice as fast as healthy bats, they will deplete their body fat reserve twice as quickly. Thus, the body fat content of a *M. lucifugus* bat with WNS is mostly likely half of that of a healthy *M. lucifugus* bat

5. Scoring Guidelines for Free-Response Question 5

(a) **Describe** the components of the logistic growth model given by $\frac{\Delta N}{\Delta t} = rN\frac{(K-N)}{K}$. **(1 point)**

Description (1 point)
• $\frac{\Delta N}{\Delta t}$ is the change in the population size over the given time
• N is the size of the original population
• r is the rate of population growth
• K is the carrying capacity.

Here is a possible response that would receive full credit:

The logistic model shows the limited increase in population size over a given time. For $\frac{\Delta N}{\Delta t} = rN\frac{(K-N)}{K}$, N is the size of the original population, r is the rate of population growth, and K is the carrying capacity.

(b) Using the information presented in the graph, **determine** the carrying capacities of the three populations relative to one another and give support for your answer. **(1 point)**

Determination (1 point)
• The higher the plateau, the higher the carrying capacity. Populations A and B have the same carrying capacity, and C has the lowest carrying capacity.

Here is a possible response that would receive full credit:

All of the populations start with the same number of individuals at the far left of the curves. Logistic growth is typified by an ultimate plateau that is equal to the carrying capacity: the higher the plateau, the higher the carrying capacity. Carrying capacity is the maximum number of individuals an environment will support. Based on the plateaus of the three curves, populations A and B have the same carrying capacity, which is higher than it is for population C.

(c) Based on the graph, **determine** the relative growth rates of the three populations. **(1 point)**

Determination (1 point)
• The steeper the curve before the plateau, the greater the growth rate. Population A has the greatest growth rate, B has the second greatest, and C has the lowest growth rate.

Here is a possible response that would receive full credit:

Growth rate affects the steepness of the exponential portion of the curve: the incline before the plateau. The quicker the population gets to its carrying capacity, the higher the growth rate. Population A has the highest growth rate, followed by B, then C.

(d) **Explain** factors that affect population growth. **(1 point)**

Explanation (1 point)
• Density-independent factors affect populations in the same way regardless of how many individuals are in the population. • Density-dependent factors affect populations differently as the number of individuals in the population increases.

Here is a possible response that would receive full credit:

Density-independent factors affect populations in the same way regardless of how many individuals are in the population at that given time. These factors tend to be abiotic factors, such as flood, drought, hurricane, or fire.

The impact of density-dependent factors on populations increases as the number of individuals in the population, and therefore the density in a given area, increases. These factors include competition among or between species, predation, or emigration.

6. **Scoring Guidelines for Free-Response Question 6**

(a) Using the data, **determine** the optimum pH of pepsin. **(1 point)**

Determination (1 point)
• The optimum pH of pepsin, which occurs at maximum activity, is approximately 2

Here is a possible response that would receive full credit:

The optimum pH occurs when an enzyme is at maximum activity. Thus, the optimum pH of pepsin is 2.

(b) **Explain** why the optimum pH is not the same for each enzyme. **(1 point)**

Explanation (1 point)
• The structure (and in turn function) of enzymes depends on the charges of the amino acids, which are influenced by pH

Here is a possible response that would receive full credit:

The optimum pH is not the same for each enzyme because the structure and function of enzymes depend on the amino acids in the chain, and these amino acids have functional groups that are influenced by pH. When pH changes, the charges on the amino acids may change due to the available H^+ and OH^- ions. Thus, the 3-D shape and function of the protein may be altered.

(c) Based on the data, **predict** the optimum pH of chymotrypsin, which has a working pH range of 5.2 to 9.8. **(1 point)**

Prediction (1 point)
• The optimum pH of chymotrypsin is about 7.5

Here is a possible response that would receive full credit:

Optimum pH occurs at the maximum enzyme activity, which is typically the midpoint of the working pH range. Since the working pH range of chymotrypsin is from 5.2 to 9.8, the optimum pH of chymotrypsin would be the average, or 7.5.

(d) **Explain** how pepsin aids in the digestion of proteins in the stomach. **(1 point)**

Explanation (1 point)
• Pepsin speeds up the reaction of protein digestion.

Here is a possible response that would receive full credit:

The stomach is very acidic, which is the optimum pH of pepsin. Pepsin lowers the activation energy required to breakdown proteins, thus speeding up the reaction. Enzymes do not get used up in a reaction, so they can catalyze millions of reactions per second.

Practice Exam 2

Section I

1 Ⓐ Ⓑ Ⓒ Ⓓ	12 Ⓐ Ⓑ Ⓒ Ⓓ	23 Ⓐ Ⓑ Ⓒ Ⓓ	34 Ⓐ Ⓑ Ⓒ Ⓓ	45 Ⓐ Ⓑ Ⓒ Ⓓ	56 Ⓐ Ⓑ Ⓒ Ⓓ
2 Ⓐ Ⓑ Ⓒ Ⓓ	13 Ⓐ Ⓑ Ⓒ Ⓓ	24 Ⓐ Ⓑ Ⓒ Ⓓ	35 Ⓐ Ⓑ Ⓒ Ⓓ	46 Ⓐ Ⓑ Ⓒ Ⓓ	57 Ⓐ Ⓑ Ⓒ Ⓓ
3 Ⓐ Ⓑ Ⓒ Ⓓ	14 Ⓐ Ⓑ Ⓒ Ⓓ	25 Ⓐ Ⓑ Ⓒ Ⓓ	36 Ⓐ Ⓑ Ⓒ Ⓓ	47 Ⓐ Ⓑ Ⓒ Ⓓ	58 Ⓐ Ⓑ Ⓒ Ⓓ
4 Ⓐ Ⓑ Ⓒ Ⓓ	15 Ⓐ Ⓑ Ⓒ Ⓓ	26 Ⓐ Ⓑ Ⓒ Ⓓ	37 Ⓐ Ⓑ Ⓒ Ⓓ	48 Ⓐ Ⓑ Ⓒ Ⓓ	59 Ⓐ Ⓑ Ⓒ Ⓓ
5 Ⓐ Ⓑ Ⓒ Ⓓ	16 Ⓐ Ⓑ Ⓒ Ⓓ	27 Ⓐ Ⓑ Ⓒ Ⓓ	38 Ⓐ Ⓑ Ⓒ Ⓓ	49 Ⓐ Ⓑ Ⓒ Ⓓ	60 Ⓐ Ⓑ Ⓒ Ⓓ
6 Ⓐ Ⓑ Ⓒ Ⓓ	17 Ⓐ Ⓑ Ⓒ Ⓓ	28 Ⓐ Ⓑ Ⓒ Ⓓ	39 Ⓐ Ⓑ Ⓒ Ⓓ	50 Ⓐ Ⓑ Ⓒ Ⓓ	
7 Ⓐ Ⓑ Ⓒ Ⓓ	18 Ⓐ Ⓑ Ⓒ Ⓓ	29 Ⓐ Ⓑ Ⓒ Ⓓ	40 Ⓐ Ⓑ Ⓒ Ⓓ	51 Ⓐ Ⓑ Ⓒ Ⓓ	
8 Ⓐ Ⓑ Ⓒ Ⓓ	19 Ⓐ Ⓑ Ⓒ Ⓓ	30 Ⓐ Ⓑ Ⓒ Ⓓ	41 Ⓐ Ⓑ Ⓒ Ⓓ	52 Ⓐ Ⓑ Ⓒ Ⓓ	
9 Ⓐ Ⓑ Ⓒ Ⓓ	20 Ⓐ Ⓑ Ⓒ Ⓓ	31 Ⓐ Ⓑ Ⓒ Ⓓ	42 Ⓐ Ⓑ Ⓒ Ⓓ	53 Ⓐ Ⓑ Ⓒ Ⓓ	
10 Ⓐ Ⓑ Ⓒ Ⓓ	21 Ⓐ Ⓑ Ⓒ Ⓓ	32 Ⓐ Ⓑ Ⓒ Ⓓ	43 Ⓐ Ⓑ Ⓒ Ⓓ	54 Ⓐ Ⓑ Ⓒ Ⓓ	
11 Ⓐ Ⓑ Ⓒ Ⓓ	22 Ⓐ Ⓑ Ⓒ Ⓓ	33 Ⓐ Ⓑ Ⓒ Ⓓ	44 Ⓐ Ⓑ Ⓒ Ⓓ	55 Ⓐ Ⓑ Ⓒ Ⓓ	

SECTION I
90 Minutes—60 Questions

Directions: Section I of this exam contains 60 multiple-choice questions to be answered in 90 minutes. Each question is followed by four suggested answers. Using the information provided and your own knowledge of biological systems, select the best answer choice and fill in the corresponding letter on your answer grid or a sheet of scratch paper.

1. A botanist observes that a mature root cell can dedifferentiate in tissue culture and give rise to a diversity of plant cells. Which of the following best explains this observation?

   (A) Root cells contain all the genes necessary to produce a variety of cells.

   (B) Each type of cell has a unique genetic blueprint.

   (C) The tissue culture transferred proteins necessary for plant differentiation to the root cell.

   (D) mRNA transcripts from the root cell are translated only after appropriate stimulation.

2. In human beings, sex-linked recessive disorders are usually carried on the X chromosome and most often affect males. Which of the following supports this claim?

   (A) Mothers always pass the disorders on to their sons

   (B) It takes only one copy of the gene to affect males

   (C) It takes only one copy of the gene to affect females

   (D) It takes two copies of the gene to affect males

Questions 3–4

The following are the net reactions of photosynthesis and aerobic cellular respiration:

$$6\,CO_2 + 6\,H_2O + \text{energy} \rightarrow C_6H_{12}O_6 + 6\,O_2$$

$$C_6H_{12}O_6 + 6\,O_2 \rightarrow 6\,CO_2 + 6\,H_2O + \text{ATP}$$

3. During aerobic cellular respiration, the C-H bond in glucose is broken and electrons are ultimately transferred to oxygen. Which of the following best explains why the concentration of electron-poor hydrogens (H^+) does not drastically change as a result?

   (A) Protons are used as building blocks for macromolecules.

   (B) Protons associate with ATP, which normally carries a negative charge.

   (C) Protons are transported out of the cell, where they are removed via diffusion.

   (D) Protons combine with oxygen anions to form water.

4. Which molecule is reduced to form glucose?

   (A) CO_2

   (B) ATP

   (C) O_2

   (D) H_2O

GO ON TO THE NEXT PAGE.

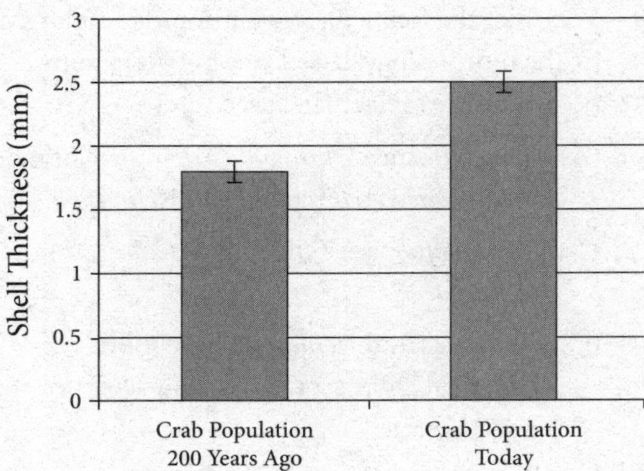

5. Crab shells are composed mainly of calcium carbonate, and their thickness determines how vulnerable crabs are to predators and environmental factors. Over the past 200 years, ocean acidification has caused a 25 percent increase in acidity. One study compared average shell thickness of a crab population 200 years ago to that of a crab population today. The data in the graph best support which of the following claims?
   (A) Average shell thickness decreased because ocean acidification lowered the pH of the seawater and caused shells to dissolve.
   (B) Average shell thickness decreased because predators selected thick-shelled crabs over thin-shelled crabs.
   (C) Average shell thickness increased because thin-shelled crabs were more easily penetrated by predators than were thick-shelled crabs.
   (D) Average shell thickness increased because ocean acidification lowered the levels of available carbonate ions for shell production.

6. *Naegleria fowleri* causes a fatal form of meningitis. The infectious form of this organism inhabits fresh water in warm climates, often in the sediment of lakes. It can infect humans when they swim in infested lakes, allowing entry through the nose. *N. fowleri* has a true membrane-bound nucleus and cellular organelles. It is a unicellular, heterotrophic organism that lacks a cell wall and moves via pseudopodia. What type of organism is it?
   (A) Bacteria
   (B) Virus
   (C) Protozoan
   (D) Fungus

7. What is the probability that a mother who is a carrier for cystic fibrosis, an autosomal recessive disorder, will have an affected child if the father is genotypically normal?
   (A) 0%
   (B) 25%
   (C) 50%
   (D) 100%

8. The sodium potassium pump is an ATPase that pumps 3 Na^+ out of the cell and 2 K^+ into the cell for each ATP hydrolyzed. Cells can use the pump to help maintain cell volume. Which of the following would most likely happen to the rate of ATP consumption immediately after a cell is moved to a hypotonic environment?
   (A) It would remain the same.
   (B) It would decrease.
   (C) It would increase.
   (D) It would increase, then decrease.

Practice

GO ON TO THE NEXT PAGE.

Questions 9–12

Two teams of scientists created the following pair of phylogenetic trees based on available data.

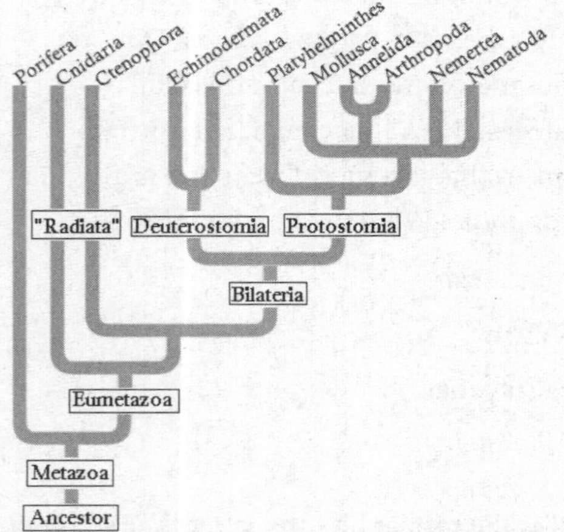

Phylogenetic tree based mainly on morphological comparisons

Phylogenetic tree based mainly on morphological comparisons

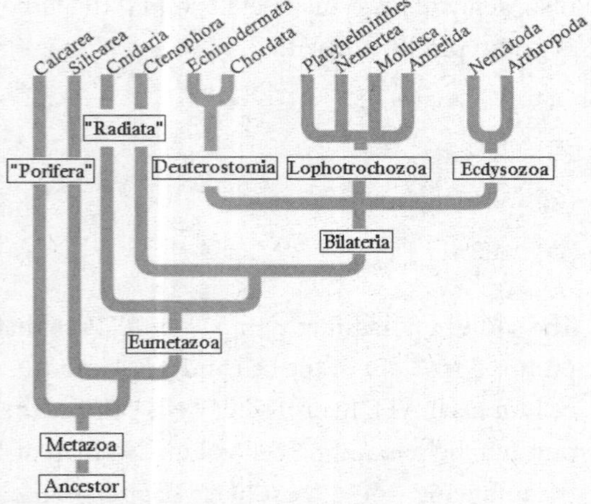

Phylogenetic tree based mainly on molecular data

9. Which of the following indicates the most substantial difference between the two phylogenetic trees?
   (A) Molecular differences among *Porifera* undetected by visual comparison
   (B) *Protostomia*'s replacement by two sister taxa in the molecular tree
   (C) The relative timeline of *Deuterostomia* divergence
   (D) The convergence of the entire animal kingdom on a common ancestor

10. Which of the following conclusions is supported by the morphology-based tree but NOT supported by the molecular-based tree?
    (A) *Annelida* and *Arthropoda* are more closely related than *Arthropoda* and *Nematoda*.
    (B) The *Porifera* are divided into *Calcarea* and *Silicarea*.
    (C) *Nemertea* and *Nematoda* have more common DNA sequences than *Mollusca* and *Nemertea*.
    (D) The *Radiata* have a more distant common ancestor than the *Bilateria*.

11. If the scientists who created the morphology-based tree relied mainly on phenotypic comparisons of adult and developing organisms, while those who created the molecular-based tree compared homologous *hedgehog* genes (a gene important for development), then which of the following statements would both teams of scientists most likely agree upon?
    (A) As the number of shared features increases so does the likelihood of independent evolution.
    (B) The fossil record is the ultimate authority in phylogenetics.
    (C) The molecular-based tree is the result of modern technology and is unlikely to be altered.
    (D) Neither phylogenetic tree completely and accurately describes actual evolutionary history.

GO ON TO THE NEXT PAGE.

12. The phylum *Cnidaria* consists of over 10,000 aquatic species, with cnidocytes (specialized cells used mainly for capturing prey) being the distinguishing characteristic. Which of the following is most likely to be classified as a Cnidarian?

    (A) An organism lacking digestive, circulatory, and nervous systems that feeds by drawing in water through pores

    (B) An organism with clearly defined sides (i.e., top/bottom, left/right)

    (C) A planktonic organism capable of responding to the environment equally from all directions

    (D) An organism that actively stalks its prey with coordinated complex movements

13. A foreign object originally located next to a tree is eventually "consumed" by the growing tree. As the tree grows taller, the height of the object in the tree is unchanged. These observations support which of the following statements?

    (A) Trees get taller by growing at the branch tips.

    (B) Vertical growth and horizontal growth are independent.

    (C) Once a certain maximum is attained, vertical height is relatively constant.

    (D) Height is increased via cell proliferation in the root system.

14. It is hypothesized that negative pressure in the phloem and cohesion among water molecules are responsible for the bulk movement of water against gravity in vascular plants. Which of the following fluid-filled cylinders best illustrates cohesion?

    (A)

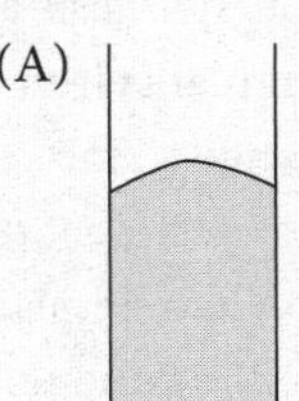

    (B)

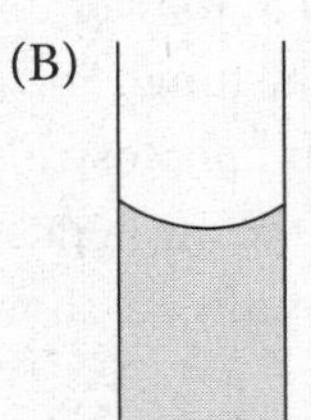

    (C)

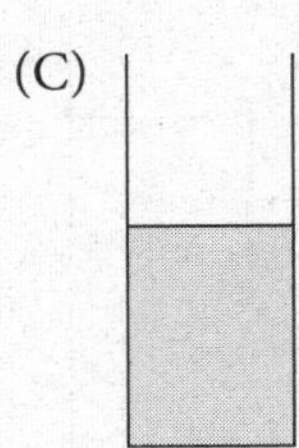

    (D)

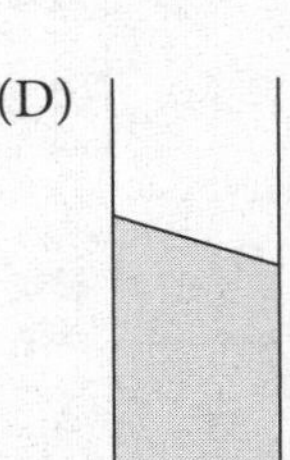

Practice

GO ON TO THE NEXT PAGE.

Questions 15–17

A scientist studying the ecology of cities found that, in developed landscapes, plant roots were not colonized by mycorrhizal fungi to the same degree that they were in a nearby nature preserve. In addition, she found that rates of photosynthesis and root respiration were much higher in plants in the preserve than for plants in city landscapes. She conducted a controlled greenhouse experiment to see what effects mycorrhizal colonization had on plant photosynthesis and respiration. Her experimental design involved growing 10 plants in soil rich in mycorrhizal fungal elements and 10 in the same soil that had been sterilized to remove the fungi. She made periodic measurements of plant photosynthesis and root respiration and calculated the mean rates for each experimental treatment. Her results are shown below.

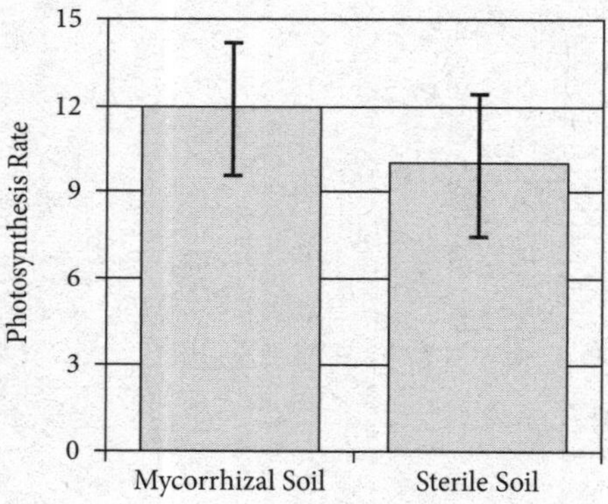

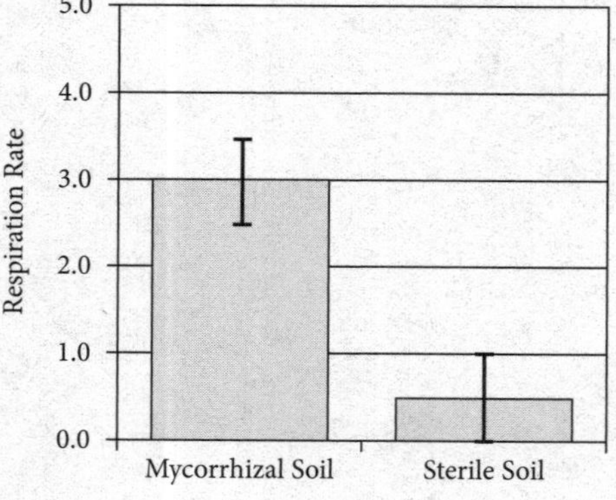

15. What conclusion can most reasonably be drawn from the data?
    (A) The presence of mycorrhizae increased photosynthesis and respiration rates significantly.
    (B) The presence of mycorrhizae increased photosynthesis but not respiration.
    (C) The presence of mycorrhizae increased respiration but not photosynthesis.
    (D) The presence of mycorrhizae had no effect on photosynthesis or respiration.

16. Which of the following provides the most plausible reason for why mycorrhizae have the influence they do on plant metabolic processes?
    (A) Mycorrhizae are important plant pathogens.
    (B) Mycorrhizae are important plant parasites.
    (C) Mycorrhizae are important plant predators.
    (D) Mycorrhizae are important plant symbionts.

17. Assuming that further experimentation showed conclusively that plants in cities had reduced rates of photosynthesis and respiration due to lack of colonization by mycorrhizae, what important biogeochemical cycle of an ecosystem would be most affected?
    (A) The nitrogen cycle
    (B) The water cycle
    (C) The hydrological cycle
    (D) The carbon cycle

GO ON TO THE NEXT PAGE.

18. In some organisms, features that have no function become vestigial and are ultimately lost. In many cave-dwelling animals, organs such as the eyes have been lost while other sense organs have increased in size. Which of the following hypotheses to explain the loss of nonfunctioning organs would NOT be considered correct according to the contemporary understanding of evolution?

(A) Mutations causing the reduction in the size of nonfunctional organs become fixed by genetic drift.

(B) There is natural selection against organs that are not used because the organs interfere with other, more important bodily functions.

(C) The development of a nonfunctional organ requires energy expenditures that would be better spent on building other tissues or maintaining other traits.

(D) All organs are maintained or eliminated as a result of how much they are used.

19. The Cdk inhibitor p16 binds to Cdk4/cyclin D complexes, which are normally responsible for allowing cells to pass through the restriction point from G_1, the first growth phase of the cell cycle, into S phase, when chromosome replication occurs. Underexpression of p16 protein could lead to which of the following?

(A) Uncontrolled cell division

(B) Cessation of mitosis

(C) Increased inhibition of Cdk4/cyclin D complexes

(D) Overexpression of p53 protein

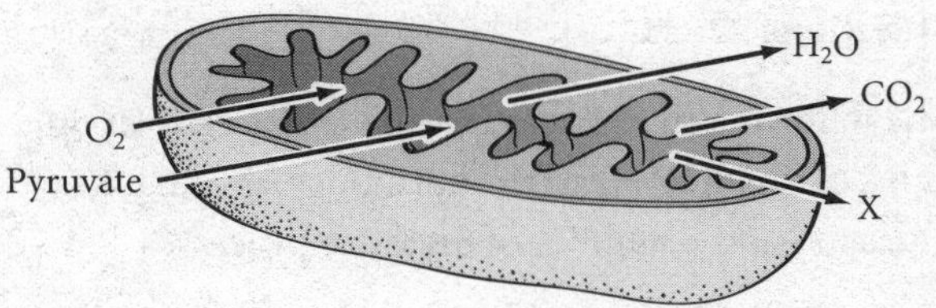

20. In the diagram pictured above, what does the letter *X* represent?

(A) Glucose

(B) $NADP^+$

(C) ATP

(D) ADP

21. Some plants can reproduce through self-pollination, in which a plant is fertilized by its own pollen. Self-pollination in plants is an example of which type of reproduction?

(A) Asexual reproduction, because a single parent is involved

(B) Asexual reproduction, because offspring are genetically identical to the parent plant

(C) Asexual reproduction, because offspring are genetically unique

(D) Sexual reproduction, because offspring are produced via fusion of gametes

Practice

GO ON TO THE NEXT PAGE.

Questions 22–24

The following diagram shows the feedback relationships between levels of the hormones insulin and glucagon and a number of digestive processes.

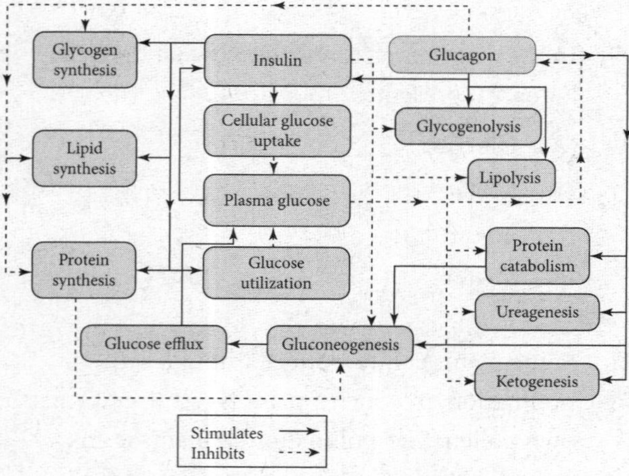

22. Based on the data in the figure above, which of the following is the most reasonable conclusion?
    (A) Protein is the body's preferred energy source
    (B) Insulin and glucagon act antagonistically
    (C) Insulin and glucagon are produced by beta and alpha cells, respectively
    (D) Brain cells are able to uptake glucose without insulin

23. Which of the following would a person with an inability to synthesize insulin be expected to show?
    (A) Low glucagon levels and low glucose levels
    (B) High glucagon levels and low glucose levels
    (C) Low glucagon levels and high glucose levels
    (D) High glucagon levels and high glucose levels

24. The respiratory quotient (RQ) is calculated as the ratio of carbon dioxide produced to the oxygen consumed for the complete combustion of a given fuel source. The RQ for carbohydrates is around 1.0, while the respiratory quotient for lipids is around 0.7. In resting individuals, what is the RQ most likely to be?
    (A) 1.2
    (B) 1.0
    (C) 0.8
    (D) 0.6

25. In a food chain that consists of grass → grasshoppers → spiders → mice → snakes → hawks, which organisms possess the most biomass within the community?
    (A) Grass
    (B) Grasshoppers
    (C) Mice
    (D) Snakes

26. Biotechnology is used for a number of applications in medicine, such as the manufacture of drugs and essential biological compounds. How are bacteria typically used to produce human insulin?
    (A) They are grown on media rich in sugar, which stimulates insulin production in the bacteria.
    (B) The DNA sequence that codes for human insulin production is inserted into the bacterial genome.
    (C) Human pancreas cells are grown in culture with bacteria and transformation occurs.
    (D) Specific bacteriophage viruses are used to produce the correct mutation in the bacterial genome.

GO ON TO THE NEXT PAGE.

27. Which of the following would slow the rate of increase in atmospheric carbon dioxide?
    (A) an increase in the rate of photosynthesis
    (B) a decrease in the rate of respiration
    (C) an increase in the carbon dioxide dissolved in the ocean
    (D) All of these would decrease the rate at which atmospheric carbon dioxide increases.

28. What is the primary danger of a population relying on an agricultural monoculture for its dietary staple?
    (A) People get tired of eating the same thing.
    (B) Children develop aversions to foods that appear too often in their diet.
    (C) Allergies develop after repeated exposure to the same food.
    (D) Genetic similarity makes an entire crop vulnerable to a single pest or pathogen.

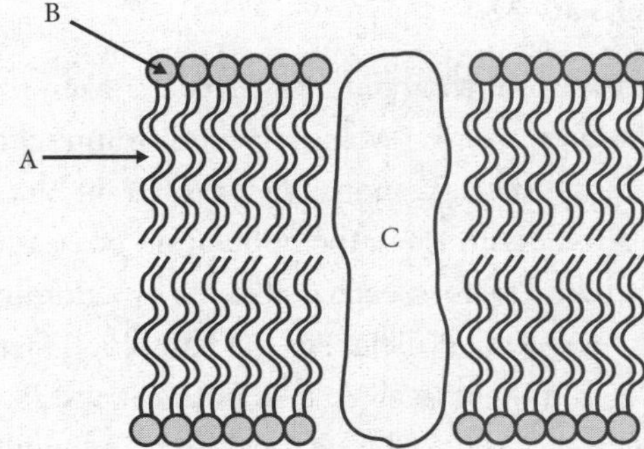

29. The diagram above depicts a cross-section of a cell membrane. Which of the following accurately describes the diagram?
    (A) The region of the membrane labeled A is nonpolar.
    (B) The region of the membrane labeled B is hydrophobic.
    (C) The structure labeled C is a complex carbohydrate.
    (D) Charged ions such as Na^+ diffuse directly through the membrane bilayer.

Practice

GO ON TO THE NEXT PAGE.

Questions 30–31

Consider the following blood group data taken from a population in Hardy-Weinberg equilibrium with respect to the alleles responsible for different blood factors. All individuals in the population possess two different blood factors, each coded for by a dominant allele and a recessive allele. For the first blood factor, allele *R* is dominant to allele *r*, so both *RR* and *Rr* individuals test as blood type R, while *rr* individuals test as blood type r. For the second blood factor, allele *F* is dominant to allele *f*, so both *FF* and *Ff* individuals test as blood type F, while *ff* individuals test as blood type f. The frequencies observed for blood type in the population are as follows:

Blood Types	Frequency
R and F	0.60
R and f	0.15
r and F	0.24
r and f	0.01

30. Based on the data provided, what are the frequencies of the *r* allele and the *F* allele, respectively?

    (A) 0.6 and 0.5

    (B) 0.25 and 0.25

    (C) 0.5 and 0.6

    (D) 0.25 and 0.84

31. Given the information above, all of the statements concerning this population are true EXCEPT for which of the following?

    (A) There are no new mutations arising among the blood type alleles

    (B) There is significant gene flow between this population and others

    (C) The population is large in size

    (D) There is no positive selection for the *R* allele

32. A certain plant is grown in darkness and observed to produce tall stems with non-expanded leaves. After being transported into daylight, the same plant develops broad, green leaves; short, sturdy stems; and long roots. Which of of following provides the best explanation for these observations?

    (A) There is minimal energy expenditure until sunlight is detected.

    (B) Sunlight stimulates the elongation of stems and proliferation of leaves.

    (C) The plant is exhibiting a specialized response to mechanical stress.

    (D) The plant normally sprouts underground.

33. A patient's parents both have a disease that is caused by a sex-linked dominant allele. This disease was passed down to the patient's mother from the patient's grandfather. The chance that the patient will have this disease is

    (A) 25%

    (B) 50%

    (C) 75%

    (D) 100%

34. Which of the following relationships is NOT an example of symbiosis?

    (A) A tick sucking blood from a dog

    (B) A clownfish living in the tentacles of a sea anemone for protection

    (C) Ants farming aphids to feed on their honeydew

    (D) Cows eating grass and leaving manure behind as fertilizer

GO ON TO THE NEXT PAGE.

35. Human white blood cells can "crawl" to damaged tissues to contact and phagocytize bacteria at the site of the damage. Which of the following processes facilitates this type of movement?

(A) Addition of phospholipids to the cell plasma membrane

(B) Rapid formation and deformation of actin filaments in the cytoskeleton

(C) Beating of cilia against vessel and tissue surfaces

(D) Whip-like motion of the cell's flagellum

36. Evolution may be defined as changes over time in the allele frequency of a population or species. All of the following are examples of evolutionary processes EXCEPT which of the following?

(A) Artificial selection in domestic dogs

(B) Populations in Hardy-Weinberg equilibrium

(C) Mutations that decrease reproductive fitness

(D) Natural selection between different colors of insects

37. Organisms that reproduce sexually exhibit zygotic, gametic, or sporic meiosis. What is one way to determine the type of life cycle of an organism?

(A) By observing embryonic development

(B) By comparing the diploid and haploid forms of the organism

(C) By determining when in the life cycle fertilization occurs

(D) By determining if gametes are multicellular or unicellular

38. Which of the following shows the correct order of hierarchy from simple to complex?

(A) Molecule → tissue → cell → organism

(B) Cell → tissue → organ → organism

(C) Organism → species → biosphere → ecosystem

(D) Molecule → cell → organism → tissue

39. Production of purple kernels in *Zea mays* (corn) is dominant over yellow kernels, and smooth kernels are dominant over wrinkled. The two traits are passed on independently of one another. Two heterozygous corn plants with purple, smooth kernels are crossed. Out of 160 of their offspring plants, how many would be expected to have yellow and smooth kernels?

(A) 120

(B) 90

(C) 60

(D) 30

40. Organisms transform energy when they ingest food, breaking it down into nutrients used to build tissues and make repairs. Each energy transfer increases the universe's level of which of the following?

(A) Order

(B) Stability

(C) Disorder

(D) Energy

Practice

GO ON TO THE NEXT PAGE.

41. According to the ABC hypothesis for the functioning of organ identity genes, three classes of genes (*A*, *B*, and *C*) are responsible for the spatial pattern of floral parts. Sepals develop from the region where only *A* genes are active. Petals develop where both *A* and *B* genes are expressed. Stamens arise where *B* and *C* genes are active and carpels arise where only *C* genes are expressed. Furthermore, it is observed that if *A*-gene or *C*-gene activity is missing, then the activity of the other spreads throughout.

    Which of the following could be the floral morphology for a mutant lacking *C*-gene activity?

    (A) Sepal-petal-stamen-carpel-stamen-petal-sepal
    (B) Carpel-stamen-carpel-stamen-carpel
    (C) Sepal-carpel-sepal
    (D) Sepal-petal-sepal-petal-sepal

42. What is the probability that the genotype *rrss* will be produced by a cross in which the genotypes of the parents are both *RrSs*?

    (A) 1/16
    (B) 1/8
    (C) 1/4
    (D) 1/2

Questions 43–44

The figure illustrates three types of point mutations involving a nucleotide base substitution.

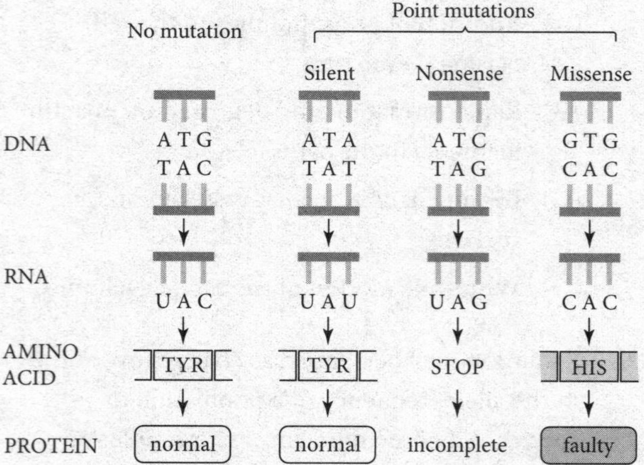

43. Hemoglobin is a tetramer, which consists of four subunits (two α and two β). Sickle cell anemia (abnormally shaped red blood cells) is caused by a mutation in the 20th nucleotide of the hemoglobin β gene, in which glutamic acid is changed for valine. Which of the following describes the result of the substitution?

    (A) The mutation has no effect on the structure of hemoglobin.
    (B) The cell stops the production of hemoglobin.
    (C) The properties of the amino acid are not altered, so hemoglobin does not lose its function.
    (D) The structure and function of hemoglobin are altered.

GO ON TO THE NEXT PAGE.

44. A frameshift mutation is caused by an insertion or deletion of one or more nucleotides in the DNA sequence. Which would most likely be the result of inserting one nucleotide instead of substituting a base in the DNA sequence?

(A) The frameshift mutation alters the R-groups in RNA.

(B) The amino acid sequence changes after the insertion.

(C) The insertion causes a missense mutation, resulting in a nonfunctional protein.

(D) The frameshift mutation affects the RNA sequence but not the amino acid sequence.

45. Enzymes are regulated in a number of different ways, including through competitive and noncompetitive inhibition. Competitive inhibitors of enzymes can be reversed by which of the following?

(A) Increasing the pH above the enzyme's optimal range

(B) Increasing the concentration of substrate

(C) Adding noncompetitive inhibitors

(D) Lowering the temperature below the enzyme's optimal range

46. Deep-diving air breathers (e.g., seals, whales, penguins) have numerous adaptations that allow them to dive to great depths and remain underwater for long periods of time. Which of the following would be LEAST likely to be found in a deep diver?

(A) Decreased O_2 storage in the lungs and increased O_2 storage in the blood

(B) High concentrations of myoglobin

(C) A large volume of blood per body mass ratio

(D) Efficient use of buoyancy to aid locomotion

47. Organisms rely upon the physical and chemical properties of water to facilitate a variety of bodily functions. For instance, human beings sweat when overheated in order to regulate temperature. Why is sweat such an effective mechanism for cooling the human body?

(A) Water has a high heat of vaporization

(B) Water has a low heat of vaporization

(C) Water dissolves salts

(D) Water adheres to heat receptors

48. Bacteriophages are a class of viruses that infect bacteria. Upon entering a bacterium, a phage can enter into one of two reproductive cycles: the lytic and the lysogenic. What is the main difference between the lytic and lysogenic cycles of phage reproduction?

(A) The lysogenic cycle alters the bacteria, while the lytic cycle does not.

(B) The lytic cycle alters the bacteria, while the lysogenic cycle does not.

(C) The lysogenic cycle kills the host cell, while the lytic cycle does not.

(D) The lytic cycle kills the host cell, while the lysogenic cycle does not.

49. Eukaryotic cells generate up to 36 ATP per molecule of glucose during aerobic respiration, but only 2 ATP per molecule during fermentation. Why does aerobic respiration produce significantly more ATP per glucose molecule than fermentation?

(A) Aerobic respiration uses glycolysis to oxidize glucose.

(B) Oxygen is necessary to release energy stored in pyruvate.

(C) Fermentation uses NAD^+ as the oxidizing agent in glycolysis.

(D) It requires energy to transport molecules into mitochondria.

GO ON TO THE NEXT PAGE.

50. A certain species of bird has beaks of variable length, with an average length of 10 centimeters and a standard deviation of 2 centimeters. It feeds primarily on a species of insect that burrows underground, with most of the insects located 8 to 10 centimeters beneath the surface. If the insect starts burrowing to a greater depth, which of the following population distributions for beak length is most likely to result?

(A) A mean of 10 centimeters with a standard deviation of 4 centimeters

(B) A mean of 8 centimeters with a standard deviation of 2 centimeters

(C) A mean of 12 centimeters with a standard deviation of 2 centimeters

(D) A mean of 10 centimeters with a standard deviation of 1 centimeter

51. An object known as the Murchison meteorite struck the Earth in Australia in 1969. Analysis of the rock shows it contains at least 50 amino acids, 19 of which are found in living organisms on Earth, as well as several nucleotides. Assume such amino acid-containing meteorites have existed throughout the history of the solar system. Of the following statements, which of the following is most reasonable to conclude of one or more meteorites?

(A) Meteorites are the sole source of all life on Earth.

(B) Meteorites cannot be the origin of amino acids on Earth.

(C) Meteorites contained complex biological polymers, such as proteins and nucleic acids.

(D) Meteorites contributed some of the precursors of life, but never contained any living organisms.

52. The *cecum* is a portion of the large intestine located near the junction of the small intestine and the large intestine. The following table lists the diets of several vertebrates, as well as the average length of the cecum as measured in 20 individuals of that species.

Species	Average Cecum Length	Diet
1	40.1 cm	herbivore
2	5.7 cm	omnivore
3	6.8 cm	carnivore
4	30.2 cm	ruminant

Based on the table, which of the following conclusions is most plausible?

(A) The cecum can become a vestigial structure in meat eaters, since it is shorter in species 2 and 3 than in species 1 and 4.

(B) The cecum evolved to have an important role in the digestion of protein, since it is shorter in species 2 and 3 than in species 1 and 4.

(C) Species 1 is more closely related to species 4 than to species 2, because 1 and 4 are both herbivores.

(D) Species 2 and 3 have a recent common ancestor, since the average cecum lengths in species 2 and 3 are approximately equal.

GO ON TO THE NEXT PAGE.

53. Most biomass pyramids show a rapid decrease in biomass as trophic level increases. In aquatic systems, however, this pattern may be reversed so that one observes a larger standing crop of consumers compared with producers. Which of the following offers the best explanation for this pattern?

    (A) Aquatic producers tend to have larger body sizes than terrestrial producers.

    (B) Water is an easier medium to live in, so aquatic organisms require less food.

    (C) Biomass in aquatic systems cannot be measured accurately.

    (D) Phytoplankton is rapidly consumed, but it has a high turnover rate.

Questions 54–56

Sickle cell anemia is caused by mutant hemoglobin DNA, which is more common in humans with African ancestry than in those with European ancestry. The sickle cell allele creates an altered mRNA codon that produces hemoglobin containing valine rather than glutamic acid. If a person inherits both alleles for the sickle cell trait, that person's hemoglobin will polymerize under low oxygen conditions (i.e., elevated physical activity). This can result in brain damage, paralysis, kidney failure, and other very serious physiological problems.

54. Based on the information provided, the mutation for sickle cell hemoglobin is most likely an example of which of the following?

    (A) A base-pair substitution

    (B) A frameshift mutation

    (C) A silent mutation

    (D) A mutagen

55. Heterozygotes for the sickle cell trait have an increased resistance to malaria. If malaria were eradicated and effective treatment for sickle cell anemia made universally available, what would be the expected effect on the sickle cell allele?

    (A) The frequency of the allele would remain roughly constant.

    (B) The frequency of the allele would decrease.

    (C) The frequency of the allele would increase.

    (D) The frequency of homozygous individuals would decrease, but the frequency of heterozygous individuals would increase.

56. Genetic mutations, such as the one found in the sickle cell anemia allele, can be found in all organisms, from the simplest prokaryotes to the most complex eukaryotes. However, these mutations have a greater impact on the genetic diversity of bacteria populations than on that of human populations for which of the following reasons?

    (A) Human sexual reproduction recombines existing alleles

    (B) Bacteria reproduce more rapidly than humans

    (C) New bacteria are generated through sexual reproduction

    (D) Genetic mutations are much rarer in humans than in bacteria

GO ON TO THE NEXT PAGE.

57. Substances that are formed as intermediates or products during a biochemical reaction are called "metabolites." Certain drugs, called "antimetabolites," possess chemical similarities that allow them to mimic metabolites and participate in normal biochemical reactions, but are different enough that they interfere with overall cellular function. These drugs are commonly used as antibacterial or anticancer agents. Which of the following statements offers the most plausible explanation of how these drugs work?

(A) The antimetabolite binds to the active site of an enzyme and directly inhibits it, acting as a competitive inhibitor.

(B) The antimetabolite binds to the active site of an enzyme, allowing the reaction to proceed and produce an unusable end product.

(C) The antimetabolite binds to the enzyme at a location that is not the active site, changing the structure and function of the enzyme.

(D) The antimetabolite binds to the metabolite, keeping it from reaching and binding to the active site of an enzyme.

58. The concept of gradualism was initially used to explain the formation of geologic features over vast stretches of time, but aspects of this idea were later incorporated into Darwin's theory of evolution. Which of the following best describes an idea shared both by geologic gradualism and Darwin's theory?

(A) Change occurs mainly through catastrophic events.

(B) Slow and continuous processes can lead to drastic changes.

(C) Certain heritable traits are gradually favored over others.

(D) Resources are limited and there is a struggle for existence among individuals.

59. In a particular species of guppy, tails can be either long or short and either feathered or straight. A mating between a short feathered-tailed female and a short straight-tailed male produces 30 short straight-tailed guppies, 42 short feathered-tailed guppies, 10 long straight-tailed guppies, and 14 long feathered-tailed guppies. Calculate the chi-square value for the null hypothesis that the short feathered-tailed guppy was heterozygous for the feathered-tail allele.

(A) 0

(B) 1.33

(C) 2.67

(D) 24

Practice

GO ON TO THE NEXT PAGE.

60. The table shows properties of water, isopropanol, and benzene.

Liquid	Molecular Formula	Boiling Point (°C)	Melting Point (°C)	Specific Heat Capacity (kJ/kg°C)
Water	H_2O	100.0	0.0	4.18
Isopropanol	C_3H_8O	82.6	–89.0	2.68
Benzene	C_6H_6	80.1	5.5	1.73

Based on the information provided in the table, which of the following best explains why living systems depend on the properties of water, rather than those of other liquids?

(A) Water is composed of hydrogen and oxygen, the two most important essential elements found in living organisms.

(B) Water's high specific heat capacity enables water to buffer temperature changes in living systems.

(C) Water's melting point at 0°C enables water to act as a solute for chemical reactions.

(D) Water's high boiling point enables living organisms to thrive at higher temperatures.

END OF SECTION I

IF YOU FINISH BEFORE TIME IS CALLED, YOU MAY CHECK YOUR WORK ON SECTION I ONLY.

DO NOT GO ON TO SECTION II UNTIL YOU ARE TOLD TO DO SO.

SECTION II
80 Minutes—6 Questions

Directions: Begin by taking a 10-minute reading period, during which time you may sketch graphs, make notes, and plan your answers. You then have 80 minutes to complete your responses to the 6 free-response questions. Questions 1 and 2 are long free-response questions that should require about 22 minutes each to answer. Questions 3 through 6 are short free-response questions that should require about 9 minutes each to answer. Read each question carefully and write your response on scratch paper. Answers must be written out in paragraphs. Outline form is not acceptable. It is important that you read each question completely before you begin to write.

1. The one gene-one enzyme hypothesis states that one gene directly produces one enzyme, which directly affects one step in a metabolic pathway (Figure 1).

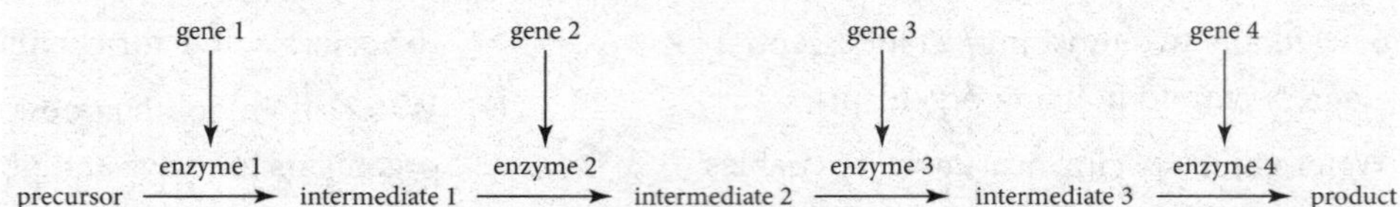

Figure 1. One gene-one enzyme hypothesis

Researchers performed an experiment to investigate the pathway of arginine (an essential amino acid). *Neurospora* mutants, each mutated in a different gene, were grown on minimal medium (sugar, salts, and vitamins) and minimal medium with an additional supplement. The results are shown in Table 1. If a mutant is supplied with the compound it is unable to produce, it will grow.

Mutant	Minimal Medium (MM)	MM + Arginine	MM + Citrulline	MM + Arginino-succinate	MM + Ornithine
1	no	yes	no	yes	no
2	no	yes	yes	yes	yes
3	no	yes	no	no	no
4	no	yes	yes	yes	no

(a) **Explain** how genetic mutations affect enzyme synthesis.

(b) **Propose** an appropriate control treatment for the experiment, and **describe** how the control treatment would increase the validity of the results.

(c) Using the data in the table, **identify** the intermediates of the arginine pathway on the template provided.

(d) **Provide** reasoning for the intermediates on the pathway.

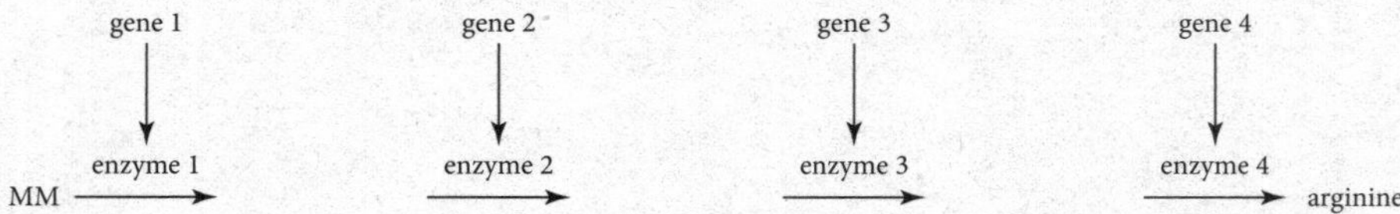

GO ON TO THE NEXT PAGE.

2. Consider the following graphs of seasonal trends in the Chesapeake Bay.

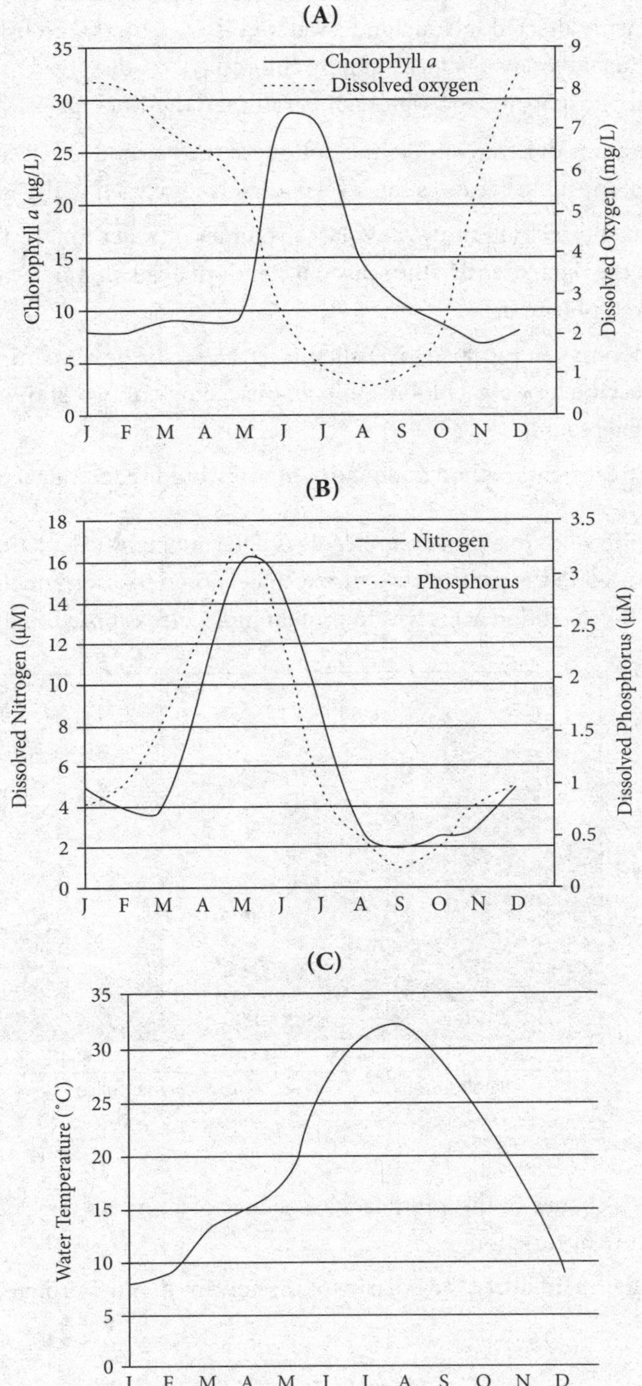

Figure 1. Seasonal trends in the Chesapeake Bay of (A) chlorophyll *a* and dissolved oxygen, (B) dissolved nitrogen and dissolved phosphorus, and (C) water temperature

GO ON TO THE NEXT PAGE.

Harmful algae blooms fueled by excess nutrients (nitrogen and phosphorus) from agricultural fields, sewage treatment plants, industrial facilities, and the atmosphere disrupt marine ecology. Examples of organisms affected include underwater grasses, oysters, and fish. The graphs show the seasonal trends of algal blooms (as measured by chlorophyll *a*), dissolved oxygen, dissolved nitrogen, dissolved phosphorus, and water temperature in the Chesapeake Bay.

(a) When algal blooms die, they sink to the bottom of the bay and are decomposed by bacteria. **Describe** what most likely causes the level of dissolved oxygen to dip in the summer months.

(b) In spring, rain and melting snow cause large volumes of water to flow into the bay. Based on an analysis of the data, **identify** the seasonal trends of algal blooms, of dissolved nitrogen, and of dissolved phosphorus.

(c) Some algae blooms can produce toxic chemicals. **Describe** how toxins may affect fish-eating birds, and **describe** how algal blooms impact the ecology of two of the listed organisms affected by algal blooms.

(d) **Propose** one treatment method and **justify** how it would reduce algal blooms.

3. The location of amino acids in a protein molecule is determined by calculating the extent by which an amino acid is buried in the protein structure when exposed to a solvent. The figure below shows the distribution of the 20 amino acids within protein molecules exposed to the same solvent.

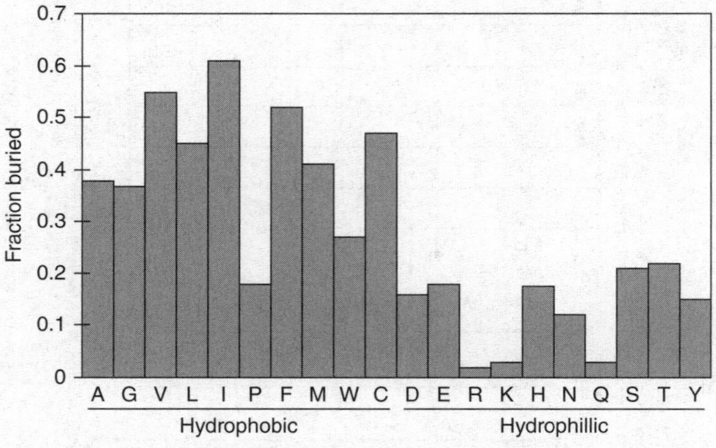

This data is adapted from *Structural bioinformatics, protein crystallography, sequence analysis & homolog modeling*.

(a) **Describe** how a change in the amino acid sequence of a protein may lead to changes in the protein structure or function.

(b) Based on the figure, **identify** the polarity of the solvent that the protein molecules were exposed to.

GO ON TO THE NEXT PAGE.

(c) A transmembrane protein spans the entire width of a cell membrane. For amino acids A F, N, and R, **predict** their location in the amino acids sequence of a transmembrane protein in the template below.

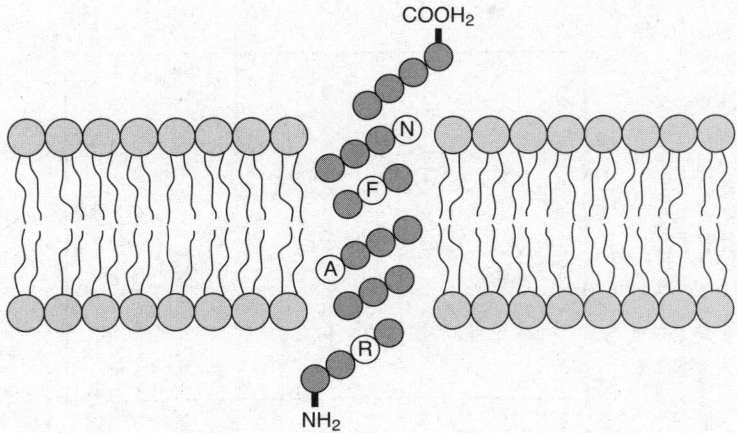

(d) **Justify** your predictions.

4. All living organisms contain genetic information that provides several functions inherent to the individual organism and to the perpetuation of its species. The frequency of two alleles, *A* and *a* (for which *A* is dominant and *a* is recessive) are shown below for an original population and after a random event.

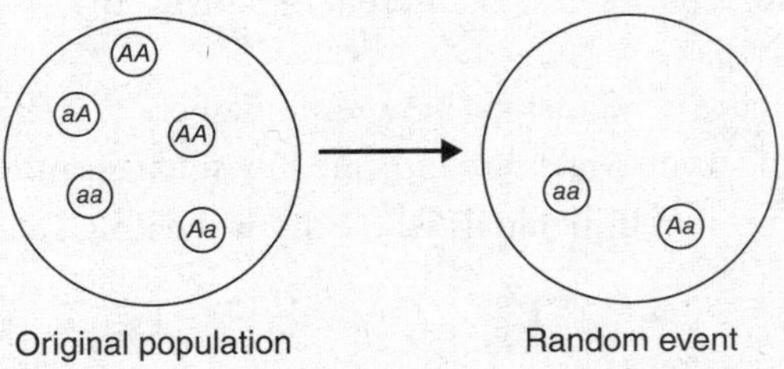

(a) **Describe** one example of genetic drift.

(b) **Explain** why a new species formed from genetic drift may not be well adapted to survive in its environment.

(c) The individuals from the random event reproduce and form a new population. Based on the data, **predict** the allele frequencies for the new population.

(d) **Justify** your prediction.

GO ON TO THE NEXT PAGE.

5. In a long-term project studying the interactions of several species of animals on an isolated island, scientists counted the number of individuals of each species visiting a site on the island over the course of several days, every summer for 100 years. The results from that study are shown in the following graph.

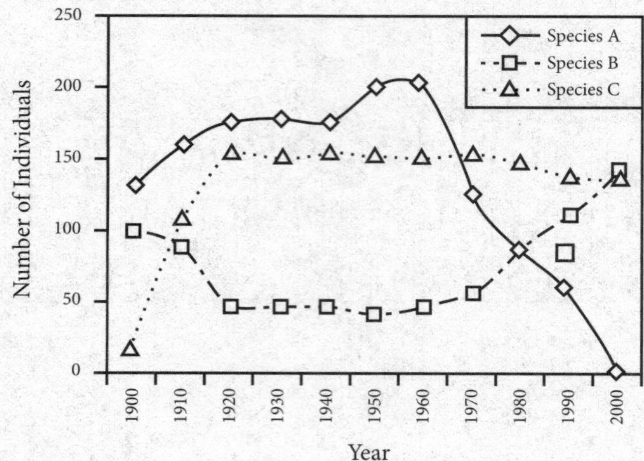

(a) **Describe** one behavioral response of organisms within a population that affects their overall fitness and may contribute to the success of the population.

(b) **Explain** the trends of population Species A and B in the figure.

(c) Based on the data, **determine** the ecological relationship between Species B and C.

(d) **Explain** how symbiosis between two different organisms drives population dynamics.

6. A pedigree shows the genetic relationships between members of a family and can be used to analyze the inheritance of a trait. Males are indicated by squares, and females are indicated by circles. Shaded shapes represent individuals who exhibit the trait. Refer to the following pedigree.

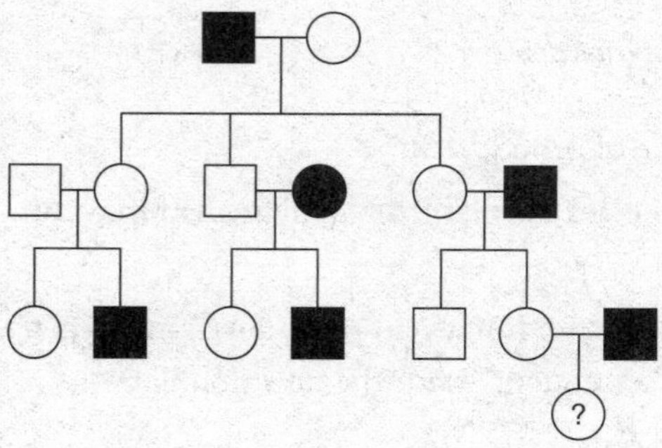

(a) **Determine** the inheritance pattern of the allele(s) represented in the pedigree.

(b) **Identify** the genotype for the grandmother at the top of the tree.

(c) **Calculate** the likelihood that a female in the fourth generation expresses the disorder.

(d) **Explain** deviations from Mendel's pattern of inheritance.

STOP—END OF EXAM

The answer key to this quiz is located on the next page.

ANSWER KEY

Section I

1. A	16. D	31. B	46. D
2. B	17. D	32. D	47. A
3. D	18. D	33. C	48. D
4. A	19. A	34. D	49. B
5. C	20. C	35. B	50. C
6. C	21. D	36. B	51. D
7. A	22. B	37. B	52. A
8. C	23. C	38. B	53. D
9. B	24. C	39. D	54. A
10. A	25. A	40. C	55. A
11. D	26. B	41. D	56. B
12. C	27. D	42. A	57. B
13. A	28. D	43. D	58. B
14. A	29. A	44. B	59. C
15. C	30. C	45. B	60. B

Section II

1. See Answers and Explanations
2. See Answers and Explanations
3. See Answers and Explanations
4. See Answers and Explanations
5. See Answers and Explanations
6. See Answers and Explanations

SCORING

Section I Number Correct: ______________

Section II Points Earned: ______________

Enter your results to your Practice Exam 2 assignment to see your 1–5 score and view detailed answers and explanations by logging in at kaptest.com.

Haven't registered your book yet? Go to kaptest.com/booksonline to begin.

PRACTICE EXAM 2 ANSWERS AND EXPLANATIONS

Section I

1. A

Cell differentiation depends upon the selective transcription of the genome, which is fully included in all of the cells of an organism. That a mature root cell can dedifferentiate is not so surprising, given the fact that it contains all the genes necessary to produce any type of cell distinctive to the species of plant. **(A)** is thus correct. (B) is factually inaccurate; all cells contain the same genetic blueprint (with at most minor differences arising from occasional mutations during development). (C) is incorrect because there is not enough information about the tissue culture to reach this conclusion. (D) is incorrect because additional stimulation is not necessary for mRNA transcripts to be translated.

2. B

In human males (who have only one X chromosome), one copy of a recessive gene will be expressed because there is no dominant copy to mask its expression. **(B)** is therefore correct. (A) is incorrect because X-linked recessive disorders are not always passed on to male offspring; a mother who is simply a carrier for the disorder could pass on her unaffected X chromosome. (C) is incorrect because a human female needs to have the recessive allele on both of her X chromosomes in order to be affected. (D) is incorrect because human males only have one X chromosome, so they only need one copy to be affected.

3. D

While neither electrons nor protons are *directly* transferred from glucose to oxygen, both are indirectly combined with oxygen to form water. The second equation shows the "big picture," wherein glucose loses hydrogen atoms and oxygen gains hydrogen atoms (and electrons) to form water. **(D)** is thus correct.

4. A

The first equation shown is photosynthesis, the process by which plants harvest the Sun's energy to fix carbon dioxide and produce glucose. An important step in the light-independent reactions of photosynthesis, the Calvin cycle, involves the reduction of fixed carbon dioxide to the three-carbon precursors of glucose. Water is oxidized during the light-dependent reactions of photosynthesis. **(A)** is correct.

5. C

Ocean acidification selects against thinner-shelled crabs. Since ocean acidification dissolves shells, shells of thinner-shelled crabs would become even thinner and more vulnerable to predators than thicker-shelled crabs. As a result, thicker-shelled crabs would have survived and produced offspring that were also thicker-shelled, leading to an increase in average shell thickness. **(C)** is thus correct. (A) is incorrect because the data provide no evidence to support the conclusion that average shell thickness decreased. Ocean acidification does cause shells to dissolve, but according to the data, the average shell thickness of the crab population today is greater than that of the crab population 200 years ago. (B) is incorrect because thin-shelled crabs would have been eaten by predators more easily since they were more vulnerable. (D) is incorrect because ocean acidification decreases the number of carbonate ions available for shell production and would result in a shift toward more thin-shelled crabs.

6. C

The key to this question is in the last sentence before the question. Any organism that is a unicellular heterotrophic eukaryote without a cell wall must be a protist. **(C)** is thus correct. Bacteria, (A), and viruses, (B), lack a nucleus and cellular organelles, and fungi, (D), have cell walls containing chitin.

7. A

If the mother is a carrier for an autosomal recessive disorder, then her genotype would be represented with one dominant allele and one recessive allele (*Aa*). The father is genotypically normal (*AA*), so 50% of the children will be carriers (*Aa*) and 50% will be homozygous dominant (*AA*). The likelihood of having an affected child is therefore 0%, **(A)**.

8. C

A cell placed in a hypotonic environment will swell unless action is taken to reduce the concentration gradient. The sodium potassium pump can be used to maintain cell volume. During each cycle of the pump, the *net* effect is the removal of one solute particle, so activating the pump (and consuming ATP) would reduce the solute gradient between the cell and the environment and mitigate the swelling. Thus, when placed in a hypotonic environment, ATP consumption will immediately increase. **(C)** is correct. (D) is incorrect because the question specifically asks about an *immediate* effect.

9. B

While there are some subtle differences between the two phylogenetic trees, the *main* difference is in the relationship among the bilaterians. The morphology-based tree divides the bilaterians into two taxa: deuterostomes and protostomes. In contrast, the tree based on molecular data omits *Protostomia*, replacing it with *Lophotrochozoa* and *Ecdysozoa*, providing a total of three branches for the bilaterians. **(B)** is thus correct. (A) and (C) do represent differences, but they are fairly small in comparison. Both trees show a convergence on a single ancestor, ruling out (D).

10. A

In the morphology-based tree, *Annelida* and *Arthropoda* have a more recent common ancestor than *Arthropoda* and *Nematoda*. However, in the molecular data-based tree, *Arthropoda* and *Nematoda* have a more recent ancestor than *Annelida* and *Arthropoda*, making **(A)** correct. (B) is incorrect because it is not a conclusion that can be drawn from the morphology-based tree. (C) can be eliminated because the morphology-based tree does not use DNA sequencing. (D) is incorrect because both trees show a more distant common ancestor for the *Radiata* than the *Bilateria*.

11. D

Phylogenetics, by its very nature—looking back into prehistory—is an imperfect science, one that will never be completely accurate. **(D)** is therefore correct. (A) is unlikely to be a true statement and less likely to be supported by both teams of scientists. (B) is incorrect because the fossil record may be "set in stone," but it is far from the ultimate authority—the most robust phylogenetic hypotheses are those evidenced by numerous lines of molecular and morphological incidence as well as by fossil evidence. (C) is unlikely to be supported by either team of scientists, who would recognize that phylogenetic trees can always be altered in light of new evidence.

12. C

The question stem mentions *Cnidaria,* and the figure implies that these organisms exhibit radial symmetry (they are "Radiata" as opposed to "Bilateria"). Because "responding to the environment equally from all directions" implies radial symmetry, **(C)** is correct.

13. A

The tree grows taller, but the height of the foreign object remains constant. This implies that the tree does not grow taller by adding building blocks at the base, but rather by extending the ends. This matches **(A)**. (B) is not a valid conclusion based on the stated observations because both types of growth are observed. (C) is not supported by the observations because the tree continues to grow taller. If the tree grew via the mechanism described in (D), then the foreign object would be expected to rise over time.

14. A

The best example of cohesion would illustrate the preference of the fluid molecules for other fluid molecules—meaning the correct answer will show the fluid sticking together and forming a convex meniscus, instead of sticking to the container and forming a concave meniscus (which would exhibit *adhesion*). **(A)** is thus correct.

15. C

Although both figures show that rates of photosynthesis and respiration tended to be higher when mycorrhizae were present, the error bars indicate that there really was no difference in terms of photosynthesis, because of the overlapping confidence intervals for the two treatments. This eliminates (A) and (B). The figures do show a difference between the two treatments in terms of respiration, ruling out (D) and making **(C)** correct.

16. D

Most land plants form mutualistic symbioses with mycorrhizae; it is thought that the two groups evolved together. Both organisms typically benefit from the association. **(D)** is thus correct. (A) is incorrect because the mycorrhizae do not infect and harm the plants, but coexist with them for mutual benefit. (B) is incorrect because the plants also benefit from the symbiosis. (C) is incorrect because the mycorrhizae do not consume the plants.

17. D

Both photosynthesis and respiration are important components of the carbon cycle. **(D)** is thus correct. The nitrogen cycle is predominantly moderated by bacteria, so (A) is incorrect. The water cycle and hydrological cycle are basically the same thing and would not necessarily be influenced by changes in photosynthesis and respiration, so (B) and (C) are also incorrect.

18. D

The pre-Darwinian biologist Jean-Baptiste Lamarck maintained that traits evolved by means of the use or disuse of acquired characteristics. For instance, a Lamarckian explanation of the long necks in giraffes would be that the ancestors of giraffes had shorter necks, but they would stretch their necks to reach food at greater heights, and that these stretched necks would be passed on to the next generation, who would stretch even more and pass on those even longer necks, and so on. **(D)** presents such a Lamarckian explanation, making it incompatible with the contemporary understanding of evolution. (A) is incorrect because genetic drift is an accepted part of the contemporary evolutionary theory. (B) and (C) both point to the role of natural selection in evolution, so they are also incorrect.

19. A

Removal of cyclin D inhibition would allow cells to progress unhindered from G_1 into S phase. This could cause, in the absence of other controls, a more rapid onset of mitosis and uncontrolled cell division. **(A)** is thus correct.

20. C

The diagram depicts a mitochondrion, where the reactions of cellular respiration take place. Pyruvate, produced from the breakdown of glucose in glycolysis, enters the mitochondrial matrix along with oxygen and NADH; the Krebs cycle takes place (as well as electron transport and oxidative phosphorylation); and ATP is produced. In fact, the letter *X* represents ATP, whose production is an essential aspect of the entire cellular respiration process. **(C)** is correct. (A) is incorrect because glucose is broken down to two pyruvate molecules before entering a mitochondrion for aerobic respiration. (B) is incorrect because $NADP^+$ is used in photosynthesis, not respiration. (D) is incorrect because ADP is a precursor to ATP.

21. D

By definition, asexual reproduction is the production of offspring genetically identical to the single parent, primarily occurring via binary fission. Sexual reproduction involves fusion of unique haploid gametes to produce genetically unique diploid offspring. Self-pollination is a form of sexual reproduction. A single parent plant produces both male and female gametes, but each is the product of meiosis and genetically unique. Fusion of two gametes produces a unique diploid zygote, which will develop into a unique adult. **(D)** is correct. (A), (B), and (C) are incorrect because self-pollination is not a form of asexual reproduction.

22. B

The most reasonable conclusion that can be drawn from the figure is that glucagon and insulin act antagonistically. Most processes in the figure that are stimulated by insulin are inhibited by glucagon (and vice versa). **(B)** is thus correct. (A) is untrue and not a reasonable conclusion that may be drawn from the figure. (C) and (D) are both true statements, but neither can be concluded from the data in the figure.

23. C

Insulin lowers plasma glucose levels. Plasma glucose inhibits glucagon. An inability to synthesize insulin would lead to increased plasma glucose levels and decreased glucagon levels, as in **(C)**.

24. C

In resting individuals, both carbohydrates and fats are utilized for energy, so the RQ would be expected to be somewhere between the RQ for carbohydrates and the RQ for lipids. **(C)** is thus correct.

25. A

Organisms at the top of the food chain generally have the least biomass, while organisms at the bottom have the greatest biomass. In this food chain, grass is at the bottom and has the greatest biomass, making **(A)** correct.

26. B

The only way a bacterium can be induced to produce a human protein such as insulin is to have the gene that codes for the production of that protein inserted into the bacterial genome. **(B)** is thus correct. (A) is clearly incorrect; bacteria are cultured on sugar-rich media routinely, but it does not confer human abilities upon them. Both (C) and (D) suggest technically feasible possibilities, except that the chances of such a beneficial transfer of genetic information occurring in either case would be virtually zero.

27. D

All three of the ways mentioned in (A), (B), and (C) would decrease the rate at which atmospheric carbon dioxide increases. While it is true that photosynthesis removes carbon dioxide from the atmosphere and that an increase in the rate of photosynthesis would decrease the rate at which atmospheric carbon dioxide increases, (A) is incorrect because it is not the only correct method among the answer choices. (B) is also a correct method because decreasing the rate of respiration would release less carbon dioxide and would thus slow the rate of atmospheric carbon dioxide increase. Again, (B) is incorrect because it is not the only correct method among the answer choices. (C) is also a true statement. Increasing the amount of carbon dioxide dissolved in the ocean would remove it from the atmosphere and thus decrease the rate at which atmospheric carbon dioxide increases. Again, since (C) is not the only correct method among the answer choices, it is not the correct answer.

28. D

Potato blight caused the destruction of much of Ireland's potato crop in the 1800s, which in part led to a famine. All potato plants were susceptible to the blight because they were so genetically uniform. Thus, the primary danger of relying on a monoculture as a primary food source is the threat posed by pests and pathogens, **(D)**. (A) may be true, but it is hardly a matter of life and death, such as a famine would be. (B) and (C) are factually inaccurate.

29. A

The region labeled A is composed of long fatty acid tails of the phospholipids that make up the cell plasma membrane. These tails are nonpolar, so **(A)** is correct. Region B is the polar, hydrophilic region of the molecule, so (B) is incorrect. The structure labeled C is a membrane-bound protein, making (C) incorrect. The lipid portion of the membrane is nonpolar and thus hydrophobic, which prevents charged ions such as Na^+ from crossing the membrane, so (D) is incorrect.

30. C

You might have been tempted to choose (D) if you simply added the frequency of individuals expressing blood type r (0.24 + 0.01) and did the same for blood type F (0.24 + 0.60). This type of addition can lead you to the frequency of the recessive *r* allele, but it does not work for the *F* allele. The frequency of the r blood type (q^2 in the Hardy-Weinberg equation, $p^2 + 2pq + q^2 = 1$) is equal to the sum of the last two rows in the table, or $0.24 + 0.01 = 0.25$. Thus, the frequency of the *r* allele (q) is the square root of 0.25, or 0.5. On the other hand, individuals with blood type F can be either homozygous or heterozygous; therefore, the first and third rows of the table include both *FF* individuals as well as *Ff* ones. Therefore, you cannot simply add 0.6 to 0.24 to calculate the frequency of the *F* allele. Instead, you must first find the frequency of the recessive *f* allele (q) and then subtract that from 1 to find the frequency of the *F* allele (p). The frequency of blood type f is the sum of the second and fourth rows, or $0.15 + 0.01 = 0.16$, so the frequency of the *f* allele is the square root of 0.16, or 0.4. The allele frequency of *F* is $1 - 0.4 = 0.6$. Therefore, **(C)** is correct.

31. B

Hardy-Weinberg equilibrium requires random mating, large population size, no natural selection, no mutations, and no gene flow. Because **(B)** suggests significant gene flow between this population and others, it is correct. The other answers are consistent with Hardy-Weinberg equilibrium.

32. D

The plant in question is grown in darkness, and certain traits are observed, but then the same plant is transported into the light and the previously observed traits are "reversed." Without light, the plant produces tall stems and non-expanded leaves—as if it were simply trying to reach something. When presented with light, the same plant develops broad, green leaves and short sturdy stems and long roots (i.e., it begins to look like a "normal plant"). The best explanation is that the plant is exhibiting normal behavior, which is explained if the plant normally sprouts underground. **(D)** is thus correct. (A) can be eliminated because the plant is observed to expend energy (tall stems and leaves) in darkness. (B) is wrong because the sunlight actually thickens the stems—the elongation is more evident in the darkness. (C) may be tempting, but cannot be concluded because there is no evidence of mechanical stress.

33. C

Call the disease allele X^A. The corresponding normal allele is X^a. If the patient's mother inherited the disease from her father (the patient's grandfather), her genotype is X^AX^a. Because the patient's father is also affected, his genotype is X^AY. The cross is, therefore, $X^AX^a \times X^AY$. Using a Punnett square, the offspring can have the following genotypes:

	X^A	X^a
X^A	X^AX^A	X^AX^a
Y	X^AY	X^aY

The patient's chances of having the disease are 75%, because 3 of the 4 offspring in the Punnett square inherit the dominant X^A allele. **(C)** is thus correct.

34. D

Symbiosis is simply defined as organisms living in close association with each other. (A) is an example of parasitism, (B) of commensalism, and (C) of mutualism, all forms of symbiosis. **(D)** is an example of predation, one organism feeding on another, which is not a form of symbiosis.

35. B

Most cell movement is associated with the cytoskeleton, so **(B)** is correct. Changes in the plasma membrane, as in (A), are associated with endocytosis and exocytosis. Cilia, as in (C), and flagella, as in (D), are not associated with the amoeboid movement of human white blood cells.

36. B

Artificial selection, as in (A), leads to changes in gene frequencies, as breeders select for some traits over others. The presence of mutations that decrease reproductive fitness, as in (C), will also lead to genetic change, as the new alleles are selected against over time. Natural selection for some colors of insects over others, as in (D), will also lead to changes in allele frequencies. In contrast, a population in Hardy-Weinberg equilibrium is one for which allele frequencies remain constant, the complete absence of evolution. **(B)** is thus correct.

37. B

Life cycles are characterized by the timing of meiosis and the characteristics of the diploid and haploid generations. Embryonic development, (A), does not reveal either of these characteristics, nor does the timing of fertilization, (C). All gametes are unicellular, ruling out (D). Only **(B)** involves an examination of characteristics of the diploid and haploid generations, making it correct.

38. B

(A) is incorrect because tissues are composed of cells. In C, biosphere should follow ecosystem, because it encompasses all other levels. (D) shows tissues being made up of organisms, but the opposite is true. Only **(B)** shows a correct order of hierarchy: cells make up tissues, which make up organs, which in turn make up organisms.

39. D

In a dihybrid cross such as the one described in the question, the expected phenotypic outcome is: 9/16 individuals will show both dominant traits, 3/16 will show one dominant and one recessive trait, 3/16 will show the *other* dominant and the *other* recessive trait, and 1/16 will show both recessive traits. The question asks about individuals showing one dominant and one recessive trait, so out of 160 offspring, 30 would be expected to fit the bill, **(D)**.

40. C

The second law of thermodynamics maintains that every change in energy or energy transfer contributes to the entropy of the universe. Entropy is also known as disorder, so **(C)** is correct. (A) and (B) are incorrect because they suggest the opposite. (D) is incorrect because it violates the first law of thermodynamics: energy can be neither created nor destroyed.

41. D

The mutant lacking the *C* gene will be incapable of producing stamens and carpels, so rule out all answers with those organs. According to the observation mentioned in the last sentence, if a mutant lacks *C*-gene activity, then *A*-gene activity will spread throughout. So expect only sepals and petals. The correct answer is **(D)**.

42. A

To determine the answer, construct a Punnett square as follows:

	RS	*Rs*	*rS*	*rs*
RS	*RRSS*	*RRSs*	*RrSS*	*RrSs*
Rs	*RRSs*	*RRss*	*RrSs*	*Rrss*
rS	*RrSS*	*RrSs*	*rrSS*	*rrSs*
rs	*RrSs*	*Rrss*	*rrSs*	*rrss*

According to the Punnett square, only 1 out of every 16 of the offspring will have the rrss genotype. Therefore, the probability of the rrss genotype is 1/16. Therefore, **(A)** is correct.

43. D

Sickle cell anemia results from a missense mutation, which affects the structure of the hemoglobin β subunit and in turn hemoglobin's function. The correct answer is **(D)**. (A) is incorrect because the point mutation causes a change in the amino acid sequence and therefore would have an effect on the structure of hemoglobin. (B) is incorrect because the mutation in the 20th nucleotide *does* translate to an amino acid, so it will not stop the formation of hemoglobin. (C) is incorrect because substituting glutamic acid with valine will create nonfunctional hemoglobin.

44. B

Unlike point mutations, which substitute a base in the DNA sequence and result in changing one amino acid, frameshift mutations can cause a change in multiple amino acids. Insertions and deletions shift the reading of the codons (sequence of three DNA/RNA nucleotides) after the mutation, thereby altering the amino acid sequence. The correct answer is **(B)**. Proteins created by frameshift mutations may be abnormally short or long and may be beneficial, non-functional, or lethal. (A) is incorrect because R-groups are found in amino acids, not RNA. (C) is incorrect because a frameshift mutation is not a missense mutation, which is a point mutation involving base substitution. (D) is incorrect because the frameshift mutation affects both RNA and amino acid sequences.

45. B

In competitive inhibition, inhibitor molecules compete with substrates to occupy the active site of an enzyme. Increasing the concentration of substrate will increase the likelihood that substrate molecules, rather than inhibitor molecules, will bind to the enzyme's active site. **(B)** is therefore correct. All of the other choices would reduce the effectiveness of an enzyme, which is what a competitive inhibitor does.

46. D

Deep-diving animals are least likely to be buoyant, to possess a tendency to float on water, since they depend on being able to stay underwater. **(D)** is thus correct. (B) is incorrect because myoglobin stores oxygen in the muscle. Deep divers tend to have high concentrations of this protein, which allows much greater storage of O_2 in their muscles than commonly seen in non-deep-diving animals. (A) and (C) are also modifications that are found in deep divers, allowing them to carry additional oxygen to their muscles and other tissues.

47. A

Liquids need to absorb heat to escape the liquid phase and be converted into a gas (the amount of heat required to do this is called the heat of vaporization). Water's high heat of vaporization enables evaporative cooling because the liquid that is converted into gas when sweat evaporates into the environment absorbs and removes heat from the body. **(A)** is therefore correct. (B) suggests the opposite. (C) is irrelevant for cooling the body. (D) is factually inaccurate.

48. D

In the lysogenic cycle, a bacteriophage inserts its DNA into the bacterial chromosome. This does not kill the host cell, but it does ensure that the viral DNA is replicated when the bacterium engages in DNA replication and cell division. In the lytic cycle, the virus exploits the cellular machinery of the bacterium to produce additional phages, causing the bacterial cell to lyse and release the new viruses. Thus, because the lytic cycle kills the host but the lysogenic does not, **(D)** is correct.

49. B

Oxidation of pyruvate, which produces a lot of ATP, occurs during aerobic respiration but not fermentation. **(B)** is thus correct. (A) and (C) are factually accurate but fail to explain the difference in ATP generation. (D) is true but misleading: molecules are not transported into mitochondria during fermentation, but only during aerobic respiration. However, the energy cost of transport across mitochondrial membranes is outweighed by the energy generated from oxidizing pyruvate.

50. C

The question stem gives the original distribution of bird beak length, and asks what would happen if the insect on which the bird feeds begins burrowing deeper into the soil. This suggests that there would be selection *for* birds with longer beaks and *against* birds with shorter beaks. It would thus be reasonable to expect that average beak length will increase over time. Only **(C)** indicates such an increase in mean length. (A) increases the variability in beak size, which would not be the result with a selective pressure in only one direction. (B) would be the result of selection for shorter beaks. (D) would result from selection against *both* birds with shorter beaks *and* birds with longer beaks.

51. D

Current hypotheses state that life on Earth was preceded by the generation of the precursors of life, such as amino acids and nucleotides. That generation thus took place under abiotic conditions. Of the stated choices, the most reasonable conclusion is **(D)**: some of the amino acids from earlier meteorites may have served as precursors of life. (A) implies that life on Earth originated on meteorites. This is a much stronger conclusion than can safely be inferred from the limited data in the question stem. (B) might seem tempting—after all, there are 20 amino acids in proteins, so the Murchison meteorite must be missing at least one of the amino acids found in proteins. That said, it is not reasonable to conclude that if meteorites did not provide *all* the necessary amino acids, then they did not provide *any* of the amino acids. (C) is incorrect because the presence of amino acid and nucleotide monomers does not guarantee the presence of more complex polymers such as proteins and nucleic acids.

52. A

The data here suggest that the cecum is significantly longer in species that consume plant matter for their diets (1 and 4) compared to those that consume at least some meat (2 and 3). Of the statements listed, the only one that logically follows is **(A)**: in organisms that eat meat, the cecum becomes unnecessary, and starts to shorten. (B) is the opposite of what would be expected, since diets that contain meat tend to be richer in proteins than those that are exclusively plant-based. The data on cecum length and diet are insufficient to draw conclusions about the relatedness of species, so rule out (C) and (D).

53. D

Heterotrophic zooplankton rapidly deplete the biomass of autotrophic phytoplankton, which allows the zooplankton to have a larger standing crop. This is sustainable because phytoplankton can reproduce very quickly to replenish consumed biomass. However, because the zooplankton are continually grazing down the phytoplankton, the phytoplankton standing crop population size remains small. **(D)** is thus correct. (A) is factually inaccurate: phytoplankton species tend to be much smaller than many terrestrial plant species. (B) is an oversimplification; aquatic organisms do not necessarily require less food than terrestrial organisms of comparable size. (C) fails to answer the question; even if measuring biomass in aquatic systems is more difficult, it does not explain the dramatic departure of their biomass pyramids from those of terrestrial systems.

54. A

The sickle cell anemia allele has adenine substituted for thymine, which codes for valine rather than the normal glutamic acid in the hemoglobin protein. Further evidence that this is a base-pair substitution includes that the mutation involves a single codon and that the protein is still produced, but with one different amino acid. **(A)** is thus correct. Usually frameshift mutations will have a much greater effect on the resulting protein, so (B) is unlikely. (C) is incorrect because silent mutations have no effect on the protein, due to the redundancy in the genetic code. Mutagens are agents that cause mutations, so (D) is also incorrect.

55. A

If malaria were eliminated, selection *for* the sickle cell allele would cease. Individuals with the allele would no longer have a greater chance of survival than those without it. At the same time, in this scenario, homozygous individuals would receive treatment that would ensure their survival through reproductive age and beyond. Thus, the allele would not be selected *against* either. With no selective pressure in either direction, the frequency of the allele would remain roughly constant, **(A)**.

56. B

Bacteria, which have short generation times, reproduce asexually and very rapidly. This allows rare individual genetic mutations that create new alleles to have a large effect on the overall genetic diversity observed in a population. **(B)** is therefore correct. While much of the genetic variation among humans is the result of sexual recombination of existing alleles, this does not explain why the impact of mutations is greater in bacteria populations, so (A) is incorrect. (C) is factually inaccurate; bacteria reproduce asexually. (D) is incorrect because it is too extreme to be accurate.

57. B

The guiding principle behind chemotherapy involves utilizing the biochemical similarities and differences between host and parasite cells. As stated in the question stem, antimetabolites, while chemically similar enough to metabolites to allow a reaction to proceed, are also chemically dissimilar enough that they eventually result in the failure of the pathway. Therefore, the correct answer will involve a mechanism that addresses downstream effects of antimetabolite use. **(B)** does exactly this. (A) and (C) focus on the antimetabolite causing the initial biochemical reaction to fail. Based on the question stem, use of antimetabolites results in a normal biochemical reaction in cells, rendering these choices incorrect. Nowhere in the question stem does it say that the antimetabolite has any affinity for the metabolite, so it is unlikely that it would bind to the metabolite, eliminating (D).

58. B

The theory of gradualism maintains that dramatic changes are the products of slow and continuous processes acting over a long time. Darwin employed this idea to explain how seemingly vast differences between species could develop, even though all organisms come from common ancestors. **(B)** describes exactly this idea. (A) describes a contrasting viewpoint known as catastrophism. (C) and (D) are tenets of Darwin's theory, but not directly connected to the idea of geologic gradualism.

59. C

The question posed asks for the chi-square value for a null hypothesis that concerns only the feathered-tail and straight-tail alleles, so ignore the role that the long and short alleles play. Out of 96 total offspring, $30 + 10 = 40$ had straight tails and $42 + 14 = 56$ had feathered tails. Using *F* to represent the feathered-tail allele and *f* to represent the straight-tail allele, a cross between an *Ff* female and an *ff* male would be expected to produce half feathered tails (*Ff*) and half straight tails (*ff*), or 48 of each. This gives a chi-square calculation as follows:

phenotype	observed	expected	obs–exp	(obs–exp)^2	(obs–exp)^2/exp
feathered tail	56	48	8	64	1.333
straight tail	40	48	−8	64	1.333
					$\chi^2 = 2.666$

Rounding to the nearest hundredth gives a chi-square value of 2.67. Therefore, **(C)** is correct.

60. B

In living systems, water acts as a solvent for chemical reactions in cells (catalyzed by enzymes that operate within an optimal temperature range) and helps transport compounds in and out of cells. Due to its high specific heat capacity (heat required to raise 1 kg of water by 1°C), water is able to absorb a large amount of energy without significantly changing its temperature, so it serves as a buffer against sudden temperature changes. Thus, the correct answer is **(B)**. (A) is incorrect because the most common essential elements found in living organisms are carbon, hydrogen, oxygen, and nitrogen; hydrogen and oxygen are no more important than the others. (C) is incorrect because water acts a solvent, not a solute, for chemical reactions. (D) is incorrect because most organisms are not able to thrive at temperatures so high that they come close to the boiling point of water.

ANSWERS AND EXPLANATIONS

Section II

1. **Scoring Guidelines for Free-Response Question 1**

(a) **Explain** how genetic mutations affect enzyme synthesis. **(2 points maximum)**

Explanation (1 point each)
• A mutation in the sequence of nucleotide bases affects the sequence of amino acids. • In enzyme synthesis, a change in amino acid sequence alters protein structure and function.

Here is a possible response that would receive full credit:

The sequence of nucleotide bases in DNA determines the sequence of amino acids in proteins. DNA is transcribed into mRNA, which is translated into amino acids. Amino acids are the primary structure in proteins and determine a protein's shape and function. Enzymes are proteins that catalyze chemical reactions. Thus, a genetic mutation affects the amino acid sequence, which in turn may affect the enzyme structure during synthesis, as well as the enzyme's function.

(b) **Propose** an appropriate control treatment for the experiment, and **describe** how the control treatment would increase the validity of the results. **(2 points maximum)**

Control Treatment (1 point)	**Description (1 point)**
• Grow wild type *Neurospora* on minimal medium and minimal medium with each supplement	• To demonstrate wild type *Neurospora* grows on each medium

Here is a possible response that would receive full credit:

A control treatment for the experiment would be to grow wild type *Neurospora* on minimal medium and minimal medium with each supplement (arginine, citrulline, argininosuccinate, and ornithine). This control would demonstrate that each medium can support growth.

(c) Using the data in the table, **identify** the intermediates of the arginine pathway on the template provided. **(3 points maximum)**

Identification (1 point for each of the 3 intermediates)
• Ornithine is the first intermediate • Citrulline is the second intermediate • Argininosuccinate is the third intermediate

Here is a possible response that would receive full credit:

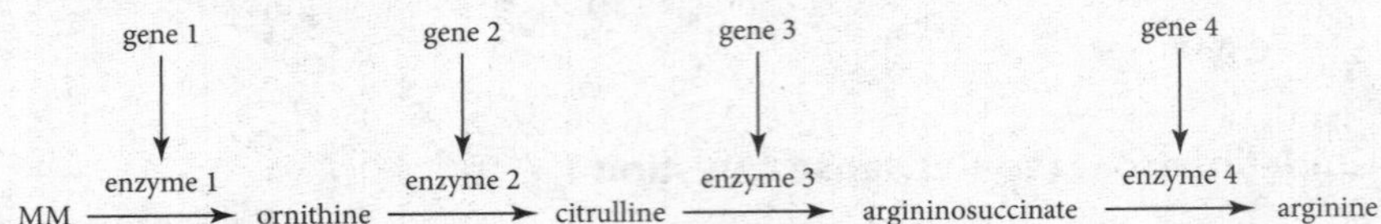

(d) **Provide** reasoning for the intermediates on the pathway. **(3 points maximum)**

Reasoning (1 point each, 3 points maximum)
• Ornithine is the first intermediate because Mutant 4 grows if given arginine, citrulline, and argininosuccinate, but not ornithine. • Citrulline is the second intermediate because Mutant 1 can grow in arginine and argininosuccinate, but not ornithine and citrulline (and ornithine is the first intermediate). • Argininosuccinate is the third intermediate because Mutant 3 only grows in minimal medium with arginine (and ornithine and citrulline are the first two intermediates).

Here is a possible response that would receive full credit:

If a mutant grows in supplemented minimal medium, it suggests that the supplement is a product of the pathway disrupted by the mutant. Mutant 4 grows if given arginine, citrulline, and argininosuccinate but not ornithine. This suggests that ornithine is the first intermediate, and the mutation impacts the function of enzyme 2, which catalyzes ornithine. Mutant 1 can grow in arginine and argininosuccinate but not ornithine and citrulline. This suggests that citrulline is the second intermediate, and the mutation impacts the function of enzyme 3, which catalyzes citrulline. Mutant 3 only grows in minimal medium with arginine, so the mutation impacted the function of enzyme 4, which catalyzes argininosuccinate. Since ornithine and citrulline are the first two intermediates, argininosuccinate is the third intermediate. Mutant 2 grows in all of the media with supplement but not minimal medium itself, so the mutation impacted the function of enzyme 1.

2. **Scoring Guidelines for Free-Response Question 2**

(a) When algal blooms die, they sink to the bottom of the bay and are decomposed by bacteria. **Describe** what most likely causes the level of dissolved oxygen to dip in the summer months. **(2 points maximum)**

Explanation (1 point each, 2 points maximum)
• Bacteria that decompose dead algal blooms consume dissolved oxygen. • Warmer water holds less dissolved oxygen; oxygen escapes to the atmosphere.

Here is a possible response that would receive full credit:

In the summer, algal blooms consume available nutrients and die when nutrients become limited. When they die, they are decomposed by bacteria. This decay process consumes oxygen in the water. In addition, according to the data, the temperature of the water increases during the summer months, which causes water to hold less dissolved oxygen. These together contribute to a decrease in the level of dissolved oxygen in the bay.

(b) In spring, rain and melting snow cause large volumes of water to flow into the bay. Based on an analysis of the data, **identify** the seasonal trends of algal blooms, of dissolved nitrogen, and of dissolved phosphorus. **(3 points maximum)**

Explanation (1 point each, 3 points maximum)
• Dissolved nitrogen in the bay increases in the spring months (March and April) due to rain and melting snow carrying nitrogen into the bay. • Dissolved phosphorus in the bay increases in the spring months (March and April) due to rain and melting snow carrying phosphorus into the bay. • Algal bloom (as indicated by chlorophyll a) increases in the months following an increase in nutrients in the bay (May through July).

Here is a possible response that would receive full credit:

According to the data, in the spring (March and April), dissolved nitrogen and dissolved phosphorus both increased in the bay. This suggests the rain and melting snow that flowed into the bay contained nitrogen and phosphorus from agricultural fields and industrial facilities. An increase in nutrients (dissolved nitrogen and phosphorus) in turn supported algal bloom, so algal blooms increased in the months following (May through July).

(c) Some algae blooms can produce toxic chemicals. **Describe** how toxins may affect fish-eating birds, and **describe** how algal blooms impact the ecology of two of the listed organisms affected by algal blooms. **(3 points maximum)**

Description (1 point each, 3 points maximum)
• Dense algal blooms block sunlight for underwater grasses. • Large concentrations of unconsumed algal blooms can interfere with the filter-feeding mechanism, so oysters starve to death. • Dead algal blooms are decomposed by bacteria, which decreases the levels of dissolved oxygen and leaves little to no oxygen for fish. • Birds that eat fish may through indirect ingestion take up toxins, which at high levels may cause death.

Here is a possible response that would receive full credit:

When algal blooms increase, they become dense and block sunlight vital for underwater grasses. Bay grasses are important because they provide food and an essential habitat and because they filter water by taking up nitrogen and phosphorus. Oysters and other filter feeders, which normally feed on algae, are unable to consume all the algae in the water. Large concentrations of unconsumed algal blooms can interfere with the filter-feeding mechanism and cause filter feeders to starve to death.

Birds are at risk from toxins produced by algal blooms through indirect ingestion. When birds consume many fish containing toxins, the toxins accumulate and the toxin level can become deadly.

(d) **Propose** one treatment method and justify how it would reduce algal blooms. **(2 points maximum)**

Proposal (1 point)	Explanation (1 point)
• Prevention OR • Mechanical (aeration, circulate, surface skimming, etc.) OR • Chemical OR • Biological	• Prevention measures reduce the amount of nitrogen and phosphorus entering the bay. OR • Mechanical methods reduce the competitive advantage of algal blooms by limiting the accessibility of nutrients at the surface. OR • Chemical methods could either bind to excess nutrients to remove them from the water or inhibit algae growth/kill algae. OR • Biological methods, such as the introduction of wetland plants, will compete for nutrients to limit algal growth.

Here is a possible response that would receive full credit:

Harmful algal blooms can be controlled using chemical and biological measures. Chemical treatment is used to inhibit algal growth or kill algae but may impact the pH of the water.

3. **Scoring Guidelines for Free-Response Question 3**

(a) **Describe** how a change in the amino acid sequence of a protein may lead to changes in the protein structure or function. **(1 point)**

Description (1 point)
• Protein structure and function depend upon a protein's tertiary structure, which is determined by the protein's primary structure.

Here is a possible response that would receive full credit:

Protein structure and function depend upon the tertiary structure of a protein, which is formed from the interactions between the R groups of amino acids. Changing the amino acid sequence of a protein may affect the interactions between the R groups, and thereby affect the overall protein structure and function.

(b) Based on the figure, **identify** the polarity of the solvent that the protein molecules were exposed to. **(1 point)**

Identification (1 point)
• The solvent is hydrophilic.

Here is a possible response that would receive full credit:

The figure shows that the fraction of hydrophobic amino acids buried is greater than the fraction of hydrophilic amino acids buried. A protein in an aqueous environment will fold so that the hydrophilic R groups are at the surface and the hydrophobic R groups are on the inside of the protein. Thus, the solvent that the protein molecules were exposed to is hydrophilic.

(c) A transmembrane protein spans the entire width of a cell membrane. For amino acids A F, N, and R, **predict** their location in the amino acids sequence of a transmembrane protein in the template below. **(1 point)**

Prediction (1 point)
• Amino acids A and F will be in the hydrophobic region of the cell membrane, and amino acids N and R will be in the hydrophilic region of the cell membrane

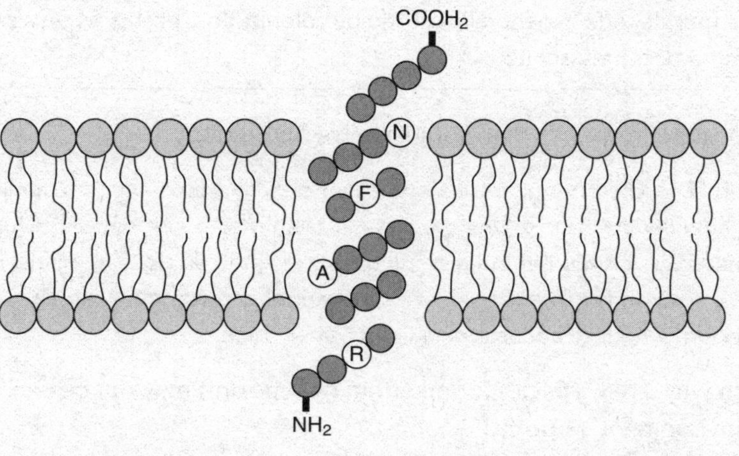

Here is a possible response that would receive full credit:

Amino acids A and F will in the hydrophobic region of the cell membrane, and amino acids N and R will be in the hydrophilic region of the cell membrane.

(d) **Justify** your predictions. **(1 point)**

Justification (1 point)
• Amino acids A and F are hydrophobic, so they will interact with the hydrophobic interior of the phospholipid bilayer; amino acids N and R are hydrophilic, so they will be in regions that associate with the aqueous environment inside and outside of the cell

Here is a possible response that would receive full credit:

Transmembrane proteins span the entirety of a phospholipid bilayer. They have hydrophobic regions that contain a high fraction of hydrophobic amino acids and hydrophilic regions that contain a high fraction of hydrophilic amino acids. Hydrophobic regions can be found in the interior of the cell membrane, while hydrophilic regions are found in contact with the aqueous intracellular and extracellular environments. Thus, amino acids A and F, which are hydrophobic, will be on the interior, and amino acids N and R, which are hydrophilic, will be on the exterior.

4. **Scoring Guidelines for Free-Response Question 4**

(a) **Describe** one example of genetic drift. **(1 point)**

Description (1 point each, 1 point maximum)
• The bottleneck effect, generally caused by a catastrophe, leaves behind a small group of individuals.
• The founder effect, generally caused by colonization, occurs when a small group breaks off and becomes isolated.

Here is a possible response that would receive full credit:

Genetic drift is the change in allele frequencies due to a random sampling process that occurs in small populations. One example is the bottleneck effect, which occurs when a natural disaster leaves behind a small group of individuals. The allele frequencies of the small group may differ from that of the original population, resulting in changes in the allele distribution that become prominent over time.

(b) **Explain** why a new species formed from genetic drift may not be well adapted to survive in its environment. **(1 point)**

Explanation (1 point)
• A new species formed from genetic drift is a result of random changes and not natural selection.

Here is a possible response that would receive full credit:

A new species formed from genetic drift is the result of a chance event and not natural selection. Thus, it may not be well adapted to survive in its environment.

(c) The individuals from the random event reproduce and form a new population. Based on the data, **predict** the allele frequencies for the new population. **(1 point)**

Prediction (1 point)
• The allele frequencies of the new population are $p = 0.25$ and $q = 0.75$.

Here is a possible response that would receive full credit:

The allele frequencies of the offspring population are $p = 0.25$ and $q = 0.75$.

(d) **Justify** your prediction. **(1 point)**

Justification (1 point)

- A testcross shows the allele frequencies of the offspring of the individuals that remain after the random event.

	a	*a*
A	*Aa*	*Aa*
a	*aa*	*aa*

p (*A*) $= 0.25$ and q (*a*) $= 0.75$

Here is a possible response that would receive full credit:

The individuals that remain after the chance effect are aa and Aa. If they were to reproduce, their offspring would be half Aa and half aa. Since A is dominant and a is recessive, the frequency of A is p and the frequency of a is q. There are 2 As and 6 as. Thus, the allele frequencies are p equals 2 over 8 or 0.25 and q equals 6 over 8 or 0.75.

5. **Scoring Guidelines for Free-Response Question 5**

(a) **Describe** one behavioral response of organisms within a population that affects their overall fitness and may contribute to the success of the population. **(1 point maximum)**

Description (1 point each, 1 point maximum)
• Cooperation within a population can provide protection from predators and acquisition of prey and resources.
• Competition decreases the availability for use by others, so the less competitive individual will be forced to go elsewhere.

Here is a possible response that would receive full credit:

Cooperation is the process of acting together for common benefits. Cooperative behavior within a population can promote survival by providing protection from predators and acquisition of prey and resources.

(b) **Explain** the trends of population Species A and B in the figure. **(1 point)**

Explanation (1 point)
• In 2015, Species A became extinct, suggesting Species B is a predator of Species A.

Here is a possible response that would receive full credit:

In the year 2015, Species A is effectively out of the picture, while Species B is increasing in number. It is reasonable to conclude that Species B is a predator of Species A, causing its virtual extinction in the ecosystem over the last 60 years.

(c) Based on the data, **determine** the ecological relationship between Species B and C. **(1 point)**

Determination (1 point)
• Over the last 60 years, Species C decreased and Species B increased, suggesting Species B is a predator of Species C.

Here is a possible response that would receive full credit:

Over the last 60 years, Species C decreased and Species B increased. It is plausible that Species B is a predator of Species C and that the expansion of Species B's population has resulted in a decline in the population in Species C. Alternatively, the slight decline in Species C over the previous 30 years could mean that it has reached its carrying capacity.

(d) **Explain** how symbiosis between two different organisms drives population dynamics. **(1 point)**

Explanation (1 point)
• A symbiotic relationship may be mutualistic, parasitic, or commensal, which affects population size and age.

Here is a possible response that would receive full credit:

In a symbiotic relationship, populations may be mutualistic, parasitic, or commensal. These relationships drive the birth and death rates, affecting the size and age of a population. In a mutualistic relationship, both populations benefit. In a parasitic relationship, one population benefits, while the other is harmed. In a commensal relationship, one population benefits, and the other is not affected.

6. **Scoring Guidelines for Free-Response Question 6**

(a) **Determine** the inheritance pattern of the allele(s) represented in the pedigree. **(1 point)**

Determination (1 point)
• The trait is sex-linked recessive. OR • The trait is autosomal recessive.

Here is a possible response that would receive full credit:

The typical inheritance pattern in sex-linkage is for a trait to be more common in males and to skip generations. Both of these occur in this pedigree. The grandfather at the top of the tree expresses the trait. If the gene in question were sex-linked, the father would pass on his X chromosome to all of his daughters. If the trait were dominant, all of his daughters would express the trait. The daughters do not express the trait, so the answer is sex-linked recessive. This pedigree could also exist for an autosomal recessive allele.

(b) **Identify** the genotype for the grandmother at the top of the tree. **(1 point maximum)**

Identification (1 point)
• The grandmother could be either a heterozygous carrier or homozygous dominant.

Here is a possible response that would receive full credit:

None of the offspring of the grandfather and grandmother exhibit the trait. Thus, the grandmother cannot be homozygous recessive. The grandmother could be either a heterozygous carrier or homozygous dominant and still produce the pedigree.

(c) **Calculate** the likelihood that a female in the fourth generation expresses the disorder. **(1 point)**

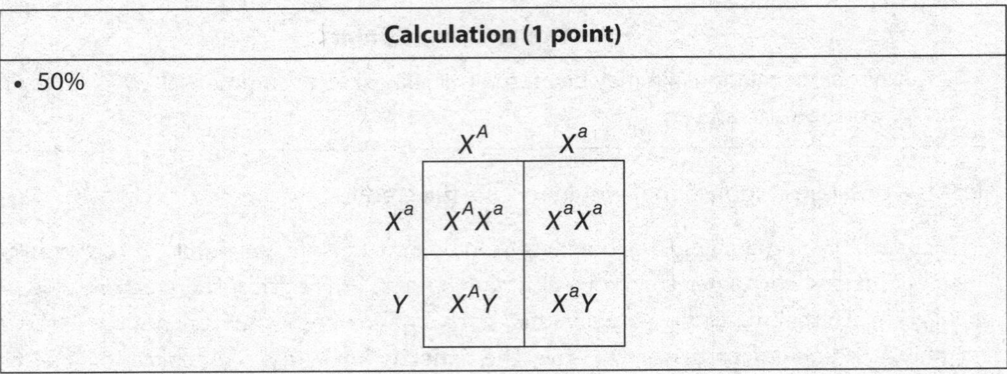

Calculation (1 point)
• 50%

	X^A	X^a
X^a	X^AX^a	X^aX^a
Y	X^AY	X^aY

Here is a possible response that would receive full credit:

The mother of the fourth generation female is heterozygous. Since the father of the fourth generation female expresses the disorder, the likelihood that the fourth generation female expresses the disorder is one-half or 50%.

(d) **Explain** deviations from Mendel's pattern of inheritance. **(1 point maximum)**

Explanation (1 point each, 1 point maximum)
• Linked genes violate the law of independent assortment, which states individual traits assort independently during gamete production. • Multiple genes violate the law of segregation, which states the offspring receives one allele for a trait from each parent. • Co-dominance or incomplete dominance violate the law of dominance, which states recessive alleles will be masked by dominant alleles.

Here is a possible response that would receive full credit:

Genes close together on the same chromosome may be linked, and thus be inherited together. This deviates from Mendel's law of independent assortment, which states that individual traits assort independently such that each gamete carries only one allele for each gene.

Practice Exam 3

Section I

1 Ⓐ Ⓑ Ⓒ Ⓓ	12 Ⓐ Ⓑ Ⓒ Ⓓ	23 Ⓐ Ⓑ Ⓒ Ⓓ	34 Ⓐ Ⓑ Ⓒ Ⓓ	45 Ⓐ Ⓑ Ⓒ Ⓓ	56 Ⓐ Ⓑ Ⓒ Ⓓ
2 Ⓐ Ⓑ Ⓒ Ⓓ	13 Ⓐ Ⓑ Ⓒ Ⓓ	24 Ⓐ Ⓑ Ⓒ Ⓓ	35 Ⓐ Ⓑ Ⓒ Ⓓ	46 Ⓐ Ⓑ Ⓒ Ⓓ	57 Ⓐ Ⓑ Ⓒ Ⓓ
3 Ⓐ Ⓑ Ⓒ Ⓓ	14 Ⓐ Ⓑ Ⓒ Ⓓ	25 Ⓐ Ⓑ Ⓒ Ⓓ	36 Ⓐ Ⓑ Ⓒ Ⓓ	47 Ⓐ Ⓑ Ⓒ Ⓓ	58 Ⓐ Ⓑ Ⓒ Ⓓ
4 Ⓐ Ⓑ Ⓒ Ⓓ	15 Ⓐ Ⓑ Ⓒ Ⓓ	26 Ⓐ Ⓑ Ⓒ Ⓓ	37 Ⓐ Ⓑ Ⓒ Ⓓ	48 Ⓐ Ⓑ Ⓒ Ⓓ	59 Ⓐ Ⓑ Ⓒ Ⓓ
5 Ⓐ Ⓑ Ⓒ Ⓓ	16 Ⓐ Ⓑ Ⓒ Ⓓ	27 Ⓐ Ⓑ Ⓒ Ⓓ	38 Ⓐ Ⓑ Ⓒ Ⓓ	49 Ⓐ Ⓑ Ⓒ Ⓓ	60 Ⓐ Ⓑ Ⓒ Ⓓ
6 Ⓐ Ⓑ Ⓒ Ⓓ	17 Ⓐ Ⓑ Ⓒ Ⓓ	28 Ⓐ Ⓑ Ⓒ Ⓓ	39 Ⓐ Ⓑ Ⓒ Ⓓ	50 Ⓐ Ⓑ Ⓒ Ⓓ	
7 Ⓐ Ⓑ Ⓒ Ⓓ	18 Ⓐ Ⓑ Ⓒ Ⓓ	29 Ⓐ Ⓑ Ⓒ Ⓓ	40 Ⓐ Ⓑ Ⓒ Ⓓ	51 Ⓐ Ⓑ Ⓒ Ⓓ	
8 Ⓐ Ⓑ Ⓒ Ⓓ	19 Ⓐ Ⓑ Ⓒ Ⓓ	30 Ⓐ Ⓑ Ⓒ Ⓓ	41 Ⓐ Ⓑ Ⓒ Ⓓ	52 Ⓐ Ⓑ Ⓒ Ⓓ	
9 Ⓐ Ⓑ Ⓒ Ⓓ	20 Ⓐ Ⓑ Ⓒ Ⓓ	31 Ⓐ Ⓑ Ⓒ Ⓓ	42 Ⓐ Ⓑ Ⓒ Ⓓ	53 Ⓐ Ⓑ Ⓒ Ⓓ	
10 Ⓐ Ⓑ Ⓒ Ⓓ	21 Ⓐ Ⓑ Ⓒ Ⓓ	32 Ⓐ Ⓑ Ⓒ Ⓓ	43 Ⓐ Ⓑ Ⓒ Ⓓ	54 Ⓐ Ⓑ Ⓒ Ⓓ	
11 Ⓐ Ⓑ Ⓒ Ⓓ	22 Ⓐ Ⓑ Ⓒ Ⓓ	33 Ⓐ Ⓑ Ⓒ Ⓓ	44 Ⓐ Ⓑ Ⓒ Ⓓ	55 Ⓐ Ⓑ Ⓒ Ⓓ	

SECTION I
90 Minutes—60 Questions

Directions: Section I of this exam contains 60 multiple-choice questions to be answered in 90 minutes. Each question is followed by four suggested answers. Using the information provided and your own knowledge of biological systems, select the best answer choice and fill in the corresponding letter on your answer grid or a sheet of scratch paper.

1. To study the effects of higher [CO_2] on plants, a researcher set up a large-scale free-air carbon dioxide enrichment (FACE) experiment by fortifying a natural prairie grassland with CO_2. The researcher compared the stomatal conductances (the higher the conductance, the more open the stomata) of grasses before and after CO_2 enrichment. The results show that over the first two years, C_3 grasses experienced a 14% change in plant height and a −36% change in stomatal conductance.

   Based on these findings, which of the following phenomena would not be affected in the grasses in the FACE experiment?

   (A) Photosynthetic rates

   (B) Water loss from the leaf surface

   (C) Water transport through the xylem

   (D) Active uptake of potassium

2. A novel antibacterial drug is isolated from a newly discovered species of fungus. While the drug is able to enter into both bacterial and human cells by endocytosis, the drug only degrades DNA in bacteria, so it is safe to use in humans. Which of the following could be a plausible explanation for why only bacteria are affected?

   (A) Bacteria have cell walls, which makes bacteria more susceptible to drugs.

   (B) DNA is protected when wrapped around histones, which only prokaryotes have.

   (C) The drug causes lysosomes, which are not found in eukaryotes, to release DNA-degrading enzymes.

   (D) Human cells have nuclear envelopes that are impermeable to the drug.

GO ON TO THE NEXT PAGE.

3. While many metabolic diseases follow typical patterns of inheritance, some have long perplexed scientists because their inheritance can't be explained by traditional genetics. A physician studying a family suffering from a metabolic disorder hypothesizes that the inheritance might be mitochondrial in origin. Knowing that mitochondria are inherited maternally, she does a pedigree analysis of the family to test her hypothesis.

   Which of the following pedigrees would support her hypothesis?

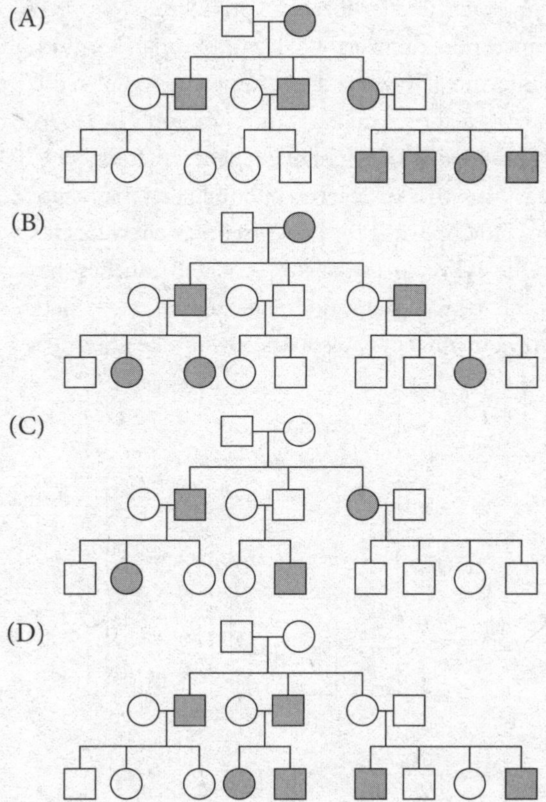

4. A student in biology class collects a sample of stream water to look at under the microscope as part of an assignment. The student sees a single-celled organism with a membrane-bound nucleus and no cell wall. Which of the following observations would also be expected in this cell?

   (A) Within the cytoplasmic region is a piece of circular, double-stranded DNA.
   (B) Glycoproteins are found only on the intracellular side of the membrane.
   (C) Ribosomes are only found free-floating in the cytosol.
   (D) Some ribosomes are membrane-bound.

Practice

GO ON TO THE NEXT PAGE.

5. The grizzled skipper butterfly (*Pyrgus centaureae*), once found throughout Minnesota, was thought to be limited to the Superior National Forest in the northernmost part of the state due to habitat loss. Recently, this endangered butterfly was found on a number of remote islands in the Boundary Waters (a wilderness area between the United States and Canada). Conservationists desiring to increase the grizzled skipper population decided to mate the two butterflies in the Boundary Waters. Larvae emerged from the butterfly eggs but died within a few hours. Repeated attempts at interbreeding have failed.

   Which statement best explains why the two populations of grizzled skipper butterflies cannot produce viable offspring?

   (A) The Boundary Waters grizzled skipper males are sterile due to the founder effect.

   (B) The Boundary Waters grizzled skipper has adapted to its wilderness home, and any offspring due to interbreeding with the Superior National Forest population results in progeny unable to survive the cold weather.

   (C) The Boundary Waters grizzled skipper is not able to produce viable offspring because interbreeding with the Superior National Forest population would increase the frequency of harmful mutations.

   (D) The Boundary Waters grizzled skipper was reproductively isolated from the Superior National Forest population and has developed into an incompatible species.

Questions 6–10

In certain areas of the world, the amount of cattle allowed per acre is based on many factors, including climate, land quality, and grass type. For example, a cow/calf pair on high-quality land only needs 1–2 acres for 12 months of grazing. However, lower-quality land may require 5–10 acres for the same pair and time frame. The development of genetically modified (GMO) crops has further complicated this relationship. While bermudagrass has been a common grass grown by many ranchers for cattle, the development of GMO haygrazer has produced a faster-growing grass. The same amount of sun, water, and fertilizer can produce up to over 100 times more dry weight of haygrazer than bermudagrass. Unfortunately, while the bermudagrass can be used at any stage for grazing, haygrazer can only be fed to cattle after it reaches a height of 18 inches. Before then, the haygrazer has toxic levels of nitrate that may cause death in cattle. So while the haygrazer may allow for more cattle to live on a smaller plot of land, ranchers must work to grow grass and rotate the cattle such that the cows are only grazing on the low nitrate grass.

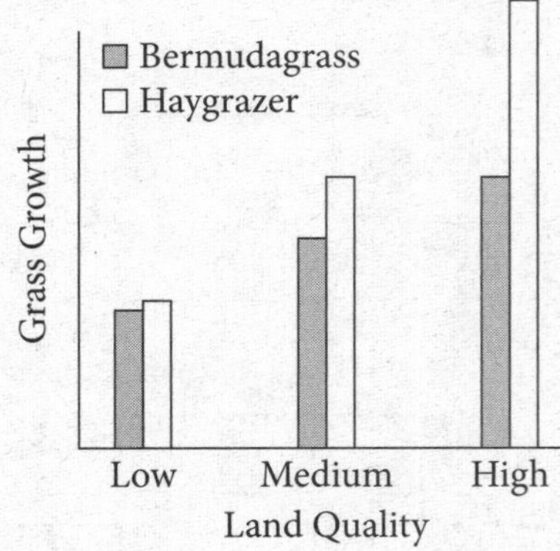

Figure 1. The effect of land quality on rate of growth of bermudagrass and haygrazer

GO ON TO THE NEXT PAGE.

6. Which of the following statements is not supported by the information in the passage?
   (A) Nitrogen-fixing bacteria are more active in haygrazer less than 18 inches tall.
   (B) Haygrazer is utilizing energy more efficiently than bermudagrass.
   (C) Slower growing bermudagrass has more non-nitrate nutrients than haygrazer.
   (D) High-quality land has more nutrients needed for grass growth.

7. Cattle are often separated based on age and size, and ranchers must calculate how much acreage is required for each herd. When predicting how much grass each herd will need, which of the following statements about cattle grass metabolism must be true?
   (A) On average, larger cattle require more energy per gram of body weight.
   (B) On average, smaller cattle require more energy per gram of body weight.
   (C) Larger cattle are more susceptible to nitrate toxicity.
   (D) Cattle obtain more energy from haygrazer than bermudagrass.

8. MicroRNA, small pieces of RNA, can be used to increase the amount of proteins that help cattle to eliminate nitrates. If microRNAs increase protein production during the translation step, what part of the cell would microRNA need to be inserted?
   (A) Cytoplasm
   (B) Rough endoplasmic reticulum
   (C) Smooth endoplasmic reticulum
   (D) Nucleus

9. During times of high rainfall, haygrazer grows much faster than bermudagrass and the haygrazer soil is dry while the bermudagrass soil remains moist. In dry conditions, both types of grass have similar growth patterns and no difference in soil water content. Which of the following explains these differences?
   (A) Bermudagrass is better than haygrazer at absorbing large amounts of water during heavy rainfall.
   (B) Haygrazer is not good at absorbing large amounts of water during heavy rainfall.
   (C) In the presence of large amounts of water, bermudagrass activates more water uptake.
   (D) In the presence of large amounts of water, haygrazer activates more water and nutrient uptake mechanisms.

10. Researchers discovered that haygrazers produce energy through photosynthesis faster during the summer season. Which of the following best describes the observed tropism in haygrazers?
   (A) Haygrazers are short-day plants that grow when exposed to less than 12 hours of daylight.
   (B) Haygrazers are short-day plants that require at least 12 hours of darkness to grow.
   (C) Haygrazers are long-day plants that need less than 12 hours of darkness to grow.
   (D) Haygrazers are day-neutral plants that do not depend upon the amount of darkness or daylight hours.

GO ON TO THE NEXT PAGE.

11. Down syndrome (trisomy 21) is one of the few examples in which an abnormal number of chromosomes results in a viable phenotype. While Down syndrome has been long understood to be caused by an extra copy of chromosome 21, approximately 5% of all infants born with Down syndrome have a normal number of chromosomes.

    Which of the following could explain the occurrence of Down syndrome with a normal chromosome number?

    (A) The third copy of chromosome 21 was lost during meiosis of the fetal cells during embryogenesis.
    (B) The third copy of chromosome 21 was lost during genetic recombination of the fetal cells during embryogenesis.
    (C) The third copy of chromosome 21 was translocated onto and fused with another chromosome during spermatogenesis.
    (D) The third copy of chromosome 21 was lost during spermatogenesis.

12.

		Species						
		1	2	3	4	5	6	7
Species	1		61	94	75	73	70	89
	2	61		55	43	42	47	37
	3	94	55		50	52	66	98
	4	75	43	50		97	92	49
	5	73	42	52	97		93	48
	6	70	59	66	92	93		53
	7	89	37	98	49	48	53	

Figure 1. The percent similarity between seven different species of birds on the Galapagos Islands

Which of the following cladograms would not be possible based on the data in Figure 1 ?

(A)

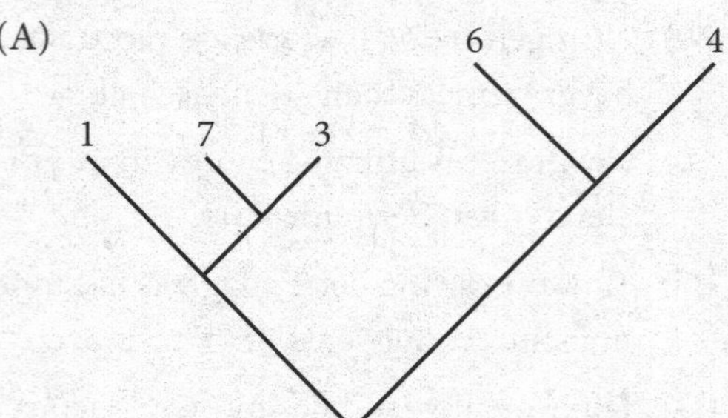

(B)

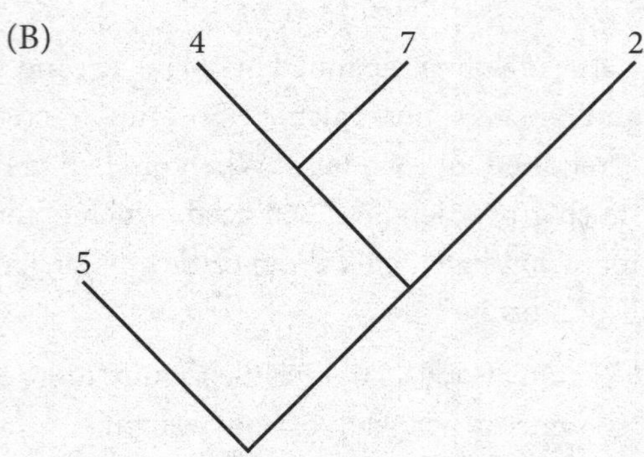

(C)

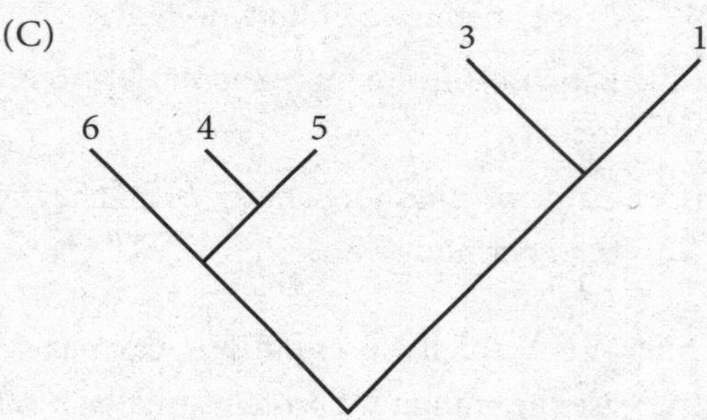

(D)

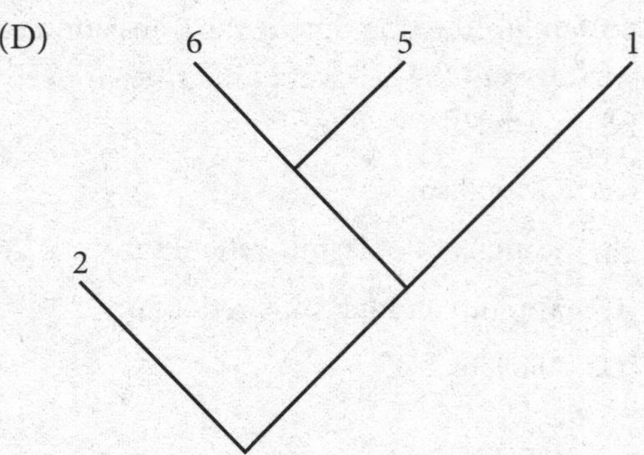

GO ON TO THE NEXT PAGE.

13. Bacteria, fungi, and protists can form biofilms, collectives of microorganisms that grow on wet surfaces underground, above ground, and underwater. Biofilms are held together by a self-produced extracellular matrix. They are large and grow slowly over days or weeks and reproduce by fragmentation or sloughing. The metabolic waste products of one organism can become the nutrients for another, which allows multiple-species biofilms to be energy-efficient.

Which of the following growth curves best describes the growth in a biofilm population?

(A)

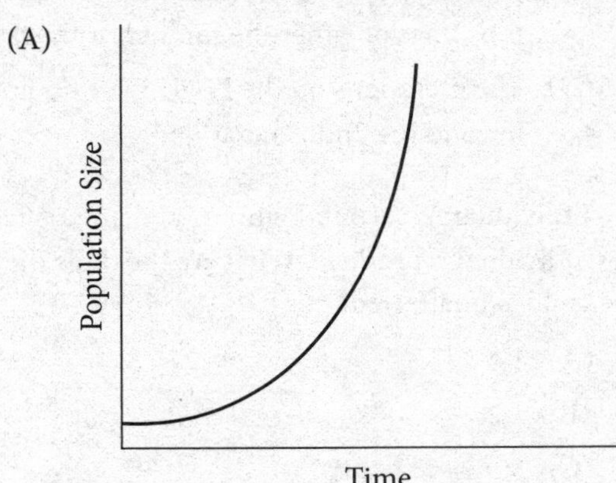

(B)

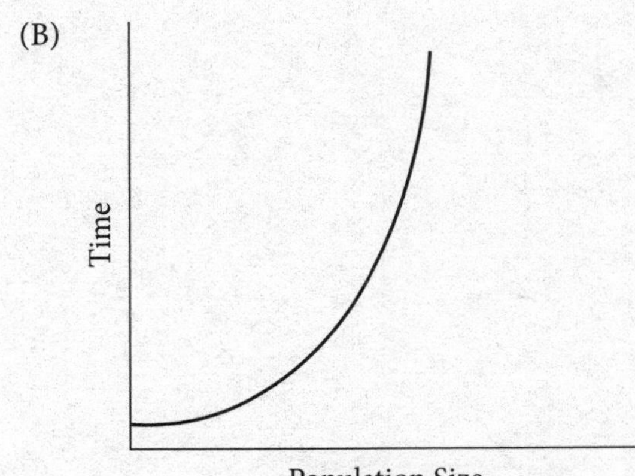

(C)

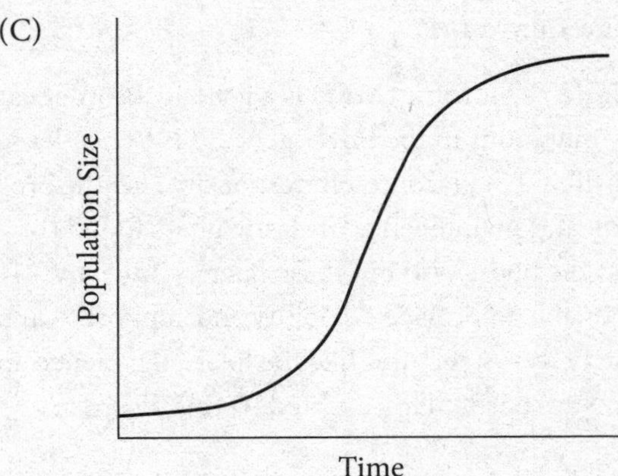

(D)

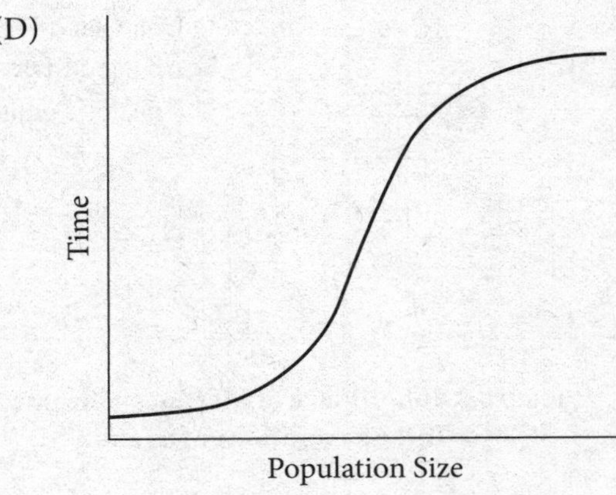

Practice

GO ON TO THE NEXT PAGE.

Questions 14–17

Marfan syndrome (MFS) is a genetic disorder caused by mutations in the FBN1 gene. FBN1 encodes for fibrillin-1, a glycoprotein responsible for the production and maintenance of elastic fibers found throughout the body. Without these fibers, connective tissue is unable to provide flexibility and support to areas of the body—structures like the heart, ligaments, and eyes are particularly affected by this disorder.

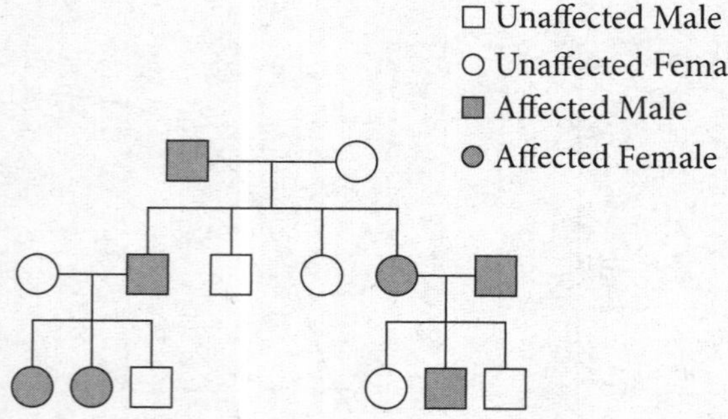

Figure 1. Inheritance of Marfan syndrome over three generations of a family

The pedigree above traces the incidence of Marfan syndrome in one family over the course of three generations. The results presented in the table indicate the prevalence of MFS in several countries around the world.

Table 1. Prevalence of MFS in select countries

Country	Prevalence
Northern Ireland	1.5/100,000
China	17.2/100,000
Scotland	6.8/100,000
Denmark	4.6/100,000

14. Based on the information presented, in order for the MFS phenotype to be present in the next generation, which of the following conclusions can be made?

(A) The mutations on the FBN1 gene are most likely passed on during meiosis from the father's chromosomes.

(B) The mutations on the FBN1 gene are passed on during meiosis from the chromosomes of both the mother and the father.

(C) The mutations on the FBN1 gene are passed on during meiosis from the chromosomes of either the mother or the father.

(D) The mutations on the FBN1 gene do not go beyond the 2nd generation.

15. Individual A, seen in Figure 1, and an unaffected individual reproduce. What are the odds their child is unaffected?

(A) 1:4

(B) 1:2

(C) 3:4

(D) 4:4

GO ON TO THE NEXT PAGE.

16. A European research team studied MFS patients ranging in age from 0 to 74 years old and found that the average age of patients at diagnosis was 19 years old. They also expect the prevalence of MFS in the country studied will increase by approximately 0.17/100,000 patients in the next several decades. Based on the data, which of the following statements does not support the researchers' expectation of increasing prevalence for MFS?

    (A) Current medical and surgical treatment of aortic disease in MFS has substantially increased the average life expectancy: up to 70 years of age.

    (B) Increased use of FBN1 genotyping provides more accurate identification of those with MFS

    (C) Females affected by MFS are able to access prenatal diagnostics and may elect to abort a fetus with the FBN1 mutation.

    (D) Diagnosis of individuals well into adulthood may indicate that those with less severe symptoms of MFS are being identified and receiving appropriate care.

17. Haploinsufficiency, in diploid organisms, may arise from a de novo or an inherited loss-of-function mutation in one allele. Marfan syndrome is one such example of haploinsufficiency. Individuals with and without MFS have two alleles that encode for the production of fibrillin-1, a protein responsible for elastic fiber creation and maintenance throughout the body. Which of the following statements is most in line with the information provided?

    (A) The amount of fibrillin-1 produced by the cells in the people without MFS is higher than those with MFS.

    (B) The amount of fibrillin-1 produced by the cells in the people without MFS is lower than with MFS.

    (C) The amount of fibrillin-1 produced by the cells in the people without MFS is the same as those with MFS.

    (D) The amount of fibrillin-1 produced by the cells is dependent on the phenotype.

GO ON TO THE NEXT PAGE.

18. Many G-protein-coupled receptors use the second messenger cyclic AMP (cAMP). The binding of adenylyl cyclase to the G_s α subunit activates adenylyl cyclase and then generates cAMP. The most well-known effector of cAMP in the cell is protein kinase A (PKA), which phosphorylates proteins at the PKA phosphorylation site. cAMP binds to two sites on each of PKA's two regulatory subunits, which causes the dissociation of PKA's two catalytic subunits. PKA can activate phosphodiesterase, which degrades cAMP.

    Which of the following is expected to directly increase cAMP levels in the cytoplasm?

    (A) Blocking the PKA phosphorylation site on cAMP.

    (B) Inhibiting phosphorylation of phosphodiesterase by PKA.

    (C) Inhibiting the dissociation of the catalytic subunits of PKA.

    (D) Blocking the binding of adenylyl cyclase to the Gs alpha subunit.

19. In a recent study, researchers classified men with many different levels of hair loss into four groups: no hair loss, slight hair loss, moderate hair loss, and severe hair loss. They analyzed a large number of genes associated with hair loss and then classified individuals by a genetic risk score, ranging from 1 to 10.

    Percentage of Hair Types Within Each Genetic Risk Group

    No hair loss
    Slight hair loss
    Moderate hair loss
    Severe hair loss

    Percentage of Men

    Genetic Risk of Male Pattern Baldness
    (1 = lowest risk, 10 = highest risk)

    Adapted from Turcotte MM, Corrin MSC, Johnson MTJ (2012) Adaptive Evolution in Ecological Communities. PLoS Biol 10(5): e1001332.

    Based on the data from the graph, does the male pattern for baldness follow a Mendelian inheritance pattern?

    (A) Yes, because individuals can be classified into distinct groups.

    (B) No, because there is continuous variation in this trait and a complex inheritance pattern.

    (C) Yes, because individuals with greater genetic risk are more likely to have severe baldness.

    (D) No, because there are only two phenotypes present.

GO ON TO THE NEXT PAGE.

20. Atrial natriuretic peptide (ANP) is secreted by myocytes in the heart's atria in response to stretching of the atrial walls, which is caused by increased blood volume. The primary function of ANP is to increase excretion of sodium and water, but it also reduces hormone secretion in the adrenal gland and has been shown to inhibit cardiac hypertrophy in heart failure.

    Which of the following statements concerning ANP's actions in the body is most likely to be true?

    (A) ANP acts in an endocrine manner to increase glomerular filtration rate and glomerular permeability in the kidney.
    (B) ANP acts in a paracrine manner to reduce aldosterone secretion by the zona glomerulosa of the adrenal cortex.
    (C) ANP acts in an exocrine manner to inhibit norepinephrine-mediated calcium influx in cardiac myocytes.
    (D) ANP acts in an autocrine manner to relax the vascular smooth myocytes in arterioles and venules.

21. In a certain desert biome, primary producers like velvet mesquite and prickly pear cactus produce 100% of the energy for the food web. Primary consumers like red harvester ants and antelope squirrels receive only 10% of that energy, and only a tenth of the energy primary consumers received is available for secondary consumers like collared lizards and grasshopper mice. The red-tailed hawk and elf owl, which are tertiary consumers, receive 0.1% of the energy that was produced by the original primary producers.

    Which of the following most completely accounts for the loss of energy at higher trophic levels?

    (A) Energy gets trapped in unusable forms like bone, is consumed by autotrophs, and is lost as heat, and the feces of organisms are passed to decomposers.
    (B) Energy is lost as heat, gets trapped in unusable forms like bone, and is consumed in metabolism, and the remains of dead organisms are passed to decomposers.
    (C) Energy gets trapped in unusable forms like bone, is consumed in metabolism, and is lost through photosynthesis, and the feces of organisms are passed to decomposers.
    (D) Energy is consumed in metabolism, gets trapped in unusable forms like fur, and is lost as heat, and the remains of dead organisms are passed to apex consumers.

Practice

GO ON TO THE NEXT PAGE.

Questions 22–25

A researcher has discovered two new proteins in the bacterium *Escherichia coli* (*E. coli*): protein X and protein Y. Protein X, whose structural gene is located in the X operon, is found to bind to its own promoter. Protein Y, whose structural gene is located in the Y operon, has an unknown function. To determine the function of protein X and protein Y with respect to the regulation of the X and Y operons, the researcher does three experiments:

Experiment 1: The researcher adds wild-type protein X at hour one to *E. coli* cells; Experiment 2: The researcher adds wild-type protein Y at hour one to *E. coli* cells; Experiment 3: The researcher adds both wild-type protein X and wild-type protein Y at hour one to *E. coli* cells.

For each experiment, the amount of protein X and protein Y was measured over time and compared to that of a non-treated control. The results for protein X levels are shown in the figure. Note: The asterisk indicates the time point at which protein X and Y are added in the experiments.

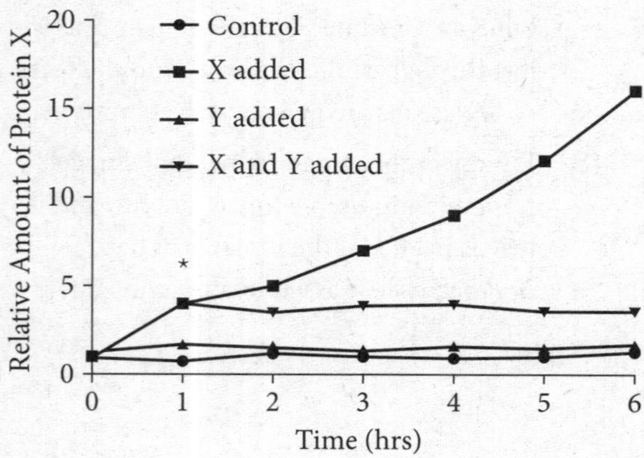

22. A point mutation is introduced into the structural gene that codes for protein X in a special line of *E. coli* cells, resulting in a mutant version of protein X that can no longer bind to its promoter. When Experiment 1 is repeated with this special line of *E. coli*, the expected outcome would most likely be:

    (A) A continuous increase in the amount of mutant protein X over time

    (B) No increase in the amount of mutant protein X

    (C) A continuous increase in wild-type protein X over time

    (D) An initial increase in mutant protein X that levels off over time

23. Based on the data, how could protein X and protein Y be regulating the X operon?

    (A) Protein X induces a positive feedback loop on itself, while protein Y inhibits transcription of the X operon.

    (B) Protein X induces a positive feedback loop on itself, while protein Y promotes transcription of the X operon.

    (C) Protein X induces a negative feedback loop on itself, while protein Y inhibits transcription of the X operon.

    (D) Protein X induces a negative feedback loop on itself, while protein Y promotes transcription of the X operon.

GO ON TO THE NEXT PAGE.

24. The researcher does a fourth experiment in which they add high amounts of lactose to *E. coli* cells. Afterward, the researcher sees an increase in protein Y levels. How does lactose increase protein Y levels?

    (A) Lactose acts as an enhancer of the Y operon.

    (B) Lactose decreases transcription of the X operon, thereby stopping protein X from acting as a repressor for the Y operon.

    (C) Lactose binds to the Y operon repressor as an inducer.

    (D) Lactose inhibits RNA polymerase from binding to the Y operon promoter.

25. In a later discovery, the researcher learns that glucose acts as a corepressor for the Y operon. How would high levels of glucose be expected to affect protein X levels?

    (A) Protein X levels would decrease, because high glucose levels would help the X operon repressor bind the X operon operator

    (B) Protein X levels would increase, because high levels of glucose would decrease the amount of protein Y

    (C) Protein X levels would decrease, because high levels of glucose would increase the amount of protein Y

    (D) Protein X levels would increase, because high levels of glucose would provide enough energy for more transcription

26. One hormone that is involved in the regulation of hunger is leptin. Released by fat cells in adipose tissue, leptin binds to receptors in the hypothalamus to inhibit the feeling of hunger. What organelle will leptin pass through before it is secreted?

    (A) Centrosome

    (B) Golgi apparatus

    (C) Smooth ER

    (D) Lysosome

27. The Allee effect is a process in which population growth in plant and animal species is limited by low population density. For example, at the beginning of the 18th century, the passenger pigeon (*Ectopistes migratorius*) was the most abundant bird in North America. Over the next 200 years, humans hunted passenger pigeons on a large scale and decimated the birds' habitat through deforestation. As the number of pigeons grew smaller and smaller, the birds struggled to minimize predation risk and to reproduce. By the early 1900s, no pigeons were left.

    Which of the following terms best describes the phenomenon undergone by passenger pigeons in the 18th and 19th centuries?

    (A) Positive feedback

    (B) Negative feedback

    (C) Homeostasis

    (D) Equilibrium

GO ON TO THE NEXT PAGE.

28. In healthy human skin cells, cyclin-dependent kinases or CDKs are enzymes that actively regulate the cell cycle from interphase through mitosis as shown in Figure 1. When the cells are exposed to a carcinogen such as nicotine, the cells' DNA may become damaged. CDKs' activity is then blocked as shown in Figure 2.

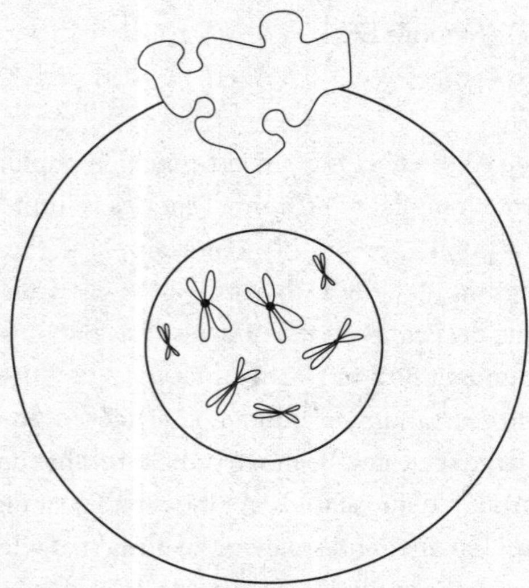

Figure 1. Healthy cell with a CDK regulating cell's activity

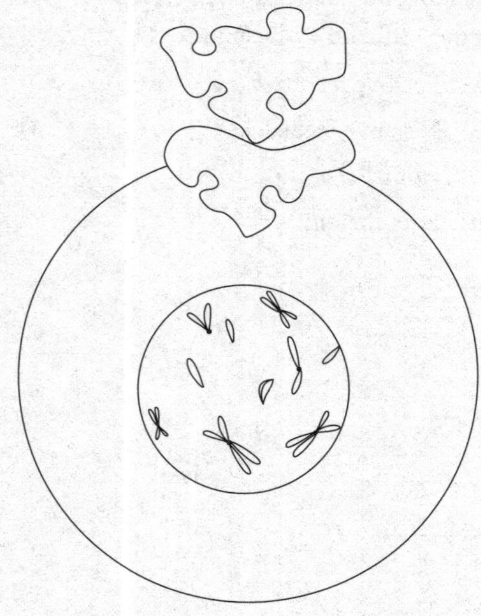

Figure 2. Cell with damaged DNA and blocked CDK

Which of the following statements explains why a cell with damaged DNA affects CDKs' activity?

(A) The damaged DNA in cells in G1 phase will trigger a response from CDK inhibitors such as p53 to repair the damaged DNA before the cells progress to G2 phase.

(B) The damaged DNA in cells in G1 phase will trigger a response from CDK inhibitors such as p53 to repair the damaged DNA before the cells progress to S phase.

(C) The damaged DNA in cells in G2 phase will trigger a response from CDK inhibitors such as p53 to repair the damaged DNA before the cells progress to G1 phase.

(D) The damaged DNA in cells in G2 phase will trigger a response from CDK inhibitors such as p53 to repair the damaged DNA before the cells progress to S phase.

29. Why would clumps of overlapping mammalian cells indicate a problem with cell cycle regulation?

(A) Normal cells use cyclin-dependent kinases for regulation.

(B) Normal cell division is stimulated by active anaphase-promoting complex.

(C) Normal cells show density-dependent inhibition and anchorage dependence.

(D) Normal cell division is stimulated by the presence of growth factor.

GO ON TO THE NEXT PAGE.

Questions 30–34

Lemurs are social creatures and are known to use different calls to communicate with their group members. Though the lemur language has yet to be fully understood, researchers have found that lemurs can communicate to other lemurs to distinguish between different but similar situations, for instance, if food or water has been located or if danger is close or far away and is approaching in the form of a fox or a bird.

A researcher has discovered two groups of lemurs in Madagascar that have developed distinct forms of communication and eating habits within their respective groups, despite living in close proximity to one another. Lemurs in Group 1 live in the northern part, while lemurs in Group 2 live in the southern part. Both groups, however, mainly subsist on a diet of plants. Table 1 indicates how each group interprets the same sounds.

Table 1.

Sound	Group 1 meaning	Group 2 meaning
Growl	Territorial warning	Territorial warning
Two clicks	Food	Water
Chirp	Greeting	Food
Shout	Danger: Far away	Greeting
Cry	Danger: Fox	Water
Chatter	Water	Danger (unspecified)
Squeal	Danger: Close	Danger: Bird

30. If a lemur from Group 1 is heard by a lemur from Group 2, which of the following sounds would put the lemur listening in the most danger?
    (A) Squeal
    (B) Cry
    (C) Two clicks
    (D) Chatter

31. Both groups of lemurs feed on the same plants found in unclaimed territory between the two groups. How might Group 1 successfully compete against Group 2 for the plants?
    (A) Lemurs from Group 1 growl at other Group 1 lemurs.
    (B) Group 1 lemurs wait until Group 2 lemurs leave to search for plants.
    (C) When Group 2 lemurs are spotted, Group 1 lemurs shout to scare Group 2 away.
    (D) Group 1 lemurs learn to chatter when they find plants.

32. A certain type of fruit, which is eaten by both groups, has suddenly become twice as abundant. Afterward, there is a noticeable decrease in the population of Group 1 lemurs. What could explain the decline in the Group 1 lemur population?
    (A) The population of Group 2 lemurs increased substantially due to the increase in food availability and outcompeted Group 1 lemurs.
    (B) The increase in food availability caused Group 2 lemurs to become overweight and slower, which led to more being eaten by predators.
    (C) Group 1 and Group 2 began to integrate with each other due to no longer needing to compete for food.
    (D) A third group of lemurs that used to exclusively eat leaves begins to eat half of all the available fruit.

Practice

GO ON TO THE NEXT PAGE.

33. The lemurs live in a place where every winter, clouds cover the sky for two months and severely limit the amount of available sunlight. How would this affect the lemurs?

    (A) The lemurs would not be affected because they would eat mostly insects instead.

    (B) Less light would lead to less plant growth, meaning both lemur populations would stagnate or decline during the winter.

    (C) Group 1 lemur populations would increase because Group 2 lemur numbers would decline due to a reduction in plant availability.

    (D) Both groups would grow larger because there would be fewer predators.

34. Another researcher claims that Group 1 and Group 2 did not split from a common group, but rather are two distinct lemur subspecies. What information would support this claim?

    (A) Group 1 and Group 2 are found to have more similarities in their languages than differences.

    (B) Viable offspring are produced when the two groups mate.

    (C) Group 1 is found to be more genetically similar to a different group of lemurs in a different region of Africa than to Group 2.

    (D) Foxes prey mostly on Group 2 lemurs, because Group 1 lemurs have comparatively darker fur that makes them less noticeable.

35. A wildlife ecologist studying deer migrations found that many members had been afflicted by an unknown wasting disease after traveling through a muddy swamp. Believing the infectious agent was transmitted through the soil, the ecologist worked with a molecular biologist to isolate DNA and RNA from soil samples. They screened for genetic sequences of known pathogens that could explain the disease. The sequencing data did not match that of any known pathogens.

    Assuming the procedures were performed correctly, what statement would support the ecologist's hypothesis that the infectious agent was transmitted through the soil?

    (A) The virus causing the wasting disease was transmitted through the air.

    (B) Bacteria in the soil caused the wasting disease, which is why viral DNA wasn't detected.

    (C) The wasting disease is caused by prions present in the soil.

    (D) The hypothesis can't be correct because all bacteria and viruses have DNA and RNA, which means the soil can't be the source of transmission.

GO ON TO THE NEXT PAGE.

36. Albinism in plants is characterized by full or partial chlorophyll deficiency. Which of the following best describes a consequence of a genetic mutation that gives rise to albinism?

(A) Albino plant seedlings will emerge and grow but encounter premature death.

(B) Young albino plants via a symbiotic relationship will harvest energy from mycorrhizae.

(C) Albino plants will increase their rate of photosynthesis to accumulate carbohydrates in the form of starch.

(D) Albino plants will perform respiration to generate ATP for cellular activities.

37. The origin of life is a frequently debated topic in the scientific community. The most established theory, the "organic soup" model, is best described by which of the following?

(A) The first cells used simple organic molecules like ATP as their source of energy.

(B) Protobionts were formed first from proteins, then nucleic acids.

(C) RNA was the first self-replicating molecule but was later mostly replaced by DNA due to stability.

(D) Simple organic molecules became more complex and became self-replicating structures, ultimately leading to the creation of the first cells.

---

38. Two experiments were performed with a container containing two chambers to determine how pillbugs react to their environment. Twenty pillbugs were placed into the leftmost chamber of the container. In the first experiment, the right chamber was moistened with water absorbed by paper towels. In the second, filter paper dampened with 1% acetic acid was placed into the right chamber.

Table 1 shows the number of pillbugs per chamber over time.

	Experiment 1		Experiment 2	
Time (min)	**Left chamber**	**Wet chamber**	**Left chamber**	**Acidic chamber**
0	20	0	20	0
5	15	5	17	3
10	9	11	10	10
15	12	8	9	11
20	13	7	7	13
25	14	6	3	17
30	10	10	1	19
35	7	13	0	20
40	11	9	0	20

Which of the following statements best summarizes the results of the experiments?

(A) The pillbugs exhibit kinesis in the first experiment and positive taxis in the second experiment.

(B) The pillbugs exhibit kinesis for both experiments.

(C) The pillbugs exhibit negative taxis in the first experiment and positive taxis in the second experiment.

(D) The pills bugs exhibit positive taxis in the first experiment and kinesis in the second experiment.

GO ON TO THE NEXT PAGE.

39.

AmB+ 25%
A+B+ 25%
25% AmBm
A+Bm 25%
Mendelian Segregation

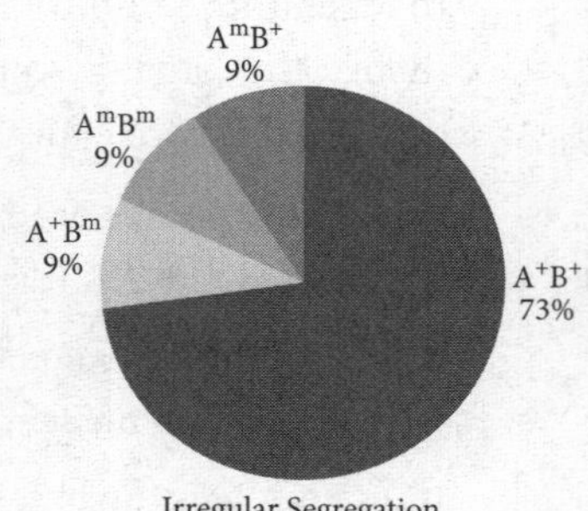

Figure 1. Proportion of offspring showing each genotype in the female A^+A^m × male B^+B^m cross (left) and female B^+B^m × male A^+A^m cross (right)

A female mouse carrying a normal and a mutant copy of gene A (A^+A^m) was mated to a male mouse carrying a normal and a mutant copy of gene B (B^+B^m), and the offspring followed the inheritance pattern predicted by Mendel's law of segregation (left side). However, when a B^+B^m female was mated to an A^+A^m male, the offspring showed irregular segregation (right side).

Which of the following is the most reasonable explanation for the irregular segregation?

(A) Unequal recombination between homologous chromosomes occurred during prophase I of meiosis.

(B) Non-random pairing of sperm and egg occurred at fertilization.

(C) Chromosomes undergoing independent assortment during oogenesis resulted in diploid gametes.

(D) Only the first sperm that reached the oocyte fertilized it.

Questions 40–44

Irrigation for agriculture can lead to salinization of soils, especially in arid and semi-arid climates. In order to study the effects of saline soils on plant development, students collected soil samples from a variety of locations. They noted the soil texture and measured electrical conductivity (EC), measured in decisiemens per meter (dS/m), which reflects the soil's salinity, or total soluble salt concentration, as shown in Table 1.

Table 1. Soil texture of each sample with corresponding electrical conductivity

Sample	Soil texture	EC (dS/m)
1	Loamy sand	1.2
2	Silt loam	2.2
3	Loam	11.2
4	Clay	5.8
5	Silty clay loam	9.7

Pea (Pisum sativum) seeds were planted in each type of soil. Following the emergence of the cotyledon leaves, each plant was placed in a sunny location and watered with 50 mL of water each morning for 60 days. The plants ($n = 10$) were then clipped close to ground level and weighed (fresh weight). Results are shown in Figure 1. The samples were then left to dry for 72 hours and then weighed again (dry weight), in order to estimate their biomass. Results are shown in Figure 2.

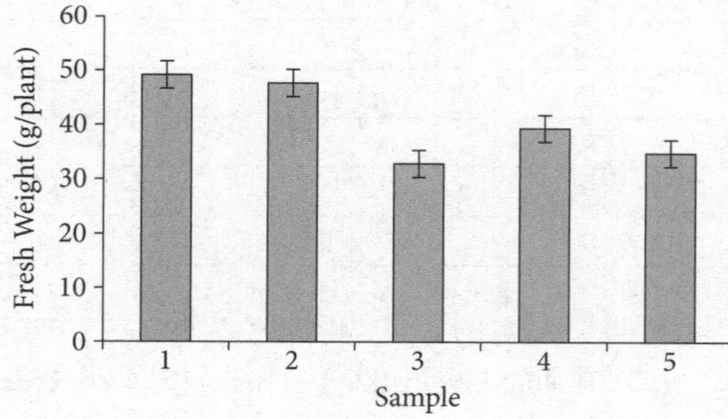

Figure 1. Mean fresh weight ± standard error of the mean for plants grown in each sample soil

GO ON TO THE NEXT PAGE.

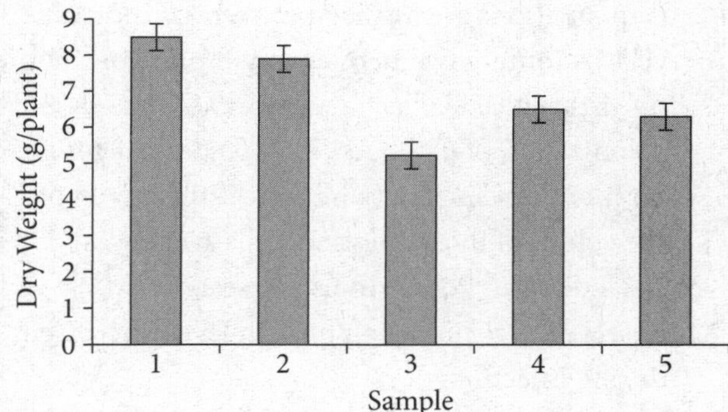

Figure 2. Mean dry weight ± standard error of the mean for plants grown in each sample soil

40. Which of the following, if true, would help support the findings of the experiment?

(A) Crops such as field beans, onions, peas, and red clover are particularly sensitive to salinity.

(B) In areas where evapotranspiration exceeds precipitation, soil salinization is very common.

(C) Peas grown in sandy loam with a high EC produce taller plants with more peas per pod than those grown in sandy loam with a low EC.

(D) Soybeans grown in clay loam with a high EC produce fewer beans per plant than those grown in clay loam with a low EC.

41. The pH of saline soils tends to be alkaline, which can affect nutrient availability in the soil. Which of the following types of substances, if added to saline soil, would be expected to improve the nutrient availability?

(A) Strongly basic

(B) Weakly basic

(C) Strongly acidic

(D) Neutral

42. Which of the following conclusions provides the best answer to the question that the students set out to investigate?

(A) Plants grown in Sample 1 soil show an 83% decrease between fresh weight and dry weight.

(B) Any soil with a saline content above 8 dS/m significantly affects the growth of pea plants.

(C) For every 10-fold increase in the EC, there is a decrease of roughly 50% in the biomass.

(D) Pea plants grown in soil with an EC above 2.2 dS/m show a significant reduction in their biomass.

43. Salinization of soil can lead to decreased crop yield (the amount of agricultural production per unit area). Which of the following provides the best explanation for this phenomenon?

(A) A shift in the osmotic pressure impairs the flow of water out of the roots.

(B) The concentration gradient driving water into the roots is reversed.

(C) The hypertonic soil makes the plant's cell walls more permeable to water.

(D) Saline soil has a lower concentration of ions than does root sap.

44. Suppose that just prior to cutting the plants, the students measured the oxygen output of each plant over a 24-hour period. If the majority of the pea plant's biomass is comprised of leaves, plants grown in which soil sample would be expected to produce the least oxygen?

(A) Sample 1

(B) Sample 2

(C) Sample 3

(D) Sample 4

Practice

GO ON TO THE NEXT PAGE.

45. In humans, an increase in blood pressure will lead to increased parasympathetic activity and decreased sympathetic activity. These changes, in turn, may affect heart rate, hormone release, and blood vessel diameter. All of these mechanisms function together to return the blood pressure to its set point when there are deviations.

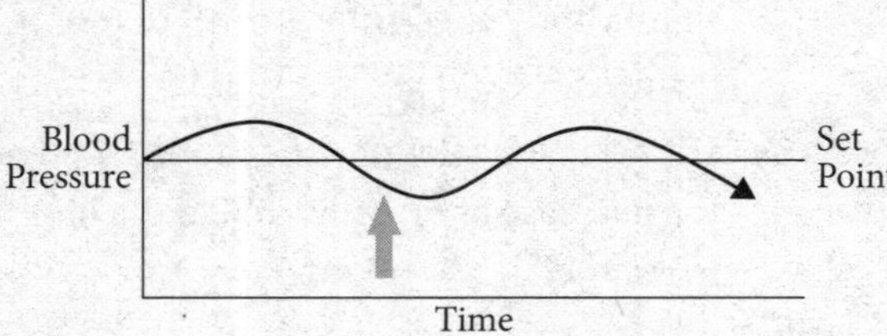

Figure 1. The relative stability of blood pressure

What two processes will the autonomic nervous system initiate at the point indicated by the arrow in Figure 1 ?

(A) Decrease heart rate and dilate blood vessels
(B) Decrease heart rate and constrict blood vessels
(C) Increase heart rate and dilate blood vessels
(D) Increase heart rate and constrict blood vessels

46. Gap junctions are formed between two adjacent cells by joining two hemichannels or connexons, one in each cell's plasma membrane. The colonization of pathogen, *Citrobacter rodentium*, on the apical surface (which faces the lumen of the colon) of mice intestinal cells causes diarrheal disease (liquid stools). Researchers observed that connexon protein levels increased during infection.

Based on the information given, which of the following is a plausible explanation for the mechanism by which *Citrobacter rodentium* causes diarrheal disease?

(A) Facilitated diffusion of small molecules and ions against their concentration gradient causes transport of water into the lumen via open unpaired hemichannels.
(B) Functionally open unpaired hemichannels of infected cells allow small molecules and subsequently water to be released from intestinal cells into the lumen.
(C) Unpaired hemichannels open to actively transport water against its concentration gradient from the intestinal cells into the lumen.
(D) The passive transport of small molecules and ions via open unpaired hemichannels from the lumen to intestinal cells draws water into the lumen via osmosis.

GO ON TO THE NEXT PAGE.

47. Stenospermocarpic seedlessness in grapes is caused by a point mutation (a change of only one nucleotide base in DNA or RNA) in a section of the chromosome responsible for seed development. Fertilization occurs, and seeds begin to develop but fail to fully mature. Grape breeders will remove these developing seeds before they die and grow them into plants by using tissue culture techniques.

Which statement gives the most likely reason the breeders are harvesting and growing these immature seeds?

(A) Harvesting immature seeds and growing them outside of the original plant enable the seed to fully develop and grow into plants that will produce seeds without the mutation.

(B) Harvesting the immature seeds from the original plant will allow the affected plant to invest all its energy in growing larger fruit.

(C) Harvesting the immature seeds allows breeders to then grow plants that will possess the seedless trait and produce a higher number of seedless offspring.

(D) Harvesting the immature seeds and growing them into plants enable breeders to cross breed them into a new species of grape.

48. To gain a better understanding of nodule regulation on the roots of legumes infected with nitrogen-fixing bacteria, researchers grew *Lotus japonicus* seedlings for 2 weeks in nitrogen-poor soil. Roots were examined, and the number of nodules infected with bacteria was counted. Shoot samples were found to have cytokinins (plant growth substances) concentrated in the leaves. Nitrogen was then added to the soil and maintained for a 2-week period. Researchers did not observe new nodule formation. Cytokinins were found concentrated in the roots.

Which of the following conclusions is supported by the observations made by the researchers?

(A) Cytokinins seem to play a role in stimulating and inhibiting colonization by nitrogen-fixing bacteria in nitrogen-poor and nitrogen-rich soil.

(B) Cytokinins do not seem to play a role in stimulating and inhibiting new colonization by nitrogen-fixing bacteria in nitrogen-poor and nitrogen-rich soil.

(C) Cytokinins seem to play a role only in inhibiting new colonization by nitrogen-fixing bacteria in nitrogen-poor soil.

(D) Cytokinins seem to play a role only in inhibiting new colonization by nitrogen-fixing bacteria in nitrogen-rich soil.

Practice

GO ON TO THE NEXT PAGE.

49. Although cheetahs survived the mass extinction at the end of the last ice age over 10,000 years ago, the total population was greatly reduced.

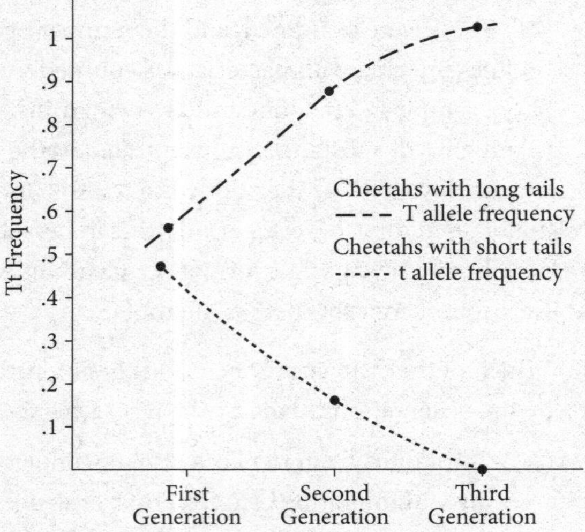

Figure 1. Change in frequency of the alleles *Tt* for tail length over three generations of cheetahs

According to Figure 1, which of the following conclusions can be made regarding the effects of genetic drift on the small population of cheetahs that survived the ice age?

(A) Long tails are superior to short tails, supported by the increasing frequency of the T allele.

(B) The frequency of alleles for long tails only has become fixed by the third generation.

(C) Both harmful and beneficial alleles are kept due to natural selection.

(D) The allele frequencies in the cheetah population before the ice age would be about the same as that of the third generation.

50. A scientist extracts DNA from the nucleus of cells and sequences it. The scientist determines that 27% of the nucleotide bases are guanine. What percentage of the bases is thymine?

(A) 23%

(B) 27%

(C) 46%

(D) 54%

Questions 51–54 refer to the following material.

Anoles are a group of lizard species from the genus *Anolis* that are widespread in the Americas and have successfully adapted to life in multiple different ecological niches. Given their many diverse forms and environments, anoles have been used to study evolutionary processes in nature.

A research group studying communication displays in reptiles found that several species of anoles have ultraviolet (UV) photoreceptors in their eyes, providing the ability to detect UV light. Furthermore, multiple studies have found that the males' *dewlap*, a large flap of skin under the chin that can be expanded, is commonly used in visual communication. In light of these findings, the researchers hypothesized that UV reflectance of the dewlap might play a role in these species' communication displays.

After identifying five closely related species of anoles from various environments, the researchers used sophisticated cameras to measure the UV reflectance of the dewlaps. The mean reflectance values for UVA and UVB wavelengths as well as the species' natural habitats are shown in Table 1.

GO ON TO THE NEXT PAGE.

Table 1. Dewlap UV reflectance in anole species

Species	Habitat	Percent reflectance UVA	Percent reflectance UVB
A	Grassland	20%	38%
B	Grassland	15%	33%
C	Grassland	9%	25%
D	Forest floor	1%	13%
E	Forest floor	1%	7%

51. A critic of the researchers claims that the differences in UV reflectance indicate that these species aren't closely related. Which of the following would best support the claim of an earlier evolutionary divergence between some species of anoles?

    (A) In regions of habitat that overlap between species D and A, fertile hybrid offspring have been found.

    (B) A large mountain range separates the five species of anoles from each other.

    (C) The dewlaps of species D and E develop from the skin on the face, whereas the dewlaps of species A, B, and C develop from the skin on the neck.

    (D) The different species of anoles frequently hunt each other and fight over territory.

52. After genomic sequencing, the researchers found that species B and D share more of the same random mutations than any of the other species. According to the molecular clock hypothesis, what does the data most likely suggest?

    (A) Species D must have diverged from species B.

    (B) Species B must have diverged from species D.

    (C) Species B or D must be the common ancestor for the other three species of anoles.

    (D) Species B and D diverged more recently from a common ancestor than from other species.

53. Which of the following statements provides the best evidence that the anoles are a separate species?

    (A) UV reflectance of the dewlap makes it difficult for related species to communicate.

    (B) One of the species is stronger and better at hunting than the others.

    (C) Anoles that breed outside of their group cannot produce viable, fertile offspring.

    (D) The separate species occupy different environments and thus have different food sources.

54. Following a wildfire, a large population of species D migrated to a grassland inhabited by species A. After several generations, it was found that most of the new juveniles from species D had UV reflectance values closer to those of species A. What is the most likely reason for this trend?

    (A) Higher UV reflectance provides a reproductive advantage, resulting in directional selection in species D.

    (B) The interbreeding of species A and D led to a new hybrid phenotype.

    (C) Higher UV reflectance provides a reproductive advantage, thus causing disruptive selection in species D.

    (D) Higher UV reflectance provides a reproductive advantage, thus causing stabilizing selection in the population.

GO ON TO THE NEXT PAGE.

55. As part of an experiment, a cell biologist treats cultured mitochondria with 2,4-dinitrophenol (DNP), a compound that dissipates the mitochondrial proton gradient, and measures the resulting oxygen concentration in the culture compared to that of untreated controls. In follow-up experiments, the cell biologist also plans to measure additional metabolic markers to evaluate DNP's effects on cellular metabolism.

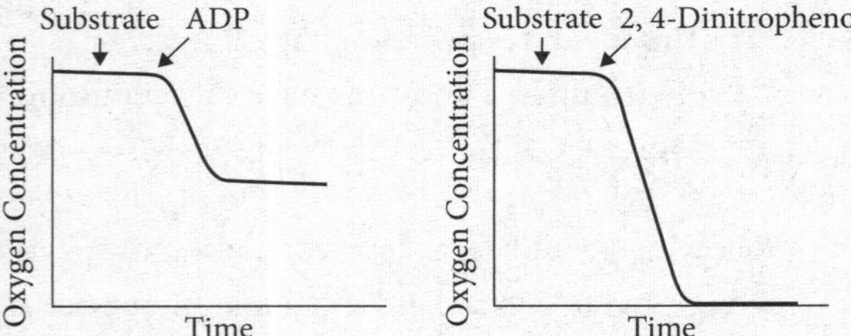

Figure 1. Oxygen concentration vs. time in cell culture with and without DNP

Based on the data represented in Figure 1, what could the cell biologist reasonably expect to find following the addition of DNP?

(A) An increase in the rate of ATP production and an increase in the concentration of NADH.

(B) A decrease in the rate of ATP production and an increase in the concentration of NADH.

(C) A decrease in the rate of ATP production and a decrease in the concentration of NAD^+.

(D) A decrease in the rate of ATP production and an increase in the concentration of NAD^+.

56. In a rare breed of African cichlids, the allele for yellow fins (*Y*) is dominant over the allele for orange fins (*y*). Yellow and orange cichlids were bred in captivity. The breeders counted 426 yellow babies and 328 orange babies. What is the chi-square value for the null hypothesis that the yellow parent was heterozygous for yellow fins?

(A) 0

(B) 3.6

(C) 6.4

(D) 12.7

57. Modern-day therapy for type 1 diabetes utilizes recombinant insulin, which is produced by transformed *Escherichia coli* bacteria. Traditionally, this process involves inserting the gene for insulin into the *lac* operon on a plasmid containing an ampicillin-resistance gene. Because antibiotic-resistant bacteria are a growing public health threat, a student tries to make insulin-producing *E. coli* using a plasmid that lacks the ampicillin-resistant gene.

Despite his careful work, why would the student's experiment be unlikely to succeed without the ampicillin-resistant gene?

(A) Ampicillin-resistance allows for the production of insulin by successfully transformed *E. coli*.

(B) Ampicillin-resistance allows for the selective growth of only the successfully transformed *E. coli*.

(C) Ampicillin-resistance allows for the insertion of the insulin gene into the *lac* operon.

(D) Ampicillin-resistance has no effect on a recombinant bacteria's ability to produce insulin via induction of the *lac* operon and his experiment should work.

GO ON TO THE NEXT PAGE.

58. Tasmanian devils almost became extinct over 10,000 years ago, except for a small population that remained on Tasmania. This founding population of devils was left with a smaller gene pool and less genetic variation than the original population. This lack of genetic variety has severely limited the devils' immune system in recognizing and fighting off pathogens. Today's devil population has been decimated by devil facial tumor disease, DFTD, an aggressive non-viral transmissible cancer. However, a group of devils from Northwest Tasmania seems to be resistant to DFTD.

Which of the following statements regarding the various processes of selection would help the Tasmanian devil evade extinction?

(A) Stabilizing selection will cause the frequency of the allele for DFTD resistance to remain constant in successive generations of devils.

(B) Disruptive selection will cause the frequency of the allele for DFTD resistance to decrease in successive generations of devils.

(C) Directional selection will cause the frequency of the allele for DFTD resistance to increase in successive generations of devils.

(D) Directional selection will cause the frequency of the allele for DFTD resistance to decrease in successive generations.

59. Scientists are studying the use of inhibitors against non-kinase enzymes in an effort to slow tumor growth. One class of enzymes of particular clinical interest includes the acetyltransferases, which are responsible for adding an acetyl group, a type of post-translational modification, onto other proteins.

From which of the following experiments could scientists determine whether an inhibitor to acetyltransferase is competitive or noncompetitive?

(A) Adding increasing amounts of substrate to the enzyme to determine V_{max} in the absence and presence of the inhibitor.

(B) Changing the temperature in 10°C increments and measuring the enzymatic rate.

(C) Varying the amount of enzyme when the inhibitor is present and measuring the V_{max}.

(D) Determining where the inhibitor binds on the substrate.

Practice

GO ON TO THE NEXT PAGE.

60.

State	Activity	Substrate binding	DNA binding
Non-phosphorylated	++++	++++	+
Phosphorylated	+	++++	++++

Table 1.

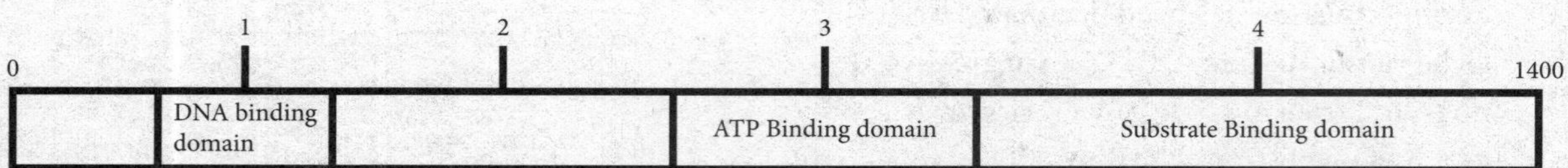

Figure 1. Primary sequence of the enzyme

Table 1 shows the enzymatic activity, substrate binding, and DNA binding affinity of an ATP-dependent enzyme in its non-phosphorylated and phosphorylated state. Figure 1 shows the primary sequence of the enzyme and its different functional domains, where the numbers 1–4 denote the possible sites of phosphorylation. Which of the following conclusions could most reasonably be made?

(A) Sites 2 and 3 are the most likely sites of phosphorylation, where phosphorylation inhibits enzymatic activity and increases DNA binding.

(B) Sites 1 and 4 are the most likely sites of phosphorylation, where phosphorylation increases DNA and substrate binding.

(C) Sites 1 and 3 are the most likely sites of phosphorylation, where phosphorylation inhibits enzymatic activity and increases DNA binding.

(D) Sites 1, 3, and 4 are the most likely sites of phosphorylation, where phosphorylation increases enzymatic activity, substrate binding, and DNA binding.

END OF SECTION I

IF YOU FINISH BEFORE TIME IS CALLED, YOU MAY CHECK YOUR WORK ON SECTION I ONLY.

DO NOT GO ON TO SECTION II UNTIL YOU ARE TOLD TO DO SO.

SECTION II
80 Minutes—6 Questions

Directions: Begin by taking a 10-minute reading period, during which time you may sketch graphs, make notes, and plan your answers. You then have 80 minutes to complete your responses to the 6 free-response questions. Questions 1 and 2 are long free-response questions that should require about 22 minutes each to answer. Questions 3 through 6 are short free-response questions that should require about 9 minutes each to answer. Read each question carefully and write your response on scratch paper. Answers must be written out in paragraphs. Outline form is not acceptable. It is important that you read each question completely before you begin to write.

1.

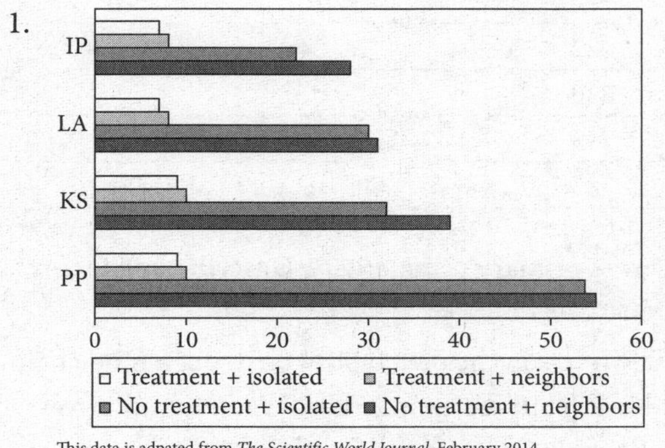

This data is adpated from *The Scientific World Journal*, February 2014.

Figure 1. Fungal Colonization Rate (%)

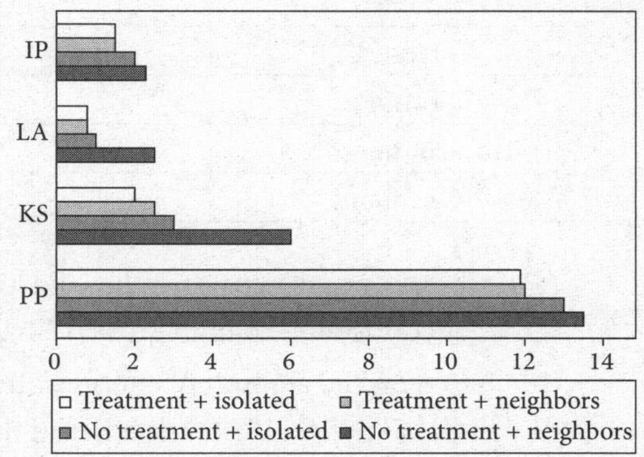

This data is adpated from *The Scientific World Journal*, February 2014.

Figure 2. Shoot Biomass (g)

Mycorrhizal fungi colonize the roots of a large variety of plants. Plants that are exposed to hostile soil may lose mycorrhizae. Researchers conducted field experiments in a marshland for four species of plant: *Ixeris polycephala, Leonurus artemisia, Kummerowia striata*, and *Polygonum pubescens*. They studied the effect of fungicide and the presence of neighboring plants on the rate of fungal colonization and shoot biomass. Forty quadrants were divided into blocks and randomly assigned factors: no treatment + neighbors, treatment + neighbors, no treatment + isolated, and treatment + isolated. Samples were taken, and soil oxygen concentration was measured for each block. The results are shown in Figures 1 and 2.

(a) **Describe** the relationship between plants and mycorrhizae.

(b) **Identify** the control group in the experiment.

(c) **Determine** the effect fungicide and the presence of neighbors have on shoot biomass.

(d) Suppose oxygen concentration promotes mycorrhizal colonization. **Predict** which plant species under what conditions would produce the greatest amount of oxygen and **justify** your prediction.

Practice

GO ON TO THE NEXT PAGE.

2. Pyramids of energy differ between ecosystems due to the effect of climate on productivity. Gross productivity is the energy organisms incorporate from the organisms at the previous trophic level, and net productivity is the energy that is available to the organisms at the next trophic level. Researchers investigated the average amount of energy at varying trophic levels in five different ecosystems. The results are shown in Table 1. The rows represent ecosystems, and the columns represent trophic levels.

Table 1. Average trophic level energy in 5 different ecosystems

	Average amount of energy at trophic level (kJ/m^2)			
Ecosystem	**A**	**B**	**C**	**D**
Temperate deciduous forest	6	620	60	6,025
Desert	0.2	22	2	212
Tundra	2	230	21	2,225
Tropical rainforest	17	1,743	172	17,450
Grassland	8	785	80	7,900

(a) **Identify** which ecosystem has the highest gross primary productivity. **Describe one factor** that contributes to high gross primary productivity.

(b) Based on the amount of energy at the trophic levels in the data table, **construct** a food chain in the template. Write the letter of the trophic level in the appropriate box and draw arrows to indicate the energy flow between levels. **Provide two reasons** for your answer.

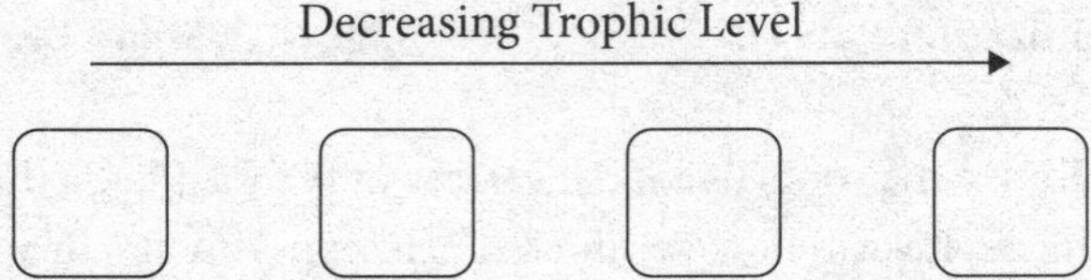

(c) **Calculate** the transfer efficiency between trophic level D and trophic level B for the desert ecosystem. **Explain** why ecosystems rarely have more than 4 or 5 trophic levels.

(d) An invasive predator was introduced in the temperate deciduous forest ecosystem. **Predict** the short-term impact on the pyramid of biomass for the ecosystem. **Justify** your prediction.

GO ON TO THE NEXT PAGE.

3. The mating of *Saccharomyces cerevisiae* (budding yeast) occurs between two haploid mating types: a and α. A culture of a-type cells only was incubated with and without the supernatant from a co-culture of a- and α-type cells. The expression of genes in the pheromone and mating pathways (*ste* 2, *ste* 3, and *ste* 12) is shown in the figure.

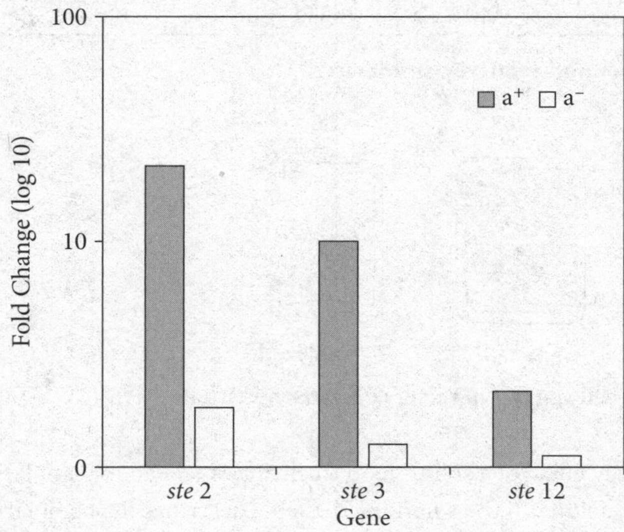

(a) **Describe** how environmental factors affect phenotypic plasticity, the ability of one genotype to exhibit different phenotypes in different environments.

(b) **Identify** the control in the experiment.

(c) In a different experiment, a culture of α-type cells was incubated with and without yeast mating pheromone. The fold change (log 10) for *ste* 3 expression for α with and without pheromone was 7 and 0.7, respectively. **Predict** the likely results of *ste* 2 expression for this experiment.

(d) **Justify** your predictions.

GO ON TO THE NEXT PAGE.

4.

Tonicity of solution	Relative osmolarity	
	Extracellular fluid	Intracellular fluid
Hypertonic	Higher	Lower
Hypotonic	Lower	Higher
Isotonic	Equal	Equal

Table 1. Tonicity and relative osmolarity

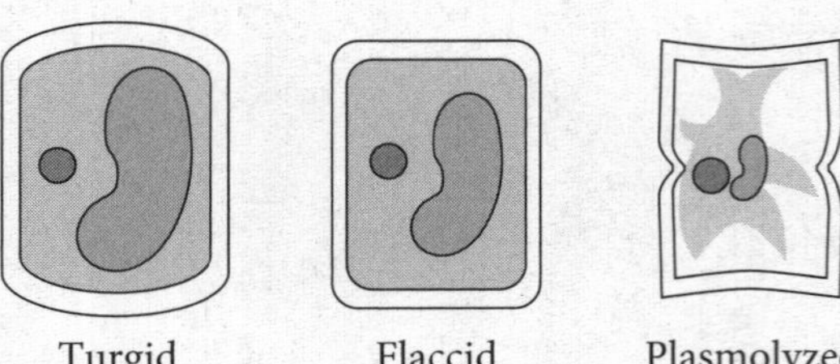

Figure 1. Plant cell states in a variety of water potentials

Tonicity describes the effect of a solution on a cell and is related to osmolarity, which is the total concentration of all solutes in the solution. The table describes the osmolarity of extracellular and intracellular fluids of cells in hypertonic, hypotonic, and isotonic solutions. Depending on the tonicity, a plant cell may be turgid, flaccid, or plasmolyzed (Figure 1).

(a) **Explain** what would happen to an animal cell placed in a hypotonic solution.

(b) **Explain** under what conditions plant cells may experience plasmolysis.

(c) A plant cell is placed in a solution with higher water potential. Based on the information, **identify** the state of the plant cell.

(d) **Justify** your conclusion.

GO ON TO THE NEXT PAGE.

5. 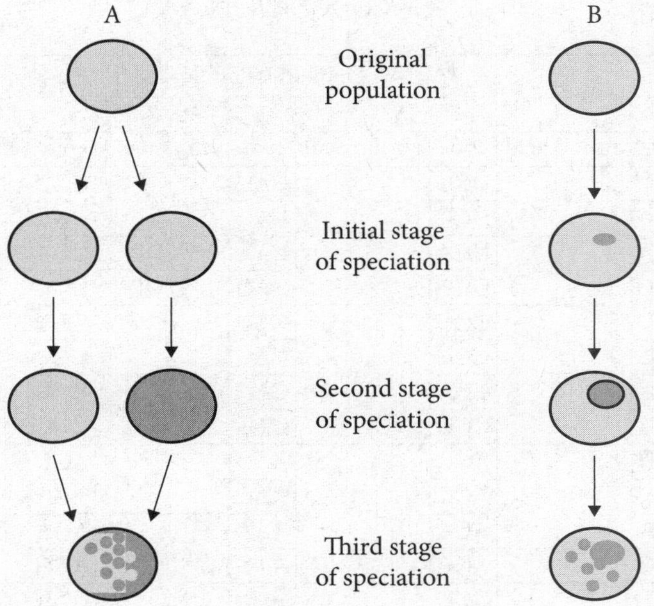

Figure 1. Allopatric and Sympatric Speciation

Speciation involves the evolutionary formation of new and genetically distinct species. Figure 1 shows two types of speciation.

(a) **Describe** the three stages of speciation shown in the figure.

(b) **Explain** the difference between allopatric and sympatric speciation.

(c) Selecting from allopatric and sympatric speciation, **identify** which type of speciation A and B in the figure represent.

(d) **Explain** whether allopatric or sympatric specification is the more common mode of speciation.

6. Point mutations that involve the substitution of a base can result in a silent, missense, or nonsense mutation. Scientists investigated three different point mutations in the genes of hemoglobin. The results are shown in the table.

Mutation location		**Before mutation**	**After mutation**
Globin	**Codon**		
beta	6	GAA	GUA
beta	39	CAG	UAG
alpha	142	UAA	CAA

GO ON TO THE NEXT PAGE.

CODON TABLE

FIRST POSITION		SECOND POSITION: U code	U amino acid	C code	C amino acid	A code	A amino acid	G code	G amino acid	THIRD POSITION
U		UUU	phe	UCU	ser	UAU	tyr	UGU	cys	U
		UUC		UCC		UAC		UGC		C
		UUA	leu	UCA		UAA	STOP	UGA	STOP	A
		UUG		UCG		UAG	STOP	UGG	trp	G
C		CUU	leu	CCU	pro	CAU	his	CGU	arg	U
		CUC		CCC		CAC		CGC		C
		CUA		CCA		CAA	gln	CGA		A
		CUG		CCG		CAG		CGG		G
A		AUU	ile	ACU	thr	AAU	asn	AGU	ser	U
		AUC		ACC		AAC		AGC		C
		AUA		ACA		AAA	lys	AGA	arg	A
		AUG	met	ACG		AAG		AGG		G
G		GUU	val	GCU	ala	GAU	asp	GGU	gly	U
		GUC		GCC		GAC		GGC		C
		GUA		GCA		GAA	glu	GGA		A
		GUG		GCG		GAG		GGG		G

(a) **Identify** the amino acid before and after the mutation at codon 142 in the alpha globin.

(b) **Describe** how the structure and function of hemoglobins are affected by the mutation at codon 142.

(c) Sickle cell anemia results from a point mutation at codon 6 in the beta globin of hemoglobin. Based on the information provided, **determine** the type of mutation that causes sickle cell anemia.

(d) Beta-thalassemia is an inherited blood disorder in which less beta globin is produced. Based on the information, **explain** how a point mutation could cause beta-thalassemia.

STOP—END OF EXAM

The answer key to this quiz is located on the next page.

ANSWER KEY

Section I

1. D	16. C	31. D	46. B
2. D	17. A	32. A	47. C
3. A	18. B	33. B	48. D
4. D	19. B	34. C	49. B
5. D	20. A	35. C	50. A
6. C	21. B	36. C	51. C
7. B	22. D	37. D	52. D
8. B	23. A	38. A	53. C
9. D	24. C	39. B	54. A
10. A	25. B	40. D	55. D
11. C	26. B	41. C	56. D
12. B	27. A	42. D	57. B
13. C	28. B	43. B	58. C
14. C	29. C	44. C	59. A
15. B	30. B	45. D	60. C

Section II

1. See Answers and Explanations
2. See Answers and Explanations
3. See Answers and Explanations
4. See Answers and Explanations
5. See Answers and Explanations
6. See Answers and Explanations

SCORING

Section I, Number Correct: __________

Section II, Points Earned: __________

Enter your results to your Practice Exam 3 assignment to see your 1–5 score and view detailed answers and explanations by logging in at kaptest.com.

Haven't registered your book yet? Go to kaptest.com/booksonline to begin.

PRACTICE EXAM 3 ANSWERS AND EXPLANATIONS

Section I

1. D

The grasses showed a negative change in stomatal conductance, meaning the stomata were open less under elevated CO_2. Stomata allow the exchange of gases and water between the atmosphere and the leaf's interior, so if the stomata are open less, there is less exchange of gases and water. Therefore, water loss from the leaf's surface would change—eliminating (B). Furthermore, since the plants are losing less water to the atmosphere, there will be a decrease in negative pressure in the xylem channels—resulting in less water traveling from the roots through the xylem—eliminating (C). Finally, the grasses would be expected to increase their photosynthetic rates (despite the lower stomatal conductance) because of the greater availability of CO_2. The fact that plant height increased supports this—eliminating (A). **(D)** is correct because, while mineral uptake occurs through the phloem via the pull of osmotic pressure, uptake of potassium occurs through passive rather than active transport.

2. D

Human cells are eukaryotic with nuclear envelopes around their genetic material that could prevent the drug from accessing the DNA even after entering the cell while bacteria are prokaryotes and only possess a nucleoid region within their cytoplasm; thus, the drug would be able to access bacterial DNA after entering the cell **(D)**. (A) is incorrect given the question has stated the drug can enter both bacterial and human cells, which makes the cell wall irrelevant to answering how the drug targets bacterial DNA. Likewise, histones are only found in eukaryotes, and lysosomes are found in both eukaryotes and prokaryotes, meaning neither (B) nor (C) could explain why the drug targets only bacterial DNA.

3. A

As stated in the question, mitochondrial inheritance is maternal, meaning that if the mother is affected by the disorder, then all offspring will be affected as well. Since males don't pass on mitochondria to offspring, an affected male will have unaffected offspring. **(A)** is the only pedigree that shows this pattern of inheritance. The pattern of inheritance shown in (B) is X-linked dominant, in (C) is autosomal recessive, and in (D) is X-linked recessive.

4. D

Since the student noticed a membrane-bound nucleus in the cell, the organism must be a eukaryote. While eukaryotes do have free-floating ribosomes, they also possess membrane-bound ribosomes in a region of the endoplasmic reticulum (ER) referred to as the rough ER. Therefore, the student would observe membrane-bound ribosomes, making **(D)** correct. The DNA of prokaryotes is kept directly within the cytoplasmic region. However, (A) would not be observed in eukaryotic cells, which house their genetic material within a nucleus. Glycoproteins are found in eukaryotes and play an important role in cell-to-cell signaling, but these proteins are located on the outer surface of the membrane, so (B) is also incorrect. (C) would also be unique to prokaryotes because the lack of membrane-bound organelles results in ribosomes only free-floating in the cytosol.

5. D

Speciation can occur when a small group breaks off from the main group. After several more generations, the new species will become reproductively isolated from the main species and speciation will have occurred. Thus, **(D)** is correct. It is highly unlikely that the male Boundary Waters grizzled skipper is sterile since that population of butterfly had not died out; therefore, (A) is incorrect. (B) can be eliminated because the Superior National Forest grizzled skipper is able to survive in the Boundary Waters long enough to mate, so there is no reason to believe the offspring would not be capable of surviving these same conditions. Lastly, if the Boundary Waters grizzled skipper were able to successfully reproduce with the Superior National Forest population, the introduction of this new population would increase genetic variation and decrease the frequency of rare harmful mutations—eliminating (C).

6. C

Nitrogen-fixing bacteria helps move nitrogen through the nitrogen cycle turning ammonia into nitrites and then to nitrates. The passage states haygrazer under 18 inches has high nitrate levels, and nitrates are produced from nitrogen-fixing bacteria. If haygrazer has high levels of nitrogen-fixing bacteria that would create high levels of nitrates, thus (A) is a true statement and can be eliminated. The passage states that same amount of Sun, water, and fertilizer will produce over 100 times more dry weight of haygrazer than bermudagrass. This means that haygrazer is doing a better job of utilizing the Sun, water, and nutrients and thus (B) is also a true statement and can be eliminated. The paragraph and the graph do not discuss the nutrient levels in the grasses except to say that the haygrazer has toxic levels of nitrates. Therefore, **(C)** is an incorrect statement and thus is the correct answer to this question. According to the graph, the high-quality land had the highest grass growth. Since grasses must obtain nutrients in order to grow, it is safe to assume that the high-quality land has larger levels of nutrients and that is responsible for the grass growth. Therefore, (D) is a correct statement and can be eliminated.

7. B

While larger animals typically require more energy in total, smaller animals require more energy per gram of body weight. The ratio between overall surface area and internal volume would increase as the size of the cattle decreases, resulting in smaller cattle giving off more heat to its surroundings—therefore **(B)** is correct. (A) is incorrect because it states the opposite; you know that larger animals need less energy per gram of body weight. There is nothing in the passage to suggest that the age or size of the cattle influences the amount of nitrate that would cause nitrate toxicity; therefore, (C) is incorrect. There is nothing in the passage to suggest that one type of grass has more caloric input than the other; therefore, (D) is incorrect.

8. B

Translation is when mRNA is used to create proteins, and the process happens in the ribosomes. Ribosomes are the "rough" part of the rough endoplasmic reticulum (ER); therefore, translation happens in the rough ER and **(B)** is the correct answer. Smooth endoplasmic reticulum does not have ribosomes and therefore (C) is incorrect. While ribosomes are in the cytoplasm, (A) does not provide enough information to assure that the microRNAs will be close to the site of translation. It is easy to think of all genetic processes happening in the nucleus, but translation only happens in the ribosome, making (D) incorrect.

9. D

The question stem states that haygrazer grows faster and has drier soil during times of heavy rainfall compared to that of bermudagrass. (A), (B) and (C) are all direct contradictions of that information. **(D)** is correct because if haygrazer absorbs more water and nutrients, then it will have drier soil and better growth.

10. C

Plant phototropism is a plant's response to light and darkness. The finding that haygrazers produce energy through photosynthesis faster during the summer season suggests that haygrazers grow better when exposed to more hours of light. The length of days is longer in the summer season, so haygrazers would be classified as long-day plants, which require less than 12 hours of darkness to grow. Thus, **(C)** is correct. (A), (B), and (D) are incorrect because haygrazers are not short-day plants, which grow better when the day length is less than 12 hours of daylight, nor are they day-neutral plants, which do not depend upon the amount of darkness or daylight.

11. C

During "crossing over" in prophase I of meiosis, sections of chromosomes can be excised and recombined with other chromosomes. Typically, this is a balanced exchange between homologous chromosomes; however, sometimes whole fragments can break off and be added to the end of a nonhomologous chromosome. This unbalanced exchange between nonhomologous chromosomes is known as translocation and can result in a trisomy or monosomy phenotype despite having a normal chromosome number. In Down syndrome cases with a normal chromosome number, the cause is what's known as a Robertsonian translocation, where the extra copy of chromosome 21 is translocated on chromosome 14, thus **(C)** is correct. During embryogenesis, fetal cells replicate by mitosis—eliminating (A)—and genetic recombination is not a part of mitosis—eliminating (B). If the extra copy of chromosome 21 were lost during spermatogenesis, then Down syndrome would not have occurred: eliminating (D).

12. B

Only **(B)** shows a cladogram that does not match the data given in the table. According to the table, species 4 is more closely related to species 5 than species 7, and thus the cladogram should show species 4 sharing a closer common ancestor with species 5, rather than with species 7. The same logic applies when looking at the last common ancestor for species 4 and species 2. The cladogram suggests species 4 shares a more recent common ancestor with species 2 than with species 5, which doesn't match the information given in the table.

13. C

Although the individual microorganisms that make up the biofilm are r-strategists, the biofilm itself acts more like a K-selected species (it doesn't reproduce quickly, it's energy-efficient, it's large, and it's organized). Therefore, it will follow a logistic, not an exponential, pattern of growth: eliminate (A) and (B). **(C)** is the correct answer, since the axes are not labeled correctly in (D).

14. C

According to the pedigree, the 1st generation father is affected (has MFS) and the mother is not. Out of their four children, one male and one female have MFS. In the 3rd generation, both of these affected siblings produce children of their own. The affected 2nd generation male produces two affected females and one unaffected male. The 2nd generation female produces three children, with one male having MFS. The pattern of inheritance presented here is autosomal dominant. The genetic mutation is not passed on through the sex chromosomes, and MFS occurs in both sexes. **(C)** is correct.

In the 2nd generation, the affected female and unaffected male produce an affected male. The 2nd generation father is not affected yet still has a son who does have MFS. This tells you that the mutation, in this case, was passed on through the mother's chromosomes. (A) is incorrect. (B) is a tempting answer choice, since either the father or mother can pass on the genetic mutation. However, here you see that only one affected parent is needed to produce offspring with the mutation. Lastly, (D) is incorrect because the pedigree shows three consecutive generations within the same family that have affected relatives at every generation.

15. B

MFS is an autosomal dominant trait. Individual A, a 3rd generation child, has MFS through an unaffected mom and an affected dad. On that basis, individual A must be heterozygous dominant. The probability of her passing on the dominant allele is $^1/_2$, and the probability of her passing on the recessive allele is also $^1/_2$. Having a child with an unaffected individual means the probability of her partner passing on a recessive allele is $^1/_1$. Therefore, there is a $(^1/_2) \times (^1/_1) = {}^1/_2$ or 1:2 chance the child will be born with two recessive alleles and be unaffected. **(B)** is correct.

16. C

Women that are affected by MFS are able to seek genetic counseling to discuss the possibility of passing on the MFS mutation to their unborn child. They are also able to screen the fetus for the presence of the FBN1 mutation, and affected fetuses may be aborted. Over time, reducing the number of those born with the FBN1 mutation would also decrease the incidence of those with MFS in the general population. **(C)** is correct. (A), (B) and (D) are all incorrect because they are all factors—better treatment, early diagnosis, genotyping, and identifying the disease in individuals with mild symptoms even in later adulthood—that would also increase the prevalence of MFS.

17. A

While those with MFS still produce fibrillin-1, one of those two alleles has a mutation that causes it to produce less than the normal amount, resulting in less than the normal amount of fibrillin-1 than those without MFS. **(A)** is correct. (B) is incorrect because it is stating the opposite. Those with MFS have inherited a genetic mutation (or spontaneous) that affects the function of one of the alleles of the gene, which encodes for fibrillin-1, such that it produces little or no fibrillin-1. (C) is incorrect because those with MFS produce less fibrillin-1 due to a genetic mutation than those that do not have MFS. (D) Production of fibrillin-1 is not dependent on the phenotype. The organism's phenotype will be influenced by the lack, overabundance, or adequate amount of fibrillin-1 being produced by the cells.

18. B

The question stem asks what will increase cAMP levels in the cell, and so **(B)** is the correct answer since PDE degrades cAMP. Inhibiting the phosphorylation of PDE by PKA will mean less PDE activity, and thus higher cAMP levels in the cell. (A) is incorrect because there is no PKA phosphorylation site on cAMP, since cAMP is not a protein. This will have no effect on cAMP levels. (C) is incorrect because the dissociation of PKA is required for PKA's catalytic actions, but not for cAMP generation, so this will not directly affect cAMP levels. (D) is incorrect because the binding of adenylyl cyclase to the Gs alpha subunit is required for cAMP generation, and so blocking this interaction would lead to less cAMP generation and thus lower cAMP levels.

19. B

The existence of two phenotypes that are distinctly different (dominant and recessive) would support the hypothesis that this is a Mendelian trait. However, as shown on the graph, there are more than two phenotypes. There is continuous variation in the population for which the researchers have produced simplified group classifications for analysis. Thus, the correct answer is **(B)**. (A) is incorrect because the groups of male pattern baldness are not distinct. (C) is incorrect because traits that are inherited in a non-Mendelian manner can also have a strong genetic basis. (D) is incorrect because four categories are used in the graph, and the level of baldness varies continuously among the men.

20. A

(A) is the correct answer because the target organ, the kidney, fits the type of action described: since ANP is produced in the heart, when it acts on the kidney, it must do so in an endocrine fashion. (B) and (D) are incorrect because these actions of ANP would be best described as endocrine. (C) is incorrect because this action of ANP would be best described as autocrine (ANP acts on the same cells that produced the peptide).

21. B

(B) is the correct answer as all the factors listed are ways that a food web can lose energy, and thus not pass the energy to higher trophic levels. (A) is incorrect because energy at higher levels is not consumed by autotrophs; autotrophs are primary producers. Any energy consumed by an autotroph is not a source of loss for the food web, since autotrophs are the producers in the food web. (C) is incorrect because energy is produced, not lost, through photosynthesis and chemosynthesis. (D) is incorrect because the remains of dead organisms are passed to decomposers, not apex consumers (apex consumers are at the highest trophic level).

22. D

When additional wild-type protein X is added to normal *E. coli* cells in Experiment 1, the level of protein X continues to go up over time. This suggests that protein X stimulates transcription of the operon through a positive feedback loop, potentially through its ability to bind the X operon promoter. In this question, the researcher is using *E. coli* cells that make a mutant version of protein X that can no longer bind the X operon promoter, indicating an inability to induce transcription. When wild-type protein X, which can bind the X operon promoter, is added to the mutant *E. coli*, an initial increase in transcription of the mutant protein X would be expected, eliminating (B) and (C). However, the amount of mutant protein X would eventually level off because the mutant protein made by the mutant *E. coli* cells would not be able to sustain a positive feedback loop, which makes (A) incorrect and matches **(D)**.

23. A

When protein X is added into the system in Experiment 1, protein X levels continue to rise afterward, suggesting protein X induces transcription of the X operon in a positive feedback loop, ruling out (C) and (D). In Experiment 2, the addition of protein Y by itself does not change protein X levels, while in Experiment 3, the addition of protein Y with protein X prevents an increase in protein X levels over time, indicating protein Y is most likely inhibiting rather than promoting the transcription of the X operon. This observation is the opposite of (B), but matches **(A)**.

24. C

Adding lactose to *E. coli* cells increases protein Y levels, suggesting lactose is increasing the transcription of the Y operon. Inhibiting RNA polymerase from binding to the promoter would instead decrease transcription, ruling out (D). Experiments 1, 2, and 3 also show that protein X does not affect protein Y levels, so even if lactose did decrease transcription of the X operon, protein Y levels would not change, eliminating (B). Inducers increase gene expression by binding to and disabling repressors, which matches **(C)**. While enhancers also increase transcription, they are DNA sequences, not disaccharides like lactose, making (A) incorrect.

25. B

If glucose levels are high, the Y operon would be expected to be repressed. If the Y operon is no longer being transcribed, protein Y levels would fall, eliminating (C). Based on the experimental results, protein Y seems to inhibit X operon transcription. Therefore, protein X levels would increase when protein Y levels decrease and no longer inhibit transcription of the X operon, making **(B)** correct. There is no indication that glucose would directly bind the X operon operator or increase transcription as described in (A) and (D), respectively.

26. B

Proteins that are destined to be released from the cell, such as peptide hormones, are synthesized by ribosomes bound to the rough endoplasmic reticulum (ER). The proteins are then moved to the Golgi apparatus before they are shuttled in vesicles to the membrane for release. Thus, **(B)** is correct. Smooth ER produces hormones, but these steroid hormones take a different route for transport. Lysosomes, which are involved in the breakdown of materials, and centrosomes, which play a part in cell division, are not involved in the secretory pathway.

27. A

A positive feedback loop amplifies change (an increase in births leads to more births or an increase in deaths leads to more deaths). The passenger pigeons, therefore, experienced a positive feedback loop: a decrease in population caused the mean fitness level of the pigeons to decrease, which led to further population decrease, and so on. Thus, **(A)** is the correct answer. Negative feedback counters change (an increase in births limits population growth).

28. B

Cells that have damaged DNA in cells in G1 phase will activate a checkpoint pathway that blocks or suppresses the activity of CDKs. While the CDK inhibitors are active, the DNA can be repaired. The cell is unable to progress to the next phase in the cell cycle, S phase, until the DNA is repaired. **(B)** is correct. (A) is incorrect because S phase, when the cell's DNA is replicated, follows G1. (C) and (D) are incorrect because mitosis follows G2.

29. C

Cells normally show density-dependent inhibition, meaning that when cells become crowded (e.g., more than a single layer in a culture dish), they stop dividing. Normal cells will also stop dividing if they are not attached to substrate, either the culture dish or extracellular tissue matrix (anchorage dependence). **(C)** is thus correct. The other answer choices are true statements but do not address the reason why clumping or overlapping cells are abnormal.

30. B

According to the chart, Group 1 and Group 2 lemurs interpret the same sounds differently. Only **(B)** presents a situation where the lemur from Group 2 would not be aware of incoming danger, instead interpreting a warning of an incoming fox as a source of water being found. In (C) and (D), the lemur from Group 2 would still be alerted to approaching danger, while a misinterpretation of two clicks would not put the listening lemur in danger, ruling out (A).

31. D

Growling at other Group 1 members as detailed in (A) would not help in competing against Group 2. Similarly, waiting for Group 2 lemurs to leave before looking for plants would not give Group 1 lemurs any competitive advantage, eliminating (B). However, if Group 1 learned to chatter when they found plants, Group 2 lemurs would interpret the chattering as a signal for danger and stay away, making **(D)** correct. (C) is incorrect because shouting, while a sign of danger for Group 1 lemurs, would be heard as a greeting by Group 2 lemurs and fail to drive them away.

32. A

This question is asking for an explanation as to why the population of Group 1 decreases despite an increase in the amount of fruit available. While (B) could be a plausible reason for a decrease in population, it refers to only Group 2, not Group 1. (C) would not be expected to result in a decrease in population. Introducing a third group of lemurs could create competition, but with no information about Group 1's population provided, there is no way to know what, if anything, was responsible for the decline. Therefore, (D) is incorrect. Only **(A)** provides a possible explanation for a decrease in the population of Group 1, where an increase in Group 2 numbers due to more food leads to Group 1 being outcompeted.

33. B

A lack of sunlight would lead to less plant growth. The passage notes that both groups of lemurs eat mainly plants, so fewer plants would be expected to negatively impact both groups, which matches **(B)** and eliminates (A), (C), and (D).

34. C

(A), which indicates close similarities between the two groups, would weaken the researcher's claim that the two groups did not come from a common group. (B) is only possible between species, which contradicts the claim that the groups are subspecies. While (D) mentions that the two groups have different fur colors, the fact that foxes prefer one group over the other does give any information on if the groups are distinct subspecies. However, if Group 1 lemurs are more genetically similar to another group, that would suggest they recently split from that group rather than Group 2, making **(C)** correct.

35. C

Common causes of disease transmission through microorganisms, such as bacteria and viruses, contain DNA and/or RNA. The absence of DNA/RNA gene signatures from known and potential pathogens suggests that if the pathogen is in fact present in the soil, its transmission must not involve nucleic acids. Prions are misfolded proteins that can transmit a disease independently of any genetic material. Thus, the scientist's methods wouldn't detect their presence, but their hypothesis would still be correct, so you can choose **(C)**. (A) is incorrect because it distorts the original hypothesis, which stated it was transmitted via the soil. (B) is incorrect because bacteria have DNA, and no DNA was detected. (D) is incorrect because viruses and bacteria aren't the only possible causes of disease, and prions would be an alternative explanation.

36. A

Chlorophyll is essential for a plant's ability to undergo photosynthesis. An albino plant that cannot produce chlorophyll is unable to use light energy to make sugars. Thus, once energy stored in the seed is exhausted, the plant will die. **(A)** is correct. (B) is incorrect because in the symbiotic relationship, the opposite occurs; plants supply sugars to the fungus while the fungus supplies water and mineral nutrients, like phosphorus, from the soil to the plant. (C) is incorrect because, albino plants with little to no chlorophyll will have impaired photosynthesis. (D) is incorrect because cellular respiration requires glucose to make ATP; without chlorophyll, plants cannot make glucose.

37. D

The organic soup model suggests that life first arose when simple organic molecules continued to increase in complexity until nucleic acids and amino acids were formed. Nucleic acids eventually joined together to create polynucleotides that eventually became self-replicating, organizing into protobionts. Further organization eventually led to the creation of the first cells. This matches **(D)**. While (A) and (B) are accurate, both are incorrect because they are components of bigger theories. (C) is incorrect because this is referred to as the RNA world hypothesis.

38. A

According to the chart, the pillbugs seem to have no preference between the left chamber and the right chamber in Experiment 1, and thus are demonstrating kinesis, or random movement in response to a stimulus. In the second experiment, however, the pillbugs are shown to preferentially move towards the acidic chamber, which describes positive taxis, or an attraction towards a stimulus. These observations match **(A)**; eliminate (B), (C), and (D).

39. B

According to Mendel's law of segregation, allele pairs separate independently of each other during gamete formation and randomly unite at fertilization. The graph for Mendelian segregation shows an equal split among the four possible genotypes. The graph for irregular segregation, however, shows a shift toward the genotype A^+B^+ in the offspring. This shift can be explained by biased fertilization, which causes some unions to be more probable than others. Thus, non-random pairing of sperm and egg at fertilization distorts the expected ratio offspring genotypes. **(B)** is correct. As for (A), unequal recombination—where some gametes end up with too much genetic information and others with too little—can lead to disorders in the fetus but would not explain the irregular segregation. The genotypes of the offspring of the two crosses indicate the same genes are present. (C) is incorrect because independent assortment is a normal occurrence, and the gametes that result from oogenesis are haploid, not diploid. (D) is incorrect because this actually describes the situation that leads to Mendelian inheritance patterns: when the sperm are all equally capable of producing viable offspring, then fertilization should be random, and offspring will show Mendelian segregation.

40. D

The experiment showed that there was some reduction in biomass, and therefore in plant growth, in higher saline soils (had a higher electrical conductivity). In order to strengthen this conclusion, the correct answer must make a link between higher salinity and reduced plant growth and development. Therefore, **(D)** is the correct answer, because it connects high salinity (high EC) with a decrease in plant development (fewer beans per plant). While (A) mentions a sensitivity of plants to salinity, it does not actually say how the plants are more sensitive to salinity. (B) mentions salinity, but not plant development. Finally, (C) connects higher salinity to increased, rather than decreased, plant growth and development (taller plants with more peas per pod).

41. C

Alkaline soil is basic (has a pH greater than 7.0), and only an acid would be expected to be able to make the pH of the soil less basic. **(C)** is the correct answer.

42. D

As stated in the passage, the point of the experiment was "to study the effects of saline soils on plant development." As such, the correct conclusion is one that incorporates both of these elements: something about the salinity of the soil, and something about the plant's development. **(D)** fits these parameters and makes an appropriate statement about the data. (A) is incorrect because the students' goal was not to compare fresh and dry weight of plants. (B) is incorrect because the students measured the electrical conductivity of the soil, not the "saline content." Finally, (C), while it connects the electrical conductivity measurement to the biomass, describes a relationship that is not supported by the data.

43. B

Under normal conditions, with non-saline soil, the driving force for the uptake of water into the roots of the plant is that the root sap is hypertonic compared to the soil; this creates a concentration gradient that drives water into the roots. With saline soil, the opposite is true: the soil is hypertonic compared to the root sap, the osmotic pressure gradient is reversed, and so the plant has to actively pump more ions into the root to take in water. Thus, **(B)** is correct. (A) and (D) express the opposite of what would actually happen, while (C) is an untrue statement because the tonicity of the fluids on either side of a cell membrane won't change the permeability of the membrane.

44. C

Since oxygen is produced as a result of photosynthesis, it is directly related to the amount of leafy biomass that a plant has. Thus, the more biomass, the more photosynthesis, and the greater the oxygen production. Since plants grown in Sample 3 soil had the lowest biomass, they would be expected to have the least photosynthetic ability, and therefore produce the least oxygen. **(C)** is the correct answer.

45. D

At the point indicated by the arrow, the blood pressure has dipped below the set point. That means that the body will act to increase the blood pressure. Vasoconstriction, or narrowing the arterioles, will increase the blood pressure (imagine taking a rubber pipe and squeezing to decrease the diameter; the fluid inside the pipe will be under more pressure). (A) and (C) can be eliminated. Furthermore, given that an increase in blood pressure increases parasympathetic and decreases sympathetic activity, a drop in blood pressure must do the opposite: decrease parasympathetic and increase sympathetic activity. An increase in heart rate is associated with increased sympathetic activity. Therefore, **(D)** is correct.

46. B

Normally, hemichannels at nonjunctions remain closed. The attachment of *Citrobacter rodentium* disrupts normal cellular function, increasing connexon protein levels during infection. This suggests that unpaired hemichannels are opened. The opening of unpaired hemichannels would allow small molecules and ions to leak out of intestinal cells into the lumen of the colon. This change in concentration gradient, in turn, would cause water to flow into the lumen, resulting in more water in the stool. **(B)** is correct. (A) is incorrect because facilitated diffusion involves the transport of molecules down their concentration gradient and not against. (C) is incorrect because transport down a concentration gradient does not require active transport. (D) is incorrect because the transport of small molecules and ions would be from the intestinal cells to the lumen.

47. C

Seeds of the affected grape plant are unable to fully develop. The seeds are aborted, leaving only a seed trace behind. In order to continue the next generation of seedless grapes, the seeds have to be harvested before they are aborted. The seeds will have the mutation for seedlessness and be able to pass on this trait to its offspring. **(C)** is correct. Growing the immature seeds outside of the original plant by using tissue culture techniques does not change the mutation in the chromosome. (A) is incorrect. (B) is incorrect because the original plant will continue to produce seedless fruit. The size of the fruit is dependent on many factors including availability of nutrients in the soil, temperature, drought conditions, etc. (D) is incorrect because the seeds are able to produce seedless offspring not another new species of grape.

48. D

Plants regulate or inhibit the formation of nodules on their roots by a chemical pathway that involves cytokinins, which are produced in the leaves and sent down to the roots; therefore, (A) and (B) are incorrect. In this experiment, the cytokinins played a role in inhibiting new nodule formation when the plants were grown in nitrogen-rich soil. The plant did not have a need for more nodules to be formed since it was able to obtain nitrogen from the nitrogen-rich soil. (C) is incorrect and **(D)** is correct.

49. B

In the first and second generations, there are two alleles for tail length, *T*, and *t*. By the third generation, only one of the alleles for tail length, *T*, remains and has become fixed, while *t*, the allele for short tails has been lost. Therefore, **(B)** is correct. Although the fixed *T* allele results in an entire third generation of long-tailed cheetahs, that does not make it a superior phenotype. Long tails may be better suited in that particular environment, but natural selection could just as easily act on short-tailed cheetahs in a different environment, so (A) is incorrect. Natural selection selected for a single allele in the population over three generations; since (C) mentions two alleles, it cannot be correct. While it is not known what the allele frequencies were in the original cheetah population, it is clear that at least two alleles, *T* and *t*, for tail length existed before the mass extinction. Since the allele for short tails exists in the original cheetah population and is lost by the third generation, then (D) must be incorrect.

50. A

In DNA, guanine pairs with cytosine and adenine pairs with thymine. If 27% of the bases are guanine, then another 27% of the bases are cytosine. Therefore, $100\% - 27\% - 27\% = 46\%$ of the bases are adenine and thymine, which each occur at the same rate. That means that $46\% \div 2 = 23\%$ of the nucleotide bases are thymine. Therefore, **(A)** is correct.

51. C

Comparative anatomy is one tool that biologists use to assess how recently two species might have shared a common ancestor. If two organisms have the same or similarly derived anatomical features, they are considered to have homologous structures. For instance, despite humans having arms and bats having wings, they share the same arm bones and are thus considered to have a common ancestor. In convergent evolution, however, distantly related organisms can evolve analogous structures that perform the same function but are structurally and developmentally distinct from each other. Thus, **(C)** is correct because it suggests the dewlaps of species D and E are analogous to the other anoles, and not homologous. (A) is incorrect because interbreeding would be evidence of a very close evolutionary relationship. (B) is incorrect because while geographic isolation can cause speciation, it's not proof of speciation. (D) is incorrect because aggression and predation are not exclusive to distantly related species.

52. D

The molecular clock hypothesis states that random genetic mutations occur at a constant rate and will steadily accumulate over time. Thus, if two species share many of the same genetic mutations, they must have a more recent common ancestor. **(D)** correctly concludes that the more similar the genetic mutations, the more recently they diverged from a common ancestor. (A) and (B) are incorrect because there's insufficient evidence to determine whether species B or D was a common ancestor for the other. (C) is incorrect because without data from the other three species, you can't determine whether or not species B or D was a common ancestor for the others.

53. C

According to Darwin's theory of evolution, organisms are said to be of different species when the differences in their behavior, environment, and biology are such that they can no longer interbreed and produce viable offspring. **(C)** precisely matches this definition of species and is thus correct. (A), (B), and (D) can all cause speciation to occur but aren't sufficient to prove individuals are indeed different species.

54. A

Based on the trends described in the question and in Table 1, it suggests that the average anole from species D is increasing its UV reflectance. According to the theory of natural selection, this must be because increased UV reflectance provides a reproductive advantage over low UV reflectance. As a result, the population must be under directional selection, which matches **(A)**. (B) is wrong because different species are unable to produce viable offspring, according to Darwin's theory of evolution. (C) is incorrect because if the population were under disruptive selection then there would be a shift to both a high and a low reflectance phenotype. Similarly, (D) is incorrect because a population experiencing stabilizing selection shifts the phenotypes away from the extremes, resulting in UV reflectance values somewhere in between species A and D.

55. D

According to the question, DNP dissipates the mitochondrial proton gradient, which is essential for generating ATP via the ATP synthase. Because there is less of a proton gradient, the rate of ATP production should decrease, which allows you to eliminate (A). While the rate of ATP production is decreased by DNP, this doesn't impair the actions of the electron transport chain, which will continue to oxidize NADH and $FADH_2$ to NAD^+ and FAD^+ and consume oxygen. Thus, you will see an increase in the concentration of NAD^+ and a decrease in NADH concentration, which eliminates (B) and (C). Therefore, **(D)** is correct.

56. D

A cross between a heterozygous yellow fish (*Yy*) and an orange fish (*yy*) would yield an offspring ratio of 1:1 between yellow- and orange-finned fish. Of the 754 fish, it would be expected that 377 would be yellow and 377 would be orange. The chi-square values are calculated as follows:

phenotype	observed	expected	obs–exp	$(obs–exp)^2$	$(obs–exp)^2/exp$
yellow	426	377	49	2,401	6.37
orange	328	377	–49	2,401	6.37
					$\chi^2 = 12.74$

Rounded to the nearest tenth, the answer is 12.7. Therefore, **(D)** is correct.

Practice

57. B

When attempting to transform bacteria with a plasmid, only a small percentage of the bacteria present will take up the plasmid. Because you only want the transformed bacteria to proliferate, you need a way to positively select for your recombinant colonies. Ampicillin-resistance is a common strategy for positive selection, and without it, the student wouldn't be able to isolate the transformed bacteria; thus, you will choose **(B)**. While ampicillin-resistance is needed to isolate the transformed bacteria, it's not necessary for the actual production of insulin, thus (A) is incorrect. Ampicillin-resistance is only required for the positive selection of transformed colonies, not the insertion of the insulin gene into the plasmid, thus (C) is wrong. Finally, while it's correct that ampicillin-resistance has no effect on insulin production via induction of the *lac* operon, ampicillin-resistance is needed to isolate transformed colonies so that insulin can be produced, thus (D) is also incorrect.

58. C

Alleles linked to phenotypes that are reproductively successful will increase in frequency, while those that are linked to phenotypes that are not reproductively successful will decrease. Directional selection will increase the phenotype that is reproductively successful, which would include those that are resistant to DFTD. The frequency of the allele for resistance to DFTD will also increase. Therefore, (D) is incorrect and **(C)** is correct. Stabilizing selection maintains the average phenotype over long periods of time, eliminating the extremes in phenotypes. Since there are few resistant DFTD devils, stabilizing selection would not increase the frequency of the allele for DFTD resistance; thus (A) is incorrect. Disruptive selection selects the extremes, so it will not favor those with an intermediate phenotype—perhaps those that recover from the disease but remain in a weakened state—to pass on the desired trait, resistance to DFTD; thus (B) is incorrect.

59. A

Competitive inhibitors do not lower the V_{max} of an enzyme, while noncompetitive inhibitors do. Therefore, the experiment described in **(A)** could be used to determine if there is a difference in V_{max} in the absence and presence of the inhibitor, and thus whether the inhibitor is a competitive or noncompetitive inhibitor. The experiments in (B) and (C) would not provide any information on how V_{max} is affected by the inhibitor. (D) is incorrect because the inhibitor is not expected to bind to the substrate.

60. C

Post-translational modifications, such as phosphorylation, can impact how an enzyme functions and interacts with other molecules. Based on the table, phosphorylation decreases enzymatic activity and increases DNA binding. You can then infer that phosphorylation would most likely be located at site 3 in the ATP-binding domain, which would affect the enzymatic rate given that the enzyme is ATP-dependent, and at site 1 in the DNA-binding domain, which would impact DNA binding. This matches **(C)** and eliminates (A) and (B). Phosphorylation does not affect substrate binding, meaning site 4 is probably not a place of phosphorylation, ruling out (D).

ANSWERS AND EXPLANATIONS

Section II

1. Scoring Guidelines for Free-Response Question 1

(a) **Describe** the relationship between plants and mycorrhizae. **(2 points)**

Description (2 points)
• Plants live in symbiosis with mycorrhizal fungi • Their relationship is mutualistic because both organisms benefit

Here is a possible response that would receive full credit:

Plants live in symbiosis with mycorrhizal fungi. Since both organisms benefit, their relationship is mutualistic. The fungi that colonize the root of the plant provide the plant with increased water and nutrient absorption, protecting the plant from abiotic (drought) and biotic (pathogen) stresses. The plant, in turn, provides the fungus nutrients in the form of sugars.

(b) **Identify** the control group in the experiment. **(2 points)**

Identification (2 points)
• No treatment + isolated

Here is a possible response that would receive full credit:

The use of fungicide and the presence of neighboring plants are being studied, so they are the independent variables. The group that does not receive treatment is the control group. Thus, the results of no treatment + isolated are a baseline to which the results of the other groups will be compared.

(c) **Determine** the effect fungicide and the presence of neighbors have on shoot biomass. **(2 points)**

Determination (2 points)
• Fungicide decreased shoot biomass • The presence of neighbors increased the shoot biomass

Here is a possible response that would receive full credit:

According to Figure 2, the lengths of the white and gray bars, which represent the use of fungicide (treatment), are shorter than those of the striped and black bars, which represent no treatment with fungicide for all four plant species. Thus, the use of fungicide decreased shoot biomass.

Figure 2 also shows for all four species that the lengths of the white bars, which represent treatment + isolated plants, are shorter than those of the gray bars, which represent treatment + neighbors. Similarly, the lengths of the striped bars, which represent no treatment + isolated plants, are shorter than those of the black bars, which represent no treatment + neighbors. Therefore, the presence of neighbors increases shoot biomass.

(d) Suppose oxygen concentration promotes mycorrhizal colonization. **Predict** which plant species under what conditions would produce the greatest amount of oxygen and **justify** your prediction. **(4 points)**

Prediction (2 points)	Justification (2 points)
• *Polygonum pubescens* (no treatment + neighbors)	• The bars with the greatest percent of fungal colonization rate correspond to the gray and black bars (no fungicide treatment)

Here is a possible response that would receive full credit:

According to Figure 1, of the four plant species, the species with the greatest fungal colonization is *Polygonum pubescens*. The data suggest that no treatment and the presence of neighbors result in greater colonization. Thus, *Polygonum pubescens* under those conditions would produce the greatest amount of oxygen.

2. Scoring Guidelines for Free-Response Question 2

(a) **Identify** which ecosystem has the highest gross primary productivity. **Describe one factor** that contributes to high gross primary productivity. **(2 points maximum)**

Identification (1 points)	Description (1 point each, 1 point maximum)
• Tropical rainforest (highest amount of energy at 17,450 kj/m^2)	• Warmer temperatures increase the rate of enzymatic light independent-reactions in photosynthesis • High precipitation increases the rate of photosynthesis from photolysis of water

Here is a possible response that would earn full credit:

Gross primary productivity is the energy producers incorporate from the sun. The ecosystem with the highest gross primary productivity is tropical rainforest, which (according to the table) has the highest amount of energy at trophic level D (17,450 kJ/m^2). High precipitation in the tropical rainforest increases the rate of photosynthesis, which converts light energy to chemical energy, as a result of the photolysis of water during non-cyclic photophosphorylation.

(b) Based on the amount of energy at the trophic levels in the data table, construct a food chain in the template. Write the letter of the trophic level in the appropriate box and draw arrows to indicate the energy flow between levels. Provide two reasons for your answer. **(4 points maximum)**

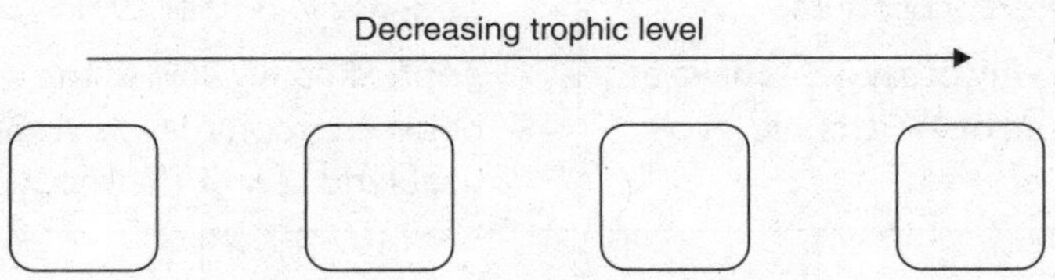

Identification (1 point each, 2 points maximum)
• All four trophic levels in appropriate boxes • All four arrows correctly drawn between trophic levels Here is a sample food chain that would receive full credit: Decreasing trophic level → D ← B ← C ← A

Reasoning (1 point each, 2 points maximum)
• On a food chain, lower trophic levels have more energy than higher trophic levels; energy decreases as it is used by organisms in each level. • Arrows on a food chain show the direction of the energy as it is transferred from one organism to another.

Here is one possible response that would earn full credit:

According to Table 1, trophic level A for all five ecosystems has the lowest amount of energy, followed by trophic level C, trophic level B, and then trophic level D, which has the highest amount of energy. On a food chain, the lower trophic levels like producers have more energy than higher trophic levels such as tertiary consumers. Thus, the four trophic levels in order of decreasing trophic levels are D, B, C, A. Arrows on a food chain represent energy flow and show the direction energy is transferred from one organism to another. Energy flows from high quantities (low trophic levels) to low quantities (high trophic levels), so A to C to B to D.

(c) **Calculate** the transfer efficiency between trophic level D and trophic level B for the desert ecosystem. **Explain** why ecosystems rarely have more than 4 or 5 trophic levels. **(2 points maximum)**

Identification (1 points)	Calculation/Explanation (1 point each)
• The transfer efficiency between trophic level D and trophic level B for a desert ecosystem • Explain why ecosystems have a maximum of 4–5 trophic levels	• Divide energy at higher trophic level by energy at lower trophic level: $22 \div 212 \approx 0.10$ or 10% • Approximately 90% of the energy is lost between trophic levels via respiration and heat (2nd Law of Thermodynamics)

Here is one possible response that would earn full credit:

The transfer efficiency is the efficiency of energy transfer from one trophic level to the next level up. The table shows for desert ecosystem, the amount of energy at trophic level D and B is 212 and 22, respectively. Energy flows from D to B. Thus, the transfer efficiency is 22 divided by 212, which is approximately 0.10 or 10%. Ecosystems rarely have more than 4 of 5 trophic levels because about 90% of the energy is lost between trophic levels via respiration and heat. The 2nd Law of Thermodynamics states that in every energy transfer, some amount of energy is lost in a form that is unusable and increases entropy.

(d) An invasive predator was introduced in the temperate deciduous forest ecosystem. **Predict** the short-term impact on the pyramid of biomass for the ecosystem. **Justify** your prediction. **(2 points maximum)**

Prediction (1 point each, 1 point maximum)	Justification (1 point each, 1 point maximum)
• Reduce the biomass of trophic level D and C, and increase the biomass of trophic level B • Reduce the biomass of trophic level D and B, and increase the biomass of trophic level C	• Invasive predator eats D; fewer D to eat B, more B to eat C • Invasive predator outcompetes D for B; D die off; fewer B to eat C

Here is one possible response that would earn full credit:

An invasive predator introduced in the temperate deciduous forests has no native predator and readily available food. If the predator feeds on organisms in trophic level D, the biomass of trophic level D will decrease. In turn, the number of organisms that organisms in trophic level D prey on will increase. Thus, the biomass of trophic level B will increase. This will cause the amount of food that organisms in trophic level B feed on to decrease, so the biomass of trophic level C will decrease.

3. **Scoring Guidelines for Free-Response Question 3**

(a) **Describe** how environmental factors affect phenotypic plasticity, the ability of one genotype to exhibit different phenotypes in different environments. **(1 point)**

Description (1 point)
• Environmental factors turn genes on and off

Here is a possible response that would receive full credit:

Phenotypic plasticity is influenced by the external environment organisms live in, as well as the organism's internal environment. External factors include temperature, light, and nutrients, and internal factors include hormones and toxins. These environmental factors can turn genes on and off, thereby influencing the development and function of an organism. For example, during unfavorable environmental conditions, some animals may enter quiescence, a state of dormancy, by turning specific genes on and/or off to slow down their metabolism. When favorable conditions return, quiescence ends.

(b) **Identify** the control in the experiment. **(1 point)**

Identification (1 point)
• The control is the a– culture.

Here is a possible response that would receive full credit:

The control is the culture of a-type cells with no supernatant from the co-culture (a–) added, meaning no pheromones were present.

(c) In a different experiment, a culture of α-type cells was incubated with and without yeast mating pheromone. The fold change (log 10) for *ste* 3 expression for α with and without pheromone was 7 and 0.7, respectively. **Predict** the likely results of *ste* 2 expression for this experiment. **(1 point)**

Prediction (1 point)
• The fold change (log 10) for ste 2 with pheromone will be 10 times greater than that without pheromone and twice that of ste 3.

Here is a possible response that would receive full credit:

The fold change (log 10) for ste 2 with and without pheromone is 14 and 1.4, respectively.

(d) **Justify** your predictions. **(1 point)**

Justification (1 point)
• The results for α-type cells will be the same trend as a-type cells.

Here is a possible response that would receive full credit:

The addition of yeast mating pheromone to the α-type cells is equivalent to adding the supernatant from the co-culture to the a-type cells. Thus, the results for α-type cells will be the same trend as a-type cells. The figure shows the expression of a+ is 10 times more than a– for all three genes; expression for ste 3 is approximately twice that of ste 12 and expression for ste 2 is approximately twice that for ste 2. Since the expression of α+ and α– for ste 3 was 7 and 0.7, the expression for ste 2 is most likely $2 \times 7 = 14$ and $2 \times 0.7 = 1.4$.

4. Scoring Guidelines for Free-Response Question 4

(a) **Explain** what would happen to an animal cell placed in a hypotonic solution. **(1 point)**

Explanation (1 point)
• The animal cell would swell and may burst.

Here is a possible response that would receive full credit:

In a hypotonic solution, water will move into the cell since the extracellular fluid has a lower solute concentration. This will cause the animal cell to swell. Since animal cells do not have cell walls like plant cells, their cell membranes cannot withstand as much pressure and may burst.

(b) **Explain** under what conditions plant cells may experience plasmolysis. **(1 point)**

Explanation (1 point)
• Hypertonic environment (e.g., plant is not watered)

Here is a possible response that would receive full credit:

Plant cells experience plasmolysis when a plant does not receive sufficient water and the extracellular fluid becomes hypertonic. In a hypertonic environment, water leaves the plant's cells, which causes the cells to lose turgor pressure, shrink, and experience plasmolysis.

(c) A plant cell is placed in a solution with higher water potential. Based on the information, **identify** the state of the plant cell. **(1 point)**

Identification (1 point)
• The plant cell is turgid.

Here is a possible response that would receive full credit:

A plant cell placed in a solution with higher water potential will be turgid.

(d) **Justify** your conclusion. **(1 point)**

Justification (1 point)
• A solution with higher water potential is a hypotonic solution, so water moves into the cell.

Here is a possible response that would receive full credit:

Since the solution has higher water potential than the plant cell, the solution is hypotonic (lower solute concentration). This indicates that water will move into the cell by osmosis from higher water potential to lower water potential until the concentrations of solute equalize. This causes the cell to swell and become turgid. The cell wall provides a barrier that prevents too much water from moving into the cell.

5. Scoring Guidelines for Free-Response Question 5

(a) **Describe** the three stages of speciation shown in the figure. **(1 point)**

Description (1 point)
• The three stages of speciation are separation, evolution, and reproductive isolation resulting in a new species.

Here is a possible response that would receive full credit:

Speciation is the formation of new and distinct species via divergent evolution. The initial stage is separation, the second stage is evolution (e.g., adaptation to new environments), and the third stage is reproductive isolation resulting in a new species.

(b) **Explain** the difference between allopatric and sympatric speciation. **(1 point)**

Explanation (1 point)
• In allopatric speciation, the original population is separated into two distinct populations by some geographic barrier. In sympatric speciation, polymorphism occurs.

Here is a possible response that would receive full credit:

In allopatric speciation, geographic isolation occurs. The original population separates and evolves into distinct populations since the geographic barrier prevents genetic interchange. In sympatric speciation, polymorphism (individuals within a population acquire distinctively different traits while in the same geographic area and become reproductively isolated from each other) occurs. Species diverge in sympatric speciation due to behavioral, temporal, and resource-based mechanisms of reproductive isolation. In sympatric speciation, groups of similar organisms can coexist in the same geographic area (or overlapping geographic areas) without interbreeding.

(c) Selecting from allopatric and sympatric speciation, **identify** which type of speciation A and B in the figure represent. **(1 point)**

Identification (1 point)
• Allopatric speciation is represented by A, and sympatric speciation is represented by B.

Here is a possible response that would receive full credit:

Allopatric speciation involves geographic barriers that physically isolate populations, which is shown in the initial stage of speciation for A in the figure. Sympatric speciation involves the emergence of a new species within the geographic range of its parent population, which is shown in the initial stage of speciation for B. Thus, A is allopatric speciation and B is sympatric speciation.

(d) **Explain** whether allopatric or sympatric specification is the more common mode of speciation. **(1 point)**

Explanation (1 point)
• Allopatric speciation is the dominant mode of speciation because geographic barriers are more common than polymorphisms.

Here is a possible response that would receive full credit:

Allopatric speciation is more common than sympatric speciation because species often develop when a few individuals in a population move to a different geographical area, whether by choice or force (for example, a storm or food shortage). Sympatric speciation is rare because it generally occurs through polyploidy (possessing more than two sets of chromosomes). It may occur among plants that can self-fertilize.

6. Scoring Guidelines for Free-Response Question 6

(a) **Identify** the amino acid before and after the mutation at codon 142 in the alpha globin. **(1 point)**

Identification (1 point)
• Codon 142 before the mutation coded for a stop codon (no amino acid) and after the mutation coded for amino acid GLN.

Here is a possible response that would receive full credit:

The point mutation at codon 142 in the alpha globin changed a U to a C. According to the tables, codon 142 before the mutation was UAA, which codes for a stop codon (no amino acid), and after the mutation was CAA, which codes for GLN.

(b) **Describe** how the structure and function of hemoglobins are affected by the mutation at codon 142. **(1 point)**

Description (1 point)
• The two alpha globins of hemoglobin may be elongated, causing hemoglobin to no longer function properly.

Here is a possible response that would receive full credit:

The mutation at codon 142 in the alpha globin changes UAA (STOP) to CAA (gln), which may create an elongated alpha chain. Hemoglobin consisting of two elongated alpha globins may no longer function properly.

(c) Sickle cell anemia results from a point mutation at codon 6 in the beta globin of hemoglobin. Based on the information provided, **determine** the type of mutation that causes sickle cell anemia. **(1 point)**

Determination (1 point)
• The mutation at codon 6 is a missense mutation.

Here is a possible response that would receive full credit:

Sickle cell anemia has adenine substituted for thymine, which codes for valine rather than the normal glutamic acid in the hemoglobin protein. Clues that this scenario describes a missense mutation include 1) the involvement of a single codon, and 2) the protein that is still produced but with a different amino acid, which affects the protein function. Silent mutations have no effect on the protein, and nonsense mutations would cause a stop where there should not be one.

(d) Beta-thalassemia is an inherited blood disorder in which less beta globin is produced. Based on the information, **explain** how a point mutation could cause beta-thalassemia. **(1 point)**

Explanation (1 point)
• The mutation is a nonsense mutation that changes the codon to a chain-terminating codon.

Here is a possible response that would receive full credit:

Beta-thalassemia is caused by a nonsense mutation. Specifically, it changes codon 39 in the beta globin from CAG (gln) to UAG (STOP). The STOP signal shortens the globin chain and makes it functionally useless.

APPENDIX

AP Biology Equations and Formulas

STATISTICAL ANALYSIS AND PROBABILITY

Mean	Standard Deviation
$\bar{x} = \frac{1}{n}\sum_{i=1}^{n} x_i$	$s = \sqrt{\frac{\sum (x_i - \bar{x})^2}{n-1}}$
Standard Error of the Mean	**Chi-Square**
$SE_{\bar{x}} = \frac{s}{\sqrt{n}}$	$\chi^2 = \sum \frac{(o-e)^2}{e}$

s = sample standard deviation (i.e., the sample-based estimate of the standard deviation of the population)
$\bar{x}$ = mean
n = size of the sample
o = observed results
e = expected results
Degrees of freedom equal the number of distinct possible outcomes minus one.

CHI-SQUARE TABLE

p value	Degrees of Freedom							
	1	2	3	4	5	6	7	8
0.05	3.84	5.99	7.82	9.49	11.07	12.59	14.07	15.51
0.01	6.64	9.21	11.34	13.28	15.09	16.81	18.48	20.09

LAWS OF PROBABILITY

If A and B are mutually exclusive, then $P(\text{A or B}) = P(\text{A}) + P(\text{B})$
If A and B are independent, then $P(\text{A and B}) = P(\text{A}) \times P(\text{B})$

HARDY-WEINBERG EQUATIONS

$p^2 + 2pq + q^2 = 1$ — p = frequency of the dominant allele in a population

$p + q = 1$ — q = frequency of the recessive allele in a population

METRIC PREFIXES

Factor	Prefix	Symbol
10^9	giga	G
10^6	mega	M
10^3	kilo	k
10^{-2}	centi	c
10^{-3}	milli	m
10^{-6}	micro	μ
10^{-9}	nano	n
10^{-12}	pico	p

Mode = value that occurs most frequently in a data set
Median = middle value that separates the greater and lesser halves of a data set
Mean = sum of all data points divided by number of data points
Range = value obtained by subtracting the smallest observation (sample minimum) from the greatest (sample maximum)

RATE AND GROWTH

Rate

$$\frac{dY}{dt}$$

Population Growth

$$\frac{dN}{dt} = B - D$$

Exponential Growth

$$\frac{dN}{dt} = r_{max} N$$

Logistic Growth

$$\frac{dN}{dt} = r_{max} N\left(\frac{K-N}{K}\right)$$

dY = amount of change

dt = change in time

B = birth rate

D = death rate

N = population size

K = carrying capacity

r_{max} = maximum per capita growth rate of population

Temperature Coefficient Q_{10}

$$Q_{10} = \left(\frac{k_2}{k_2}\right)^{\frac{10}{T_2 - T_1}}$$

Primary Productivity Calculation

mg O_2/L × 0.698 = mL O_2/L

mL O_2/L × 0.536 = mg carbon fixed/L

(at standard temperature and pressure)

T_2 = higher temperature

T_1 = lower temperature

k_2 = reaction rate at T_2

k_1 = reaction rate at T_1

Q_{10} = the factor by which the reaction rate increases when the temperature is raised by ten degrees

Water Potential (Ψ)

$\Psi = \Psi_P + \Psi_S$

Ψ_P = pressure potential

Ψ_S = solute potential

The water potential will be equal to the solute potential of a solution in an open container because the pressure potential of the solution in an open container is zero.

The Solute Potential of the Solution

$\Psi_S = -\,iCRT$

i = ionization constant (this is 1.0 for sucrose because sucrose does not ionize in water.)

C = molar concentration

R = pressure constant (R = 0.0831 liter bars/mole K)

T = temperature in Kelvin (°C + 273)

SURFACE AREA AND VOLUME

Volume of a Sphere

$V = \frac{4}{3}\pi r^3$

Volume of a Rectangular Solid

$V = lwh$

Volume of a Right Cylinder

$V = \pi r^2 h$

Surface Area of a Sphere

$A = 4\pi r^2$

Surface Area of a Cube

$A = 6s^2$

Surface Area of a Rectangular Solid

$A = \Sigma$ (surface area of each side)

r = radius

l = length

h = height

w = width

s = length of one side of a cube

A = surface area

V = volume

Σ = sum of all

Dilution (used to create a dilute solution from a concentrated stock solution)

$C_i V_i = C_f V_f$

i = initial (starting)

C = concentration of solute

f = final (desired)

V = volume of solution

Gibbs Free Energy

$\Delta G = \Delta H - T\Delta S$

ΔG = change in Gibbs free energy

ΔS = change in entropy

ΔH = change in enthalpy

T = absolute temperature (in Kelvin)

pH $= -\log_{10}[H^+]$

Glossary

abiotic

Nonliving, as in the physical environment

activators

Regulatory proteins that switch genes on

active immunity

Protective immunity to a disease in which the individual produces antibodies as a result of previous exposure to the antigen

adaptation

A behavioral or biological change that enables an organism to adjust to its environment

adaptive radiation

The production of a number of different species from a single ancestral species

adenine

A purine nitrogenous base that pairs with thymine in DNA and uracil in RNA

adenosine phosphate

Adenosine diphosphate (ADP) and adenosine triphosphate (ATP), which are energy-storage molecules

adipose

Fatty tissue, fat-storing tissue, or fat within cells

aerobic

Requiring free oxygen from the atmosphere for normal activity and respiration

aerobic catabolism

Metabolic breakdown of complex molecules into simple ones through the use of oxygen; results in the release of energy

allele

One of two or more types of genes, each representing a particular trait; many alleles exist for a specific gene locus

allopatric speciation

Evolution of species that occurs in separate geographic areas

alternation of generations

The description of a plant life cycle that consists of a diploid, asexual, sporophyte generation and a haploid, sexual, gametophyte generation

anaerobic

Living or active in the absence of free oxygen; pertaining to respiration that is independent of oxygen

anaerobic catabolism

Metabolic breakdown of complex molecules into simple ones without the use of oxygen; results in the release of energy

analogous

Describes structures that have similar function but different evolutionary origins (e.g., a bird's wing and a moth's wing)

anaphase

The stage in mitosis that is characterized by the migration of chromatids to opposite ends of the cell; the stage in meiosis during which homologous pairs migrate (anaphase I); and the stage in meiosis during which chromatids migrate to different ends of the cell (anaphase II)

androgen

A male sex hormone (e.g., testosterone)

Animalia

Kingdom that includes all extinct and living animals

antibiotic

An antipathogenic substance (e.g., penicillin)

appendage

A structure that extends from the trunk of an organism and is capable of active movements

Archaea

Domain comprised of an ancient group of microorganisms (prokaryotes) that are metabolically and genetically different from bacteria; they came before the eukaryotes

asexual reproduction

The production of daughter cells by means other than the sexual union of gametes (as in budding and binary fission)

ATPase

Adenosine triphosphatase; enzyme that catalyzes the hydrolysis of ATP to ADP, thereby releasing energy

autosomal genes

Non-sex-linked genes

autosome

Any chromosome that is not a sex chromosome

autotroph

An organism that utilizes the energy of inorganic materials, such as water and carbon dioxide or the Sun, to manufacture organic materials; plants are examples of autotrophs

Bacteria

Domain of single-celled organisms that reproduce by fission and can be spiral, rod, or spherical shaped; often pathogenic organisms that rapidly reproduce

base-pair substitution

When one base pair is incorrectly reproduced and substituted for another base pair

binary fission

Asexual reproduction; in this process, the parent organism splits into two equal daughter cells

biological species concept (BSC)

Definition of a species as a naturally interbreeding population of organisms that produce viable, fertile offspring

biome

A habitat zone, such as desert, grassland, or tundra

biotic

Living, as in living organisms in the environment

Calvin cycle

Cycle in photosynthesis that reduces fixed carbon to carbohydrates through the addition of electrons (also known as the "dark cycle")

carbohydrate

An organic compound to which hydrogen and oxygen are attached; the hydrogen and oxygen are in a 2:1 ratio; examples include sugars, starches, and cellulose

carbon cycle

The recycling of carbon from decaying organisms for use in future generations

carbon fixation

Conversion of carbon dioxide into organic compounds during the Calvin cycle, the second stage of photosynthesis; known as a "dark reaction"

carrying capacity

The number of organisms an environment can support

catabolism

Metabolic breakdown of complex molecules into simple ones, releasing energy

cell

Smallest structural unit of an organism

cell wall

A wall composed of cellulose that is external to the cell membrane in plants; it is primarily involved in support and in the maintenance of proper internal pressure; fungi have cell walls made of chitin, and some protists also have cell walls

chemiosmosis

The coupling of enzyme-catalyzed reactions

chi-square analysis

Test to see if a theory is backed up by experimental results

chlorophyll

A green pigment that performs essential functions as an electron donor and light entrapper in photosynthesis

chloroplast

A plastid containing chlorophyll

chromatid

One of the two strands that constitute a chromosome; chromatids are held together by the centromere

chromatin

A nuclear protein of chromosomes that stains readily

chromosome

A short, stubby rod consisting of chromatin that is found in the nucleus of cells; contains the genetic or hereditary component of cells (in the form of genes)

circadian rhythms

Daily cycles of behavior

cleavage

The division in animal cell cytoplasm caused by the pinching in of the cell membrane

codominance

A form of inheritance in which neither of the alleles is completely recessive or dominant, so both alleles are expressed equally in the phenotype

codon

Three adjacent nucleotides that signal to insert an amino acid into the genetic code or end protein synthesis

coenzyme

An organic cofactor required for enzyme activity

cofactor

An ion or molecule that helps to bind an enzyme to a substrate

commensalism

A relationship in which an organism lives symbiotically with a host; this host neither benefits nor suffers from the association

communities

Groups of interacting organisms that live in the same geographic area under similar environmental conditions

competition

The process of striving for limited resources

complementary base pairs

Pairing of purines and pyrimidines in DNA and RNA

concentration gradient

Difference in concentration of a solute between two areas of a solution

conditioning

The association of a physical, visceral response with an environmental stimulus with which it is not naturally associated; a learned response

consumer

Organism that consumes food from outside itself instead of producing it (primary, secondary, and tertiary)

cooperation

The process of acting together for common benefits

coupled reaction

Chemical reaction in which energy is transferred from one side of the reaction to the other through a common intermediate

cristae

Inward folds of the mitochondrial membrane

crossing over

The exchange of parts of homologous chromosomes during meiosis

cytokinesis

A process by which the cytoplasm and the organelles of the cell divide; the final stage of mitosis

cytoplasm

The living matter of a cell, located between the cell membrane and the nucleus

cytosine

A pyrimidine nitrogenous base that is present in nucleotides and nucleic acids; it is paired with guanine

cytoskeleton

The organelle that provides mechanical support and carries out motility functions for the cell

dark reactions

Processes that occur after the light reactions of photosynthesis (during carbon fixation), without the presence of light

Darwin, Charles Robert (1809–1882)

Naturalist who came up with the theory of evolution based on natural selection

decomposers

Organisms that feed on and break down dead plant or animal matter

degree of freedom (d.f.)

Independent statistical category; the number of categories of observation minus one

deletion

The loss of all or part of a chromosome

density-dependent factors

Effects that increase with population density and smaller population size

density-independent factors

Effects that are independent of population size

deoxyribose

A five-carbon sugar that has one oxygen atom less than ribose; a component of DNA (deoxyribonucleic acid)

differentiation

The process in which a cell becomes specialized by changing from one cell type to another

diffusion

The movement of particles from one place to another as a result of their random motion

dihybrid

An organism that is heterozygous for two different traits

dihybrid cross

A hybridization between two traits, each with two alleles

diploid

Describes cells that have a double set of chromosomes in homologous pairs ($2N$)

directional selection

Favors organisms within a population that have one extreme variation of a trait

disruptive (diversifying) selection

Favors organisms within a population with extreme variations of a trait over those with moderate variations

DNA

Deoxyribonucleic acid; found in the cell nucleus, its basic unit is the nucleotide; contains coded genetic information; can replicate on the basis of heredity

domains

Biological classification of prokaryotes and eukaryotes into Bacteria, Archaea, and Eukarya

dominant

A dominant allele suppresses the expression of the other member of an allele pair when both members are present; a dominant allele exerts its full effect regardless of the effect of its allelic partner

ecological succession

The orderly process by which one biotic community replaces another until a climax community is established

ecology

The study of organisms in relation to their environment

ecosystem

Ecological community and its environment

egg (ovum)

The female gamete; it is nonmotile, large in comparison to male gametes, and stores nutrients

electron transport chain

A complex carrier mechanism located on the inside of the inner mitochondrial membrane of the cell; releases energy and is used to form ATP

endocytosis

A process by which the cell membrane is invaginated to form a vesicle that contains extracellular medium

endoplasmic reticulum

A network of membrane-enclosed spaces connected with the nuclear membrane; transports materials through the cell; can be smooth or rough

energy flow

The movement of energy throughout the trophic levels of an ecosystem

enhancers

DNA sequences that increase the level of transcription of a gene on the same chromosome

enzyme

An organic catalyst; typically, a protein

epidermis

The outermost surface of an organism

Eukarya

Domain containing all eukaryotic organisms

eukaryote

Organism consisting of one or more cells with genetic material in membrane-bound nuclei

evolution, theory of

Theory that organisms have developed over time to produce current biomes

exocrine

Pertaining to a type of gland that releases its secretion through a duct (e.g., the salivary gland or the liver)

exocytosis

A process by which the vesicle in the cell fuses with the cell membrane and releases its contents to the outside

exons

DNA that is transcribed to mRNA and codes for protein synthesis

extinction

The termination of an organism or species

extracellular matrix

Material found outside of the cell

F1

The first filial generation (first offspring)

F2

The second filial generation; offspring resulting from the crossing of individuals of the F1 generation

facilitated diffusion

A type of passive transport in which a membrane-embedded protein allows the movement of impermeable ions or molecules down their concentration gradients

fats

Solid, semi-solid, or liquid organic compounds composed of glycerol, fatty acids, and organic groups

feedback mechanism

The process by which a certain function is regulated by the amount of the substance it produces

fertilization

The fusion of the sperm and egg to produce a zygote

fitness

The ability of an organism to contribute its alleles and, therefore, its phenotypic traits to future generations

food web

The interaction of feeding levels in a community, including energy flow throughout the community

Fungi

Kingdom of eukaryotic organisms that lack vascular tissues and chlorophyll, possessing chitinous cell walls; reproduction occurs through spores

gamete

A sex or reproductive cell that must fuse with another of the opposite type to form a zygote, which subsequently develops into a new organism

gametogenesis

The formation of gametes

gametophyte

The haploid sexual stage in the life cycle of plants (alternation of generations)

gene

The portion of a DNA molecule that serves as a unit of heredity; found on the chromosome

gene expression

Conversion of information from a gene to mRNA to a protein

gene flow

The process of moving genes between populations as a result of the movement of individual organisms

genetic code

A four-letter code made up of the DNA nitrogenous bases A, T, G, and C; each chromosome is made up of thousands of these bases

genetic drift

Random evolutionary changes in the genetic makeup of a (usually small) population

genotype

The genetic makeup of an organism without regard to its physical appearance; a homozygous dominant and a heterozygous organism may have the same appearance but different genotypes

genus

In taxonomy, a classification between species and family; a group of very closely related species (e.g., *Homo, Felis*)

geographic barrier

Any physical feature that prevents the ecological niches of different organisms (not necessarily different species) from overlapping

geographic isolation

Isolation due to geographic factors (e.g., islands are geographically isolated)

glycolysis

The anaerobic respiration of carbohydrates

Golgi apparatus

Membranous organelles involved in the storage and modification of secretory products

gravitropism

Directional growth according to the gravitational field; roots grow downward with gravity, while the shoots of plants grow up toward sunlight

growth curve

Growth of an organism or population plotted over time

guanine

A purine nitrogenous base that is a component of nucleotides and nucleic acids; it links up with cytosine in DNA

habitat

The environment a community or organism lives in

haploid

Describes cells (gametes) that have half the chromosome number (*N*) typical of the species

Hardy-Weinberg equilibrium

In a randomly breeding population, gene frequency and genotype ratios remain constant over generations of organisms

hemizygous

Describes an organism that possesses only a single copy of a gene, which it expresses (e.g., human males are hemizygous for X-linked traits)

heterotroph

An organism that must get its inorganic and organic raw materials from the environment; a consumer

heterozygous

Describes an individual that possesses two contrasting alleles for a given trait (e.g., *Tt*)

homeostasis

The dynamic process by which an organism maintains a stable internal environment

homologous

Describes two or more structures that have similar forms, positions, and origins despite the differences between their current functions; examples are the arm of a human, the flipper of a dolphin, and the foreleg of a horse

homozygous

Describes an individual that has the same gene for the same trait on each homologous chromosome (e.g., *TT* or *tt*)

homozygous dominant

Having two dominant alleles of the same gene; dominant alleles are expressed in a heterozygous as well as a homozygous genotype

homozygous recessive

Having two recessive alleles of the same gene; recessive alleles are only expressed when a gene is homozygous recessive

hormone

A chemical messenger that is secreted by one part of the body and carried by the blood to affect another part of the body, usually a muscle or gland

hormone-receptor system

Chemical messengers (hormones) travel throughout the body and are read by receptor proteins, which respond to the message each hormone codes for

hybrid

An offspring that is heterozygous for one or more gene pairs

hydrophilic

Having an affinity for water, "water loving"

hydrophobic

Repelling water, "water fearing"

hypertonic

Describes a fluid that has a higher osmotic pressure than another fluid it is compared to; it exerts greater osmotic pull than the fluid on the other side of a semipermeable membrane; hence, it possesses a greater concentration of particles, and acquires water during osmosis

hypotonic

Describes a fluid that has a lower osmotic pressure than a fluid it is compared to; it exerts lesser osmotic pull than the fluid on the other side of a semipermeable membrane; hence, it possesses a lesser concentration of particles, and loses water during osmosis

incomplete dominance

Genetic blending; each allele exerts some influence on the phenotype (for example, red and white parents may yield pink offspring)

Independent Assortment, Law of

Genes independently sort and do not affect the sorting of other genes in the formation of gametes in diploid organisms

indeterminate growth

Growth without a termination point

inducers

Molecules that bind to repressors and/or activators to increase gene expression

induction

Initiating enzyme production or genetic transcription

ingestion

The intake of food from the environment into the alimentary canal

innate behaviors

Instinctive responses encoded in the genes of organisms

insertion

Addition of one or more nucleotides to a chromosome, usually by mutation

intercellular

Between cells

intermembrane space

Space between the outer and inner membranes of a mitochondrion

interphase

The cellular phase between meiotic or mitotic divisions

intracellular

Within a cell

intron

Part of a gene that is located between exons and is removed before the translation of mRNA; does not code for protein synthesis

invasive species

Nonnative species that cause harm to the environment, the health of native species, or the local economy

isolation

The separation of some members of a population from the rest of their species; prevents interbreeding and may lead to the development of a new species

isotonic

Describes a fluid that has the same osmotic pressure as a fluid it is compared to; it exerts the same osmotic pull as the fluid on the other side of a semipermeable membrane; hence, it neither gains nor loses net water during osmosis and possesses the same concentration of particles before and after osmosis occurs

keystone species

A species that increases overall diversity of an ecosystem and whose removal would affect the ecosystem's balance

kinesis

Movement of an organism in response to light

kingdom

Second-highest taxonomic classification of organisms, after domain

Krebs cycle

Process of aerobic respiration that fully harvests the energy of glucose; also known as the citric acid cycle

K-selected

Organisms in more stable environments that tend to produce fewer offspring and invest more energy in rearing offspring

lactase

An enzyme that breaks down the sugar lactose

lactic acid fermentation

A type of anaerobic respiration found in fungi, bacteria, and human muscle cells

Lamarck, Jean Baptiste de (1744–1829)

Naturalist who studied evolution and classified invertebrate organisms; Lamarck produced several theories of evolution including *Philosophie Zoologique*; best known for his incorrect theory of inheritance of acquired characteristics

learned behaviors

Responses acquired/lost through interaction with the world or through teaching

light reactions

Photosynthetic reactions that occur in the presence of light

linkage

Occurs when different traits are inherited together more often than they would have been by chance alone; it is assumed that these traits are linked on the same chromosome

lipase

An enzyme that catalyzes the hydrolysis of lipids

lipid

An organic compound that contains hydrocarbons and includes fats, oils, waxes, and steroids

lysosome

An organelle that contains enzymes that aid in intracellular digestion

meiosis

A process of cell division whereby each daughter cell receives only one set of chromosomes; the formation of gametes

membrane

Thin structure connecting or separating structures or regions of an organism

Mendelian laws

Laws of classical genetics established through Mendel's experiments with peas; include laws of segregation and independent assortment

metabolism

A group of life-maintaining processes that includes nutrition, respiration (the production of usable energy), and the synthesis and degradation of biochemical substances

metaphase

A stage of mitosis; chromosomes line up at the equator of the cell

micronutrients

Vitamins or minerals essential for growth and metabolism in an organism

minerals

Naturally occurring inorganic elements essential in the nutrition of organisms

mitochondria

Cytoplasmic organelles that serve as sites of respiration; rod-shaped bodies in the cytoplasm known to be the center of cellular respiration

mitosis

A type of nuclear division that is characterized by complex chromosomal movement and the exact duplication of chromosomes; occurs in somatic cells

molecular clock hypothesis

Hypothesis that genetic mutations occur in a genome at a linear rate

Monera

The kingdom of bacteria (no longer used under the three domain system)

monohybrid

An individual that is heterozygous for only one trait

monohybrid cross

Cross involving a single trait and two alleles

monomer

A basic molecule that can covalently bond to other monomers to form long chains called polymers

monosaccharide

A simple sugar; typically, a five- or six-carbon sugar (e.g., ribose or glucose)

mRNA (messenger RNA)

RNA that transfers genetic information from the cell nucleus to ribosomes, serving as a template for protein synthesis

multiple allele inheritance

A form of inheritance in which more than two alleles code for a certain trait, resulting in more than two phenotypes

mutagen

Agent that induces mutations; often carcinogenic

mutation

Changes in genes that are inherited

mutualism

A symbiotic relationship from which both organisms involved derive some benefit

NAD $^{+}$

An abbreviation of nicotinamide adenine dinucleotide; a respiratory oxidation-reduction molecule

NADP $^{+}$

An abbreviation of nicotinamide adenine dinucleotide phosphate; an organic compound that serves as an oxidation-reduction molecule, used in photosynthesis

natural selection

Process by which organisms best adapted to their environment survive to pass their genes on through offspring; idea pioneered by Charles Darwin

negative control

Regulation in which the regulatory protein is a repressor and transcription is inhibited (switched off)

niche

The functional role and position of an organism in an ecosystem; embodies every aspect of the organism's existence

nitrogenous bases

The five purine and pyrimidine bases found in nucleic acid—adenine, thymine (in DNA only), cytosine, guanine, and uracil (in RNA only)

novel structures

Cellular structures used primarily for reproduction

nuclear membrane

A membrane that envelopes the nucleus and separates it from the cytoplasm; present in eukaryotes

nucleolus

A dark-staining small body within the nucleus; composed of RNA

nucleotide

An organic molecule consisting of phosphate joined with a five-carbon sugar (deoxyribose or ribose) and a purine or a pyrimidine (adenine, guanine, uracil, thymine, or cytosine)

nucleus

An organelle that regulates cell functions and contains the genetic material of the cell

nutrient

A substance that can be metabolized by an organism to provide energy and build tissue

ontogeny

The origin and subsequent growth of an organism, from embryo to adult

operon

A series of linked genes that regulate a biological function

organ system

A group of organs that work together to perform a specific task in the body, such as the circulatory system distributing blood throughout the body

organelle

A specialized structure that carries out particular functions for eukaryotic cells; examples include the plasma membrane, nucleus, and ribosomes

organic molecules

Most molecules that contain carbon (C); note that there are exceptions (e.g., carbon dioxide is not considered organic)

Origin of Species, The

Charles Darwin's book in which he expressed his theory that species evolve through natural selection

osmosis

The diffusion of water through a semipermeable membrane, from an area of greater water concentration to an area of lesser water concentration

osmotic pressure

Pressure exerted by the flow of water through a semipermeable membrane, which separates a solution into two concentrations of solute

oxidation

The addition of oxygen to or removal of hydrogen or electrons from a compound; half of a redox (oxidation or reduction) process

oxidative phosphorylation

Formation of ATP from energy released during the oxidation of various substances and substrates, as in aerobic respiration and the Krebs cycle

oxygen

A nonmetallic element essential in animal and plant respiration

parapatric speciation

Occurs when limited interbreeding and negligible genetic exchange takes place between two populations

parasitism

A relationship in which an organism lives on or within a host; this host suffers from the association

pedigree

A family tree depicting the inheritance of a particular genetic trait over several generations

pepsin

An enzyme that breaks down proteins into smaller peptides

pH

A symbol that denotes the relative concentration of hydrogen ions in a solution: the lower the pH, the more acidic a solution; the higher the pH, the more basic a solution; pH is equal to $-\log [H^+]$

phenotype

The physical appearance of an individual, as opposed to its genetic makeup

phloem

The vascular tissue of a plant that transports organic materials (photosynthetic products) from the leaves to other parts of the plant

phospholipids

Phosphorus-containing lipids composed of two fatty acids and a phosphate group modified with simple organic molecules

photoperiodism

Response to seasonal changes in the length of day or night

photophosphorylation

The synthesis of ATP using radiant energy absorbed during photosynthesis

photosynthesis

The process by which light energy and chlorophyll are used to manufacture carbohydrates out of carbon dioxide and water; an autotrophic process using light energy

phototropism

Plant growth stimulated by light (stem: +, toward light; root: –, away from light)

phylogeny

The study of the evolutionary descent and interrelations of groups of organisms

phylum

A category of taxonomic classification that is ranked above class; kingdoms are divided into phyla

Plantae

Kingdom containing all extinct and living plants

plasma membrane

The cell membrane

pollen

The microspore of a seed plant

pollination

The transfer of pollen to the micropyle or to a receptive surface that is associated with an ovule (such as a stigma)

polymer

A large molecule that is composed of many similar molecular units (e.g., starch)

polymorphism

The individual differences of form among the members of a species

polyploidy

A condition in which an organism may have a multiple of the normal number of chromosomes ($4N$, $6N$, etc.)

polysaccharide

A carbohydrate that is composed of many monosaccharide units joined together, such as glycogen, starch, and cellulose

population

All the members of a given species inhabiting a certain locale

positive control

Regulation in which the regulatory protein is an activator and transcription is promoted (turned on)

postzygotic barriers

Mechanisms that prevent the development of a zygote into a fertile adult offspring

predation

Relationship in which one organism feeds on another

prezygotic barriers

Mechanisms that prevent the formation of a zygote, leading to reproductive isolation

primary consumers

Organisms that eat primary producers; herbivores

primary producers

First level of the food chain; use light to produce energy through photosynthesis

prokaryote

Unicellular organism lacking organelles, specifically a nucleus

promoter

Initial binding site on an operon for RNA polymerase

prophase

A mitotic or meiotic stage in which the chromosomes become visible and during which the spindle fibers form; synapsis takes place during the first meiotic prophase

protein

An organic compound that is composed of many amino acids; contains C, H, O, and N

protein synthesis

The creation of proteins, coded for by nucleic acids

protobionts

Metabolically active protein clusters that inaccurately reproduce; possible evolutionary precursors to prokaryotic cells

Protoctista

Kingdom composed of eukaryotic microorganisms and their immediate descendants, such as slime molds and protozoa

punctuated equilibrium

Evolution characterized by long periods of virtual standstill interspersed with periods of rapid change

Punnett square

A diagram that maps out the possible crosses between genes of breeding organisms

receptor cells

Cells throughout the body that contain receptor proteins and are activated by chemical signals, producing a systemic response

recessive

Pertains to a gene or characteristic that is masked when a dominant allele is present

repressors

Regulatory proteins that switch genes off

reproduce vegetatively/vegetative reproduction

Asexual reproduction

respiration

A chemical action that releases energy from glucose to form ATP

respiratory system

System of organs involved in the intake and exchange of oxygen and carbon dioxide between an organism and its environment

ribose

A pentose sugar that occurs in nucleotides, nucleic acids, and riboflavin

ribosome

An organelle in the cytoplasm that contains RNA; serves as the site of protein synthesis

RNA

An abbreviation of ribonucleic acid, a nucleic acid in which the sugar is ribose; a product of DNA transcription that serves to control certain cell activities; acts as a template for protein translation; types include mRNA (messenger), tRNA (transfer), and rRNA (ribosomal)

r-selected

Species that monopolize rapidly changing environments and produce many offspring in a short amount of time

rubisco

Plant protein that accepts oxygen in place of carbon dioxide and fixes carbon in photosynthetic organisms

secondary consumers

Organisms that eat primary consumers

Segregation, Law of

Genes come in pairs in diploid organisms and each gamete gets one gene at random from each gene pair

selective advantages

Characteristics that are good for survival and/or mating

selective disadvantages

Characteristics that are bad for survival and/or mating

selectively permeable

Membranes that allow some substances and particles to pass through, but not others

self-pollination

The transfer of pollen from the stamen to the pistil of the same flower

sex chromosome

There are two kinds of sex chromosomes, X and Y; XX signifies a female and XY signifies a male; there are fewer genes on the Y chromosome than on the X chromosome

sex linkage

Occurs when certain traits are determined by genes on the sex chromosomes

sexual reproduction

Reproduction by the fusion of a male and female gamete to form a zygote

sexual selection

Selection driven by the competition for mates, in relation to natural selection

somatic cell

Any cell that is not a reproductive cell

specialization

The process of a cell adapting to do a particular function

species

A group of populations that can interbreed to produce fertile, viable offspring

sperm (spermatozoon)

A male gamete

spindle

A structure that arises during mitosis and helps separate the chromosomes; composed of tubulin

spore

A reproductive cell that is capable of developing directly into an adult

sporophyte

An organism that produces spores; a phase in the diploid-haploid life cycle that alternates with a gametophyte phase

stabilizing selection

Selection that maintains the same mean in a phenotypic distribution by removing individuals from both phenotypic extremes

sterols

Polycyclic compounds (lipids), such as cholesterol, that play an important role in lipid metabolism

stomata

Pores in a leaf through which gas and water vapor pass; a small opening on the surface of a membrane; a mouthlike opening in an organism

substrate

A substance that is acted upon by an enzyme

symbiosis

A type of relationship between two different organisms (e.g. mutualism, commensalism, parasitism)

sympatric speciation

Speciation due to behavioral, temporal, or ecological factors; species live in the same geographic area but do not interbreed

synapsis

The pairing of homologous chromosomes during meiosis

TATA box

Promoter sequence that specifies where transcription begins

taxis

The responsive movement of an organism toward or away from a stimulus, such as light

taxonomy

The science of classification of living things

telophase

A mitotic stage in which nuclei reform and the nuclear membrane reappears

terminator

Sequence of nucleotides that signals the end of synthesis of a protein or nucleic acid, as well as the end of translation or transcription

terrestrial plants

Plants that live and grow on land

test cross

The breeding of an organism with a homozygous recessive to determine whether an organism is homozygous dominant or heterozygous for a given trait

tetrad

A pair of chromosome pairs present during the first metaphase of meiosis

thermoregulation

The ways in which organisms regulate their internal heat

thylakoid space

An inner compartment formed from the connected spaces between flattened sacs called thylakoids in chloroplasts

thymine

A pyrimidine nitrogenous base in nucleic acids and nucleotides; pairs with adenine in DNA

tissue

A mass of cells that have similar structures and perform similar functions

transcription

The first stage of protein synthesis, in which the information coded in the DNA base is transcribed onto a strand of mRNA

transcriptional level

The level at which many genes are primarily regulated

translation

The final stages of protein synthesis in which the genetic code of nucleotide sequences is translated into a sequence of amino acids

translocation

The transfer of a piece of chromosome to another chromosome

transpiration

The evaporation of water from leaves or other exposed surfaces of plants

tRNA (transfer RNA)

RNA molecules that transport amino acids to ribosomes

uracil

A pyrimidine nitrogenous base found in RNA (but not in DNA); pairs with adenine

vacuole

A space in the cytoplasm of a cell that contains fluid

vitamin

An organic nutrient required by organisms in small amounts to aid in proper metabolic processes; may be used as an enzymatic cofactor; because it is not synthesized, it must be obtained prefabricated in the diet

xylem

Vascular tissue of the plant that aids in support and carries water

zygote

A cell resulting from the fusion of gametes